ENVIRONMENTAL HEALTH - PHYSICAL, CHEMICAL AND BIOLOGICAL FACTORS

SOIL NUTRIENTS

Environmental Health - Physical, Chemical and Biological Factors

Additional books in this series can be found on Nova's website
under the Series tab.

Additional E-books in this series can be found on Nova's website
under the E-book tab.

ENVIRONMENTAL HEALTH - PHYSICAL, CHEMICAL AND BIOLOGICAL FACTORS

SOIL NUTRIENTS

MOHAMMAD MIRANSARI
EDITOR

Nova Science Publishers, Inc.
New York

For permission to use material from this book please contact us:
Telephone 631-231-7269; Fax 631-231-8175
Web Site: http://www.novapublishers.com

LIBRARY OF CONGRESS CATALOGING-IN-PUBLICATION DATA

Soil nutrients / editor: Mohammad Miransari.
p. cm.
Includes index.
ISBN 978-1-61324-785-3 (hardcover)
1. Soils and nutrition. 2. Plant nutrients. 3. Soil microbiology. I. Miransari, Mohammad.
S596.5.S65 2011
579'.1757--dc23

2011017281

Published by Nova Science Publishers, Inc. † New York

CONTENTS

Preface		**vii**
Chapter 1	Potassium and Sugar Accumulation in Plant Cells *M. S. Krasavina and N. A. Burmistrova*	**1**
Chapter 2	The Role of Redox Conditions on Soil Nutrients Availability *Artur Banach and Zofia Stępniewska*	**21**
Chapter 3	Role of Soil Organisms in Determining Nutrient Dynamics *Jagdamba Prasad Thapliyal and Sutapa Bose*	**55**
Chapter 4	Soil Micronutrients: Role in Plant Growth and Integrated Nutrient Management Systems *A. A. Deshmukh, N. N. Nawani, M. B. Khetmalas and P. N. Razdan*	**73**
Chapter 5	Role of Microbial, Enzymatic and Pedofauna Activities in Making Good the Adaptation of *Trachypogon* to Phosphorus Depleted Environments *D. López-Hernández, M. Toro and J. C. López-Gutiérrez*	**93**
Chapter 6	Solubilization of Mineral and Rocks: Microorganisms Acting to Improve Soil Fertility *Maria Leonor Lopes-Assad, Márcia Maria Rosa-Magri and Sandra Regina Ceccato-Antonini*	**127**
Chapter 7	Role of Microorganisms in Nitrogen Cycling in Soils *Egamberdieva Dilfuza*	**159**
Chapter 8	Manganese Deficiency and Toxicity in Soils and Mechanisms Adopted by Plants in Order to Face these Nutritional Disorders: A Review *T. Chatzistathis, I. Therios and D. Alifragis*	**177**
Chapter 9	Sustainable Use of Rhizospheric Microflora in Reclamation of Soil Health *Devendra Kumar Choudhary*	**199**

Chapter 10 The Use of Vermicompost in Sustainable Agriculture: Impact on Plant Growth and Soil Fertility **211**
Cristina Lazcano and Jorge Domínguez

Chapter 11 Optimum Soil Nutrients: A *Sine Qua Non* for Sustainable Agriculture for Food and Nutrition Security **235**
Emmanuel Uzoma Onweremadu and Angela Chika Udebuani

Chapter 12 Chemical Soil Quality and Mycorrhiza **267**
Marcela Claudia Pagano and Aba C. Persiano

Chapter 13 Micronutrient Availability in Crop Soils of the Pampas Region, Argentina **277**
Silvana Irene Torri, Susana Urricariet and Raúl Silvio Lavado

Chapter 14 Soil Microbes and Soil Nutrients **297**
M. Miransari, H. Omidi, A. Korde Naij, M. Amini and S. Maleki

Chapter 15 Uptake of Nitrogen by Arbuscular Mycorrhizal Fungi **311**
Mohammad Miransari, M. Habibi, A. M. Naji, M. H. Fotokian and H. Amirshekari

Index **321**

Preface

Soil nutrients are among the most important factors affecting plant growth and yield production as well as the environment. They are subjected to variations by different parameters such as climate, soil and plant. In the book "Soil Nutrients" the effects of different parameters on soil nutrients is presented in 15 chapters by researchers and scientists from across the globe. The materials are suitable for students, researchers and scientists. A broad range of topics related to soil nutrients is covered by the book. It is appropriate for soil and plant scientists, plant biologists, ecologists, microbiologists, plant physiologists, biotechnologists, biogeochemists and environmentalists.

With best regards,
Prof. Dr. Mohammad Miransari
Department of Soil Science,
College of Agricultural Sciences,
Shahed University, Tehran, Iran
http://www.researcherid.com/rid/B-1105-2010
http://miransari.webs.com
http://sites.google.com/site/miransari1/

In: Soil Nutrients
Editor: Mohammad Miransari
ISBN: 978-1-61324-785-3

Chapter 1

POTASSIUM AND SUGAR ACCUMULATION IN PLANT CELLS

M. S. Krasavina and N. A.Burmistrova*
Institute of Plant Physiology Russian Academy of Science, Moscow, Russia

ABSTRACT

Experimental data indicating interchangeability of potassium and sugars in the maintenance of plant cell turgor pressure are presented. Possible ways of such interchangeability are discussed on the basis of consideration of some mechanisms engaged in potassium transmembrane transport.

INTRODUCTION

Among various chemical elements used in agriculture (N, P, K), potassium takes an especial place. As distinct from nitrogen and phosphorus, it is not included in the cell molecules, remaining in a free state, and this determines its specific role in cell physiology.

Since the beginning of agricultural practice, farmers empirically selected potassium concentrations in the soil solution optimal for yield production and the ratio between it and other nutrients. The systemic analysis of potassium role in the synthesis and accumulation of compounds of interest in plant organs started only since the mid-1900s. Sucrose is one of such compounds. The fields with sugar-accumulating crops, sugar beet and sugarcane primarily, occupy large areas.

To understand the connection between processes of sugar accumulation and potassium metabolism, we firstly consider briefly processes affected by potassium and metabolic disturbances arising at its deficit.

---

* E-mail: krasavina.m@mail.ru

THE ROLE OF POTASSIUM IN THE CELLS AND METABOLISM DISTURBANCES AT ITS DEFICIT

Although potassium physiological role is not completely elucidated, some aspects of its influence on plant cell physiology are known (see Armengaud et al., 2004). Its basal functions are activation of many enzymes, the maintenance of the membrane electric potential, and turgor generation (Britto, Kronzucker, 2008). More than 60 plant enzymes require potassium for their functioning (Maathuis, 2009). Among them, enzymes catalyzing ion transfer across the membrane, redox processes, and hydrolytic processes. Potassium ions are involved in the maintenance of cell-wall structure and in the regulation of their state (Suelter, 1970; Evans, Wildes, 1971; Sorger et al, 1965).

Thus, by activation of phosphofructokinase and pyruvate kinase, potassium is involved in the initial and terminal steps of glycolysis. Potassium is required for functioning of malic enzyme, an important component of tricarboxylic acid cycle. Phosphotransacetylase and acetylkinase, catalyzing the formation of acetyl-CoA by a transfer of acetyl residue from acetylphosphate or acetyladenylate to CoA, are K-dependent enzymes. These reactions provide for fatty acid synthesis and organic acid oxidation in the Krebs cycle. Thus, both glycolysis and tricarboxylic acid cycle require potassium presence (see Kursanov, Vyskrebentzeva, 1966).

Potassium involvement in the syntheses of proteins and nucleic acids is its important function (Mengel et al., 1981). Potassium role in protein synthesis is primarily determined by its requirement for NO_3^- uptake and nitrogen assimilation, one of the elements most important for plant life and protein synthesis. In addition, enzymes activating amino acids for ribosomal protein synthesis are also K-dependent (Evans, Wildes, 1971; Mengel et al., 1980). It is believed that potassium is required in ribosomes for peptide bond formation.

Potassium deficit inhibits the pentosophosphate cycle required for the synthesis of the purine ring and then purine nucleotides and nucleic acids. Some reactions resulting in the synthesis of adenine nucleotides require potassium presence (Vyscrebentseva, Krasavina, 1966). Thus, potassium activates glutamine synthase catalyzing the condensation of glutamate and ammonia to form glutamine. Glutamine is involved in the formation of N-formylglycineamide ribonucleotide, which is conversed into 5-aminoimidazole ribonucleotide in the reaction catalyzed by K-dependent amidoimidazole ribonucleotide synthase. Another stage of inosite monophosphate synthesis, e.g., a conversion of 5-aminoimidazole-4-carboxamide ribonucleotide into 5-formamideimidazole-4-carboxamide ribonucleotide, is also requires potassium presence. The analysis of the nucleotide composition of pumpkin plants deficient in potassium demonstrated a substantial decrease of adenosine derivatives in such plants (Vyscrebentseva, Krasavina, 1966).

Potassium is also required for functioning of pyruvate kinase, the enzyme catalyzing the transfer of phosphoric acid from pyruvate to ADP, and that of pyrophosphatase (Sorger et al., 1965; Sugiyama et al., 1968) providing energy for many transport processes on the tonoplast, in particular for proton transport into the vacuole. Potassium deficit suppresses the activity of H^+-ATPase, pumping protons from the cytoplasm, which is a source of energy for the creation of the active component of the membrane potential and transmembrane electrochemical H^+ gradient driving the uptake of many ions and organic compounds. Potassium activates ATP formation in the reactions of oxidative phosphorylation. Therefore, at potassium deficit, the pool of macroergic phosphates drops sharply (Kursanov, Vyscrebentseva, 1966, 1967). This phenomenon and also retardation of electron transport in the mitochondrial electron-transport chain results in the reduction of the cell energetic status.

Some enzymes of the carbohydrate metabolism could also require potassium presence. For example, potassium activates amylase and invertase, and also hexokinase catalyzing sugar phosphorylation. Activation of ADPG-starch synthase determined the ratio between soluble sugars and starch in plant tissues. By facilitating monosaccharide conversion into sucrose, potassium also shift the ratio between groups of soluble sugars. In the presence of potassium, the total content of carbohydrates increases.

Being cofactor of many enzymatic reactions, potassium is required for the reactions of protein, lipid, and carbohydrate metabolism. The content of potassium affects enzyme transcriptional and post-transcriptional regulation. Potassium binding with enzymes occurs in its dehydrated form, evidently, via coordination with six oxygens that may derive from carboxyl, carbonyl, and hydroxyl groups and from water molecules. Such binding is selective for potassium (Britto, Kronzucker, 2008; Maathuis, 2009)

By now, some genes were identified, which expression depends on potassium presence (Marshner, 1995). They are genes for NO_3^- transporters, ribosome proteins, and sugar translocators. It was demonstrated that, at potassium deficit, expression of genes related to cell senescence (Gepstein et al., 2003) and genes involved in the synthesis of jasmonic acid is increased. The genes encoding peroxidase are also activated (Kim et al., 2010).

Many intracellular processes are determined by the presence of high concentrations of free potassium ions, which is the main movable ion in the cell. Its high chemical activity (a_K), as determined with selective microelectrodes, does not essentially differ from its concentration determined chemically; its intracellular concentrations are very high, attaining about 100 mM. Due to the high potassium activity in the cytoplasm and the high membrane permeability for potassium, it is a basic cation determining the value of the diffusion component of the membrane potential (Cheeseman, Hanson, 1979).

An important potassium function is the maintenance of electrically neutral state within the cell. Potassium serves as a counter ion for organic and inorganic anions and neutralizes negative charges of macromolecules, for instance, proteins and nucleic acids.

Potassium is the main osmotically active cation in the plant cell creating the osmotic potential for cell expansion. Potassium translocation between the cells determines some plant movements, for example, seismonastic organ movements, gravi- and chemotropisms, mimosa leaf folding and unfolding, tendril movements, trap closure in carnivorous plants (Hedick, Sievers, 1980; Moran et al., 2007). Potassium translocation along the pollen wall determines its capacity to germinate. Using a potassium-sensitive fluorescent dye, it was shown that, in dormant pollen grains, potassium is evenly distributed under the whole wall, whereas during pollen maturation, it moves toward the aperture (Rehman, Yun, 2006). Potassium is responsible for rapid pollen hydration (Rehman et al., 2004), its germination, and pollen tube growth (Fan et al., 2001).

Potassium transport into stomatal guard cells and out of them determines the size of stomata aperture. Potassium flow into stomatal guardian cells reduces their negative water potential, resulting in osmotic water uptake and turgor increase. Potassium exit from the stomatal guardian cells shifts the water potential toward more positive values, water exits from the cells, turgor pressure drops, and stomata close (Dietrich et al., 2001; Fisher 1968; Talbott, Zeiger, 1996).

The high turgor pressure maintains cell structure and induces cell expansion (Dolan, Davies, 2004). The plants containing sufficient amount of potassium are capable of more efficient usage of soil water as compared with plants experiencing potassium deficit. Therefore, in the first case, less water is required for good yield ensuring.

Aforementioned features of potassium functioning determine physiological processes occurring with its participation and disturbed at its deficit (Jackson, Volk, 1968). This element is known to participate in carbon assimilation in the process of photosynthesis, to control stomatal conductivity and mesophyll resistance to intercellular carbon transport in the leaves (Graham, Ulrich, 1972). At potassium deficit, photosynthesis is suppressed, stomatal aperture is diminished, and stomatal resistance to CO_2 is increased. As a result, PEP-stimulated CO_2 fixation is suppressed. An inhibition of electron transport in chloroplast membranes hinders photosynthetic energy production and the reactions of phosphorylation. The rate of photosynthesis and photorespiration reduce markedly, whereas dark respiration is activated (People, Koch, 1979).

At potassium starvation, a disturbance in mitochondrial respiration and ATP production were also noted. In this case, photosynthesis is inhibited and respiration is activated, and as a result, plant productivity decreases. The two-phasic response of plant respiration to potassium deficit was observed. Thus, in the beginning of starvation, the rate of respiration was lower than in plants supplied with sufficient potassium level; however, the extent of coupling with phosphorylation was high and the pentosophosphate cycle was active. That means that no metabolism disturbance still occurs in this period (Kursanov, Vyskrebentzeva, 1966, 1967). The main changes induced by potassium shortage are membrane hyperpolarization and changes in the cell ionic balance (Krasavina, 1966; Pertl et al., 2010). In this time period, adaptive processes are evidently activated; they are directed to the maintenance of potassium concentration in the cytoplasm, potassium uptake from soil due to the activation of potassium transporters, and secretion of toxic sodium ions. At the deeper potassium deficit, the membrane potential decreases and metabolism is disturbed. Respiration becomes low-productive: its rate increases sharply but its coupling reduces and ATP production drops. This results in the inhibition of the nucleic acids, protein, and lipid syntheses (Kursanov, Vyskrebentzeva, 1966; Shin, Schachtman, 2004).

Similar two-phasic response was observed for the nucleotide composition in potassium-deficient plants (Vyskrebentseva, Krasavina, 1966): short-term deficit enhanced the synthesis of purine derivatives, whereas the lengthening of starvation period suppressed this process, i.e., changes in the nucleotide content repeated changes in the activity of the pentosophosphate cycle providing synthetic processes with reduction equivalents (NADPH) and substrates (riboso-5-phosphate).

Potassium dominant role in the maintenance of turgor and water homeostasis is important for pressure-driven solute transport in the xylem and phloem (Mengel, 1980; Wyse et al., 1986; Cakmak et al., 1994). Suppression of assimilate efflux from the leaf was one of the first disturbances noted at potassium deficit (Amir, Reinhold, 1971; Mengel, Haeder, 1977). Hartt (1969, 1970) found that translocation of ^{14}C-labeled assimilates from the sugarcane leaf was retarded before a decrease in the photosynthetic carbon fixation or activation of photorespiration. Assimilate efflux was suppressed even under light intensities, which did not allow carbon fixation at all. Similar conclusion was made by Huber (1984), who observed the retardation of assimilate efflux from soybean leaves. In this work, changes in the activity of sucrose phosphate synthase were also noted. Efflux retardation and enhanced sucrose synthesis resulted in sucrose accumulation in the leaves.

THE SYSTEMS OF POTASSIUM TRANSPORT

Soil is the only source for potassium in the plant. The concentration of potassium in soil varies from 0.1 to 1 mM. Several factors affect potassium content and availability: (a) soil type, e.g., its structure and density (Kuchenbuch et al., 1986), pH, the presence of competing cations (Na, NH_4) (Spalding et al., 1999), and (b) climatic conditions (temperature regime, water supply (Libersbach et al., 2004). In soil, potassium cation is dehydrated and produces coordinate bonds with oxygen atoms. Such potassium is not available for plants (Maathuis, 2006, 2009). During its life, plants repeatedly experience potassium deficit and must adapt to these conditions.

The development of systems for active potassium uptake is one of the ways for plant adaptation to potassium shortage in soil. Plants require much potassium for their functioning. To maintain the high potassium concentration (on the average of about 100 mM in the cytoplasm), plant possesses transport systems capable of potassium uptake against the electrochemical gradient (ECG). Epstein et al. (1963) developed the main positions concerning the pathways and mechanisms of potassium uptake from soil. They demonstrated functioning of two transport systems in plant roots, which function in different ranges of potassium concentration in medium.

The first one is a high affinity transport system. It is characterized by the saturation kinetics and functions at external potassium concentrations up to approximately 1 mM. This system provides for potassium uptake against ECG; such transport demands coupling with the energy source (Kühn, Grof, 2010). The second system is switched on at the higher potassium concentrations in medium; it manifests a low affinity for potassium and depends linearly on potassium concentration in medium.

In recent years, molecular identification of these systems is performed (Very, Sentenac, 2003; Ashley et al., 2006; Gierth, Mäser, 2007; Grabov, 2007). Some families of potassium transporters function at low potassium concentrations. The high affinity transport system (HATS) is most studied (Maathuis, Sanders, 1994; Gierth, Mäser, 2007). ECG of H^+ created by the proton pump in the plasma membrane provides this system with energy. Proton enters into the cell along this gradient, and K^+ entrance is coupled with proton transport. Potassium flow into the cell and back is correlated with the intracellular potassium content. In most experiments, it is shown that ammonia and sodium suppress this system functioning.

The genes encoding HATS transporters are identified (Maser et al., 2001). Mutations in these genes inhibited potassium (Rb^{86}) uptake and suppressed root growth.

Four main families of HATS transporters are recognized.

1. HAK/KUP/KT – K^+/H^+ symporters, for example, AtHAK5 (Qi et al., 2008).
2. HKT family, i.e., the family, one function of which is a K^+/Na^+ symport at low sodium concentrations in medium and sodium transport at high sodium concentrations. The role of these transporters in potassium uptake is unclear; their role in sodium uptake is more evident. It seems likely that the TRK/HKT transporters

control sodium homeostasis in the cell and thus change the potassium status. Calcium affects the ratio between potassium and sodium transport exerted by this system, increasing selectivity for potassium (Epstein, 1961; Rus et al., 2004).

3. The members of the CPA family are also involved in potassium uptake. AtCHX-13 is most active under potassium deficit, determining potassium transport across the plasma membrane. Overexpression of its gene reduces sensitivity to potassium shortage (Zhao et al., 2008). The members of the CPA1 family mediate Na^+/H^+ exchange on both plasma membrane and intracellular membranes. Thus, NHX1 participates in the intracellular Na^+/H^+ exchange in the tonoplast vesicles, thus affecting potassium compartmentation. The *CPA2* gene encodes the K^+/Na^+-antiporter at potassium starvation and sodium-induced stress.
4. Poorly studied Shaker-type transporters (Lebaudy et al., 2007).

The second system of potassium transport, the low affinity transport system (LATS), is mediated by potassium channels (Schachtman, Shin, 2007; Gierth, Maser, 2007; Jung et al., 2009). This system is thermodynamically passive and transport potassium along the ECG depending on the value of the membrane potential. Most studied member of this family, AKT1, provides for potassium uptake by plant roots at both high and low potassium concentrations in medium (Hirsch et al., 1998, Spalding et al., 1999).

Low- and high-affinity potassium transport systems function on the tonoplast as well. Thus, five genes of high-affinity KUP family of tonoplast transporters were identified, including that encoding OsHAK10, which is supposed to mobilize vacuolar potassium for the maintenance of its cytoplasmic concentration at potassium shortage in medium. The members of tandem-pore potassium channels with a low affinity to K^+, TPC1, were found in the tonoplast.

Some data enlighten the regulatory ways for these transport systems. For example, AKT1 phosphorylation activates the channel via interaction with CIPK23 protein kinase. By reaction with calcium-binding proteins (calcineurin B-like CBL1 and CBL9}, activated AKT1 opens the channel (Li et al., 2006, Xu et al., 2006, Lee et al., 2007). AKT1 dephosphorylation by a 2C-type protein phosphatase inactivates the channel.

Another way to regulate conductivity of the potassium channel by calcium could be its interaction with guanine nucleotide-binding proteins (G-proteins). The degree of stomata opening is controlled just by this way (Fan et al. 2008). The AKT1 channel conductivity is changed due to its interaction with the cytoskeleton (Mills, Mandel, 1994). Finally, change in the plasma membrane potential could also affect gene expression for potassium transporters (Nieves-Cordones et al., 2008).

It was reported that some phytohormones, for example, jasmonic acid and ethylene, and polyamines could also regulate potassium channel functioning (Armengaud et al., 2004, Jung et al., 2009).

Activation of transport systems can not cope completely with potassium deficit, especially during long-term starvation. The additional way of plant adaptation is some changes in their morphology (Lebaudy et al., 2007; Maathuis, 2009). For example, at potassium deficit, root lengthening, enhanced growth of root hairs, and lateral root formation were frequently observed. These changes increase the root absorbing surface.

Plant adaptation also includes potassium redistribution between intracellular organelles. For example, potassium transport from the vacuole to the cytosol induced by potassium deficit maintains potassium homeostasis in the cytoplasm required for plant life (Li et al., 2006, Gierth, Mäser, 2007).

It is unclear how the cells appraise external potassium availability and which signal switch on the cascade activating potassium transport. It is known that potassium channels response to the transmembrane potassium gradient; however, it is not clear whether they function as sensors. According to Nieves-Cordones et al. (2008), the plasma membrane potential affects expression of genes encoding K^+ transporters. It is supposed that cytoplasmic calcium, calcium-binding proteins of the plasma membrane, cell wall proteins, including extensins and arabinogalactans, are involved in the recognizing signals about potassium status (Armengaud et al., 2004, Xu et al., 2006, Li et al., 2006). It is also supposed that reactive oxygen species might be involved in the perception of signals about potassium deficit (Shin, Schachtman, 2004). Thus, a possibility for regulation of AtHAK5 expression by the cell redox status was observed (Kim et al., 2010). Potassium deficit enhanced expression of peroxidase *RCI3* gene. The RCI3 protein is synthesized in the endoplasmic reticulum and transported toward the cell wall. The authors suggested that peroxidase could be a component of the K^+ signaling pathway in the roots.

YIELD

Thus, potassium is required for carbon dioxide assimilation, sucrose synthesis in source leaves, and photoassimilates transport toward sinks along the phloem. Plant supplement with sufficient amounts of potassium enhances photosynthesis and assimilate transport. Under the influence of potassium, starch, sucrose, monosaccharides, and proteins accumulate in sink organs, and this accelerates their growth and results in the high yield of main crops (Jung et al, 2009; Pettigrew, 1999, 2008)

Sufficient potassium supply not only increases the yield volume but also improves vegetable and fruit quality, and many corresponding examples are described in agronomic literature. For example, potassium fertilizers enhance vitamin C synthesis and the content of vitamin C in potato. Potassium facilitates nitrogen absorption by stimulating amino acid translocation from vegetative shoots into grains, and this increases the content of gluten and other proteins, which improve baking properties of the flour. By stimulation of carbohydrate synthesis and transport into plant reproductive organs, potassium increased markedly the content of starch in potato tubers, in the seeds of rice, wheat, soybean, sesame, and in some forage crops. The contents of sucrose in sugar beet roots and pectic compounds in fruits and berries were increased as well. Potassium application increased the content of oil in the seeds of sesame, soybean, rapeseeds, peanut, and cotton, the content of organic acids and carotene in fruits and vegetables; it prolonged the time of their storage. Potassium increased the content of proteins. In addition, it increased production of large and middle-sized potato tubers and prevented a decrease in the reduction in tuber weight after yield harvesting. Potassium shortage resulted in hexose accumulation and reduced the content of starch in potato tubers.

DISTRIBUTION OF POTASSIUM AND SUGARS IN PLANTS

When the plant is supplied with sufficient amount of potassium, there is no K^+- gradient within the cell. Thus, the X-ray microanalysis showed that potassium concentration in the cytoplasm and vacuole of differentiated root cortex cells did not essentially differ. However, before and at initial stages of cell elongation, potassium concentration in the vacuoles reduced slightly (Krasavina, Burmistrova, 1991). In this zone, phloem unloading occurs and sucrose enters the cells (Figure 1).

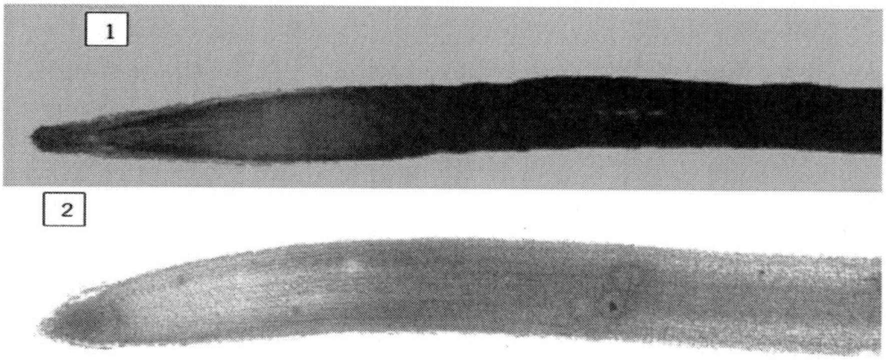

Note:

1. Fluorescence of carboxyfluorescein coming to the root apex along the phloem from cotyledons treated with 5 (6) carboxyfluorescein diacetate.
2. Cytochemical detection of invertase. Enzyme activity was assayed by blue color of reduced nitroblue tetrazolium. Cited from Feshchenko at al.,2004.

Figure 1. Phloem unloading and invertase activity in the apex of melon seedings.

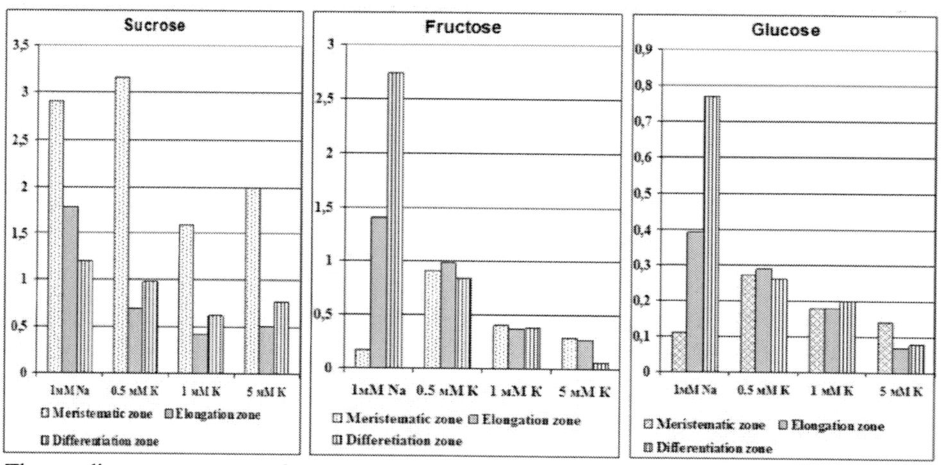

Note: The seedings were grown for 3 days in the thermostat (25°C) and 3 days on 0.5 Knop solutions with 0.5-5 mM potassium (K).

Figure 2. Sugar content in the root of pumpkin seedings grown on nutrient solution differing in potassium content (mg/g fr wt).

It was suggested that sucrose not used in the cytoplasm deposits into the vacuoles, replacing movable potassium. In the beginning of cell elongation, vacuolar invertase hydrolyzing sucrose into the two hexose molecules is activated, resulting in the turgor pressure increase facilitating cell elongation. Such change in the potassium concentration in the vacuole is in agreement with the notion that potassium distribution between the vacuoles and cytoplasm is determined by the medium composition (Grieth, Masci, 2007). When to grow maize or pumpkin seedlings on media differing in potassium concentration (0.2, 1.0, or 5 mM), potassium concentration in the high vacuolated differentiation zone reduced at potassium shortage in medium; and it increased at the higher potassium concentration (5 mM). Potassium concentration of 1 mM was evidently optimal for seedling growth: it was sufficient for metabolic processes in the cytoplasm and for its accumulation in the vacuoles in amounts providing for osmoregulatory processes (Figure 2).

Such event sequence is confirmed by the estimates of the total potassium content in the root cells at different stages of their differentiation. We analyzed the content of major cations and soluble sugars in 1-cm corn root segments positioned at different distanced from the root tip (Figure 3).

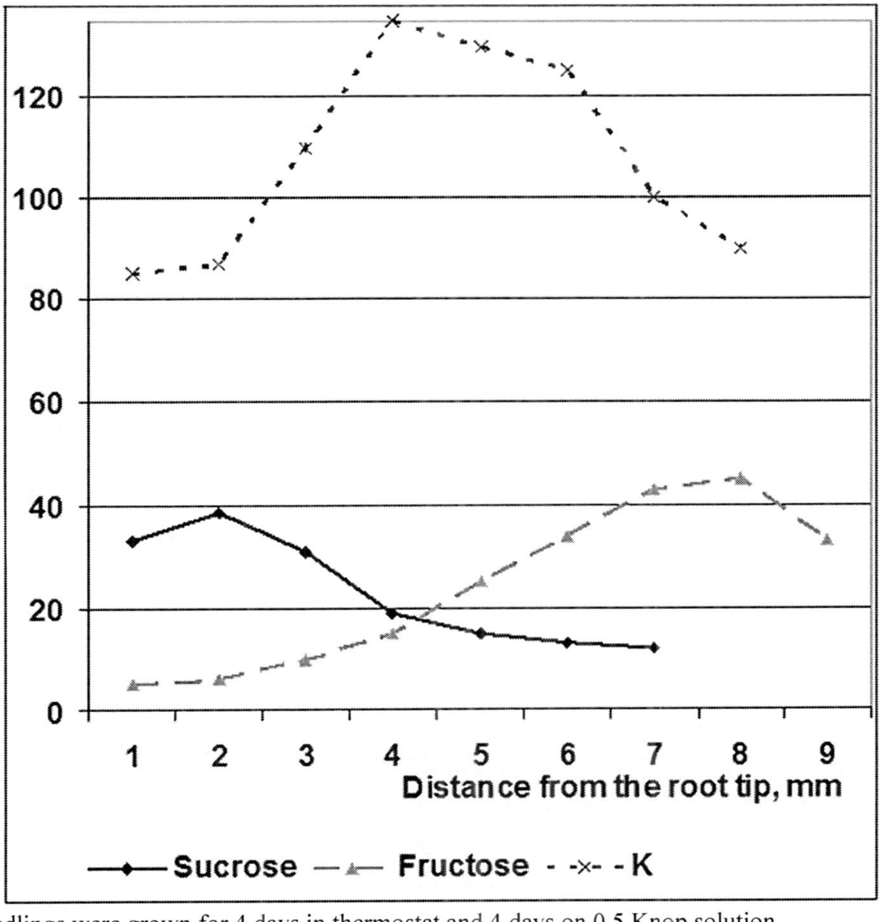

Note: seedlings were grown for 4 days in thermostat and 4 days on 0.5 Knop solution.

Figure 3. Sugars and potassium distribution along *Zea mays* root (μmol/g fr wt).

The content of potassium increased with the increase of the distance to 4--5 mm where the cells completing differentiation were located. These cells had large vacuoles occupying the most part of the cell volume. Therefore, the total content of compounds in this zone actually reflects their content in vacuoles. At the same time, the highest sucrose content was observed at a distance of approximately 2 mm from the root tip where the cells contain small vacuoles, only beginning to enlarge. In this case, the cytosol contributes much to the total content of compounds. Radial potassium distribution also indicates a dependence of potassium content on the extent of cell differentiation and vacuole development. In the beginning of differentiation, vacuoles completed their development in the cortical cells but the cells in the stele contained small vacuoles. X-ray microanalysis showed that the content of potassium was higher in the more vacuolated cortical cells then in the stellar cells with small vacuoles (Figure 4).

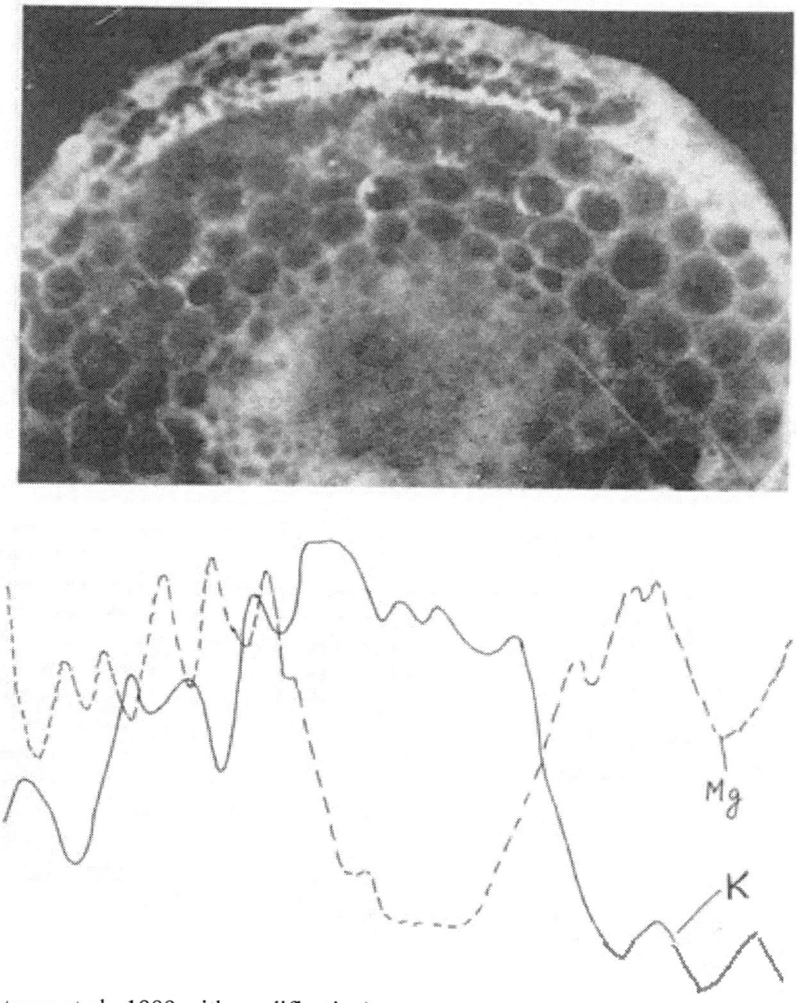

(From Burmistrova et al., 1999 with modification).

Figure 4. Frozen hydrated cryosection of differentiating pumpkin root, potassium and magnesium distribution through the centre of the root.

When maize plants were grown on medium, where potassium was replaced by sodium, the potassium content in the root differentiated zone declined, the content of sucrose increased insignificantly, but the content of hexoses increased markedly as compared with plants on potassium-containing medium (Table 1). These data indicate that the maintenance of the high potassium content in the cytosol required for metabolic processes can occur not only due to more active uptake of trace potassium amounts from medium but also due to its exit from vacuoles by H^+/K^+-symport (Walker et al., 1996). In this case, other osmotically active compounds, sugars in our example, can be involved in the osmotic potential maintenance; in fact, sugar content in vacuoles increased. It seems likely that a capacity of plants to maintain the constant potassium concentration in the cytoplasm due to its depletion in the vacuoles provides for plant tolerance to potassium deficit. In this case, other cations can participate in the regulation of osmotic pressure (Marschner et al., 1981, Subbarao et al, 1999). Under salinity or heavy metal stresses, such capacity to replace vacuolar potassium with other elements could reflect also nonspecific plant tolerance.

Similar interchangeability of osmotically active compounds determining turgor pressure was observed also during fruit maturation. For example, during maturation of grapevine berries, sugar content increased simultaneously with potassium depletion (Tsitsilashvili et al. 1984). The reverse relations between potassium and sugar contents were manifested still more brightly in the sugar beet tap roots. The two periods were observed during development of these tap roots (Semenov, 1986). In our experiment, during the early stages of root growth, sucrose accumulation was accompanied by a sharp decrease in the potassium content (Figure 5); there was no substantial changes in magnesium and sodium contents.

Table 1. Cation and sugar contents in the root zones of *Zea mays* seedlings grown on different solutions during 5 days, μmol/g fr wt

Treatment	Root zones	K^+	Na^+	Mg^{++}	Sucrose	Fructose	Glucose
Initial seedlings	1	69,6±1,7	1,4±0,4	29,7±2,1	13,7±3,3	3,6±1,1	2,7±0,5
	2	77,2±5,3	1,4±0,2	20,6±2,9	9,1±0,9	30,0±8,6	22,8±7,1
	3	49,4±5,3	0,9±0,2	10,0±0,4	6,5±0,7	26,6±4,0	31,7±2,9
KCl	1	79,8±8,2	6,6±0,6	41,8±5,3	15,0±1,3	1,3±0,2	3,8±0,6
	2	110,2±8,7	2,2±1,0	39,4±5,8	5,3±1,3	12,0±1,3	17,0±2,3
	3	93,9±5,2	1,4±0,6	27,9±4,1	3,4±0,6	4,9±0,9	10,9±1,7
NaCl	1	83,8±9,2	5,4±1,5	45,1±3,7	13,0±1,9	4,3±0,2	2,7±0,4
	2	101,1±1,3	6,0±1,1	59,2±1,6	7,6±1,2	12,7±3,5	17,5±3,6
	3	48,2±8,7	3,4±0,9	48,2±0,6	5,2±0,1	9,6±2,5	20,7±4,6

Note: Maize roots were germinated in the thermostat for 3 days and then transferred to 1 mM KCl or 1 mM NaCl. In 4 days, roots were subdivided into three zones: (1) meristem (0--1.5 mm from the tip); (2) elongation (1.5--3.0 mm from the tip); and (3) differentiation (5.0--10.0 mm from the tip). Cations were determined by flame photometry after tissue ashing in HNO_3; glucose was determined by the glucoseoxidase method; sucrose and fructose, by the Roe method. Five experiments were performed using 30--40 plants for each treatment.

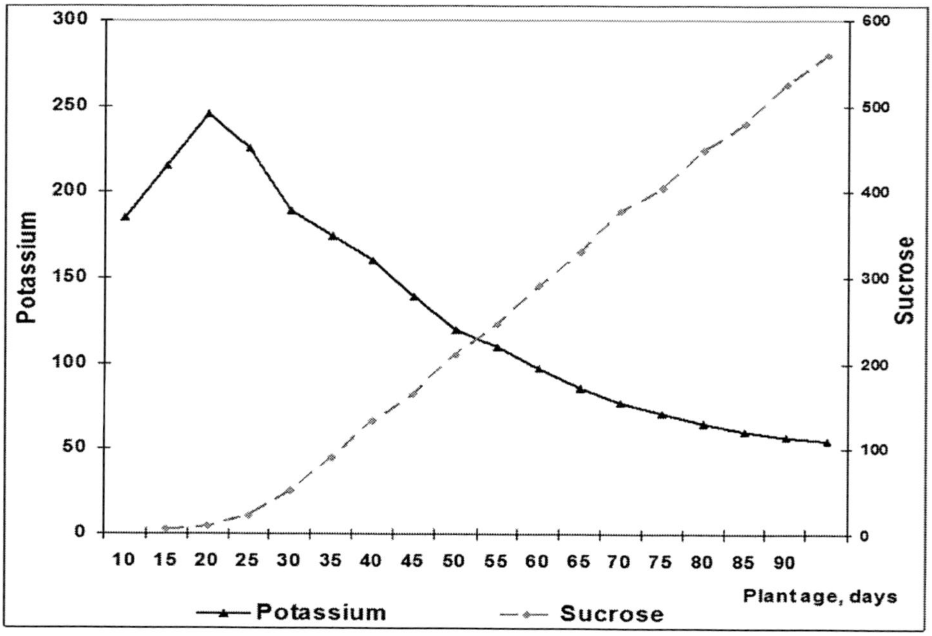

Figure 5. Potassium and sucrose content in sugar beet taproot (µmol/g fr wt).

During this period there was a balance between potassium and sucrose contents, so that the osmotic potential of the sap squeezed from the roots remained unchanged. During further plant growth, osmoregulation in its tap root tissues was disturbed. Potassium concentration attained its lowest value (about 25 µmol/g fr wt), whereas the content of sucrose continued to increase substantially. This resulted in the marked increase in the root tissue osmotic pressure. In this period, the plant copes with such situation otherwise, e.g., by morphological changes. If the sugar beet plant was transferred to darkness, the sucrose content was reduced and potassium content was increased, especially in young plants. The osmotic potential in such plants was unchanged, but in older plants it became lower (Table 2).

Similar changes in the potassium content were observed during development of red beet (Salyaev et al., 1983) and red radish root (Figure 6): it decreased sharply in the period of most active sugar accumulation in the vacuolar sap. A difference between these plant species was in the composition of accumulating sugars: sucrose predominated in the red beet roots and monosaccharides, fructose mainly, in the radish roots.

Table 2. Effect of 3 day long darkness on potassium and sucrose contents in sugar beet taproots (µmol/g fr wt)

Index	Sugar beet age			
	37 days		72 days	
	light	dark	light	dark
Potassium	101,0±5,4	163,4±15,2	90,9±5,2	72,8±5,3
Sucrose	208±17,5	63,2±8,5	319,1±19,8	116,8±9,4
Osmotic potential, mOsm	410±10,2	390±10,7	500±12,8	262,4±11,6

Note: During 37 or 72 days plants were kept in the greenhouse under natural illumination; thereafter, some plants were transferred to darkness for 4 days.

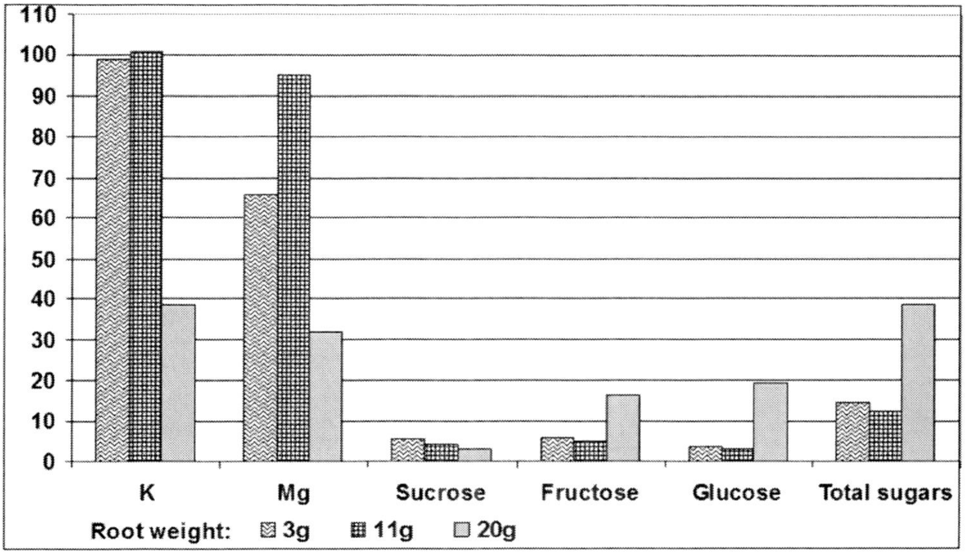

Note: It is clearly seen that potassium content was reduced with the increase in the root weight (from 3 to 20 g fr wt), whereas the content of hexoses increased.

Figure 6. Sugars and cations content in red radish taproot (μmol/g fr wt).

Like in the case of differentiated maize roots, in the process of sugar accumulation in plant storage organs, potassium translocation is evidently occurs but, in this case, between different plant parts. Thus, simultaneously with diminishing potassium concentration in the sap of grapevine berries, it increased in the fruit-stems. Similar pattern was observed between the sugar beet leaves and petioles. The highest potassium concentrations in the leaf blades were characteristic of the youngest and senescing leaves. In active source leaves, the content of potassium in the leaf blades was markedly reduced, and the content of potassium in petioles exceeded substantially that in the leaf blades. It might be that, like in grapevine, sugar accumulation under conditions of active photosynthesis induced potassium redistribution between the leaf blade and petiole. When plants were shaded for several days, potassium concentration in petioles reduced and the difference between the blades and petioles was smoothed. It might be that a decrease in the sugar content in the leaf blade occurring in darkness and, as a consequence, assimilate efflux retardation induced potassium reverse transport from petioles to leaf blades or enhanced potassium transport toward the sugar-depleted leaf blades.

A comparison of different grapevine cultivars differing in the sugar accumulation showed a stringent reverse correlation between the contents of sugar and potassium in the berries (Tsitsilashvili et al., 1984). Thus, mature berries of cultivars with high sugar content contained low potassium concentrations in berries and the high potassium concentrations in the fruit-stems. In contrast, berries with the low sugar contents were enriched in potassium, but its content in the fruit-stem was low. One more observation is of interest: cultivars accumulated the highest sugar amounts in mature berries were characterized by the highest potassium content during fruit active growth. Based on these observations, the authors supposed that berry capacity to accumulate sugars was largely determined by the initial potassium content in immature berries. Therefore, during the period of active berry growth, stable plant supplying with potassium is of importance. Just in this period, potassium fertilizers are most efficient. At the same time, organic acid accumulate as well, which decrease the pH of the cell sap. It might be that this results in the proton pump activation exchanging protons for potassium. During berry maturation, acidity reduced and the mechanisms are activated providing for the active sugar transport into the berries coupled with potassium export.

We analyzed the sugar and potassium content in the tap roots of sugar beet of different cultivars. It was shown that sugar beet cultivars behavior totally different from that of grape berry: cv. Yanash with the highest sugar content and cv. Mangold with the lowest sugar content contained approximately similar potassium concentration in the beginning of the tap root growth, and its changes during root growth were similar as well.

Thus, the fact of interchangeability of potassium and sugars in the maintenance of the osmotic potential is well established. However, the mechanisms involved are not clear. According to the Saftner and Wyse opinion, stated still in the 1980, sucrose is transported in the vacuoles of parenchymal cells in the sugar beet tap root by the active mechanism functioning on the tonoplast (Saftner, Wyse, 1980). This transport is driven by the electrochemical gradients for potassium and proton directed oppositely: potassium enters the vacuole along its ECG, whereas proton exit from the vacuole along its ECG. The K^+/H^+ antiporter is coupled with sucrose transport into the vacuole. Another antiporter was detected in the tonoplast directly binding proton exit and sucrose uptake. However, in both cases, potassium and sucrose both penetrate the vacuole.

Another suggestion concerns plants grown under potassium deficit in medium. In this case, potassium concentration in the cytoplasm is maintained at expense of vacuolar potassium. Therefore, initially potassium concentration in the vacuole reduces and turgor pressure could be reduced as well. Such a decrease could be recognized by turgor-sensitive transport mechanisms. The penetration of exogenous sugars into the cells is known to depend on turgor. Sucrose in the root intercellular space could appear due to its leakage from the phloem, and also it could be produced by the microorganisms in the rhyzosphere. Most researchers believe that plasmalemmal H^+-ATPase plays a key role in cell osmotic adjusting (Reinhold et al., 1984). At declining turgor pressure, this ATPase is activated due to interaction with 14-3-3 proteins (Gonzales-Reyes et al., 1994; Shabala, Lew, 2002; Oren-Shamir et al., 1989). Activation of H^+-ATPase leads to plasma membrane hyperpolarization (Pertl et al., 2010). Another reason for the increase in the membrane potential due to the increase in the transmembrane potassium gradient due to its lower concentration in the apoplast than in the cytosol. Plasma membrane hyperpolarization was really observed in the initial period of potassium starvation (Krasavina, 1974, 1977). ATPase activation and an increase in the membrane potential could enhance sucrose uptake when it occurs via the H^+/sucrose antiport (Walker et al., 2000). Such explanation is also not quite suitable for understanding sugar accumulation in the cells coupled with potassium depletion because ATPase activation and membrane hyperpolarization should induce absorption of apoplastic potassium by the cell (or its retaining in the cytoplasm).

A new approach for the revealing of dependency between sugar accumulation in the cells and a decrease in the content of potassium in them is evident when to take into account the possibility of effects of reactive oxygen species (ROS) on sugar uptake by the roots of potassium-deficient plants (Jung et al., 2009). It was shown that in such plants as soon as in several hours, i.e., simultaneously with activation of the system of potassium uptake, ROS accumulated (Shin, Schachtman, 2004; Jung et al., 2009). This occurred in the root zone most active in potassium uptake and its transport to the shoot, i.e., in the apical area of the absorption zone (Moritsugu, 1993). It seems likely that induction of ethylene synthesis just in this zone (Jung et al., 2009) caused ROS accumulation. Both ethylene and ROS activated K-transporter HAK5, and this resulted in activation of potassium uptake and an increase in its intracellular content. Treatment of potassium-deficient plants with hydrogen peroxide induced high-affinity systems of potassium uptake. In addition to the HAK5 K-channel, hydrogen peroxide activated KEA5 transporter (K^+/H^+-antiporter) and WRKY9 transcription factor (Shin, Schachtman, 2004). ROS accumulation was correlated with stimulation of sucrose and glucose uptake (Gonzales-Reyes et al., 1994). Plant treatment with the natural electron acceptor AFR (intermediate form of redox pair ASC/DHA) enhanced absorption of sucrose, glucose, and fucose evidently due to H^+-ATPase activation and, as a consequence, activation of H^+/sugar-antiporter on the tonoplast. Ascorbate and dehydroascorbate suppressed sugar uptake. This indicates that activation of the membrane redox system is involved not only in potassium uptake by plant roots but also sugar uptake, i.e., it increases the content of osmotically active compounds and enhance root elongation.

More reasonable seems to consider phloem sap as the only source of osmotic substances for sink cells. Phloem sap is known to contain by an order of magnitude higher sucrose than potassium concentration. In fact, potassium concentration in the phloem sap is about 100 mM, whereas that of sucrose in the phloem of source leaves could attain 1 M. When transported to sinks, sucrose concentration declines but not below 300 mM. Thus, even when we suppose that potassium concentration is not changed along the phloem, the phloem sap in sinks contains more sucrose than potassium. This ratio should be higher in plants deficient in potassium.

The osmotic pressure may be regulated by mechanosensory ionic mechanisms, including stretch-activated and stretch-inhibited ionic channels. This mechanism of osmoregulation is best studied in prokaryotes. Stretch-dependent ionic channels were also detected for K, Cl, and Ca in plants (Cosgrove, Hedrich, 1991; Shabala et al., 2000; Shepherd et al., 2002; Shabala, Lew, 2002; Okazaki et al., 2002; Martinac, 2004). In addition to specific stretch-sensitive ionic channels, other channels could manifest mechanosensory properties, for example, Shaker-type K^+-channels, the members of the superfamilies MEC/DEG (mechanosensory abnormal/degenering) or TRP (transient receptor potential). Although mechanosensory channels are found in membranes of many plant cells, they are poorly studied and their functions are unclear. It seems likely that they function under increased turgor pressure when the membrane is stretched (Zhang et al., 2007).

Thus, although some transport systems for both potassium and sugars were identified and corresponding genes were found, the interaction between potassium and sugar absorption remain puzzling. On the one hand, potassium is involved in some reactions of sugar biosynthesis in the process of photosynthetic carbon fixation. It is required for assimilate phloem loading and their long-distance transport along the phloem toward sinks. On the other hand, increase in the potassium content in medium suppresses sugar transport into the cells (Pitman et al., 1971). Moreover, during fruit and storage root maturation, potassium present in the vacuoles exit from the cells. However, even these scarce data, which causes are still not unveiled by the researchers, could be beneficial for farmers. At the early stages of growth, plants demand to be richly supplied with potassium to assure growth processes and accumulate potassium in young fruits. Therefore, in this period, the plant should be completely provided with potassium. Taking into account a scarcity of many soils and their depletion in potassium, potassium fertilizers should be used in the beginning of plant development or even before sowing. In the period of fruit or storage root development, the content of potassium drops simultaneously with sugar accumulation; therefore, there is no need in potassium fertilizers in this time. Such agricultural technology turned out to be encouraging on the vineyards (Tsitsilashvili et al., 1984).

References

Amir, S; Reinhold, L. Interaction between K+-deficiency and light in 14C-sucrose translocation in bean plants. *Physiol .Plant*, 1971 24, 226-231.

Armengaud, P; Breiting, R; Amtmann, A. The potassium-dependent transcriptome of Arabidopsis reveals a prominent role of jasmonic acid in nutrient signaling. *Plant Physiol.,* 2004 36, 2556-2576.

Ashley, MK; Grant, M; Grabov, A. Plant responses to potassium deficiencies: a role for potassium transport proteins. *J. Exp. Bot.*, 2006 57, 425-436.

Britto, DT; Kronzucker, HJ. Cellular mechanisms of potassium transport in plants. *Physiol. Plant.,* 2008 133, 637-650.

Burmistrova, NA; Krasavina, MS; Pogorelov, AG. Use of X-ray microanalysis for study of cation distribution in potassium deficient pumpkin roots. *Acta Bot. Polon.*, 1991 60, 303-312.

Burmistrova, NA; Krasavina, MS; Pogorelov, AG. Use of X-ray microanalysis for study of highly hydrated cell compartments (plant vacuoles). XIII Intern Congr X-ray Optics Microanal, 1992 Manchester, p.21.

Cakmak, I; Hengeler, C; Matschner, H. Changes in phloem export of sucrose in leaves in response to phosphorus, potassium and magnesium deficiency in bean plants. *Exp. Bot.,* 1994 45б, 1251-1257.

Cosgrove, DJ; Hedrich, R. Stretch-activated chloride, potassium, and calcium channels coexisting in plasma membranes of guard cells of *Vicia faba* L. *Planta,* 1991 186, 143-153.

Dietrich, P; Sanders, D; Hedrich, R. The role of ion channels in light-dependent stomatal opening. *J. Exp. Bot.,* 2001 52б, 1959-1967.

Dolan, L; Davies, J. Cell expansion in roots. *Curr. Opin. Plant. Biol.*, 2004 7, 33-39.

Epstein, E; Rains, DW; Elzam, OE. Resolution of dual mechanisms of potassium absorption by barley roots. *Proc. Natl. Acad. Sci. USA*, 1963 49, 684-692.

Epstein, E. The essential role of calcium in selective cation transport by plant cell. *Plant Physiol.*, 1961 36, 437-444.

Evans, HJ; Wildes, RA. Potassium and its role in enzyme activation. In: Potassium in biochemistry and physiology. International Potash Institute. Berne. Switzerland. 1971, 13-39.

Fan, L-M; Wang, Y-F; Hong W; Wu, W-H. In vitro Arabidopsis pollen germination and characterization of the inward potassium currents in Arabidopsis pollen grain protoplasts. *J. Exp. Bot.*, 2001 52, 1603-1614.

Fan, L-M; Zhang, W; Chen, J-G; Taylor, P; Jones, AM; Assmann, SM. Abscisic acid regulation of guard-cell K+ and anion channels in Gß- and RGS-deficient Arabidopsis lines. *Proc. Natl. Acad. Sci. USA*, 2008 105, 8476-8481.

Feshchenko, NF; Krasavina, MS; Burmistrova, NA; Nosov, AV. Phloem unloading and activity of enzymes hydrolyzing sucrose in the apex of the growing root. *Doklady Biochem. Biophys.*, 2004 399,705-707.

Fischer, RA. Stomatal opening: role of potassium uptake by guard cells. *Science,* 1968 169,784-785.

Gepstein, S; Sabehi, G; Carp, M-J; Hajouj, T; Nesher, MFO; Yariv, I; Dor, C; Bassani, M. Large-scale identification of leaf senescence-associated genes. *Plant J.,* 2003 36, 629-642.

Gibon, Y; Pyl, ET; Sulpice, R; Lunn, JE; Höhne, M; Günther, M; Stitt, M. Adjustment of growth, starch turnover, protein content and central metabolism to a decrease of the carbon supply when Arabidopsis is grown in very short photoperiods. *Plant Cell Environ.*, 2009 32, 859-874.

Gierth, M; Mäser, P; Schroeder, JI. The potassium transporter AtHAK5 functions in K(+) deprivation-induced high-affinity K(+) uptake and AKT1 K(+) channel contribution to K(+) uptake kinetics in Arabidopsis roots. *Plant Physiol.*, 2005 137, 1105-14.

Gierth, M; Mäser, P. Potassium transporters in plants--involvement in K+ acquisition, redistribution and homeostasis. *FEBS Lett.*, 2007 581, 2348-2356.

Gonzalez-Reyes, JA; Hidalgo, A; Caler, JA; Palos, R; Navas, P. Nutrient Uptake Changes in Ascorbate Free Radical-Stimulated Onion Roots. *Plant Physiol.*, 1994 104, 271-276.

Grabov, A. Plant KT/KUP/HAK potassium transporters: single family – multiple functions. *Ann. Bot.*, 2007 99, 1035- 1041.

Graham, RD; Ulrich, A. Potassium deficiency-induced changes in stomatal behavior, leaf water potentials, and root system permeability in *Beta vulgaris* L. *Plant Physiol.*, 1972 49, 105-109.

Hartt, CE. Effect of potassium deficiency upon translocation of C in attached blades and entire plants of sugarcane. *Plant Physiol.*, 1969 44, 1461-1469.

Hartt, CE. Effect of potassium deficiency upon translocation of C in detached blades of sugarcane. *Plant Physiol.*, 1970 45, 183-187.

Hirsch, RE; Lewis, BD; Spalding, EP; Sussman, MR. A role for the AKT1 potassium channel in plant nutrition. *Science*, 1998 280, 918-921.

Hodick, D; Sievers, A. On the mechanism of trap closure of Venus flytrap (*Dionea muscipula* Ellis). *Planta*, 1989 179, 32-42.

Huber, SC. Biochemical basis for effects of K-deficiency on assimilate export rate and accumulation of soluble sugars in soybean leaves. *Plant Physiol.*, 1984 76, 424-430.

Jackson WA, Volk RJ. Role of potassium in photosynthesis and respiration. In The role of potassium in agriculture. *Amer. Soc. Agron, Madison*, 1968 P 109-145

Jung, J-Y; Shin, R; Schachtman, DP. Ethylene mediates response and tolerance to potassium deprivation in Arabidopsis. *Plant Cell,* 2009 21, 607-621.

Kim, MJ; Ciani, S; Schachtman, DP. A peroxidase contributes to ROS production during Arabidopsis root response to potassium deficiency. *Mol. Plant,* 2010 3, 420-427.

Krasavina, MS. On an equipotentiality of cortex root cells. *Fiziologia Rastenii*, 1974 21, 5-11 (in Russian).

Krasavina, MS. Potassium deficiency and permeability of plant cell membrane. *Fiziologia Rastenii,* 1977 24, 429-430 (in Russian).

Krasavina, MS; Burmistrova, NA. Potassium distribution in root cells in relation to cell differentiation and environmental conditions. In: *Root ecology and its practical application.* Vienna, 1991 25-28.

Kuchenbuch, R; Claassen, N; Jungk. A. Potassium availability in relation to soil-moisture. 1. Effect of soil-moisture on potassium diffusion, root-growth and potassium uptake of onion plants. *Plant Soil,* 1986 95, 221-231.

Kühn, C; Grof, CP. Sucrose transporters of higher plants. *Curr. Opin. Plant Biol,* 2010 13, 288-298.

Kursanov, AL; Vyskrebentzeva, EI. Metabolism of plant in conditions of potassium deficiency. *Agrochemia,* 1967 1, 65-77. (in Russian)

Kursanov, AL; Vyskrebentzeva, EI. The rope of potassium in plant metabolism and the biosynthesis of the compounds which determine the quality of agricultural products. *6th Internat. Potash Congr. Brussels,* 1966, p.1-22.

Lebaudy, A; Véry, AA; Sentenac, H. K+ channel activity in plants: genes, regulations and functions. *FEBS Lett,* 2007 581, 2357-2366.

Lee, SC; Lan, WZ; Kim, BG; Li, L; Cheong, YH; Pandey, GK; Lu, G; Buchanan, BB; Luan, S. A protein phosphorylation/dephosphorylation network regulates a plant potassium channel. *Proc. Natl. Acad. Sci. USA,* 2007 104, 15959-15964.

Li, L; Kim, BG; Cheong, YH; Pandey, GK; Luan, S. A Ca(2+) signaling pathway regulates a K(+) channel for low-K response in Arabidopsis. *Proc. Natl. Acad. Sci. USA,* 2006 103, 12625-12630.

Libersbach, H; Steingrobe, B; Claassen, N. Poos regulate ion transport in the rhizosphere to counteract reduced mobility in dry soil. *Plant Soil,* 2004 260, 79-88.

Maathuis, FJ. Physiological functions of mineral macronutrients. *Curr. Opin. Plant Biol.,* 2009 12, 250-258.

Maathuis, FJ. The role of monovalent cation transporters in plant responses to salinity. *J. Exp. Bot.,* 2006 57, 1137-1147.

Maathuis, FJM; Sanders, D. Mechanism of high affinity potassium uptake in roots of *Arabidopsis thaliana. Proc. Natl. Acad. Sci. USA,* 1994 91, 9272-9276.

Marschner, H. Mineral nutrition of higher plants. London: Academic Press. Harcourt Braceand Company Publishers; 1995.

Martinac, B. Mechanosensitive ion channels: molecules of mechanotransduction. *J. Cell Sci.,* 2004 117, 2449-2460.

Maser, P; Thomine, S; Schroeder JI. Phylogenetic relationships within cation transporter families of Arabidopsis. *Plant Physiol.,* 2001 126, 1646-1667.

Mengel, K. Effect of potassium on the assimilate conduction to storage tissue. *Berichte Deutsch. Bot. Gesel.,* 1980 93, 353-362.

Mengel, K; Haeder, HE. Effect of potassium supply on the rate phloem sap exudation and the composition of phloem sap of *Ricinus communis*. *Plant Physiol,* 1977 59, 282-284.

Mengel, K; Secer, M; Koch, K. Potassium effect on protein formation and amino acid turnover in developing wheat grain. *Agron. J.,* 1981 73, 74-78.

Mills, JW; Mandel, LJ. Cytoskeletal regulation of membrane transport events. *FASEB J,* 1994 8, 1161-1165.

Moran, N. Osmoregulation of leaf motor cells. *FEBS Lett,* 2007 581, 2337-2347.

Nieves-Cordones, M; Miller, AJ; Alemán, F; Martínez, V; Rubio, F. A putative role for the plasma membrane potential in the control of the expression of the gene encoding the tomato high-affinity potassium transporter HAK5. *Plant Mol. Biol.*, 2008 68, 521-32.

Okazaki, Y; Ishigami, M; Iwasaki, N. Temporal relationship between cytosolic free Ca(2+) and membrane potential during hypotonic turgor regulation in a brackish water charophyte *Lamprothamnium succinctum. Plant Cell Physiol.*, 2002 43, 1027-1035.

Oren-Shamir, M; Pick, U; Avron, M. Involvement of the plasma membrane ATPase in the osmoregulatory mechanism of the alga Dunaliella salina. *Plant Physiol.*, 1989 89, 1258-63.

Peoples, TR; Koch, DW. Role of potassium in carbon dioxide assimilation in *Medicago sativa* L. *Plant Physiol.*, 1979 63, 878-881.

Pertl, H; Pöckl, M; Blaschke, C; Obermeyer, G. Osmoregulation in Lilium pollen grains occurs via modulation of the plasma membrane H+ ATPase activity by 14-3-3 proteins. *Plant Physiol.*, 2010 154, 1921-1928.

Pettigrew, WT. Potassium deficiency increases specific leaf weight and leaf glucose levels in field-grown cotton. *Agron. J.*, 1999 94, 962-968.

Pettigrew, WT. Potassium influences on yield and quality production for maize, wheat, soybean and cotton. *Physiol. Plant.*, 2008 133, 670-681.

Pitman, MG; Mowat, J; Nair, H. Interactions of processes for accumulation of salt and sugar in barley plants. *Austr. J. Biol. Sci.*, 1971 24, 619-631.

Qi,Z; Hampton, CR; Shin, R; Barkla, BJ; White, PJ; Schachtman, DP. The high affinity K+ transporter AtHAK5 plays a physiological role in planta at very low K+ concentrations and provides a caesium uptake pathway in Arabidopsis. *J. Exp. Bot.*, 2008 59, 595-607.

Rehman, S; Rha, ES; Ashraf, M; Lee, KJ; Yun, SJ; Kwak, YG; Yoo, NH; Kim, J-K. Does barley (*Hordeum vulgare* L.) pollen swell in fractions of a second? *Plant Sci.*, 2004 167, 137-142.

Rehman, S; Yun, SJ. Developmental regulation of K accumulation in pollen anthers, and papillae: are anther dehiscence, papillae hydration, and pollen swelling leading to pollination and fertilization in barley (*Hordeum vulgare* L.) regulated by changes in K concentration? *J. Exp. Bot.*, 2006 57, 1315-1321.

Reinhold, L; Seiden, A; Volokita, M. Is modulation of the rate of proton pumping a key event in osmoregulation? *Plant Physiol.*, 1984 75, 846-849.

Rus, A; Lee, BH; Munoz-Mayor, A; Sharkhuu, A; Miura, K; Zhu, JK; Bressan, RA; Hasegawa, PM. AtHKT1 facilitates Na+ homeostasis and K+ nutrition in planta. *Plant Physiol.*, 2004 136, 2500-2511.

Saftner, RA; Wyse, RE. Alkali cation/sucrose co-transport in the root sink of sugar beet. *Plant Physiol.*, 1980 66, 884-889.

Salyaev, RK; Pusanova, NA; Lobocho, CB. Metabolite distribution in red beet tap root. *Inform Bull. SIFIBR*, 1983 15-17 (in Russian).

Schachtman, DP; Shin, R. Nutrient sensing and signaling: NPKS. *Annu. Rev. Plant Biol.*, 2007 58, 47-69.

Semenov, IL. The storage organ of sugar beet: are there two stages of sugar accumulation? *Fiziologia Rastenii*, 1986 33, 1090-1095 (in Russian).

Shabala, SN; Lew, RR. Turgor regulation in osmotically stressed Arabidopsis epidermal root cells. Direct support for the role of inorganic ion uptake as revealed by concurrent flux and cell turgor measurements. *Plant Physiol.*, 2002 129, 290-299.

Shepherd, VA; Beilby, MJ; Shimmen, T. Mechanosensory ion channels in charophyte cells: the response to touch and salinity stress. *Eur. Biophys. J.*, 2002 31, 341-355.

Shin, R; Schachtman, DP. Hydrogen peroxide mediates plant root response to nutrient deprivation. *Proc. Nat. Acad. Sci. USA*, 2004 101, 8827-8832.

Sorger, GJ; Ford, RE; Evans, HJ. Effects of univalent cations on the immunoelectrophoretic behavior of pyruvic kinase. *Proc. Natl. Acad. Sci. USA*, 1965 54, 1614–1621.

Spalding, EP; Hirsch, RE; Lewis, DR; Qi, Z; Sussman, MR; Lewis, BD. Potassium uptake supporting plant growth in the absence of AKT1 channel activity: inhibition by ammonium and stimulation by sodium. *J. Gen. Physiol.*, 1999 113, 909-918.

Suelter, CH. Enzymes activated by monovalent cations. *Science*, 1970 168, 789-795.

Talbott, LD; Zeiger, E. General role of potassium and sucrose in guard cell osmoregulation. *Plant Physiol.*, 1996 111, 1051-1057.

Tsitsilashvili, OK; Okanenko, AC; Bershtein, BI; Bibilashvili, PD; Derbeneva, TG; Namgaladse, RA; Sikharulidze, LM. Changes of potassium concentration in grape buches during sugar accumulation in grapes. *Fiziologia Rastenii*, 1984 31, 469-474 (in Russian).

Véry, AA; Sentenac, H. Molecular mechanisms and regulation of K+ transport in higher plants. *Annu. Rev. Plant Biol.*, 2003 54, 575-603.

Vyskrebentseva, EI; Krasavina, MS. Effect of potassium deficiency on the free nucleotide content in pumpkin roots. *Fiziologia Rastenii*, 1966 13, 433-445 (in Russian).

Walker, DJ; Leigh, RA; Miller, AJ. Potassium homeostasis in vacuolated plant cells. *Proc. Natl. Acad. Sci. USA,* 1996 93, 10510-10514.

Walker, NA; Zhang, WH; Harrington, G; Holdaway, N; Patrick, JW. Effluxes of solutes from developing seed coats of *Phaseolus vulgaris* L. and *Vicia faba* L.: locating the effect of turgor in a coupled chemiosmotic system. *J. Exp. Bot.*, 2000 51, 1047-1055.

Wyse, RE; Zamski, E; Tomos, AD. Turgor regulation of sucrose transport in sugar beet taproot tissue. *Plant Physiol.*, 1986 81, 478-481.

Xu, J; Li, HD; Chen, LQ; Wang, Y; Liu, LL; He, L; Wu, WH. A protein kinase, interacting with two calcineurin B-like proteins, regulates K+ transporter AKT1 in Arabidopsis. *Cell,* 2006 125, 1347-1360.

Zhang, W; Fan, LM; Wu, WH. Osmo-sensitive and stretch-activated calcium-permeable channels in *Vicia faba* guard cells are regulated by actin dynamics. *Plant Physiol.,* 2007 143, 1140-1151.

Zhao, J; Cheng, N-H; Motes, CM; Blancaflor, EB; Moore, M; Gonzales, N; Padmanaban, S; Sze, H; Ward, JM; Hirschi, KD. AtCHX13 is a plasma membrane K+ transporter. *Plant Physiol.,* 2008 48, 796-807.

In: Soil Nutrients
Editor: Mohammad Miransari
ISBN: 978-1-61324-785-3

Chapter 2

THE ROLE OF REDOX CONDITIONS ON SOIL NUTRIENTS AVAILABILITY

Artur Banach* and Zofia Stępniewska
The John Paul II Catholic University of Lublin, Institute of Biotechnology, Department of Biochemistry and Environmental Chemistry, Lublin, Poland

ABSTRACT

This chapter will focus on the role of redox potential (Eh) on nutrient availability (N, P) of different soils under changing redox conditions. In addition, other elements involved in biogeochemical nutrient cycling such as Mn, Fe and S will be discussed.
Soil microbial activity leads to the changes of nutrient availability what is connected with soil aeration state. The change of soil aeration state leads to a shift of the activity of different microbes affecting biogeochemical cycling of elements. There is a subsequent reduction of nitrates, oxidized manganese and iron minerals, followed by the reduction of sulfate and, finally carbon dioxide under anaerobic conditions due to oxygen depletion (Stumm and Morgan, 1970; Ponnamperuna, 1972; Gliński and Stępniewski, 1985; Laanbroek, 1990; Schlesinger, 1997; Smolders et al., 2006). Under such conditions the accumulation of reduced products could lead to adverse effects for the organisms. Moreover, end-products such as nitrous oxide and methane are emitted into atmosphere what strengthen greenhouse effect.

Nitrogen is an important, potentially growth-limiting nutrient in floodplains and the input of nitrate may lead to the development of fast growing species instead of target vegetation. N is much more mobile than P, and is lost to the atmosphere by denitrification and anammox (Paul and Clark, 1996). It is very powerful electron acceptor and its presence may hamper P mobilization due to buffering of soil Eh (Lucassen et al., 2004; Banach et al., 2009).

Phosphorus is another important nutrient for plants; its availability has a strong impact on plant species composition as it can be a growth-limiting factor. When P availability increases it can cause eutrophication and dominance of fast-growing species whereas natural, characteristic species disappear. In the soil it is bound to Fe, Al and Ca

---

* The John Paul II Catholic University of Lublin, Institute of Biotechnology, Department of Biochemistry and Environmental Chemistry, Al. Kraśnicka 102, 20-718 Lublin, Poland. E-mail: abanach@kul.lublin.pl.

minerals as well as present in organic form. The Fe-bound P fraction is a redox sensitive and P may easily become mobilized from it leading to very high concentrations in both pore water and surface water. It is a result of the reduction of Fe minerals (Patrick and Khalid, 1974; Tiemeyer et al., 2005; Stępniewska et al., 2006; Banach et al., 2009) and by sulfate reduction. Sulfide is a product of the latter process and it interferes the Fe and P cycle by iron reduction (Caraco et al., 1989; Sperber, 1958; Roden and Edmonds, 1997; Lamers et al., 2002). This phenomenon is called internal eutrophication (Roelofs, 1991; Smolders et al., 2006).

Availability of macro- and micronutrients and their transformation related either directly or indirectly to the soil redox state are discussed.

1. Introduction

The availability of main nutrients such as C, N, P, K and others e.g. Fe, Mn, Cu, Zn and S (called macro- and microelements) is crucial for numerous living organisms. They are present in each soil phase in different forms depending on soil conditions. The soil organisms take part in circulation of elements increasing or decreasing their availability and using them for their metabolism.

All elements cycle at a global scale since forming of the Earth showing certain regularity, however, they may undergo some fluctuations or change during transient periods of rapid change. This may result from sudden events like high tectonic or volcanic activity but also from the alternations done by humans, for example it is known that industrialization resulted in the increase of CO_2 concentrations in the atmosphere. These changes modify elements pools and fluxes in each compartment of the environment (Schlesinger, 1997).

Biosphere responds to above mentioned alternations for example by switching its metabolism or changing its activity. In addition, it may also impact biogeochemical cycling of elements (Schlesinger, 1997) affecting the availability of nutrients and other elements in the soil. In that way soil quality, which can be discussed in terms of its fertility, functions and health would change. The last term is a combination of biological, chemical, and physical features necessary for long-term, sustainable agricultural productivity with minimal environmental impact (Arias et al., 2005). The study of nutrient and element cycles in relation to organisms is a topic of biogeochemistry (Lincoln et al., 1998).

One of the factors influencing the availability of nutrients in the soil is its redox state describing both soil aeration state as well as its degree of reduction. Under aerobic conditions most of nutrients is immobilized in the soil and decomposition of soil organic matter is occur aerobically. Inundation affects strongly biogeochemical soil processes and causes first, fast oxygen depletion and, consequently, shifts of microbial activity to anaerobic processes leading to more reduced conditions. The reduction of oxidized compounds such as nitrates, Fe/Mn (hydr)oxides and sulfate affects the bigeochemical cycling of nutrients and other elements (Gliński and Stępniewski, 1985; Laanbroek, 1990; Schlesinger, 1997).

It is possible to estimate the aeration status of the soil using redox potential (Eh) measurements. It was shown that there is a direct but non-constant relationship between Eh and oxygen content in the soil (Armstrong, 1967). Other authors demonstrated critical values of Eh for certain microbial processes in the soil. Patrick (1960) showed that denitrification can occur if Eh<338 mV (pH 5.1) whilst Yamane and Sato (1964) have found that methane production form flooded paddy fields will not occur until Eh drop below -200 mV.

Redox potential is also useful in the prediction of the consequences of soil flooding (e.g. mobilization of nutrients and toxic substances = CTB, Chemical Time Bomb, gas emission), effectiveness of fertilizers, the environmental impact of agriculture in expected climate change (Stępniewski et al., 2005). For that reason the estimation of soil redox potential is important to determine the availability of nutrients and other elements in the soil (Patrick and Mikkelsen, 1971). There are many relationships between redox state and the forms in which nutrients and elements occur in the soil. The changes of hydrology have a profound effect on the biogeochemical cycling of nutrients in soil. We will describe them in the following sections in this chapter.

2. Soil Redox Processes

Soil aeration state is a main factor determining the availability of nutrients in the soil. It is a part of oxygenology, a discipline defined by Stępniewski and Stępniewska (1998) explaining the role of the oxygen in the environment.

Soil Microbial Processes

The process of breakdown of organic matter can be divided into two main stages – first (decomposition) involves degradation of high-molecular weight organic compounds (e.g. polysaccharides, proteins) to shorter compounds (e.g. alcohols, acids) which can be transferred further to CO_2 or CH_4. The second process is the release of inorganic components such as N, P and S etc. from organic matter (mineralization). In order to gain the energy for respiration microorganisms need both, donors of electrons (which is mainly organic matter) and their acceptors, i.e. substances on each electron is transported (mainly oxidized inorganic compounds) what results in generation of energy for metabolism. The availability of both donors and acceptors of electrons is a crucial element for sustaining the microbial activity. There are several substances which can accept electrons (see below) and they differ in the oxidative strength; several groups of microbes have different preferences to them. Thus, the availability of certain electron acceptor is a factor controlling the soil processes. However, the process of decomposition is mediated by consortium of different microorganisms and different paths of degradation overlap. The end-products of decomposition and mineralization are consumed by other organisms what sustains the decay of organic matter (Drever, 1997; Holmer and Storkholm, 2001; Wetzel, 2001).

Prolonged inundation retard decomposition with less CO_2 production and more CH_4 whist the reducing conditions are moderate (e.g. during short periods of flooding) N_2O may be emitted (Yu et al., 2006).

Role of Oxygen

Oxygen is an essential element for life being the most powerful terminal electron acceptor it plays a crucial role in all biochemical and chemical processes. The respiration process demands oxygen, in the soil it is a combination of microbial and plant roots respiration as chemical processes can be neglected (Stępniewski et al., 2005).

Under aerobic conditions oxygen pool (present in gas and liquid phase) is used for the respiration and it is refilled by diffusion from the atmosphere. Depth of oxic layer in the soil depends on soil porosity (Eg), level of water table and the presence of plants roots. The presence of bigger pores allows more space for both gas and liquid phase. When Eg amounts 24% there is sufficient diffusion of O_2 to the soil and roots and this state is called *normoxia.* A minimum air-filled pore space at this state is 10 % (v/v) (Paul and Clark, 1996). Moreover, there is a positive, curvilinear relationship between diffusion factor and Eg (Stępniewski et al., 2005). At a lower porosity there are contemporary aerobic and anaerobic pockets in the pores and this state is called *hypoxia.* The lack of porosity (Eg=0) is an anoxic state and means the absence of molecular oxygen (Bennicelli, 1999). The solubility of gases in water is different, depending on its type, temperature and pressure. For example O_2 and N_2 are these which have very low solubility. In addition, diffusion of them is also much lower in water than in the air being 10 000 times slower in the water in relation to air. This can lead to fast depletion of oxygen after soil submergence and N-limitation in N-fixing systems (Paul and Clark, 1996). Distribution of oxygen rapidly decreases to zero below depth of 1 cm in flooded soil (Figure 1).

Gas transport and exchange in the soil is based on diffusion in the direction of decreasing concentration. This process can occur within whole profile (macrodiffusion), in particular parts and between water films surrounding plants roots (microdiffusion). In general, gas diffusion is based on the first Fick's law describing diffusion flow (f_x) as a proportional to the gradient of concentration (dC/dx) and to diffusion coefficient (D):

$$f_x = - \quad (1)$$

In the soil gas diffusion is lower than in atmosphere depending also on gas, temperature, pressure and the amount of air-filled pores (Stępniewski et al., 2005).

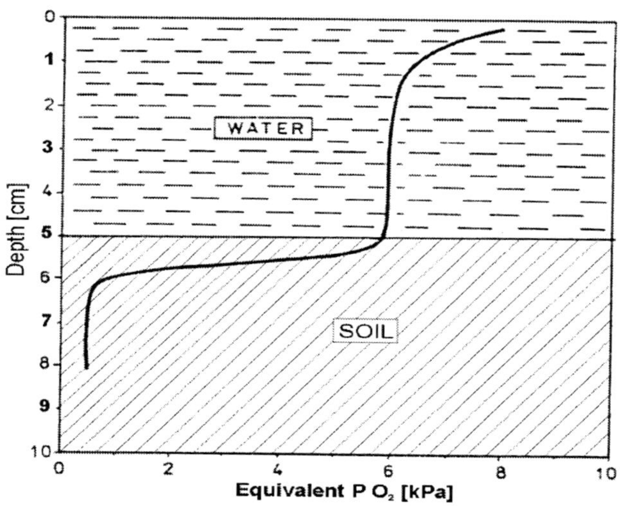

Figure 1. Oxygen distribution in flooded soil (after Patrick and Mikkelsen, 1971).

Plants have the ability to transport oxygen via roots, through aerenchyma tissue. This feature called radial oxygen loss (ROL) is very important under anaerobic conditions allowing plants to survive. The release of oxygen could be beneficial allowing more aerated conditions in the rhizosphere. Barrier for ROL, however, could also help plant due to saving oxygen pool (Gliński and Stępniewski, 1985). The method of determination the potential availability of O_2 for plant roots has been proposed in 1952 by Lemon and Erickson (1952). He designed the electrode which simulated O_2 uptake by plant root. The principle of the method is based on the measurement of the amount of oxygen diffusing to the surface of platinum electrode ("a plant root") where it is reduced electrochemically (due to its polarization to -0.65V versus saturated calomel electrode, SEC). The intensity of oxygen flux to that electrode indicates the maximum of oxygen which would be available for plant root place in the same spot in the soil. Microdiffusion of oxygen, called Oxygen Diffusion Rate (ODR) is calculated as follows:

$$ODR = 8.29 \cdot 10^{-5} \frac{i}{A} [\mu g\ m^{-2} \tag{2}$$

where i is current intensity (μA) and A the area of Pt electrode in mm^2. The values of ODR in the soil vary between 0 and 200 $\mu g\ O_2\ m^{-2}\ s^{-1}$ (Gliński and Stępniewski, 1985; Stępniewski et al., 2005; Bennicelli et al., 2006).

ODR is affected by several factors, first is water content which is negatively correlated, i.e. when it decreases ODR would increase to a certain level after which it also decreases. This factor has a strong effect on many soil parameters including its redox state; it affects the amount of free soil pores, gas diffusion, pH, osmotic potential of microbial cells and the availability of nutrients (Schimel et al., 2007). Soil texture is another factor, microdiffusion of oxygen is higher in coarse than in soils with fine texture. It was shown by Brzezińska et al. (2005) that water content in the soil of 40% (v/v) lowered ODR to 30 $\mu g\ m^{-2}\ s^{-1}$ which is the minimal ODR value necessary for root growth (Gliński and Stępniewski, 1985; Stępniewski et al., 2002; Włodarczyk et al., 2008; Wolińska and Stępniewska, 2011).

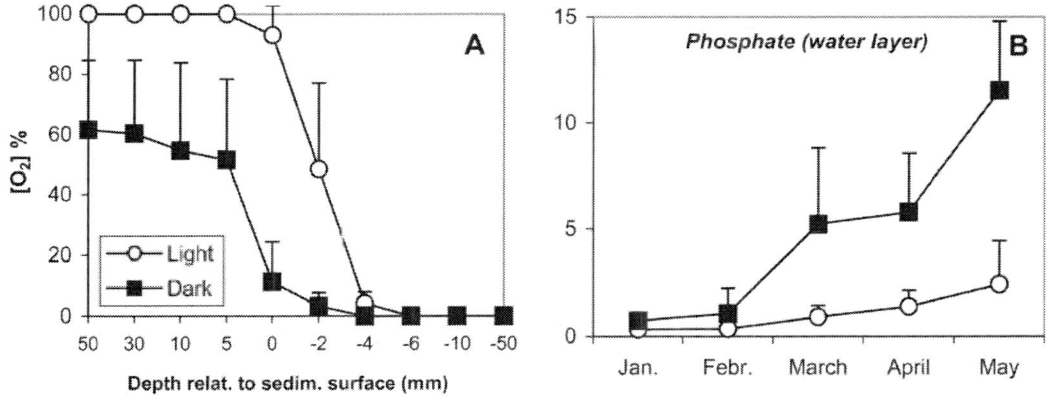

Figure 2. The effect of irradiation (open symbols) on concentrations of oxygen (A) and phosphate (B) in sediment and water layer in relation to dark conditions (black squares) (from Smolders et al, 2006, modified).

Under aerobic conditions other compounds remain in oxidized state resulting in immobilization of nutrients e.g. P in iron(hydr)oxides whilst nitrogen availability decreases due to coupled nitrification-denitrification. Ammonium is nitrified in root zone; produced nitrate diffuses to deeper layer where it is denitrified (Smolders et al., 2006). Thus, the oxygenation of the previously submerged sediment may be beneficial reducing the availability of unwanted substances. Similar effect could be achieved by irradiation, what stimulates growth of benthic algae which produce oxygen oxidizing surface layer of sediment affecting P availability (Figure 2).

Thus, aerobic conditions are much more favorable reducing the availability unwanted, often toxic compounds such as heavy metals, ammonium, phosphate and greenhouse gases (N_2O, CH_4).

Redox Processes

Inundation changes drastically conditions in the soil leading to its reduced state. Pores become completely filled with water and gas diffusion from the atmosphere declines as described above.

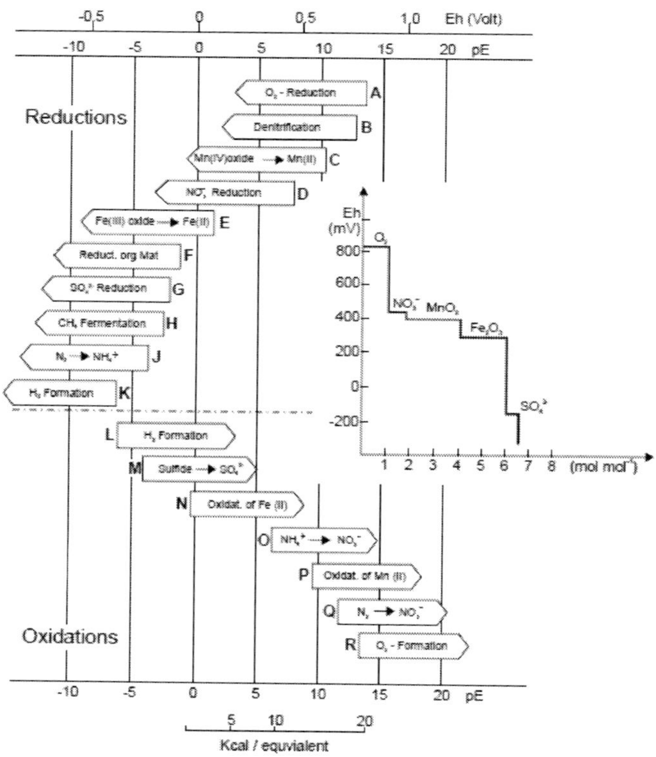

Figure 3. The cascade of redox reactions occurring during wetting and rewetting of soils. Inserted graphs shows the Eh levels of particular compound and their oxygen equivalents (mol mol^{-1}) i.e. number of moles of a given compound substituting one mole of O_2 (after Stępniewski et al., 2005).

As a result pool of gases, especially O_2 depletes quickly resulting in deficiency and anaerobic conditions. It is estimated that there is a shift from aerobic to anaerobic processes when oxygen concentration are less than 1% (Paul and Clark, 1996). This process could take even few hours for the microorganisms to consume all remain oxygen pool (Ponnamperuma, 1972). Under such conditions microorganisms use alternative electron acceptors (other oxidized compounds) resulting in a cascade of the reactions (Tab. 1; Figure3; Ponnamperuma, 1972; Gliński and Stępniewski, 1985; Laanbroek, 1990; Schlesinger, 1997; Stępniewski et al., 2005; Banach, 2010).

These reactions overlap under more or less anaerobic conditions indicating the presence of anaerobic microsites in the soil. Aerobic bacteria prepare an environment for anaerobes by consuming O_2, thus allowing them to develop (Paul and Clark, 1996). Under aerobic conditions these reactions are reversed leading to oxidized end-products (Figure 3). Each reaction is governed by the certain community of microorganisms that compete for substrates (organic matter and electron acceptors). In that way the availability of the substrates is often a key-factor for the determination of the redox state of a soil (Laanbroek, 1990). After O_2 depletion (due to aerobic respiration) denitrifying bacteria consumes nitrates. When it is depleted the reduction of Mn^{4+} starts followed by Fe^{3+} reduction. Some bacteria (e.g. *Shewanella putrefaciens*) can couple the reduction of both compounds. It was also found that there are overlapping zones where denitrification and Mn^{4+} reduction occurs by facultative anaerobes. There is not so much evidence about such zones for manganese and iron as soil bacteria prefer Mn^{4+} more than Fe^{3+} and as long as Mn^{4+} is present, the Fe^{3+} reduction will not begin. After Mn^{4+} reduction, other redox reactions are performed by obligate anaerobes (e.g. *Clostridium*). Next steps are sulfate reduction and methanogenesis (Laanbroek, 1990; Schlesinger, 1997).

The above showed oxidation-reduction reactions generate different redox potentials (Eh). The idea of this estimator is based on the rule that chemical reactions have a tendency of decrease the free energy (ΔG) to obtain equilibrium when the sum of free energies of the products equals to that of the remaining reactants. The change of free energy, (measured either in calories/Joules or volts, can written as follows:

$$\Delta G = \Delta G^0 + RT \ln \frac{(}{(} \qquad (3)$$

where (*Red*) and (*Ox*) are the activities of the reduced and oxidized species and ΔG^0 is the change of free energy when the activities have unity. *R* is universal gas constant (8.31447 J $mole^{-1}$ K^{-1}) and *T* – absolute temperature [K]. After conversion the calories into volts ($\Delta G = -nEF$) the equation has a form (Nernst equation, Bohn, 1971):

$$Eh = E_0 + \frac{RT}{nF} \ln \frac{(}{(} \qquad (4)$$

where Eh is the voltage of reaction, measured against hydrogen electrode. E_0 is the voltage when both (*Ox*) and (*Red*) are unity, *n* is the number of to the reaction concerned electrons and *F* – Faraday's constant (96500 C $mole^{-1}$) (Ponnamperuma, 1972).

As soil is heterogeneous environment there are many parallel reactions take place which are connected with different electron acceptors resulting in the final Eh which is called mixed redox potential (Bohn, 1971).

Eh is a quantitative measure of the tendency of environment to oxidize or reduce substances. When Eh is high and positive, it estimates strongly oxidizing systems (Eh>500 mV) whilst negative and low values characterize strongly reducing sites (-400 till 200 mV). For each redox reaction there is its characteristic value of Eh what can be useful in determination of both aeration state of the soil and the availability of nutrients and other elements. It is also possible to define stages of soil reduction based on Eh values (Ponnamperuma, 1972; Stępniewski et al., 2005; Table 1).

Eh changes rapidly depending on oxidation-reduction reactions in the soil. These reactions are influenced by several factors such as temperature (which affects soil microbial activity), pH and the presence of substrates. The rapid decrease of Eh results from the release of reduced substances accompanying O_2 depletion before Mn(IV) and Fe(III) compounds can mobilize their buffer capacity. The presence of each electron acceptor can stabilize Eh for a certain time; high nitrate concentration sustains positive potential for several weeks after submergence (e.g. 1-2 weeks, Banach et al., 2009a). High concentrations of Mn(IV) maintain positive Eh even for 6 weeks, whilst low levels of Mn(IV) and Fe(III) cause low and negative soil redox potential (Ponnamperuma, 1972).

Soil pH is strongly connected with redox potential, there is a negative relationship between Eh and pH and the change of pH of 1 unit leads to change of Eh of 59 mV (Figure 4). The effect of oxygen concentration and temperature on Eh is weaker as 10-fold lower concentration of O_2 lowers Eh by 15 mV and a rise of temperature by 1°C decreases Eh by 16 mV (Stępniewski et al., 2005). In order to compare Eh between different soils the following conversion is performed (Bell, 1969):

$$\mathrm{Eh7} = \mathrm{Eh} - 0.059\ (7 - \qquad (5)$$

Soil submergence causes first pH lowering followed by it increase and the stabilization at range of 6.7-7.2. The overall effect is a tendency to neutralize pH – in acidic soils it increases whilst in calcareous it lowers (Ponnamperuma, 1972).

Table 1. Stages of soil anaerobiosis (from Takai and Kamura, 1966; Gliński and Stępniewski, 1985 and Schlesinger, 1997, modified)

Stage of reduction	Degree of reduction	Reaction (redox pair in bold)	Initial Eh [mV]	Eh7 (20°C) [mV]
I	1	$O_2 + 4H^+ + 4\bar{e} \rightarrow 2H_2O$	600-500	812
	2	$NO_3^- + 2H^+ + 2\bar{e} \rightarrow NO_2^- + H_2O$	600-500	747
	3	$MnO_2 + 4H^+ + 2\bar{e} \rightarrow Mn^{2+} + 2H_2O$	600-400	526
	4	$Fe(OH)_3 + 3H^+ + \bar{e} \rightarrow Fe^{2+} + 3H_2O$	500-300	-47
II	5	$SO_4^{2-} + 8H^+ + 8\bar{e} \rightarrow S^{2-} + H_2O$	0 to -190	-221
	6	$2H^+ + 2\bar{e} \rightarrow H_2$	-150 to -220	
	7	$CO_2 + 8H^+ + 8\bar{e} \rightarrow CH_4 + 2H_2O$	-150 to -190	-244

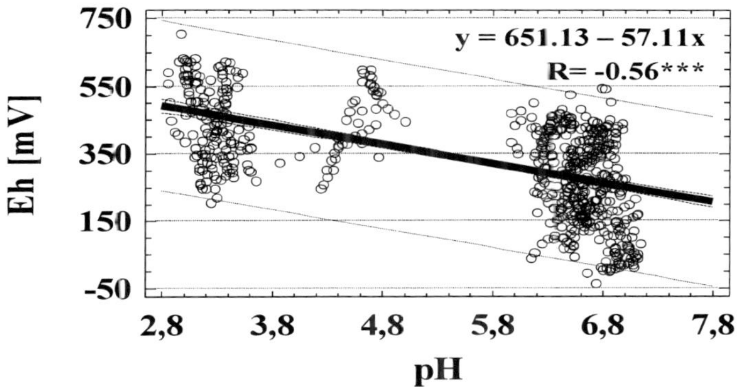

Figure 4. The relationship between Eh and pH for peaty soil (n=329) (after Szafranek-Nakonieczna, 2009).

Another parameter used for describing redox state is pE showing electron activity, which can be easy transformed into Eh. It is simply calculated as a negative logarithm of electron activity (Ponnamperuma, 1972; Stępniewski et al., 2005):

$$pE = -\log[\bar{e}] = Eh/0. \text{ at } 25°C \quad (6)$$

Under natural conditions pE ranges from -6 to +12 and Eh from -600 to +400 mV whilst pH from 3 to 9 (Figure 5).

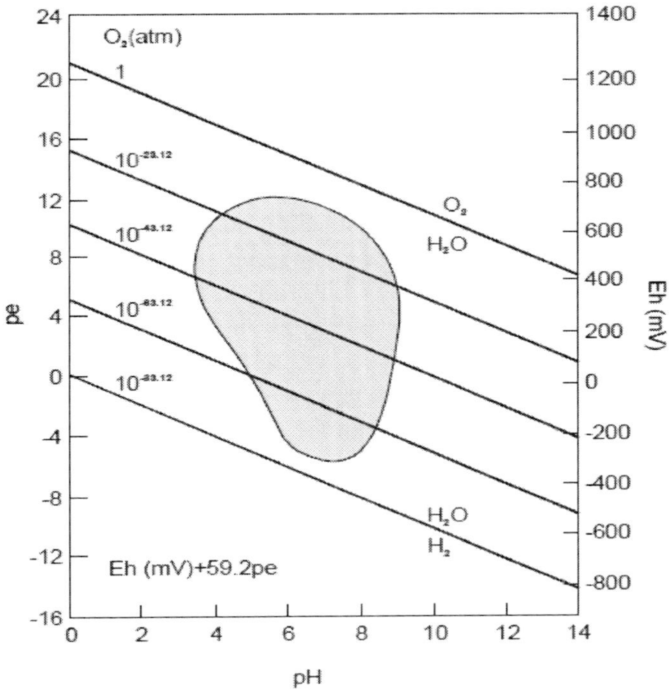

Figure 5. The range of Eh occurring in natural aqueous environments (shaded) (after Schlesinger, 1997).

The measurement of Eh is based on estimation difference of potential between two electrodes expressing the reduction state of the soil. It expresses the disequilibrium of its suite of constituents relative to a standard electrode. This electrode contains hydrogen gas overlaying the solution of known H^+ concentration. The second is inert platinum electrode, placed into soil, connected to a standard electrode. It takes the potential from the soil and Eh is determined against the reference electrode (hydrogen) with the high impedance voltmeter (e.g. pH-meter) as a voltage required to prevent interconversion of H^+ and H_2 at the standard electrode. In practice saturated calomel electrode (SEC) or more often silver-silver chloride electrode, calibrated against hydrogen electrode are as used as reference (Ponnamperuma, 1972; Schlesinger, 1997). Modern electrodes comprise of both Pt and reference electrodes in one compartment are available.

The principle of Eh measurement can be shown on the example when O_2 is present. Under such conditions molecular oxygen accepts electrons at the Pt electrode as follows:

$$O_2 + 4H^+ + 4\bar{e} \rightarrow 2H_2O \quad (7)$$

Next, the electrons are generated at the hydrogen electrode:

$$2H_2 \rightarrow 4H^+ + 4\bar{e} \quad (8)$$

and the voltmeter records a high voltage or redox potential. When O_2 is absent other electron acceptors are involved generating lower voltage (Schlesinger, 1997). The final value of Eh is calculated by adding the potential of reference electrode which is +247 mV for SEC and +238 mV for Ag/AgCl (20°C) to recorded values. When presenting Eh data it is important to mention the temperature and pH of the measurement.

Redox status of the soil can be also expressed by the ability of the soil components to stabilize Eh at a given level what is called redox capacity or redox resistance (R_c) (Ponnamperuma, 1972; Stępniewska et al., 1996; Stępniewski et al., 2005). Redox resitatnce is defined as the quantity of strong reductant which must be added in order to reduce Eh by 1 mV (Nightingale, 1958). There are two levels of Eh used for the characterization of R_c – 300 (the beginning of Mn(IV) and Fe(III) reduction) and 400 mV (the beginning of NO_3^- reduction). It is also defined as a time when Eh drops to level of 400 mV (t_{400}) or 300 mV (t_{300}) under flooded conditions. First (t_{400}) is used as an indicator of nitrate stability and the initiation of denitrfication. Stępniewska et al. (2000) demonstrated the dependence denitrification rate on redox state of the soil (Figure 6) and when t_{400} was lower than 2 days the process decreased. The status of progress in the redox process is expressed by the Eh, and the ability of the soil to maintain its Eh is a measure of soil resistance to reduction (Stępniewska, 1994; Stępniewska et al., 2000).

The form in which component is available strongly depends on Eh and pH of the soil. There are Eh-pH (or pE-pH) diagrams describing in which form of certain compound is present under exact Eh (pE) and pH values. They are very useful for prediction the oxidation state of them nutrients and other elements and to determine their availability (Stumm and Morgan, 1970; Sposito, 1989; Schlesinger, 1997). All of this emphasizes that Eh is a powerful tool for the characterization soil properties and for the prediction the consequences of altered aeration state of the soil.

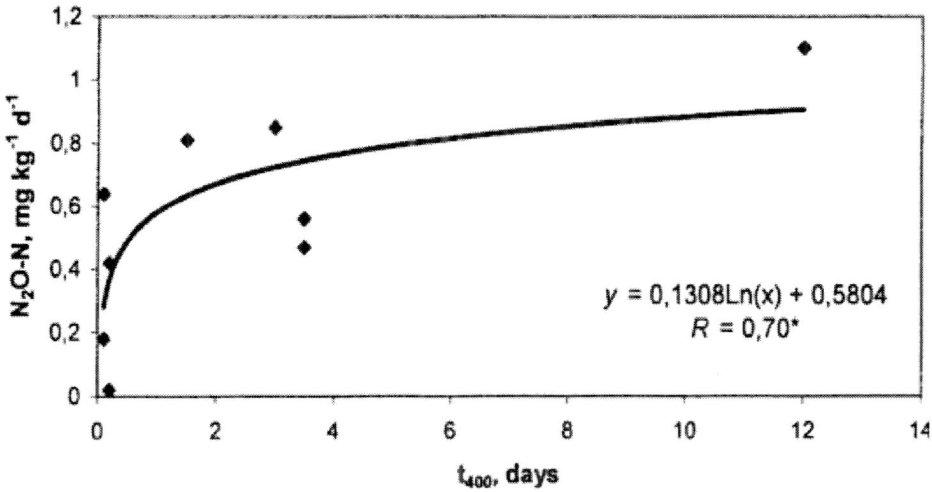

Figure 6. Denitrification rate at 20°C as a function of redox resistance (t_{400}) (after Stępniewska et al., 2000).

3. Soil Nutrients

There are group of elements such as H, B, C, N, O, Mg, P, S, Cl, K, Ca, Mn, Fe, Cu, Zn and Mo which are essential for plant growth. The most important, called macronutrients are H, C, N, O, P and K. Next, B, Cl, Mn, Fe, Cu, Zn and Mo are called micronutrients and the last Mg, S and Ca are secondary nutrients. These elements play important role in the metabolism of living organisms. For that reason, it is important to determine their availability for plants and other organisms. The element is bioavailable when it is present in the form that plants can absorb and when it is absorbed it affects the life cycle of the plant (Sposito, 1989).

Main nutrients are used by biota for building their bodies and providing their activities. Many organic compounds e.g. proteins constitute of these elements. The metabolism of organisms is also linked with the presence of nutrients (described be the ratios between them) as there is a linkage between cycles of main nutrients (C, N and P). Phosphorus may stimulate N-fixation in bacteria, the N and P affects the photosynthesis and so organic C production (Schlesinger, 1997). Nitrate, which is more powerful electron acceptor than Fe^{3+} or SO_4^{2-} may lower the mobilization of PO_4^{3-} due to hampering the reduction of less powerful electron acceptors involved in internal P mobilization acting as redox buffer (see blow, Figure 7).

The way of soil use may affect buffering capacity as shown by Banach et al. (2009a) where one soil contained much higher concentrations of nitrate resulting in a longer period of higher redox state and lower P availability (Figure 7, 8).In that way nitrate may reduce risk of eutrophication as it was proved by several experiments (Lucassen et al., 2004a). Nitrate reducing bacteria can grow anaerobically and use Fe as electron donor producing Fe(III) (Straub et al., 1996; Nielsen and Nielsen, 1998; Weber et al., 2001) what increases P binding capacity of soils and sediments.

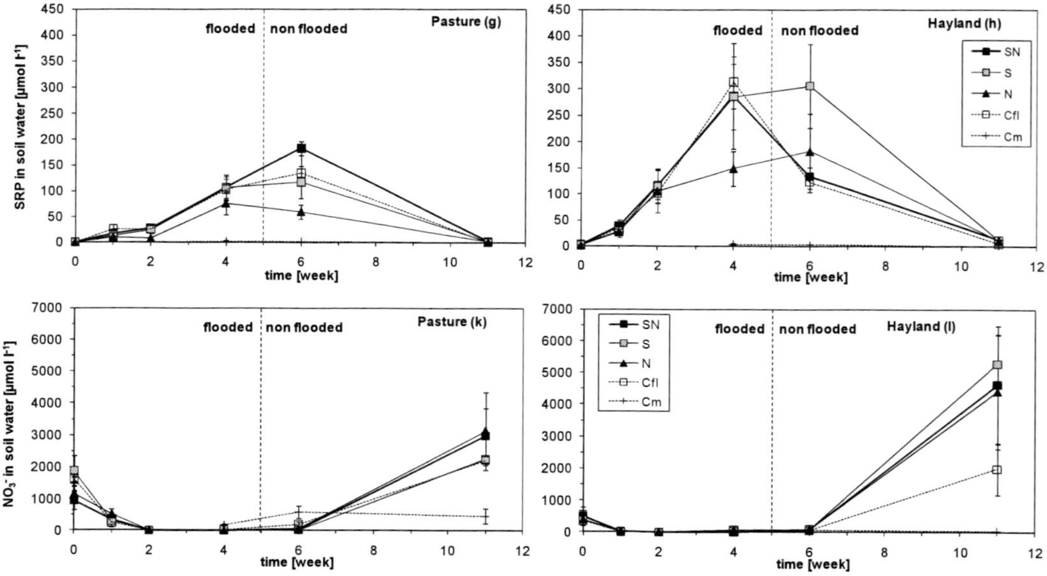

Figure 7. Changes of concentrations of SRP (soluble reactive phosphate) and NO_3^- in soil pore water during one month inundation of sods from meadows (uncultivated – Pasture and cultivated – Hayland). Floodwater was enriched in nitrate and sulfate (SN), sulfate (S), nitrate (N) and nutrient-poor (Cfl). As a control non-flooded combination (Cm) was used (from Banach et al., 2009a, modified).

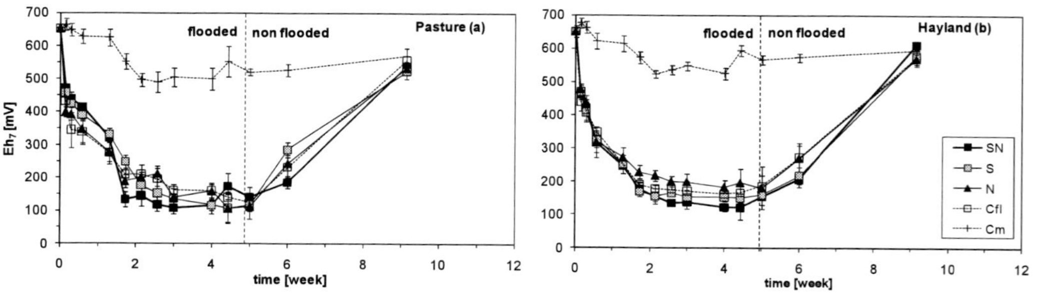

Figure 8. Redox potential during flooding sods (2 meadows) under different water qualities (from Banach et al., 2009a, modified, see Figure 6 for more details).

Carbon

Carbon is a predominant constituent of all life forms; it is involved in most of biogeochemical reactions. Its transformation in the environment directly or indirectly affects the fate of other nutrients, especially nitrogen and phosphorus. Carbon is an element which is incorporated into plants via photosynthesis when atmospheric CO_2 is reduced to alkyl compounds further transferred into various organic compounds and incorporated into cells. There are two main natural transformations of C – photosynthesis and respiration.

After death of organisms their residuals undergo process called humification resulting in the formation of organic material called humus. It consists of organic compounds having 50-

55% C, 4.5 % N, 1% S and different contents of P and metals. As humus decomposes very slowly it accumulates in the soil as soil organic matter (SOM). During process decomposition of SOM, C is again converted into inorganic products such as CO_2 and CH_4. In addition, anthropogenic organic substances increase a pool of soil C and next the fluxes of CO_2 and CH_4 into atmosphere affecting greenhouse effect.

During decomposition microbes use oxygen (aerobic process) or in the absence of O_2, alternative electron acceptors such as nitrate (V), manganese (IV) and iron (III) and the process leads to formation of CO_2 as end-product of either aerobic or facultative anaerobic respiration. Aerobic conditions boost decomposition due to high energy associated with aerobic respiration. The end-products are oxidized compounds such as CO_2, NO_3^- and SO_4^{2-}. Under more reduced conditions sulfate and CO_2 serve as electron acceptors and methane is end-product of anaerobic respiration as well as reduced compounds like ammonium and sulfide. The CH_4 formation occurs in submerged soils of wetlands, peatlands and bogs. Sediments, rice fields, sewage treatment plants are also it sources (Tab. 2). Soil inundation (e.g. irrigation or flooding) may also lead to temporal CH_4 emission depending on the availability of electron acceptors and donors. In general, oxygen presence inhibits facultative anaerobic respiration whilst nitrate and sulfate hamper methanogenesis (Ponnamperuma, 1972; Paul and Clark, 1996).

The production of methane is also related to redox status of the soil. Yu et al. (2001) showed different critical values of Eh for CH_4 emissions raging from -150 to -215 mV. Moreover, lower values of Eh resulted in higher production of CH_4 showing a risk of higher emission under strongly reduced soil conditions (Figure 9).

Table 2. Global sources and sinks of methane (Paul and Clark, 1996)

	($Tg\ CH_4\ yr^{-1}$)	Range
Sources		
Wetlands	115	100-200
Termites	20	10-50
Ocean	10	5-20
Freshwater	5	1-25
CH_4 hydrate	5	0-5
Coal mining, natural gas and oil industry	100	70-120
Rice paddies	60	20-150
Enteric fermentation[a]	80	65-100
Animal wastes[a]	25	20-30
Domestic sewage treatment	25	No data
Landfills	30	20-70
Biomass burning	40	20-80
Sinks		
Atmospheric removal	470	420-520
Removal by soils	30	15-45
Atmospheric increase	32	28-37

[a] In or from domesticated animals.

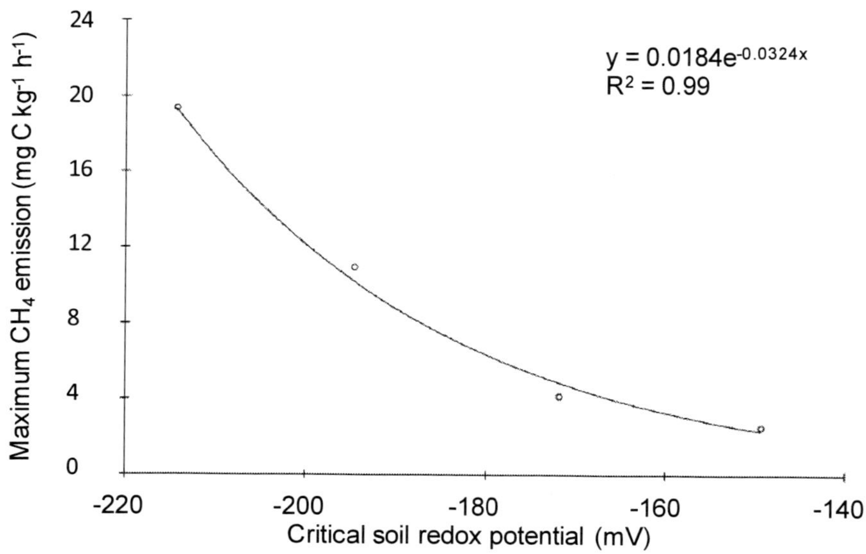

Figure 9. Relationship between critical Eh and maximum CH_4 production (after Yu et al., 2001, modified).

Nitrogen

Nitrogen is an important nutrient for organisms which can be a main element limiting rate and net primary production. It exists in different oxidation states (from -3 to +5) in opposite to phosphorus (almost all the time at +5) and these forms are change easily mainly due to the activity of soil organisms what makes N easily cycling in the ecosystems. There are two main compounds in which N is available for organisms – nitrates (V) and ammonium (NH_4^+). Nitrates have good solubility in the water and thus are well transported. Ammonium has ability to volatilize (as NH_3) and to fixation by minerals and SOM. The end-products are often gaseous forms such as N_2 and N_2O returned to the atmosphere. Nitrogen is present also in organic compounds such as nucleic acids, proteins and microbial cell wall components.

There are several sources of nitrogen in the soil – atmospheric N_2 (the most abundant form) fixed by symbiotic (e.g. *Rhizobium*) or non-symbiotic organisms (e.g. *Azotobacter*, *Clostridium*, algae), NO_3^- and NH_4^+ from fertilizers, microbial, plant and animal residues (mineralization process). The process of mineralization refers to degradation of SOM and organic residues to mineral forms. In a case of N mineralization ends with ammonium (ammonification) form which can be either immobilized in the soil or taken up by organisms, depending on their requirements.

The main source of ammonium is degradation of organic compounds such as amino acids, purines and urea, only less than 1% of it comes from nitrate reduction (see below). Ammonification is faster under anaerobic conditions due to lower N immobilization and there is wide variety of organisms which are involved (fungi are known to be the most vigorous). Temperature is a factor which strongly affects this process in flooded soil, the increase from 26 to 40°C may double the ammonium production (Ponnamperuma, 1972).

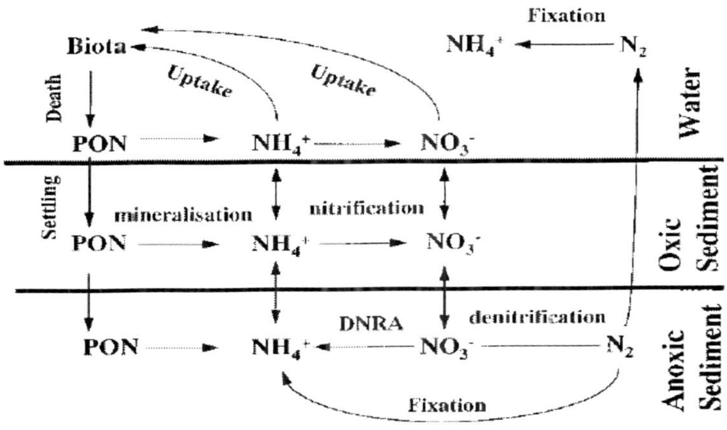

Figure 10. The nitrogen cycle in the aquatic systems (PON, particulate organic nitrogen; after Baldwin and Mitchell, 2000).

In aquatic systems N transformation differ depending on aeration state of the sediment (Figure 10). In oxic part of the sediment produced ammonium may be nitrified, taken up or volatilized. Under anoxic conditions, nitrate can be either denitrified to N_2 or converted to ammonium on dissimilatory pathway which has similar reactions of reduction such as for assimilatory process described below but other enzymes are involved (Baldwin and Mitchell, 2000).

The processes of N transformation are primarily biological, assimilation of N involves the incorporation ammonium into amino acids, often proceeded by nitrate reduction (see below):

$$NO_3^- \rightarrow NO_2^- \rightarrow NH_2OH \rightarrow NH_4^+ \rightarrow R\text{-}NH_2 \quad (9)$$

Produced ammonium is incorporated into α-ketoglutaric acid to form of glutamic acid. Next, there are several reactions resulting in numerous products including keto acids.

Ammonium may have also another fate, i.e. it could be left on the exchange complex where it may be replaced by cations from soil solution. However, it can be tapped into interlayer space of clay and under dry conditions immobilized. When is present in unabsorbed state, e.g. decaying vegetation, manure it can volatilize at high rate as ammonia. Another fate of NH_4^+ is its role as an energy source for some autotrophs in the nitrification process.

Nitrification is a two-stage process of the NH_4^+ conversion (oxidation) to NO_3^- through NO_2^-. This is governed by chemoautotrophic and heterotrophic bacteria. There are different groups of microbes involved in each stage; *Nitroso*-group for ammonium oxidation to nitrite (e.g. *Nitrosomonas europaea*) and *Nitro*-group for nitrite conversion to nitrate (e.g. *Nitrobacter*). Both processes are closely related and there is hardly any nitrite accumulation in the soil. First, oxidation of ammonium occurs as described by the following reaction catalyzed by ammonia mono-oxygenase:

$$NH_4^+ + 2H^+ + 2\bar{e} + O_2 \rightarrow NH_2OH + H_2O + H^+ \quad (10)$$

The enzyme catalyses also CH_4 oxidation but this process is faster when performed by methanotrophs than *Nitrososmonas*. Above reaction (10) is inhibited irreversibly by acetylene. Next step is oxidation of hydroxylamine to nitrite:

$$NH_2OH + H_2O \rightarrow HONO + 4\bar{e} + 4H^+ \quad (11)$$

The second step is oxidation of nitrite:

$$HNO_2 + H_2O \rightarrow HNO_3 + 2H^+ \quad (12)$$

During this stage gaseous products such as NO and N_2O may be produced. Heterotrophic bacteria such as *Arthrobacter globiformis*, *Aerobacter aerogenes*, *Mycobacterium pheli*, *Streptomyces griseus*, *Thiosphaera* and *Pseudomonas* spp. can produce nitrate from both organic and inorganic sources. Fungi appear to be major organisms involved, e.g. *Aspergillus flavus*, *Penicillium* spp. and *Cephalosporium* spp. Amines and amides may be used to a substituted hydroxylamine followed by the oxidation of nitroso and nitro compounds:

$$R\text{-}NH_2 \rightarrow R\text{-}NHOH \rightarrow R\text{-}NO \rightarrow R\text{-}NO_3 \rightarrow NO_3^- \quad (13)$$

Acetylene does not inhibit heterotrophic nitrification. The process of nitrification is regulated by several factors such as pH, aeration state, moisture and temperature. The optimum pH for agricultural soils is between 6.6 and 8.0, more acidic conditions hamper second step of nitrification. Oxygen is necessary for organisms and its availability regulates nitrification. As moisture affects O_2 availability as mentioned previously, it has also the influence on ammonium oxidation hampering it. The optimal temperature ranges between 30 and 35°C, nitrification slows below 5 and above 40°C. The interaction of these last 3 factors reveals the seasonal differences in the rate of nitrification. In temperate climates it is greatest in spring and fall and slowest during summer and winter. The process is beneficial for plants as they easily assimilate NO_3^- and it is used as an index of soil fertility. Nitrification, however, has also negative consequences – as nitrate is mobile it could lost from the soil due to denitrification leading to atmospheric contamination and leaching what results in elevated concentration of this anion in groundwater, rivers and lakes. The latter may have adverse effects such as health problems (infant and animal methemoblobinemia or formation carcinogenic nitrosoamines) and environmental effects such as euthrophication due to extensive development of plants and algae.

There are several ways in which nitrate may be treated when present in the soil. First, it can be taken up by organisms and used for amino acids synthesis (assimilatory reduction to ammonium followed by assimilation as described above). Secondly, it may be reduced to gaseous end-products such as NO_x and N_2 (denitrification). Next, nitrate may undergo intermediate (to NO_x and N_2, chemodenitrification) or complete reduction to ammonium under anaerobic conditions (dissimilatory reduction, DNRA). It is also possible that it may be leached into deeper layer of the soil or groundwater, removed by run-off or accumulated in the site.

The microbial NO_3^- reduction, called denitrification involves two main enzymes: nitrate- (NR) and nitrite reductase (NiR) and it is stimulated in NO_3^- presence and inhibited in the presence of NH_4^+. In general, the course of the reaction can be summarized as follows:

$$2NO_3^- \xrightarrow{NR} 2NO_2^- \xrightarrow{NiR} 2NO \xrightarrow{NOR} N_2O \xrightarrow{NO} \qquad (14)$$

After nitrite reduction, nitric oxide reductase (NOR) catalyses the reduction of NO to N_2O followed by the final reduction to N_2 (nitrous oxide reductase, NOS, is involved, Figure 11). All of these enzymes are susceptible to O_2 and low pH losing their activities. In addition acetylene and sulfide inhibit NOS activity.

This process is conducted by numerous bacterial species from groups of organotrophs, lithotrophs, phototrophs and diazotrophs. The requirements for denitrification are: the absence of O_2, the presence nitrate of other N oxides, a electron donor (redactor) such as organic C (for organotrophs), S, HS^- or NH_4^+ (for lithotrophs) and the denitrifying organisms. The moisture also play a role as when there is more than 60% of water-filled pores the reduction starts whilst nitrification starts to be hampered. The optimal pH is 6-8, and the increase of temperature leads to higher denitrification. This process occurs at Eh 100-400 mV (Pidello et al., 1996, Figure 12).

Chemodenitrification is a chemical reduction of nitrite to N_2O or NO by nonenzymatic pathway, e.g.:

$$R\text{-}NH_2 + HNO_2 \rightarrow R\text{-}OH + H_2O + N_2 \qquad (15)$$

$$NO_2^- + NH_4^+ \rightarrow N_2 + 2H_2O \qquad (16)$$

$$3HNO_2 \rightarrow HNO_3 + H_2O + 2NO \qquad (17)$$

Produced NO may escape to the atmosphere and next form HNO_3 which could contribute in the formation of acid rains and ozone.

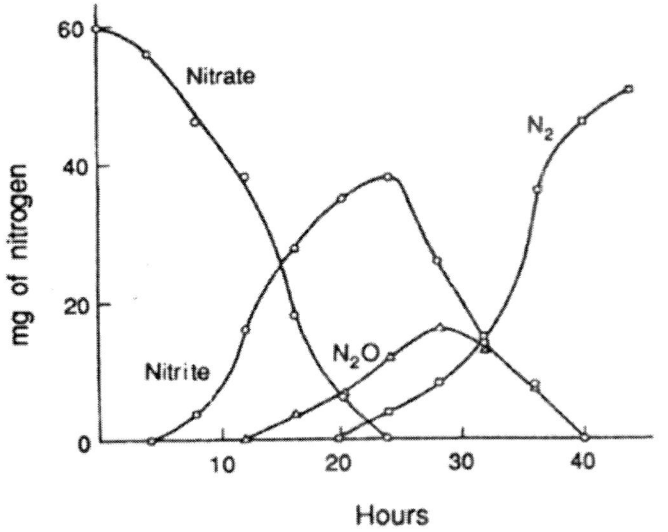

Figure 11. Sequential amounts of products formed during denitrification (after Paul and Clark, 1996).

N_2O is a radiative gas, contributing in the global warming effect. Even though, it has stronger warming potential than CO_2, its concentration in the atmosphere is much lower. It can be transferred into NO causing the effects mentioned above. In addition, it plays a role in CH_4 oxidation (Paul and Clark, 1996). It is estimated that about 70-90% of N_2O emission originates from cultivated soils which is produced both during nitrification and denitrification processes. Fertilization (ammonium addition) causes higher emission than from unfertilized soils, also the addition of nitrate and oxidizable C source enhances the process (Azam et al., 2002).

Aeration conditions play an important role in N_2O formation. High fluxes occur under incomplete denitrification, N_2O reduction to N_2 will not occur as long as more efficient electron acceptors are available (e.g. nitrate). For that reason high emission of nitrous oxide occur mainly during intermediate aeration. Such treated soils show higher emission than anaerobic and aerobic soils. The critical Eh for N_2O absorption and production ranges from 250-300 mV and pH 6.5-7.5 (Smith et al., 1983; Gliński et al., 2000; Stępniewska et al., 2000, Jungkunst et al., 2004, Figure 12). Study of Yu et al. (2001) showed more narrow range of 120-250 mV.

Stępniewska and Szmagara (2000) demonstrated the critical values of Eh for either sorption or emission of N_2O for several mineral soils. When Eh was lower than on the average 350 mV N_2O was sorbed whilst higher Eh values led to the emission of this gas.

Denitrification is a way of losing N from the soil whilst DNRA preserves nitrogen in the system. It is important if both processes nitrification and denitrification are coupled, i.e. they occur spatially close to each other, because produced nitrate is quickly reduced do nitrogen gas rather than accumulated in biomass. Such situation takes place in the sites on the border between oxic and anoxic conditions e.g. in the rhizosphere (Baldwin and Mitchell, 2000).

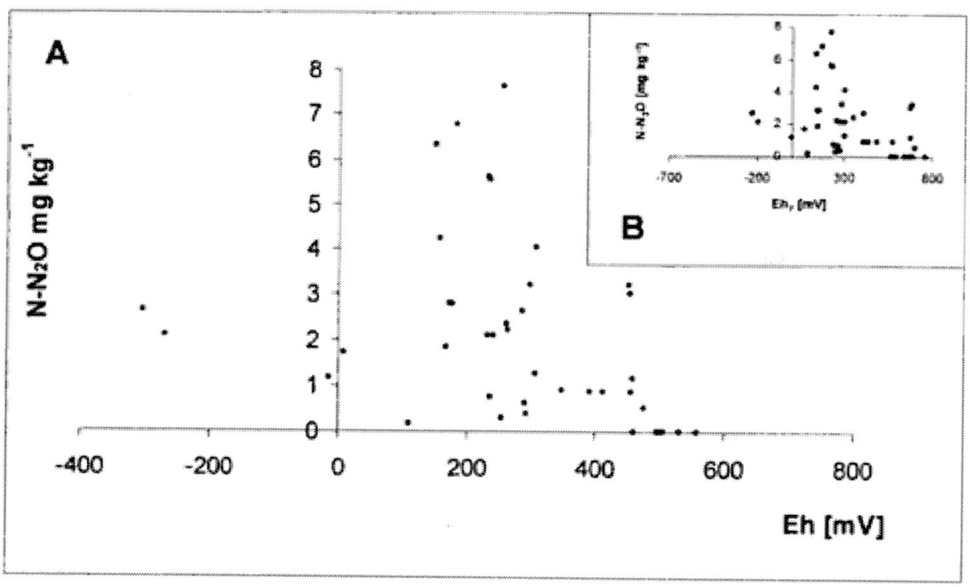

Figure 12. Cumulative denitrification at 20°C as a function of Eh (A) and Eh7 (B) (after Stępniewska et al., 2000).

Some organisms have the ability of fixation atmospheric nitrogen in a process called biological nitrogen fixation due to possessing enzyme nitrogenase. It is a reduction of N_2 to NH_3 requiring reductants of a very high electron activity (very low pE) originating mainly photosynthesis and anaerobic respiration. The organisms which are capable for N_2-fixation are blue-green algae (in flooded soils) and nitrogen-fixing bacteria in bulk soils. The examples of such cyanobacteria are the members of genera *Anabaena*, *Aulosira*, *Calothrix*, *Clostridium*, *Nostoc*, *Ocillatoria*, *Phormidium* and *Tolypotrix*. Some of them are associated with plants in symbiosis whilst other exist alone (Ponnamperuma, 1972). Some of these organisms exist alone in soil (free-living, asymbiotic organisms) whilst other like *Rhizobium* and *Frankia* form symbiotic associations with the roots of higher plants (Paul and Clark, 1996; Schlesinger, 1997).

Phosphorus

Phosphorus is another crucial element for living organisms. It is an essential component of DNA, ATP and phospholipids of cell membranes. It controls the availability of C and N. P occurs mostly in connection with oxygen (i.e. as orthophosphate, PO_4^{3-}) and sustain its valence of +5 in opposite to N. Phosphorus is present predominantly in earth's crust in a form of different minerals (Al, Ca or Fe), mainly apatite. The ocean's sediments constitute the biggest pool of P in comparison to other sources. Erosion and run-off of human wastes to the oceans supply continuous inflow on P into oceans. The weathering of minerals results in the mobilization of P from this pool. However, dissolved P in the solution undergoes several reactions with other minerals (see below) and become less available. In the soil there is small P fraction which is bioavailable and thus, organisms use mainly P which originates from degradation of organic compounds (Paul and Clark, 1996; Schlesinger, 1997).

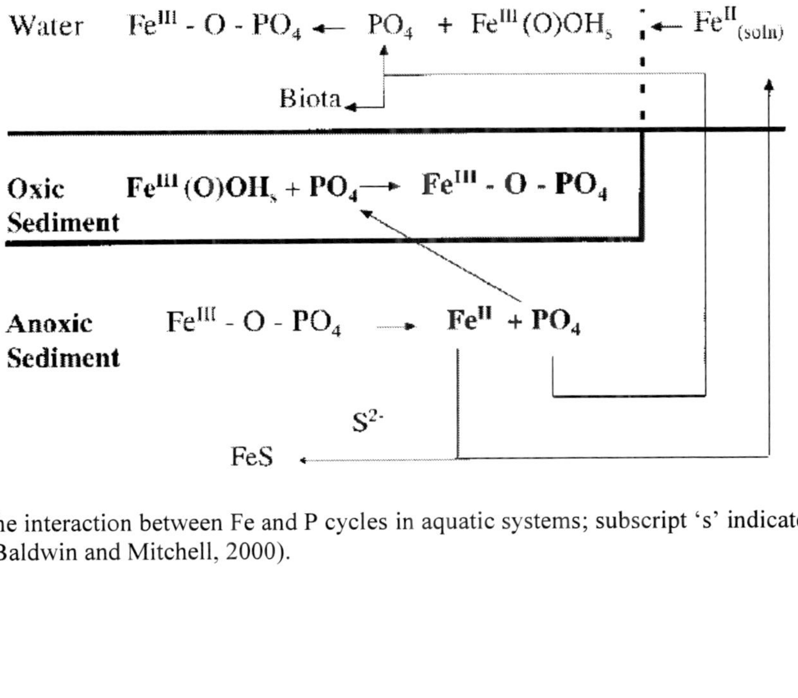

Figure 13. The interaction between Fe and P cycles in aquatic systems; subscript 's' indicates a solid chase (after Baldwin and Mitchell, 2000).

In aquatic environments phosphorus occurs as a Soluble- (Soluble Reactive P, SRP or Molybdate Reactive P, MRP or Filterable Reactive P, FRP or Dissolved Reactive P, DRP depending on the analytical method), Total Dissolved P (TDP) and Total P fraction (TP). Biologically available is SRP whilst from dissolved only part of organic (TDP) and sorbed to particles (including TP) may be easy available. Phosphorus is a key nutrient for aquatic system limiting growth of algae and macrophytes. Its high availability stimulates primary production what in some cases have negative effects. For that reason P availability is attracting the attention many researches worldwide. The external sources of P in waters are discharges of waste water, run-off from urban, agricultural and natural areas whilst precipitation is negligible (Bergametti et al., 1992; House, 2003).

The cycle of P is closely related to Fe cycle as Fe bound P fraction is one of the most important in the soil and it is redox sensitive. Under anoxic conditions iron reduction (and dissolution of Fe minerals) leads to the release of PO_4^{3-} which can be taken up by organisms or sequestered by amorphous Fe particles in oxic part of the soil or sediment (Figure 13). There are several mechanisms responsible for above mentioned P mobilization including microbial iron and sulfate reduction, polyphosphate hydrolysis and other.

Iron (hydr)oxides serve as electron acceptor for certain microbes (e.g. *Pseudomonas*) as mentioned previously. The reduction of these compounds results in release of any P ions associated with the reduced material. Part of this released phosphate may form insoluble complexes with Fe(II) and later, it could be mobilized again by the sulfide originating from sulfate reduction. This process is common in aquatic systems, sulfate serves as electron acceptor and after microbial reduction its end-product sulfide interacts with Fe-P cycle leading to additional P mobilization (Baldwin and Mitchell, 2000).

The availability of P in soil and water may be influenced by soil aeration state due to solubility of redox sensitive Mn and Fe minerals. In addition, organic P may be also released as SOM serves as electron donor and P-carbonates may dissolve due to increased pH. The factor involved in P mobilization are temperature, biological activity and the availability of electron acceptors (Scalenghe et al., 2010).

Increased availability of both phosphorus and nitrogen has serious environmental consequences, as both are important for plant growth. High concentrations of both nutrients cause development of fast growing species which may cover the whole lake overcompeting other species. Dense cover of lemnids, cayanobateria or algae causes reduced diffusion of oxygen and light penetration into water layer affecting submerged vegetation. This phenomenon is called eutrophication and it is one of the important threats for the biodiversity of lakes and wetlands nowadays. This process may result from not only internal input of nutrients from external sources but, what is more striking, also from internal mobilization of N and P inside the system. The phenomenon of so-called internal eutrophication has been well documented for many aquatic systems, especially by works of Roelofs and his group (e.g. Roelofs, 1991; Smolders and Roelofs, 1993; Smolders and Roelofs, 1995; Lamers et al., 1998; Lamers et al., 2002; Lucassen et al., 2004b; Loeb et al., 2007; Geurts et al, 2009; Banach et al., 2009a, b) and discussed in detail by Lamers et al (2006) and Smolders et al. (2006). There are two main factors involved in increased availability of P in submerged soils (anaerobic) – the P saturation in the soil (expressed as bioavailable P, Olsen P; Olsen et al., 1954) and the composition of floodwater. The first results from the way of land cultivation, e.g. fertilization, mowing etc. leading to the accumulation of nutrients and the latter is closely related with the quality of flooding water which may originate from rivers or precipitation.

From the point of view redox processes iron and sulfur are the most important elements in availability of phosphorus under changed hydrological conditions (i.e. aeration status of the soil). Soils and sediments tend to buffer P by either the release of it into solution low in P or to sorb it from high P loaded solutions. The redox state of the soil affects this significantly as under oxic conditions P mobilization is hamper whilst anaerobic conditions cause Fe reduction and P release as shown above (Ann et al., 2000). Banach et al. (2009a) showed a linear relationship between concentration of mobilized P and Fe in the soil (Figure 14). Similar relationship was demonstrated by Moore Jr. and Reddy (1994) for sediments of Lake Okeechobee and by Stępniewska and Szafranek (2004) for renaturalized and degraded peatland. This emphasizes a key role of Fe in P kinetics under different redox conditions what was confirmed by several studies (Patrick and Khalid, 1974; Zak et al., 2004; Loeb et al., 2007).

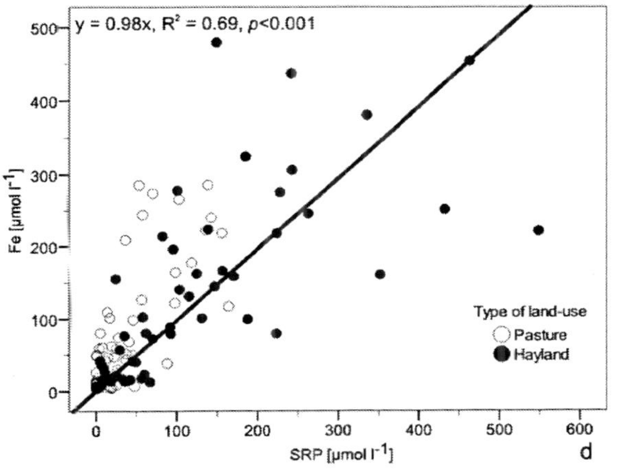

Figure 14. Relation between concentration of Fe and SRP in soil water during inundation of meadows differing in way of cultivation. Pasture – unculivated meadow, Hayland – fertilized and cultivated meadow (from Banach et al., 2009a, modified).

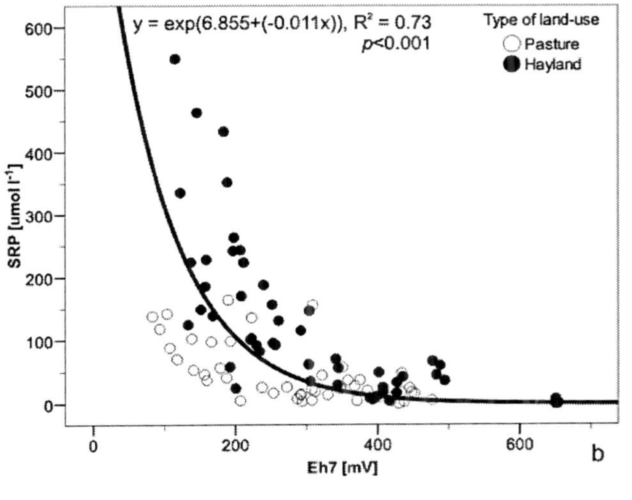

Figure 15. Relation between concentration of Eh and SRP in soil water during inundation of meadows differing in way of cultivation (from Banach et al., 2009a, modified).

There are several factors which could serve as a tool for the prediction and estimation the mobilization of phosphorus under different hydrological conditions. The ratio between soil Fe and P may be used as indicator of P saturation. Several studies showed that the value below 8-12 mol mol^{-1} indicates strong mobilization of phosphate (Jensen et al., 1992; Ramm and Scheps, 1997; Banach et al., 2009a). Another indicator of P mobilization to water layer is Fe to PO_4^{3-} ratio in soil solution which values below 1-3 mol mol^{-1} indicate the release of Fe-bound phosphorus (Smolders et al., 2001; Zak et al., 2004; Geurts et al., 2008). Leob et al. (2008a) have found that amount of mobilized P could be estimated by the phosphate saturation of the Fe binding sites in the soil which are reflected by the pool of amorphous iron. The contents of both Fe and P may differ between soils depending on their use. Banach et al. (2009a) showed that fertilized meadows were characterized by twice higher content of Olsen P than in uncultivated meadow and the fraction of Fe-bound P was also higher for fertilized meadow. Inundation of both soils led to strong mobilization of PO_4^{3-} what was clearly related to iron reduction as Eh dropped to about 100 mV when the concentration of P was the highest (Figure 15). Stępniewska and Szafranek (2004) showed even lower values of Eh around -200 mV when P mobilization was strong for peaty soils and similar relationship was observed.

The second element, involved in the availability of P is sulfur. It is present at different oxidation states in chemical compounds ranging from +6 (sulfate, SO_4^{2-}) through solid S (0) till reduced sulfides (-2, HS^-). Most of S is stored in rocks mainly as pyrite (FeS_x). During their weathering a large amounts of S are transferred into oceans as sulfate. Organisms take it up and accumulate as a reduced form in proteins. Under anaerobic conditions sulfur reducing bacteria (SRB) use SO_4^{2-} as electron acceptor and there is also S-based photosynthesis. Sulfur is also emitted (e.g. from coal burning) into atmosphere as SO_2 which takes part in acid rains formation (Schlesinger, 1997).

The mechanism of sulfate reduction under anaerobic conditions can be described as follows (at pH 6.4):

$$SO_4^{2-} + 2CH_2O \leftrightarrow HS^- + HCO_3^- + CO_2 + H_2O \qquad (18)$$

The main products, sulfide and bicarbonate (HCO_3^-) are responsible for the increased mobilization of nutrients.

Decomposition is stimulated under alkaline conditions, inlet of alkaline substances, e.g. bicarbonate, increases the rate of this process by the inhibition of decay-neutralizing acids. As process of decomposition also has a HCO_3^- as one of products it leads to higher availability of phosphate and ammonium. This factor, called alkalinity plays an important role in nutrient availability what was confirmed by positive correlations between both nutrients and bicarbonate found by Smolders (1995, Figure 16) and other researchers (Roelofs, 1991; Smolders and Roelofs, 1995; Lamers et al., 1998; Leob et al., 2008b; Banach et al., 2009b).

Higher concentrations of phosphate result not only from increased decomposition but also from the competition between HCO_3^- and PO_4^{3-} for anion adsorption sites. Alkalinization may also occur due to internal generation of biocarbonate during reduction of NO_3^-, Fe(III) and SO_4^{2-} as shown above (so-called internal alkalinization, Figure 17). This emphasizes the combined key role of sulfate and alkalinity in increased availability P in waters (Roelofs, 1991; Smolders et al., 2006).

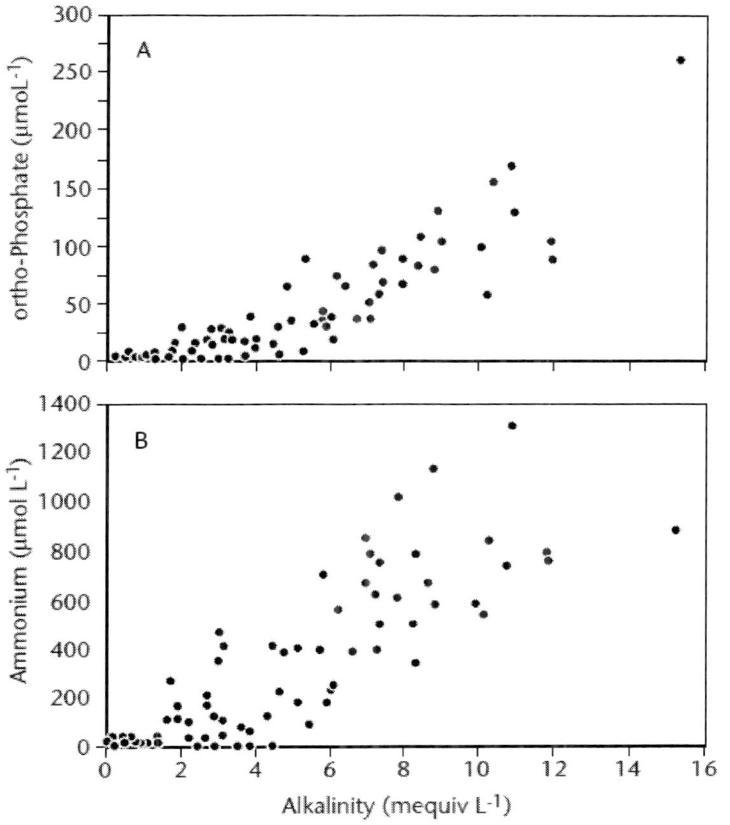

Figure 16. The relationship between alkalinity and (A) PO_4^{3-} or (B) NH_4^+ concentrations in sediment pore water for 70 randomly selected fen sediments in the Netherlands (after Smolders, 1995).

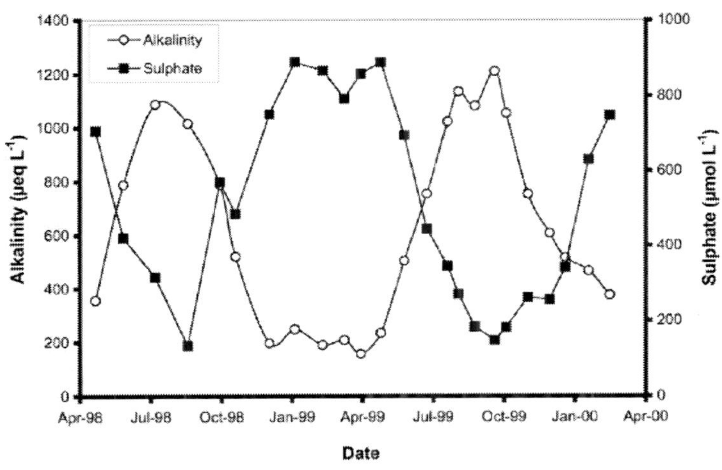

Figure 17. Changes the alkalinity and sulfate levels over time in the surface water of Dutch fen showing the effect seasonal of sulfate reduction on alkalinity (after Smolders et al., 2006).

The reduction of sulfate simulates decomposition of organic matter as it is an alternative electron acceptor. Next, it increases the alkalinity what further enhances decomposition. This together with inference of produced sulfide with Fe-P cycles mobilizes P. In addition, there is a competition between SO_4^{2-} and PO_4^{3-} for anion adsorption sites. Produced sulfide reduces Fe(III) in minerals such as strengite, iron (hydr)oxide phosphate and in humic iron phosphate complexes resulting in the release of attributed phosphate (Sperber, 1958; Patrick and Khalid, 1974) and forms insoluble iron sulfides (pyrite) decreasing the availability of free iron for P immobilization (Caraco et al., 1989). When soil is iron-rich and concentrations of sulfate are low then it can store P. High concentration of sulfate and prolonged period of anaerobic conditions favors P mobilization due to depletion of free Fe pool what eventually ends with severe eutrophication and accumulation of toxic, reduced compounds negatively affecting vegetation (Smolders and Roelofs, 1993; Lameres et al., 1998; Zak et al., 2006; Banach et al., 2009a, b; Figure 18).

Nitrate may interfere above mentioned process as it could oxidize pyrite resulting in higher levels of sulfate and lower nitrate (low NO_3^-:SO_4^{2-} ratio) what causes eutrophication. However, if concentration of nitrate are higher than produced sulfate (high NO_3^-:SO_4^{2-} ratio) it prevents further mobilization of phosphate by being more favorable electron acceptor increasing pool of Fe(III) for P binding (Lucassen et al., 2004a; Smolders et al., 2006).

The P binding depends also on pH, this together with Eh decides in which form P occurs in the soil what can be predicted using Eh-pH diagram (Figure 19).

It shows that under reduced conditions at pH 5.5 and lower the availability of phosphate is the highest as Fe and Ca minerals are dissolved under such conditions. Ca minerals are stable under neutral and alkaline conditions when precipitation P with Ca is observed (Moore and Reddy, 1994). The presence of Ca and Al compounds may influence the P availability. During P release due to Fe reduction it is possible that moderately stable Ca and Al minerals could reduce P availability by re-adsorbing part of P whilst Mg-bound P may be more easier mobilized as it is less stable than Ca-bound P fraction (Pant et al., 2002).

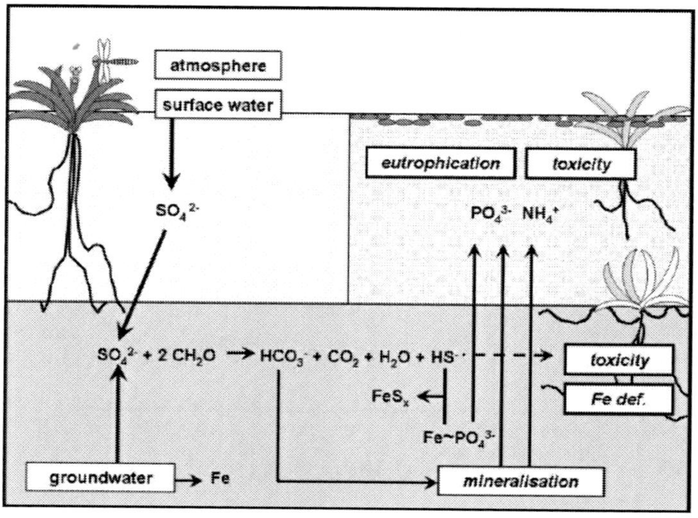

Figure 18. The hypothesized model explaining the disappearance of aquatic vegetation due to combined effects of sulfate and alkalinization resulting in internal eutrophication and toxicity (after Smolders et al., 2006).

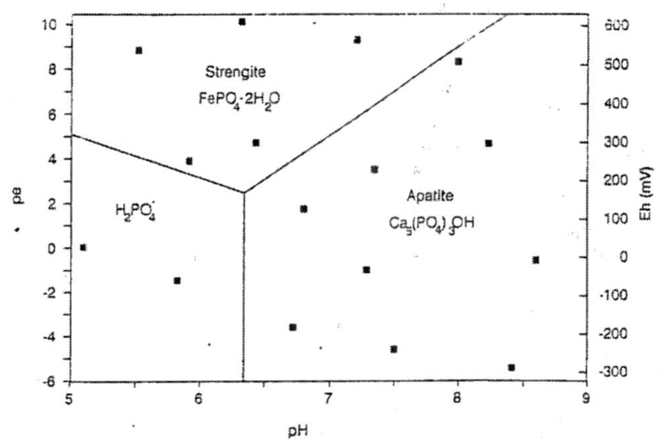

Figure 19. The pe(Eh)-pH diagram showing the stability of Fe- (strengite) and Ca-bound P (apatite) under different redox and pH conditions (after Moore and Reddy, 1994).

Alternation of soil microbial conditions affects the levels of nutrients in water layer what has severe ecological consequences such as internal eutrophication as explained previously. In addition, there are natural fluctuations of nutrient cycling. During photosynthesis CO_2 or HCO_3^- is reduced in cells resulting in the precipitation of P with Ca under higher pH. The same process is observed in green algal suspensions (*Chlorococcum* spp.) and biofilms where organisms create a higher pH for Ca-P co-precipitation (House, 2003).

Other Elements and Compounds

Because many elements posses few oxidation states they also undergo transformations as a result changed redox conditions. By construction of Eh-pH diagrams (Figure 20) for each element it is possible to predict the form of element and its availability. It is very important especially in case of heavy metals (e.g. Pb, Cr) which may be toxic for living organisms.

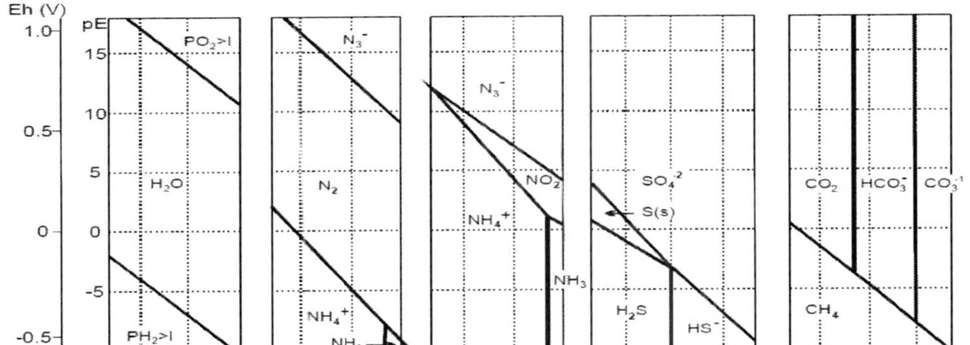

Figure 20. The examples of Eh(pe)-pH diagram for O_2 and H_2 (a), nitrogen (b and c), sulfur (d) and carbon (e) compounds at 25°C (after Stępniewski et al., 2005).

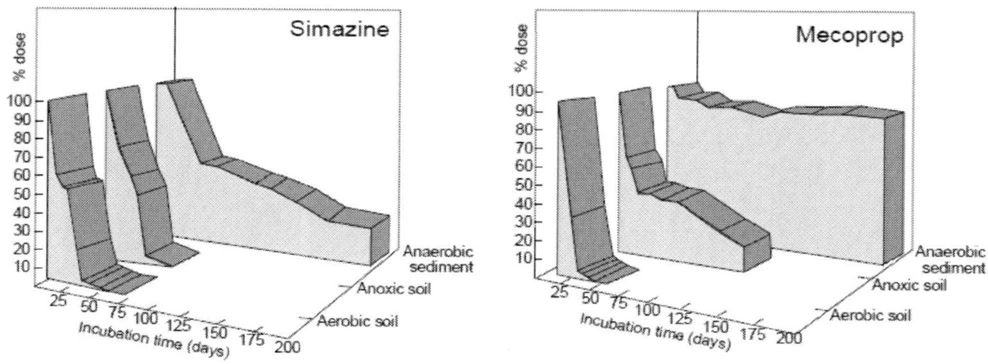

Figure 21. Stability od 2 pesticides in soils either aerobic (Eh=330 mV) or anoxic (Eh=180 mV) and anaerobic lake sediment (Eh=-120 mV) (after Vink, 1997).

Soil aeration state affects also the stability of organic compounds, e.g. pesticides. Vink in his PhD Thesis (1997) demonstrated the effect of different soil aeration condition on the stability of simazine and mecoprop showing that higher redox state led to faster degradation of selected pesticides (Figure 21).

4. Effects of Redox Conditions of Soil Organisms

The role of soil conditions for all living organisms is obvious; its properties such as: structure, porosity, moisture content, pH, temperature, aeration state and the contents of substrates (both organic and inorganic) shape the biodiversity of soil organisms creating different habitats. For that reason the knowledge about soil components is essential to evaluate its functions for living organisms (Nannipieri et al., 2003).

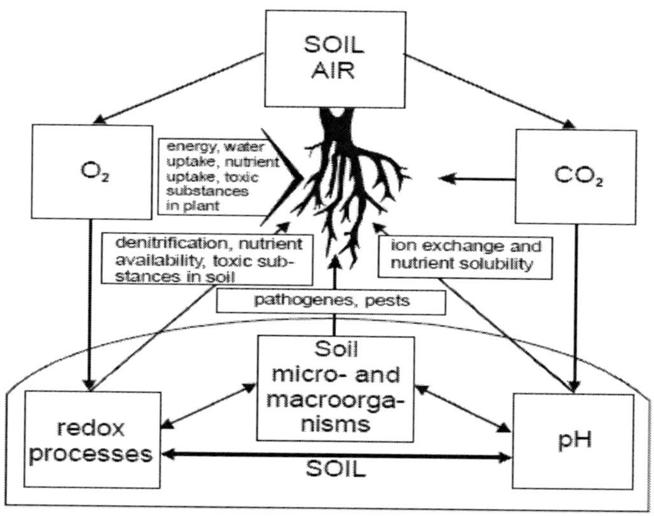

Figure 22. The schematic direct and indirect effect of soil aeration state on plants (after Gliński and Stępniewski, 1985).

Soil aeration state is very important factor for living organisms as it drives the availability of substrates for them. Plant species and some microbes depend on oxygen presence and key-nutrients such as N, P and K. Thus, the change of Eh affects their growth and the capacity of absorb substrates and unwanted substance, e.g. wastes (Ponnamperuma, 1972). The presence or absence of soil pores and water films affects gas transport (including oxygen). Soil is influenced by dry-wet cycles as a result of climatic changes. There are observed more frequent and severe flood events caused by the intensification of hydrological cycle (increased evapotranspiration, more dry summers with intense rainstorms) probably due to climate change (Milly et al., 2002; Christensen and Christensen, 2003; Huntington, 2006) what strongly affects soil organisms (Gleeson et al., 2008; Xiang et al., 2008). When soil organisms lose their activity it affects soil quality and health (Arias et al., 2005).

In a case of microorganisms determination their activity is based on measurement of activities of some enzymes involved in their metabolism. Dehydrogenases, DHA (EC 1.1.1.), enzymes from oxidoreductase class are good estimators of biological activity of soils giving information about the influence of environmental conditions on biological activity. They conduct a broad range of reactions responsible for decomposition, i.e. dehydrogenation of organic matter. Catalase (EC 1.11.1.6) is enzyme present in all aerobic organisms except for obligatory anaerobes and thus it is estimator of soil aeration status (Włodarczyk et al., 2001; Moreno et al., 2007; Wolińska and Stępniewska, 2011). Wolińska and Bennicelli (2010) demonstrated that DHA activity was negatively correlated with aeration parameters such as Eh or ODR. Flooded conditions lead to more enzyme produced and higher DHA activity what speeds up mineralization (Zhang et al., 2009). Work of Włodaczyk et al. (2001) showed that catalase activity was positively correlated with soil redox state.

Vegetation is influenced by many factors, from which oxygen availability is very important (Figure 22). The alternation of aeration state of soil affects plant species, the problem is related with limited oxygen diffusion as explained previously. Under submerged conditions plants are dependent on oxygen pool stored in the soil. After it depletion they start suffer from oxygen stress ending eventually with death of the roots and finally the whole plant. Oxygen deficiency and anaerobic condition have many negative effects on vegetation (Gliński and Stępiewski, 1985; Gibbs and Greenway, 2003). First, there is a shift of plant metabolism into less efficient anaerobic fermentation leading to fast depletion of carbonate reserves (Bailey-Serres and Voesenek, 2008). Next, the accumulation of reduced, potentially phytotoxic compounds such as ammonium, Fe(II), Mn(II) and sulfide may cause growth retardation, reduction in leaf size, wilting of shoots and necrosis (Roelofs, 1991; Snowden and Wheeler, 1993; Smolders, 1995; Lamers et al., 1998; Lucassen et al., 2000, 2002).

In response to above mentioned stress factors plants have developed several strategies to cope with them (Stępniewski et al., 2005). In order to save energy, lower energy demand may be established (Geigenberger, 2003). Next, shoot elongation may help plant to restore contact of leaves with the atmosphere (Banga et al., 1995; Voesenek et al., 2004) whilst aerenchyme tissue formation improves internal gas transport (Armstrong, 1972; Visser et al., 1996; McDonald et al., 2002; Colmer, 2003). In a case when water surface cannot be reached, underwater photosynthesis may improve both carbonate and oxygen concentrations (Laan and Blom, 1990; Mommer and Visser, 2005). Plant survival under these unfavorable conditions strongly depends on the balance between costs and benefits of these acclimations which are determined by flooding characteristics and soil processes (van Eck et al., 2004, 2005; Voesenek et al., 2004; van der Welle et al., 2007).

Conclusions

The availability nutrients and other elements in soil strongly depends on its characteristics. The main drivers of the nutrient cycling are microorganisms which demand certain conditions for their optimal activities. Their requirements are: temperature, pH, moisture, the presence of adequate levels of electron donors and acceptors and, what more important, appropriate aeration status of the soil. The latter may change quickly and influence other soil properties. For that reason soil redox state plays a key role in the microbial activity and biogeochemical cycling of elements.

In order to estimate redox conditions in the soil, there are several parameters exist. First of all, it is possible to determine oxygen availability either in terms of concentration or diffusion rate. However, determination of Eh is much easier and gives more information about soil processes. This parameter informs about the aeration state of the soil, the dominant oxidation-reduction process which tells about the forms in each all nutrients and elements are present what explains their availability.

The determination of soil aeration is important as nowadays there are numerous alternations of hydrology which differs in timing and frequency for example flood events which pose one of the important environmental problems. There are also restoration measures involving re-wetting previously desiccated wetlands or allowing temporary change of aeration state of the soil or sediment. The knowledge about the soil response to changed hydrology demands the analysis of biogeochemical processes under different conditions and it will help in the prediction environmental consequences of planed measures and occurring events.

References

Ann Y., Reddy K. R., Delfino J. J. (2000) Influence of redox potential on phosphorus solubility in chemically amended wetland organic soil. *Ecological Engineering* 14: 169-180.

Arias M. E., González-Pérez J. A., Ball A. S. (2005) Soil health – a new challenge for microbiologists and chemists. *International Microbiology* 8: 13-21.

Armstrong W. (1967) The relation between oxidation-reduction potentials and oxygen diffusion levels in some waterlogged organic soils. *Journal of Soil Science* 18: 27-34.

Armstrong W. (1972) A re-examination of the functional significance of aerenchyma. *Physiologia Plantarum* 27: 173–177.

Azam F., Müller C., Weiske A., Benckiser G., Ottow J. C. G. (2002) Nitrification and denitrification as sources of atmospheric nitrous oxide – role of oxidizable carbon and applied nitrogen. *Biology and Fertility of Soils* 35: 54-61.

Bailey-Serres J., Voesenek L. A. C. J. (2008) Flooding stress: acclimations and genetic diversity. *Annual Review of Plant Biology* 59: 313–339.

Baldwin D. S., Mitchell A. M. (2000) The effects of drying and re-flooding on the sediment and soil nutrient dynamics of lowland river–floodplain systems: a synthesis. Regulated Rivers: *Research and Management* 16: 457-467.

Banach A. M., Banach K., Visser E. J. W., Stępniewska Z., Smits A. J. M., Roelofs J. G. M., Lamers L. P. M. (2009a) Effects of summer flooding on floodplain biogeochemistry in Poland; implications for increased flooding frequency. *Biogeochemistry* 92: 247–262.

Banach A. M., Banach K., Peters R. C. J. H., Jansen R. H. M., Visser E. J. W., Stępniewska Z., Roelofs J. G. M., Lamers L. P. M. (2009b) Effects of long-term flooding on biogeochemistry and vegetation development in floodplains; a mesocosm experiment to study interacting effects of land use and water quality. *Biogeosciences* 6: 1–15.

Banach A. M. (2010) Biogeochemical constraints on combined flood control, water storage and ecological restoration in Central European river floodplains. Małopolski Przełom Wisły (Poland) as a case study. PhD Thesis, Nijmegen, *The Netherlands*, 141 pp.

Banga M., Blom C. W. P. M., Voesenek L. A. C. J. (1995) Flood-induced leaf elongation in Rumex species: effects of water depth and water movements. *New Phytologist* 131: 191–198.

Bell R. G. (1969) Studies on the decomposition of organic matter in flooded soil. *Soil Biology and Biochemistry* 1(2): 105-116.

Bennicelli R. P. (1999) Effect of soil aeration on plant biomass production. *Polish Journal of Soil Science* 33(2): 3-11.

Bennicelli R. P., Stępniewska Z., Banach K., Banach A. M., Szafranek A. (2006) Comparison of Aeration Status Measurements by Clark Sensor (DO) and ODR-Meter during Azolla caroliniana Willd. Growth in the Presence of Cd(II) and Hg(II). *Water, Air, and Soil Pollution* 180(1-4): 29-37.

Bergametti G., Remoudaki E., Losno R., Steiner E., Chatenet B. (1992) Source, transport and deposition of atmospheric P over the northwestern Mediterranean. *Journal of Atmospheric Chemistry* 14: 501–513.

Bohn H. L. (1971) Redox potentials. *Soil Science* 112: 39–45.

Bronstert A. (2003) Floods and climate change: Interactions and impacts. *Risk Analysis* 23(3): 545–557.

Brzezińska M., Włodarczyk T., Stępniewski W., Przywara G. (2005) Soil aeration status and catalase activity. *Acta Agrophysica* 5: 555-565.

Caraco N. F., Cole J. J., Likens G. E. (1989) Evidence for sulphate-controlled phosphorus release from sediments of aquatic systems. *Nature* 341: 316-318.

Christensen J. H., Bronstert, 2003; Christensen O. B. (2003) Climate modelling: severe summertime flooding in Europe. *Nature* 421: 805–806.

Colmer T. D. (2003) Long-distance transport of gases in plants: a perspective on internal aeration and radial oxygen loss from roots. *Plant, Cell and Environment* 26: 17–36.

Drever J. I. (1997) The Geochemistry of Natural Waters, Surface and Ground Water Environments, 3rd ed., Prentice Hall, Upper Saddle River, NJ, USA.

van Eck W. H. J. M., van de Steeg H. M., Blom C. W. P. M., de Kroon H. (2004) Is tolerance to summer flooding correlated with distribution patterns in river floodplains? A comparative study of 20 terrestrial grassland species. *Oikos* 107: 393–405.

van Eck W. H. J. M., Lenssen J. P. M., Rengelink R. H. J., Blom C. W. P. M., de Kroon H. (2005) Water temperature instead of acclimation stage and oxygen concentration determines responses to winter floods. *Aquatic Botany* 81: 253–264.

Geigenberger P. (2003) Response of plant metabolism to too little oxygen. *Current Opinion in Plant Biology* 6: 247–256.

Geurts J. J. M., Smolders A. J. P., Verhoeven J. T. A., Roelofs J. G. M., Lamers L. P. M. (2008) Sediment Fe:PO4 ratio as a diagnostic and prognostic tool for the restoration of macrophyte biodiversity in fen waters. *Freshwater Biology* 53: 2101–2116.

Geurts J. J. M., Sarneel J. M., Willers B. J. C., Roelofs J. G. M., Verhoeven J. T. A., Lamers L. P. M. (2009) Interacting effects of sulphate pollution, sulphide toxicity and eutrophication on vegetation development in fens: A mesocosm experiment. *Environmental Pollution* 157: 2072-2081.

Gibbs J., Greenway H. (2003) Mechanisms of anoxia tolerance in plants. I. Growth, survival and anaerobic catabolism. *Functional Plant Biology* 30: 1–47.

Gleeson D. B., Herrmann A. M., Livesley S. J., Murphy D. V. (2008) Influence of water potential on nitrification and structure of nitrifying bacterial communities in semiarid soils. *Applied Soil Ecology* 40: 189-194.

Gliński J., Stępniewski W. (1985) Soil aeration and its role for plants. Boca Raton, FL: CRC Press, 229 pp.

Gliński J., Stępniewska Z., Stępniewski W., Ostrowski J., Szagara A. (2000) A contribution to the assessment of potential denitrification in arable mineral soils of Poland. *Journal of Water and Land Development* 4: 175-183.

Holmer M., Storkholm P. (2001) Sulphate reduction and sulphur cycling in lake sediments: a review. *Freshwater Biology* 46: 431–451.

House W. A. (2003) Geochemical cycling of phosphorus in rivers. *Applied Geochemistry* 18: 739–748.

Huntington T. G. (2006) Evidence for intensification of the global water cycle: review and synthesis. *Journal of Hydrology* 319: 83-95.

Jensen H. S., Kristensen P., Jeppesen E., Skytthe A. (1992) Iron:phosphorus ratio in surface sediment as an indicator of phosphate release from aerobic sediments in shallow lakes. *Hydrobiologia* 235/236: 731–743.

Jungkunst H. F., Fiedler S., Stahr K. (2004) N2O emissions of a manure Norway spruce (Picea abies) stand in the Black Forest (southwest Germany) as differentiated by the soil pattern. *Journal of Geographical Research* 109: 1-11.

Laan P., Blom C. W. P. M. (1990) Growth and survival responses of Rumex species to flooded and submerged conditions: the importance of shoot elongation, underwater photosynthesis and reserve carbohydrates. *Journal of Experimental Botany* 41: 775–783.

Laanbroek H. J. (1990) Bacterial cycling of minerals that affect plant growth in waterlogged soils: a review. *Aquatic Botany* 38: 109–125.

Lamers L. P. M., Tomassen H. B. M., Roelofs J. G. M. (1998) Sulfate-induced eutrophication and phytotoxicity in freshwater wetlands. *Environmental Science and Technology* 32: 199–205.

Lamers L. P. M., Falla S-J., Samborska E. M., Van Dulken I. A. R., Van Hengstum G., Roelofs J. G. M. (2002) Factors controlling the extent of eutrophication and toxicity in sulfate-polluted freshwater wetlands. *Limnology and Oceanography* 47: 585–593.

Lamers L. P. M., Loeb R., Antheunisse A. M., Miletto M., Lucassen E. C. H. E. T., Boxman A. W., Smolders A. J. P., Roelofs J. G. M. (2006) Biogeochemical constraints on the ecological rehabilitation of wetland vegetation in river floodplains. *Hydrobiologia* 565: 165–186.

Lemon E. R. and Erickson A. E. (1952). The measurement of oxygen diffusion in the soil with a platinum microelectrode. *Soil Science* 16: 160.

Lincoln, R.J., Boxshall, G.A. and Clark, P.F. (1998) A dictionary of ecology, evolution and systematic - 2nd edn. Cambridge University Press, Cambridge.

Loeb R., van Daalen E., Lamers L. P. M., Roelofs J. G. M. (2007) How soil characteristics and water quality influence the biogeochemical response to flooding in riverine wetlands. *Biogeochemistry* 85: 289–302.

Loeb R, Lamers L. P. M., Roelofs J. G. M. (2008) Prediction of phosphate mobilisation in inundated floodplain soils. *Environmental Pollution* 156: 325–331.

Loeb R, Lamers L. P. M., Roelofs J. G. M. (2008b) Effects of winter versus summer flooding and subsequent desiccation on soil chemistry in a riverine hay meadow. *Geoderma* 145: 84-90.

Lucassen E. C. H. E. T., Smolders A. J. P., Roelofs J. G. M. (2000) Increased groundwater levels cause iron toxicity in Glyceria fluitans (L.). *Aquatic Botany* 66: 321–327.

Lucassen E. C. H. E. T., Bobbink R., Smolders A. J. P., van der Ven P. J. M., Lamers L. P. M., Roelofs J. G. M. (2002) Interactive effects of low pH and high ammonium levels responsible for the decline of Cirsium dissectum (L.) *Hill. Plant Ecology* 165: 45–52.

Lucassen E. C. H. E. T., Smolders A. J. P., Van der Salm A. L., Roelofs J. G. M. (2004a) High groundwater nitrate concentrations inhibit eutrophication of sulphate-rich freshwater wetlands. *Biogeochemistry* 67: 249–267.

Lucassen E. C. H. E. T., Smolders A. J. P., Van de Crommenacker J., Roelofs J. G. M. (2004b) Effects of stagnating sulphaterich groundwater on the mobility of phosphate in freshwater wetlands: a field experiment. *Archiv für Hydrobiologie* 160: 117–131.

McDonald M. P., Galwey N. W., Colmer T. D. (2002) Similarity and diversity in adventitious root anatomy as related to root aeration among a range of wetland and dryland grass species. *Plant, Cell and Environment* 25: 441–451.

Milly P. C. D., Wetherald R. T., Dunne K. A., Delworth T. L. (2002) Increasing risk of great floods in a changing climate. *Nature* 415: 514–517.

Mommer L., Visser E. J. W. (2005) Underwater photosynthesis in flooded terrestrial plants: a matter of leaf plasticity. *Annals of Botany* 96: 581–589.

Moore P. A. Jr., Reddy K. R. (1994) Role of Eh and pH on Phosphorus Geochemistry in Sediments of Lake Okeechobee, Florida. *Journal of Environmental Quality* 23: 955-964.

Moreno J. L., Aliaga A., Navarro S., Hernendez T., Garcia C. (2007) Effects of atrazine on microbial activity in semiarid soil. *Applied Soil Ecology* 35: 120-127.

Nannipieri P., Ascher J., Ceccherini M.T., Landi L., Pietramellara G, Renella G. (2003) Microbial diversity and soil functions. *European Journal of Soil Science* 54: 655-670.

Nielsen J. L., Nielsen P. H. (1998) Microbial nitrate-dependent oxidation of ferrous iron in activated sludge. *Environmental Science and Technology* 32: 3556–3561.

Nightingale E. R. Jr. (1958) Poised oxidation–reduction systems. *Analytical Chemistry* 30: 267–272.

Olsen S. R., Cole C. V., Watanabe F. S., Dean L. A. (1954) Estimation of available phosphorous in soils by extraction with sodium bicarbonate. United States Department of Agriculture Circular 93: 9, Washington, DC.

Pant H. K., Nair V. D., Reddy K. R., Graetz D. A. Villapando R. R. (2002) Influence of Flooding on Phosphorus Mobility in Manure-Impacted Soil. *Journal of Environmental Quality* 31: 1399-1405.

Patrick W. H. (1960) Nitrate Reduction Rate in a Submerged Soil as Affected by Redox Potential. *Transactions of the Seventh International Congress on Soil Science* 3: 494-500.

Patrick W. H., Mikkelsen D. D. (1971) Plant Nutrient Behaviour in flooded soil. In: Fertilizer Technology and Use. 2nd ed., Soil Science Society of America, Madison, Wisc., 187.

Patrick W. H., Khalid R. A. (1974) Phosphate release and sorption by sediments soils and sediments – effect of aerobic and anaerobic conditions. *Science* 186: 53–55.

Paul E. A., Clark F. E. (1996) Soil Microbiolgy and Biochemistry. Academic Press, San Diego, CA, 340 pp.

Pidello A., Menendez L., Perotti E. B. R. (1996) Saccharid compounds as soil redox effectors and their influence on potential N2O production. *Biology and Fertility of Soils* 23: 173-176.

Ponnamperuma F. N. (1972) The Chemistry of Submerged Soils. *Advances in Agronomy* Vol. 24, 29-88.

Ramm K., Scheps V. (1997) Phosphorous balance of a polytrophic shallow lake with the consideration of phosphorous release. *Hydrologia* 342: 43–53.

Roelofs J. G. M. (1991) Inlet of alkaline river water into peaty lowlands: effects on water quality and Stratiotes aloides L. stands. *Aquatic Botany* 39: 267–293.

Scalenghe R., Edwards A. C., Barberis E., Ajmone Marsan F. (2010) The influence of pulsed redox conditions on soil phosphorus. *Biogeosciences Discussion* 7: 9009-9037.

Schimel J., Balser T. C., Wallenstein M. (2007) Microbial stress-response physiology and its implications for ecosystem function. *Ecology* 88: 1386-1394.

Schlesinger W. H. (1997) Biochemistry. Analysis of global change. Academic Press, San Diego, CA, 588 pp.

Smith C. J, Wright M. F., Patrick W. H. (1983) The effect of soil redox potential and pH on the reduction and production of nitrous oxide. *Journal of Environmental Quality* 12(2): 186-188.

Smolders A. J. P., Roelofs J. G. M. (1993) Sulphate mediated iron limitation and eutrophication in aquatic systems. *Aquatic Botany* 46: 247–253.

Smolders A. J. P. (1995) Mechanisms involved in the decline of aquatic macrophytes in the Netherlands; in particular of Stratiotes aloides L., PhD thesis, University of Nijmegen, Nijmegen.

Smolders A. J. P., Roelofs J. G. M. (1995) Internal eutrophication, iron limitation and sulphide accumulation due to the inlet of river Rhine water in peaty shallow waters in the Netherlands. *Archiv für Hydrobiologie* 133: 349–365.

Smolders A. J. P., Lamers L. P. M., Moonen M., Zwaga K., Roelofs J. G. M. (2001) Controlling phosphate release from phosphate enriched sediments by adding various iron compounds. *Biogeochemistry* 54: 219–228.

Smolders A. J. P., Lamers L. P. M., Lucassen E. C. H. E. T., van der Velde G., Roelofs J. G. M. (2006) Internal eutrophication: How it works and what to do about it – a review. *Chemistry and Ecology* 22(2): 93–111.

Snowden R. E., Wheeler B. D. (1993) Iron toxicity to fen plant species. *Journal of Ecology* 81: 35–46.

Sperber J. I. (1958) Release of phosphate from soil minerals by hydrogen sulphide. *Nature* 181: 934.

Sposito G. (1989) The Chemistry of Soils. Oxford University Press, NY, 277 pp.

Stępniewska Z. (1994) Soil redox resistance as a factor of nitrate stability in the soil. In: Migration and Fate Pollutants in Soils and Subsoils. Therory and Practice. *CNR, Quaderni* 96: 9.1-9.7.

Stępniewska Z., Bennicelli R. P., Weiss U., Włodarczyk T., Stahr K. (2000) Denitrification rate in soils as affected by their redox potential. *Journal of Water and Land Development* 4: 163-173.

Stępniewska Z., Stępniewski W., Gliński J., Ostrowski J. (1996) Atlas of the redox properties of arable soils in Poland. Lublin-Falenty, 106 pp.

Stępniewska Z., Szafranek A. (2004) Changes in phosphate level in some peats being transformed in different degrees. Environment Protection Engineering 30(4): 121-126.

Stępniewska Z., Szmagara A. (2000) The influence of nitrates (V) level on N2O emission from selected soils in model investigations. *Acta Agrophysica* 38: 209-219.

Stępniewski W. and Stępniewska Z. (1998) Oxygenology as a new discipline in the environmental sciences – a proposal for discussion. *International Agrophysics* 12: 53-55.

Stępniewski W., Horn R., Martyniuk S. (2002) Managing soil biophysical properties for environmental protection. *Agriculture. Ecosystems and Environment* 88: 175-181.

Stępniewski W., Stępniewska Z., Bennicelli R. P., Gliński J. (2005) Oxygenology in outline. Institute of Agrophysics PAS, Lublin, 121 pp.

Straub K. S., Benz M., Schink B., Widdel F. (1996) Anaerobic, nitrate dependent microbial oxidation of ferrous iron. *Applied and Environmental Microbiology* 62: 1458–1460.

Stumm V. and Morgan J. J. (1970) Aquatic Chemistry, Wiley-Interscience, USA, 514-565.

Szafranek-Nakonieczna A. (2009) The influence of environmental conditions on the activity of methanotrophs and methanogens in selected peat deposits from Poleski National Park. PhD Thesis, Lublin, 367 pp (in Polish).

Takai Y. and Kamura T. (1966) The Mechanism of Reduction in Waterlogged Paddy Soil. *Folia Microbilogica* 11(4): 304-313.

Voesenek L. A. C. J., Rijnders J. H. G. M., Peeters A. J. M., van de Steeg H. M., de Kroon H. (2004) Plant hormones regulate fast shoot elongation under water: from genes to communities. *Ecology* 85: 16–27.

Vink J. P. M. (1997) Pesticide biotransformation and fate in heterogeneous environments. Van Zee tot Land 63. PhD Thesis, Wageningen, *The Netherlands*, 77-93.

Visser E. J. W., Blom C. W. P. M., Voesenek L. A. C. J. (1996) Flooding-induced adventitious rooting in Rumex: morphology and development in an ecological perspective. *Acta Botanica Neerlandica* 45: 17–28.

Weber K. A., Picardal F. W., Roden E. E. (2001) Microbially catalyzed nitrate-dependent oxidation of biogenic solid-phase Fe(II) compounds. *Environmental Science and Technology* 35: 1644–1650.

van der Welle M. E. W., Roelofs J. G. M., op den Camp H. J. M., Lamers L. P. M. (2007) Predicting metal uptake by wetland plants under aerobic and anaerobic conditions. *Environmental Toxicology and Chemistry* 26: 686–694.

Wetzel R. G. (2001) Limnology, Saunders W. B., Philadelphia, PA.

Włodarczyk T., Gliński J., Stępniewski W., Stępniewska Z., Brzezińska M., Kuráž V. (2001) Aeration properties and enzyme activity on the ex ample of Arenic Chernozem (Tišice). *International Agrophysics* 15: 131-183.

Włodarczyk T., Stępniewski W., Brzezińska M., Przywara G. (2008) Impact of different aeration conditions on the content of extractable nutrients in soil. *International Agrophysics* 22: 371-375.

Wolińska A., Stępniewska Z. (2011) Microorganisms abundance and dehydrogenase activity as a consequence of soil reoxidation process. In: Soil Tillage and Microbial Activities. Miransari M. eds. Research Signpost, India, 1-33.

Xiang S. R., Doyle A., Holden P. A., Schimel J. P. (2008) Drying and rewetting effects on C and N mineralization and microbial activity in surface and subsurface California grassland soils. *Soil Biology and Biochemistry* 40: 2281-2289.

Yamane I., Sato K. (1964) Decomposition of glucose and gas formation in flooded soil. *Soil Science and Plant Nutrition* 13: 94-100.

Yu K. W., Wang Z. P., Vermoesen A., Patrick Jr W. H., Van Cleemput O. (2001) Nitrous oxide and methane emissions from different soil suspensions: effect of soil redox status. *Biology and Fertility of Soils* 34: 25-30.

Yu K., Faulkner S. P., Patrick Jr W. H. (2006) Redox potential characterization and soil greenhouse gas concentration across a hydrological gradient. *Chemosphere* 62: 905-914.

Zak D., Gelbrecht J., Steinberg C. E. W. (2004) Phosphorus retention at the redox interface of peatlands adjacent to surface waters in northeast Germany. *Biogeochemistry* 70: 357–368.

Zak D., Kleeberg A., Hupfer M. (2006) Sulphate-mediated phosphorous mobilization in riverine sediments at increasing sulphate concentration, River Spree, NE Germany. *Biogeochemistry* 80: 109–119.

Zhang L., Zhije W. U., Chen L., Jiang Y., Dongpo L. I. (2009) Kinetics of catalase and dehydrogenase in main soils of Northeast China under different soil moisture conditions. *Agricultural Journal* 4: 113-120.

In: Soil Nutrients
Editor: Mohammad Miransari
ISBN: 978-1-61324-785-3

Chapter 3

ROLE OF SOIL ORGANISMS IN DETERMINING NUTRIENT DYNAMICS

***Jagdamba Prasad Thapliyal*[1] *and Sutapa Bose*[2*]**
[1]Department of Applied Sciences and Humanities, G.L.A. University, Chaumuhan, Mathura, Uttar Pradesh, India
[2]Department of Environmental Medicine, New York University, Tuxedo, NY, U. S.

ABSTRACT

Soil provides habitat to a diverse array of organisms including archaea, bacteria, fungi, protozoans, algae and several taxa of invertebrates, which may represent about 2 millions species or as much as 23% of the currently known biodiversity. On the basis of their size, these organisms are broadly categorized as microflora, microfauna, mesofauna and macrofauna. The community of these organisms plays central role in various processes that determine soil structure, nutrient dynamics and soil fertility. A number of studies from tropical regions have fairly documented the major functions associated to various groups of soil organisms. Microflora, mainly represented by bacteria and fungi, is reported to catabolise decomposition of organic matter, and mineralise and immobilise nutrients in soil. Microfauna, mainly represented by protozoans and nematodes, are known for regulating bacterial and fungal populations and altering nutrient turnover. Mesofauna, mainly represented by collembola and mites, are reported to regulate fungal and microfaunal populations, alter nutrient turnover and fragment plant residues. Macrofauna, represented by annelids, arthropods and molluscs are mainly known for their contribution in fragmenting plant residues and stimulating microbial activity. These organisms also indirectly influence soil nutrients by altering soil structure, which chiefly includes humification, formation of soil aggregates, creation of biopores and other biogenic structures. Understanding the contribution of soil organisms in soil nutrient dynamics has been a key research area for ecologists and soil biologists for many decades and still a number of processes have not been adequately explored. It has been widely acknowledged that for achieving sustainability in food production, basic processes of soil fertility including nutrient dynamics have to be understood and managed at the farm

* Corresponding autour's E-mail: sutaparai@gmail.com.

level. Soil organisms offer potential option for optimizing these processes. Studies involving soil nutrients and below ground community bear special importance for tropical region since farmers of these areas are economically weak and they are largely dependent on natural processes for managing fertility of their soils.

INTRODUCTION

Soil provides habitat to a fairly diverse communities of microorganisms, plants and animals. These organisms play major role in soil organic matter turnover, soil nutrient cycling and mineralization. In fact, their contribution in various processes occurring in the top few centimeters of the earth surface are critically important for sustenance of life in terrestrial ecosystem [1,2,3].

Soil organisms, often termed as soil biota or below ground biodiversity, include archaea, bacteria, fungi, protozoans, algae and invertebrate animals. Among invertebrates, there are various taxonomic groups including nematodes, annelids, arthropods and mollusks. The diversity of these organisms is amazingly very high [4]. According to recent estimations, soil organisms may represent over 2 million species or as much as 23% of the total diversity of living organisms that have been described till date [5]. In some ecosystems, the local diversity of soil organisms may be enormously high: far above that of aboveground plants or animals [3].

The soil organisms exert vital functions with respect to soil structure and nutrient dynamics in tropical ecosystems [6, 7]. Belowground biodiversity because of its interrelations with soil quality could be a key factor in determining stability and resilience of ecosystem [8]. A direct relationship exists between species richness of these organisms and the intensity of some fundamental processes such as decomposition, nutrient storage, primary production and water availability [9, 10].

Generally, the diversity of soil biota is described in a few size classes with specific spheres of influence. Major groups under different size classes of soil organisms are: Microflora, Microfauna (<0.2 mm), Meso-fauna (0.2 – 2 mm) and Macrofauna (>2 mm) [11]. Ecological functions associated to these groups have been discussed in many studies, although such studies for tropical soils are limited in number [12].

Microflora, mainly represented by bacteria and fungi, is reported to catabolise decomposition of organic matter, and mineralise and immobilise nutrients in soil [13, 14]. Microfauna, mainly represented by protozoans and nematodes, are known for regulating bacterial and fungal populations and altering nutrient turnover [15, 16]. Mesofauna, mainly represented by collembolan, mites and enchytraeids, are reported to regulate fungal and microfaunal populations, alter nutrient turnover and fragment plant residues [11, 17] Macrofauna, represented by annelids, arthropods and molluscs are mainly known for their contribution in fragmenting plant residues and stimulating microbial activity. Soil macrofauna also indirectly influence soil nutrients by altering soil structure which chiefly includes humification, formation of soil aggregates, creation of biopores and other biogenic structures [18, 19, 20].

Soil biotic communities are affected negatively by agricultural intensification [21] but the precise significance of these changes to agro-ecosystem functions is less well established [22]. Though soil biota have remained in major focus of ecological investigation for the last

2-3 decades, such studies are largely confined to temperate soils and their precise role in ecosystem is yet not fully understood for tropical soils [23, 24]. Our ability to manage the ecosystem processes for sustainable agriculture is explicitly limited in tropical agro-ecosystem by a fundamental lack of knowledge of biota present in soil, of the basic biology and ecology of different species and their role in ecosystem functioning [25].

The present paper reviews the current status of ecological knowledge about soil organisms in terms of their diversity, distribution and roles in ecosystem. Role of soil biota in nutrient dynamics, litter decomposition and other soil processes have been discussed for individual group as well as for the soil biota as a whole. Though studies from all over the world have been included, studies from tropical regions have been particularly emphasized.

Ecological Studies on Soil Organisms

A number of studies have documented the diversity, distribution and ecological roles of different groups of soil organisms [1, 2, 3, 19, 26]. Traditionally soil organisms were portrayed as decomposers constituting a single trophic level. In fact, the soil food web is as complex as the aboveground web, with multiple trophic levels and intricate connections to its aerial counterpart [27, 28, 29]. Studies show that the diversity of these organisms is amazingly very high [4, 30, 31, 32, 33] and sometimes exceeding that of aboveground plants or animals [3]. (Lavelle et al., 2006). Soil animals are estimated to constitute as much as 23% of the currently known biodiversity [5]. It is still an enormous task to identify these organisms at species level [34].

Soil organisms started receiving major attention of ecological investigation only for the last 2-3 decades. Perception of their presence and functional importance in soil is, however, as old as the interest for their taxonomic study. Aristotle in 350 BC called earthworm 'the intestine of the earth'. Despite early works by Gilbert White [35] and Charles Darwin [36, 37] on the earthworms and other soil invertebrates, studies on soil biota had received inadequate attention from pedologists for a long time. Thereafter, the majority of research work on soil fauna was carried out on European and North American soils. Many good studies on soil biota have been carried out between 1940 and 1980 [1].

After 1980s, considerable efforts were devoted to the description and quantification of the direct and indirect effects of soil invertebrates and other groups of soil biota on major processes of soil ecosystem, e.g., regulation of microbial activities, the dynamics of soil organic matter, nutrient cycling and the formation and maintenance of physical structures [38, 39, 40, 41]. The effects of soil invertebrate activities on soil fertility and plant growth have been considered in a number of studies [42, 43, 44, 45, 46].

In the 1970s the International Biological Programme produced a large amount of quantitative data on the abundance of soil invertebrates and their role in ecosystem [47]. (UNESCO in the 1990s sponsored a programme on "Tropical Soil Biology and Fertility". The programme resulted in the development of a series of scientific literature on tropical soils. Today, soil biology has become a well-organized international scientific discipline. Since 1955, studies on soil biota are regularly supplemented and enriched through a number of research journals and periodicals. Applied Soil Ecology, Biology and Fertility of Soils,

Pedobiologia and Soil Biology and Biochemistry are some of the major international journals exclusively devoted for soil biological studies.

Functional Classification of Soil Organisms

As soil organism comprise thousand of species with wide range of ecological strategies, attempts have been made to construct functional classification systems. Generally, the diversity of soil biota is described in a few size classes with specific spheres of influence. Major groups under different size classes of soil organisms are: Microflora, Microfauna (<0.2 mm), Meso-fauna (0.2 – 2 mm) and Macrofauna (>2 mm) [11]. Further, in order to fully appreciate structural and functional diversity presented by these organisms, and to understand effects of human induced stress and disturbance on their activities, the soil biota must be studied at taxonomic level [2].

Soil biota covers a range of trophic levels due to diversified food habits. In soil food webs they are often allocated to various functional groups based on their feeding habit [48, 49]. The primary consumers of the soil food web are the microbes (bacteria, fungi, actinomycetes, and algae) and they are primarily responsible for breaking down and mineralizing complex organic substances. Soil animals (microfauna, mesofauna and macrofauna) constitute secondary and higher level of consumers. On the basis on their functions in ecosystem, soil animals are often be distinguished in three functional groups: micro-predators, litter transformers and ecosystem engineers [50].

Micropredators contain the smallest invertebrates, protozoa and nematodes. They do not produce organo-mineral structures [3, 34], and their principal effect is to stimulate the mineralization of soil organic matter [51]. Litter transformers include mesofauna and some macrofauna and these are involved in litter decomposition [50]. When these invertebrates re-ingest their excretions, which serve as incubators for microflora, they assimilate metabolites liberated by microbial actions. The ecosystem engineers are those organisms that produce biogenic structures through which they can modify the availability or accessibility of resources for other organisms [52, 53]. The macrofauna are also divided into three groups depending on their respective niche in the ecosystems: the epigeics, the anecics and the endogeics [54, 55].

Microflora

Microflora includes bacteria, fungi (mycorrhizal and non-mycorrhizal), actinomycetes and algae which are the primary consumers of the soil food web. These are primarily responsible for breaking down and mineralizing complex organic substances. Out of these, fungi and bacteria are the most abundant groups [15, 56]. The recent development of culture-independent molecular tools has allowed ecologists to start to explore the true diversity of bacteria and fungi in soils and other habitats [57, 58].

Soil bacteria are extremely abundant and diverse organisms. They are ubiquitous in their distribution and inhabit the most extreme environments and soils. It has been estimated on the basis of analysis of soil DNA that about 10,000 genetically distinct prokaryotic types can be present in a handful of soil [59], and there may be as many as $4x10^6$ prokaryotic taxa in soil [60]. Their abundance is equally astounding as there can be as many as 10^{10}–10^{11} bacteria per gram of soil [61].

The rich taxonomic diversity of bacteria reflects itself in form of their immense metabolic diversity to break down different substrate including xenobiotic compounds that are toxic to most other organisms in soil [12]. Heterotrophic bacteria play a major role in decomposition of soil leaf litter and organic residue [38]. Certain groups of bacteria also perform specific functions in nutrient cycling. For example, symbiotic nitrogen-fixing bacteria of the genus Rhizobium infect roots of legumes, greatly increasing N supply to the system. Some free-living, non-symbiotic heterotrophic bacteria also perform this process of N fixation, for example, those of the genera Azotobacter and Clostridium. This in turn has great influence for plant productivity and community structure [12].

Fungi are distinct from bacteria in that they are eukaryotic and generally produce filamentous hyphae that can penetrate and explore microhabitats of the soil [56]. The primary role of fungi is to decompose organic materials via the production of a vast array of extracellular enzymes. Due to their large biomass, which can often more than that of other microbes, plants, and animals, they represent a significant portion of the ecosystem nutrient pool [12]. For example, in grassland, there can be as much as 250 kg/ha dry fungal hyphae within the surface 5 cm soil, containing as much as 8 kg of N and 2 kg of P [62]. By degrading relatively permanent pools of organic matter in soil, these extensive mycelial networks can exploit 'vacant' resource patches in ecosystems [63].

There are also other important roles of fungi. They act as plant pathogens, they bind soil particles together, thereby enhancing soil structural stability, and they provide a food source to microbial-feeding fauna. Some fungi, the mycorrhizae, form mutualistic associations with plant roots [12]. The term mycorrhiza literally means fungus-root, derived from the Greek mykes and rhiza, which respectively mean fungus and root. Some 95% of vascular plants belong to genera that are associated with mycorrhizal fungi. Benefits to the plant include enhanced nutrient acquisition, especially of P, and consequently increased growth and improved fitness. Plants that have mycorrhizas also tend to be more competitive and better able to tolerate environmental stress than non-mycorrhizal ones [56, 64].

MICROFAUNA

Microfauna are the small soil organisms (<0.2mm body width) which live in water filled pore space in soil. The two most abundant groups of microfauna are the protozoa and nematodes. Microfauna can affect the nutrient mineralization directly by excreting mineral nutrients and releasing nutrients from microbial biomass and indirectly by causing shifts in the microbial community structure and growth rates. They can increase microbial turnover by reducing microbial competition and increasing nutrient availability. They form food webs of great functional importance in specific microhabitats, such as rhizosphere system [16, 65].

Soil protozoa are single-celled eukaryotic organisms. Protozoan abundance varies greatly in soil, both spatially and temporally. Their densities is about 10^4–10^6 /g soil in surface soil, declining to as low as 100/g soil in deeper soil horizons where organic matter and prey availability are low [66]. Protozoa also show good taxonomic diversity. Over one thousand species of ciliate fauna have been recorded globally [67]. As protozoa require a water film for locomotion and feeding, their activity is limited to the water-filled pore space in soil. They can, however, withstand drying of the soil and other adverse conditions by rapidly forming resistant structures called cysts. They can also rapidly emerge from cysts when conditions become favourable, for example, after rains in desert ecosystems [68].

The importance of protozoa in terrestrial ecosystems is largely due to their feeding activities. Most of protozoans are thought to feed on bacteria. Other groups include fungal-feeders, predatory protozoa and saprophytic protozoa [12]. These feeding activities have major consequences for ecosystems, affecting nutrient and energy flow both directly and indirectly. Their direct effect stems from the fact that a large proportion (60%) of ingested nutrients are excreted into the soil environment, enhancing the availability of nutrients for plant uptake, whereas their indirect effect relates to their impact on the activity and composition of microbial communities. They enhance microbial growth, graze bacteria and provide resource to nematodes and mesofauna [17].

Nematodes are multicellular animals. Like protozoa, soil nematodes (roundworms) require a water film around soil particles in which to move, feed, and reproduce; they also have similarly evolved mechanisms to survive unfavourable conditions (cryptobiosis, a complete cessation of metabolism, and dormant juvenile stages). Nematodes are the most numerous multicellular animals in soil, reaching densities as high as 30 million m^{-2} in Oak forest and 10 million m^{-2} in highly productive lowland pasture [69]. Even in soils that are polluted with heavy metals, densities can reach 260,000 m^{-2} [62]. Diversity of nematodes in soil is fairly high. Bloemers et al. [70] recorded a total of 431 species from 24 sites in a Cameroon forest. Many species of nematodes also inhabit soils of less fertile ecosystems [71, 72].

Nematodes in soil may be grouped according to the type of food that they consume, based on the morphology of their mouthparts. The most common feeding groups of nematodes are plant-feeders, bacterial-feeders, fungal-feeders, omnivores, and predatory nematodes [73]. Like protozoa, the role of nematodes in soil systems relates to their feeding activities. Direct functions include excretion of nutrients into soil while indirect roles include altering the size, composition, and activity of the microbial community. They also have important roles in the inoculation of new substrates by phoretic transport (i.e. microbes adhering to the surface of larger organisms) or excretion of viable microbes. Plant-parasitic nematodes are increasingly being shown to have important direct effects on plant communities [12].

MESOFAUNA

Mesofauna (0.2–2 mm) include mites (Acari), springtails (Collembola) and enchytraeids (small Oligochaeta). Collembola and acari are collectively called as microarthropods. They are often studied together as their sampling or extraction methods fairly overlap. The Collembola (also known as springtails) are small (up to 5 mm long), wingless insects, with six abdominal segments and biting mouth-parts which can be withdrawn into the head. Acari (commonly known as mites) have sac-like bodies, which may be divided by a furrow into anterior and posterior sections [12]. The enchytraeids (Oligochaeta), commonly known as potworms, are smaller than other soil annelids like earthworms. They are about 10–50 mm long [74].

Numerically, microarthropods are the most abundant non-aquatic faunal group in the soils of most ecosystems. Densities of soil microarthropods can be as high as 300,000 m_2 in permanent grassland with dense root systems and high organic matter content [17]. In tropical grasslands the biomass of mites can be 2–5 times that of Collembola whereas in temperate grasslands, the biomass of mites and Collembola is often reported to be similar [47]. Culik et al., [75] recorded collembolan densities of the orchards ranging from 14023 to 56001 Collembola per m2 in Brazil. Microarthropod also represent rich diversity in different ecosystems [75, 76].

Mites (Acari) consume plant litter, animal carcasses, nematodes, bacteria, fungi and some insects. They may be parasites on animals of higher taxa, or can serve as prey to macrofauna. Sometimes they act as micro-ecosystem engineers. Springtails (Collembola) graze microflora and microfauna especially in rhizosphere, consume litter and carcasses and act as micropredators of nematodes and fungi. They serve as prey for macrofauna, host parasites, and also act as micro-ecosystem engineers [11].

In the presence of microarthropods, mass loss of newly senesced litter increases, amounts of inorganic nitrogen may be greater, and primary productivity can be enhanced [77]. Much of the regulation which they impose on decomposition is through their trophic interactions with the microbial community [78]. Alterations to the microarthropod assemblage structure have been shown to have significant effects on soil respiration and the leaching of a wide range of nutrients from laboratory microcosms [79].

Enchytraeids (Oligochaeta) is less widely studied group of mesofauna particularly in tropical soil. These animals are considered to be 'keystone species' in acid mor soils, such as those of Boreal forest, tundra, and heathland, where they can reach densities of up to 400,000 m^{-2}, constituting some 75% of the total faunal biomass. Much lower densities of 4000 m^{-2} are found in agricultural and mull forest soils, where earthworms typically dominate the faunal biomass [74]. Their taxonomic diversity has been documented by a limited number of studies and it is relatively lesser as compared to that of microarthropods [80, 81, 82].

The main food source of enchytraeids is fungal mycelium, but they also feed on decaying organic matter and microbes associated with it. Enchytraeids are involved in fragmentation of plant litter. They enhance microbial growth, help in dispersion of microorganisms and act as bioturbators [74]. Enchytraeids like collembola and mites are mobile throughout the pores, cracks and interfaces of litter and soil. They represent 'external rumen' type of digestion which consists of stimulating microbial activities in their fecal pellets and re-ingesting this material once microorganisms have released assimilable elements [11].

MACROFAUNA

The macrofauna (>2 mm body breadth) include earthworms, ants, termites, millipedes, centipedes, beetles and other arthropods. The macrofauna are the most conspicuous fauna of soil, and also the most documented in terms of their biology and impact on soil fertility. They include a wide array of taxonomic groups that cross several trophic levels [20, 83]. Macro-invertebrates are considered to play an important role in the decomposition of plant material

[18, 84, 85, 86, 87], with particularly strong effects at fine spatial and temporal scales [88]. By the processes of fragmentation, macro-invertebrates can influence decomposition through their effects upon fungal and bacterial populations [89]. Some groups of macrofauna are also known as 'ecosystem engineers' since they alter ecosystem dynamics through the modification, maintenance and creation of habitats (biogenic structures) for other organisms in the ecosystem [52, 53]. Earthworms, termites and ants generally act as ecosystem engineers/bioturbators.

Earthworms are probably the best known of the macrofauna and there is a huge literature on their biology and importance for soil fertility in natural and agricultural systems [90, 91, 92]. Earthworm play major role in fragmentation of litter. They enhance microbial growth, host protozoa and other parasites and disperse microorganisms and algae. Tropical soils sustain higher earthworm density as compared to temperate soils. A total of 51 exotic and 151 native species commonly occur in tropical agro-ecosystems [93].

Isoptera or termites are also considered as keystone species in the tropics and subtropics. Their biology and important roles in these ecosystems have been fairly documented. Termites feed mainly on decaying organic matter and roots, and are best known for their dramatic effects on soil nutrient cycling and the physical architecture and hydrology of soils, through their mound building and gallery excavating activities [94, 95, 96].

Ants act as bioturbators, enhance microbial growth and act as keystone species for fauna and plants associated with ant-hills. They have a variety of feeding habits, being predators, microbial- and plant-feeders, and opportunistic omnivores. In the biomes where ants are abundant, they affect many soil processes through their nesting activities. Their nests greatly affect critical ecosystem processes such as nutrient cycling and water redistribution [97, 98, 99, 100].

Other groups of macrofauna perform miscellaneous functions in soil and litter and participate in detritus food chain. Coleopterans (beetles) are involved in above ground and root herbivory as well as predation on many pest animals. They also assist in decomposition of plant and animal matter. Larval stages of Coleoptera and Lepidoptera are often involved in above ground and root herbivory. Larval stages of Diptera are mainly saprophagous. Arachnids (spiders, scorpions and pseudo-scorpions) and Chilopods (centipedes) act as predators to animals of lower taxa. Diplopods (millipedes) feed on plant residue and thus facilitate in litter decomposition. Orthoptera (e.g., grasshoppers, crickets and cockroaches) graze on above ground herbaceous cover. Mollusca (snails), Isopoda (woodlice), Thysanuria (silverfish), Demaptera (earwigs), Hemiptera (bugs), Homoptera and few other taxa also dwell in soil or litter layer and perform various functions [101, 102, 103, 104, 105].

General Linkages between Soil Biota and Soil Nutrient Dynamics

Impact soil biota on soil physico-chemical properties and various soil processes have been discussed above in terms of specific roles of different groups of soil biota. Studies show that a direct relationship exists between species richness; the intensity of some fundamental processes such as respiration, decomposition, nutrient storage, and primary production; and water [9, 106, 107].

In most terrestrial ecosystems, between 50 and 99% of aboveground net primary productivity is not consumed by herbivores but instead enters the decomposer subsystem as plant litter [108]. The decomposition of this litter has important ecosystem-level implications through regulating build up of soil organic matter, releasing of nutrients for plant growth and influencing the flux of CO_2 from the soil [28, 109]. Soil biota including microbes and many invertebrates interacting with them remarkably affect the rate of decomposition [110, 111, 112]. Decomposer organisms consist of complex community of soil biota including microflora and soil fauna [51] Fungi and bacteria are the active decomposer organisms responsible for the early phase of litter decay and nutrient release by rapidly increasing in numbers. Actinomycetes play major role at the end of the decomposition, when lignin content becomes high [113]. These are assisted by soil fauna, which enhance the biodegradation and humification of organic residues in several ways. The macrofauna, such as oligochaeta (earthworms), Gastropoda, Diplopoda, Isopoda, Coleoptera, and Hymenoptera (ants), and the mesofauna, such as oribatid mites (the most abundant group in Acari) and Collembola, are important litter fragmenters or detritivores [112].

Earthworms increase the decomposition of organic residues and the release of nutrients by comminuting plant residues, by incorporating the organic matter into the soil [114, 115] and by producing casts enriched with microflora and enzymes [116, 117, 118]. Termites are known to be efficient in digesting cellulose and in some cases also lignified substances [119, 120]. Millipedes break down plant litter and mix it with mineral soil, which they ingest [115].

Litter decomposition rates can be affected by absolute abundance as well as the diversity of fauna [79, 111, 118]. Soil macro fauna contribute relatively more to the decomposition of low quality residues because they stimulate microbial activity. Effect of fauna on the breakdown of the high quality residues is small, probably because these plant residues are easily decomposed by microorganisms without faunal stimulation of their activity [115, 121]. Macrofaunal predators may initiate trophic cascades that affect rates of decomposition, particularly through the fungal channel of the decomposer food web [122, 123, 124].

The soil macrofauna exerts a vital function with respect to soil structure and nutrient cycling in tropical ecosystems [6, 7]. Belowground biodiversity because of its interrelations with soil quality could be a key factor in determining stability and resilience of these ecosystems [8]. The loss of biodiversity in an ecosystem may lead to change in ecosystem functions. Landuse intensification is reported to reduce above and belowground biodiversity, which if accompanied by extinction of species may cause losses of certain key functions and reduce the ability of agricultural system to withstand unexpected periods of stress [125, 126, 127].

Studies show that loss of diversity of the soil macro-invertebrates might lead to degradation process through its impact on the maintenance of the soil structure and fertility parameters [128]. The impact of these land use changes on belowground biodiversity is less well established for tropical agroecosystems [22]. Moreover our ability to manage the ecosystem processes for sustainable agriculture is explicitly limited in tropical agro-ecosystem by a fundamental lack of knowledge of biota present in soil, of the basic biology and ecology of different species and their role in ecosystem functioning.

CONCLUSIONS

Studies on soil organisms over the past three decades have given many useful insights about the linkages of soil communities with various ecological processes. It is increasingly being recognized that below ground biodiversity, which remained neglected for long in past, could be a key factor in determining stability and resilience of ecosystem. Our understanding about the role of various groups of soil biota in ecosystem is yet not comprehensive. Except for some well studied taxa like earthworms, ants and termites, most of the groups of soil organisms like soil microbial diversity have not yet been adequately studied for their diversity, their precise role in ecosystem and their response to human induced disturbances.

Precise understanding about the role of different groups of soil biota is of great significance particularly for tropical regions where land degradation is prevalent due to high population pressure and soil biota can play vital role in maintaining soil structure and nutrient cycling. Moreover, farmers of these areas are economically weak in general, and they are largely dependent on natural processes for managing fertility of their soils. For these areas, soil fertility management policies and programs must be devised while considering the contribution of soil biota in various soil processes such as nutrient cycling.

REFERENCES

[1] Hole, FD. Effects of animals on soil. *Geodeoma*, 1981, 25: 75-112.

[2] Brussaard, L; Behan-Pelletier, VM; Bignell, DE. Biodiversity and ecosystem functioning in soil. *Ambio*, 1997, 26: 563-570.

[3] Lavelle, P; Decaens, T; Aubert, M; Barot, S; Blouin, M; Bureau, F; Margerie, P; Mora, P; Rosi, JP. Soil invertebrates and ecosystem services. *European Journal of Soil Biology*, 2006, 42: S1-S15.

[4] Giller, PS. The diversity of soil communities, the 'poor man's tropical rainforest'. *Biodiversity and Conservation*, 1996, 5: 135-168.

[5] Decaens, T; Jimenez, JJ; Gioia, C; Measey, GJ; Lavelle, P. The values of soil animals for conservation biology. *European Journal of Soil Biology*, 2006, 42(1): S23-S38.

[6] Oades, JM. The role of biology in the formation, stabilization and degradation of soil structure. *Geoderma,* 1993, 56: 377-400.

[7] Blanchart, E; Lavelle, P; Braudeau, E; Bissonnais, YL; Valentin, C. Regulation of soil structure by geophagous earthworm activities in humid savannas of Cote d'Ivoire. *Soil Biology and Biochemistry*, 1997, 29(3/4): 431-439.

[8] Saxena, KG; Maikhuri, RK., Rao, KS; Ramakrishnan, PS. Soil biodiversity, ecological processes and sustainable management of natural resources: where do we stand? In: Ramakrishnan, P.S., Saxena, K.G., Swift, M.J., Rao, K.S., Maikhuri, R.K. (Eds.), Soil biodiversity, ecological classes and landscape management. Oxford and IBH Publishing Co. Pvt. Ltd. New Delhi 2005.

[9] Asner, GP; Seastedt, TR; Townsend, AR. The decoupling of terrestrial carbon and nitrogen cycles: human influences on land cover and nitrogen supply are altering natural biogeochemical links in the biosphere. *Bioscience,* 1997, 47: 226-234.

[10] Pimm, SL; Sudgen, AM. Tropical diversity and global change. *Science*, 1994, 263: 933-934.

[11] Swift, MJ; Heal, OW; Anderson, MH. Decomposition in Terrestrial Ecosystems. Blackwell Scientific, Oxford, UK 1979.

[12] Bardgetts, RD. The Biology of Soil: A Community and Ecosystem Approach. Oxford University Press, 2005, pp 242.

[13] Boddy, L. Saprotrophic cord-forming fungi: meeting the challenge of heterogeneous environments. *Mycologia*, 1999, 91, 13–32.

[14] Johnson, D; Booth, RE; Whiteley, AS; Bailey, MJ; Read, DJ; Grime, JP. Plant community composition affects the biomass, activity and diversity of microorganisms in limestone grassland soil. *European Journal of Soil Science*, 2003, 54: 671–678.

[15] Clarholm, M. Possible roles of roots, bacteria, protozoa and fungi in supplying nitrogen to plants. In: Fitter, A.H., Atkinson, D., Read, D.J., Usher, M.B. (Eds). Ecological Interactions in Soils. Special Publication No. 4, British Ecology Society. Oxford, UK: Blackwell 1985.

[16] Ingham, RE; Elliott, ET; Moore, JC; Rose, SL; Reid, CPP; Morley, CR. The detrital food web in a short grass prairie. *Biology and Fertility of Soils*, 1987, 3: 57-68.

[17] Bardgett, RD; Griffiths, B. Ecology and biology of soil protozoa, nematodes and microarthropods. In: Modern Soil Microbiology (eds. J.D. van Elasas, E.Wellington, J.T. Trevors), Marcell Dekker, NY, 1997, pp. 129–163.

[18] Prins, AJ. Morphological and biological notes on some South African arthropods associated with decaying organic matter. Part 1. Chilopoda, Diplopoda, Arachnida, Crustacea, and Insecta. *Annals of the South African Museum*, 1983, 92: 53-112.

[19] Barros, E; Grimaldi, M.; Sarrazin, M; Chauvel, A; Mitja, D; Desjardins, T; Lavelle, P. Soil physical degradation and changes in macrofaunal communities in Central Amazon. *Applied Soil Ecology*, 2004, 26: 157-168.

[20] Doblas-Miranda, E; Sanchez-Pinero, F; Gonzalez-Megias, A. Soil macroinvertebrate fauna of a Mediterranean arid system: Composition and temporal changes in the assemblage. *Soil Biology and Biochemistry*, 2007, 39: 1916-1925.

[21] Swift, MJ. Special Issue on Agricultural Intensification, Soil Biodiversity and Agro-ecosystem Function. *Applied Soil Ecology*, 1997, 6(1): 1-108.

[22] Giller, KE; Beare, MH; Lavelle, P; Izac, AMN; Swift, MJ. Agricultural intensification, soil biodiversity and agro-ecosystem function. *Applied Soil Ecology*, 1997, 6: 3-16.

[23] Bhadauria, T; Saxena KG. Role of Earthworms in Soil Fertility Maintenance through the Production of Biogenic Structures. Applied and Environmental Soil Science, Vol 2010, Article ID 816073, 7 pages, doi:10.1155/2010/816073.

[24] Prasad J. Soil Macro-Invertebrate Communities in a Mid-Altitude Landscape of Western Himalayas. Unpublished Ph.D. Thesis, Jawaharlal Nehru University, New Delhi, 2008.

[25] Jimenez, JJ; Thomas, RJ. Nature's Plow: Soil Macro-invertebrate Communities in the Neotropical Savannas of Colombia. CIAT Publication, Colombia, 2001.

[26] [Lawton, J.H., Bignell, D.E., Bloemers, G.F., Eggleton, P., Hodda, M.E. Carbon flux and diversity of nematodes and termites in Cameroon forest soils. *Biodiversity and Conservation,* 1996, 5: 261-273.

[27] Wall, DH; Moore, JC. Interactions underground-soil biodiversity, mutualism, and ecosystem processes. *Bioscience*, 1999, 49: 109-117.

[28] Wardle, DA; Yeates, GW; Williamson, W; Bonner, KI. The response of a three trophic level soil food web to the identity and diversity of plant species and functional groups. *Oikos,* 2003, 102: 45-56.

[29] [Scheu, S; Setala, H. Multitrophic interactions in decomposer food-webs. In: Tscharntke, T., Hawkins, B.A. (Eds.), Multitrophic Level Interactions. Cambridge University Press, Cambridge, 2002, pp. 223-266.

[30] Tiedje, JM. Approach to the comprehensive evaluation of Prokaryote diversity of a habitat. In: Allsop, D., Colwell, R.R., Hawksworth, D.L. (Eds.), Microbial Diversity and Ecosystem Function. UNEP and CAB International, Wallingford UK, 1995, pp 73-82.

[31] Schaefer, M; Schauermann, J. The soil fauna of beech forests: a comparison between a mull and a moder soil. *Pedobiologia,* 1990, 34: 299-314.

[32] Barros, E; Curmi, P; Hallaire, V; Chauvel, A; Lavelle, P. The role of macrofauna in the transformation and reversibility of the soil structure of an oxisol in the process of forest to pasture conversion. *Geoderma,* 2001, 100: 193-213.

[33] Mathieu, J; Rossi, JP; Grimaldi, M; Mora, P; Lavelle, P; Rouland, C. A multi-scale study of soil macrofauna biodiversity in Amazonian pastures. *Biology and Fertility of Soils,* 2005, 40: 300–305.

[34] Lavelle, P; Bignell, D; Lepage, M; Wolters, V; Roger, P; Ineson, P; Heal, OW; Dhillion, S. Soil functions in a changing world: the role of invertebrate ecosystem engineers. *European Journal of Soil Biology*, 1997, 33: 159-193.

[35] White, G. The Natural History and Antiquities of Seibome. McMillan, London, 1789.

[36] Darwin, CR. On the formation of mould. Transactions of Geological Society of London, 1840, 5: 505-509.

[37] Darwin, CR. The Formation of Vegetable Mould through the Action of Worms, with Observations on their Habits. John Murray, ,London, UK, 1881.

[38] Anderson, JM; Ineson, P. Interactions between soil arthropods and microbial population in carbon, nitrogen and mineral nutrient fluxes from decomposing leaf litter. In: Lee, J.A., McNeill, S., Rorison, I.H. (Eds.), Nitrogen as an Ecological Factor. Blackwell, Oxford, 1983, pp. 413-442.

[39] Coleman, DC. Through a ped darkly: an ecological assessment of root-soil-microbial-faunal interactions. In: Atkinson, D., Fitter, A.H., Read, D.J., Usher, M.B., (Eds.) Ecological Interactions in Soil; Plants, Microbes and Animals. Blackwell Scientific Publications, Oxford, UK, 1985, pp. 01-21.

[40] Andren, O; Lindberg, T; Paustian, K; Rosswall, T (Eds.). Ecology of Arable Land: Organisms, Carbon and Nitrogen Cycling. Munksgaard, Copenhagen, Denmaerk, 1990 pp. 84-125.

[41] Blanchart, E. Restoration by earthworms (Megascolecidae) of the macro-aggregate structure of a destructured savanna soil under field conditions. *Soil Biology and Biochemistry,* 1992, 24(12): 1587-1594.

[42] Stockdill, SMJ. Earthworm improve pasture growth. *New Zealand Journal of Agriculture*, 1959, 98: 227-233.

[43] Stockdill, SMJ. Effects of introduced earthworm on the productivity of New Zealand pastures. *Pedobiologia,* 1982, 24: 29-35.

[44] Setala, H; Laakso, J; Mikola, J; Huhta, V. Functional diversity of decomposer organisms in relation to primary production. *Applied Soil Ecology*, 1998, 9: 25-31.

[45] Spain, AV; Lavelle, P; Mariotti, A. Stimulation of plant growth by tropical earthworm: *Soil Biology and Biochemistry*, 1992, 24: 1629-1633.

[46] Pashanasi, B; Lavelle, P; Alegre, J. Effects of inoculation with endogeic earthworm Pontocolex corethrurus on soil chemical characteristics and plant growth in a low input agricultural system of Peruvian Amazonia. *Soil Biology and Biochemistry*, 1996, 28(6): 801-810.

[47] Petersen, H; Luxton, M. A comparative analysis of soil fauna populations and their role in decomposition processes. *Oikos,* 1982, 39(3): 287-288.

[48] Ponsard, S; Arditi, R. What can stable isotopes (δ15N and δ13C) tell about the food web of soil macro-invertebrates? *Ecology*, 2000, 81, 852–864.

[49] Scheu, S; Falca, M. The soil food web of two beech forests (Fagus sylvatica) of contrasting humus type: stable isotope analysis of macro- and a mesofauna dominated community. *Oecologia* 123, 2000, 285–296.

[50] Lavelle, P. Diversity of soil fauna and ecosystem function. *Biological International*, 1996, 33: 3-6.

[51] Couteaux, MM; Mouseau, M; Celerier, ML; Bottner, P. Increased atmospheric CO2 and litter quality: decomposition of sweet chestnut leaf with animal food webs of different complexities. *Oikos,* 1991, 6: 54-64.

[52] Jones, C.G., Lawton, J.F., Shachak, M. 1994. Organisms as ecosystem engineers. *Oikos,* 69: 373-386.

[53] Jones, CG; Lawton, JH; Shachak, M. Positive and negative effects of organisms as physical ecosystem engineers. *Ecology,* 1997, 78(7): 1946-1957.

[54] Lavelle, P; Dangerfield, M; Fracoso, C; Eshenbrenner, V; Lopez-Hernandez, D; Pashanasi, B; Brussaard, L. The relationship between soil macrofauna and tropical soil fertility. In: Woomer PL, Swift MJ (Eds.), The biological management of tropical soil fertility. TSBF, Wiley-Sayce, *Chichester,* 1994, pp. 137-168.

[55] Bouche, MB. Strategies Lombriciennes. In: Lohm, U., Persson, T. (Eds.), Soil Organisms as Components of Ecosystems. *Ecology Bulletin*, 1977, 25: 122-132.

[56] Harley, JL. Fungi in ecosystems. *Journal of Applied Ecology* 1971, 8: 627–642.

[57] O'Donnell, AG; Colvan, SR; Malosso, E; Supaphol, S. Twenty years of molecular analysis of bacterial communities in soils and what have we learned about function? In: Biological Diversity and Function in Soil (eds. R.D. Bardgett, M.B. Usher, and D.W. Hopkins), Cambridge University Press, Cambridge, UK, 2005.

[58] Robinson, CH; Miller, JP; Deacon, LJ. Biodiversity of saprotrophic fungi in relation to their function: do fungi obey the rules? In: Biological Diversity and Function in Soil (eds. R.D. Bardgett, M.B. Usher, D.W. Hopkins), Cambridge University Press, Cambridge, UK, 2005.

[59] Torsvik, V; Ovreas, L; Thingstad, TF. Prokaryotic diversity: magnitude, dynamics, and controlling factors. *Science,* 2002, 296, 1064–1066.

[60] Curtis, TP; Sloan, WT; Scannell, JW. Estimating prokaryotic diversity and its limits. *Proceedings of the National Academy of Sciences USA*, 99, 2002, 10494–10499.

[61] Honer-Devine, MC; Carney, KM; Bohannan, BJM. An ecological perspective on bacterial diversity. Proceedings of the Royal Society London (B) 271, 2004, 113–122.

[62] Bardgett, RD; Speir, TW; Ross, DJ; Yeates, GW; Kettles, HA. Impact of pasture contamination by copper, chromium and arsenic timber preservative on soil microbial properties and nematodes. *Biology and Fertility of Soils*, 1994, 18, 71–79.

[63] Boddy, L. Saprotrophic cord-forming fungi: meeting the challenge of heterogeneous environment, *Mycologia,* 1999, 91: 13-32.

[64] Johnson, D; Vandenkoornhuyse, PJ; Leake, JR; Gilbert, L; Booth, RE; Grime, JP. Plant communities affect arbuscular mycorrhizal fungal diversity and community composition in grassland microcosms. *New Phytologist*, 2003, 161: 503–515.

[65] Trofymow, JA; Coleman, DC. The role of bacterivorous and fungivorous nematodes in cellulose and chitin decomposition in the context of a root/rhizosphere/soil conceptual model. In: Freckman, D.W. (Ed.), Nematodes in Soil Ecosystems. Austin, Texas, USA: University of Texas, 1982.

[66] Schnurer, J; Clarholm, M; Rosswall, T. Fungi, bacteria and protozoa in soil from 4 arable cropping systems. *Biology and Fertility of Soils* , 1986, 2: 119–126.

[67] Foissner, W. Global soil ciliate (Protozoa, Ciliophora) diversity: a probabilitybased approach using large samples collections from Africa, Australia and Antarctica. *Biodiversity and Conservation*, 1997, 6: 1627–1638.

[68] Parker, LW; Santos, PF; Phillips, J; Whitford, WG. Carbon and nitrogen dynamics during decomposition of litter and roots of a Chihuahuan desert annual Lepidium lasiocarpum. *Ecological Monographs*, 1984, 54: 339–360.

[69] Yeates, GW; Bardgett, RD; Cook, R; Hobbs, PJ; Bowling, PJ Potter, JF. Faunal and microbial diversity in three Welsh grassland soils under conventional and organic management regimes. *Journal of Applied Ecology*, 1997, 34: 453–471.

[70] Bloemers, GF; Hodda, M; Lambshead, PJD; Lawton, JH; Wanless, FR. The effects of forest disturbance on diversity of tropical soil nematodes. *Oecologia* 1997, 111, 575–582.

[71] Wall, JW; Skene, KR; Neilsen, R. Nematode community and trophic structure along a sand dune succession. *Biology and Fertility of Soils*, 2002, 35: 293–301.

[72] Virginia, RA; Wall, DH. How soils structure communities in the Antarctic Dry Valleys. *BioScience,* 1999, 49: 973–983.

[73] Yeates, GW; Bongers, T; De Goede, RGM; Freckman, DW; Georgieva, SS. Feeding habits in soil nematode families and genera - an outline for ecologists. *Journal of Nematology*, 1993, 25: 315–331.

[74] Didden, WAM. Ecology of terrestrial Enchytraeidae. *Pedobiologia,* 1993, 37: 2–29.

[75] Culik MP; Martins DS; Ventura JA. Collembola (Arthropoda: Hexapoda) Communities in the Soil of Papaya Orchards Managed with Conventional and Integrated Production in Espírito Santo, *Brazil. Biota Neotropica*, Sep/Dec 2006, vol 6 (3).

[76] Siepel, H. Biodiversity of soil microarthropods: the filtering of species. *Biodiversity and Conservation* 1996, 5:251–260.

[77] Setala, H; Huhta, V. Soil fauna increase Betuala pendula growth: laboratory experiments with coniferous forest floor. *Ecology,* 1991, 72(2): 665-671.

[78] Moore, JC; Hunt, HW. Resource compartmentation and the stability of real ecosystems. *Nature,* 1988, 333, 261–263.

[79] Heneghan, L; Coleman, DC; Zou, X; Crossley Jr. D.A; Haines, BL. Soil microarthropod contributions to decomposition dynamics: tropical-temperate comparisons of a single substrate. *Ecology,* 1999, 80: 1873-1882.

[80] Didden, WAM; De Fluiter, R. Dynamics and stratification of Enchytraeidae in the organic layer of a Scots pine forest. *Biology and Fertility of Soils*, 1998, 26: 305–312.

[81] Standen, V. Production and diversity of enchytraeids, earthworms and plants in fertilized hay meadow plots. *Journal of Applied Ecology*, 1984, 21: 293–312.

[82] Stefan, V. Soil Enchytraeidae from the Cerna Valley. Fourth Symposium on Soil Biology; Rumanian National Society of Soil Science. Editura Ceres, Bucuresti, 1977, pp. 277–283.

[83] Doblas-Miranda, E; Sanchez-Pinero, F; Gonzalez-Megias, A. Different microhabitats affect soil macroinvertebrate assemblages in a Mediterranean arid ecosystem. *Applied soil Ecology,* 2009, 41: 319-335.

[84] Silva, SI; MacKay, WP; Whitford, WG. The relative contributions of termites and microarthropods to fluff grass litter disappearance in the Chihuahuan Desert. *Oecologia,* 1985, 67: 31-34.

[85] Garay, I; Mollon, A; Flogaitis, E. Studies on the litter of an elm-oak mixed forest. II. Macroarthropods succession during decomposition. *Acta Oecologia*, 1986, 7: 263-288.

[86] Curry, JP; Byrne, D. Role of earthworms in straw decomposition in a winter cereal field. *Soil Biology and Biochemistry*, 1997, 29: 555-558.

[87] Paoletti, MG; Hassall, M. Woodlice (Isopoda: Oniscidea): their potential for assessing sustainability and use as bioindicators. *Agriculture, Ecosystems and Environment,* 1999, 74: 157-165.

[88] Lavelle, P; Blanchart, E; Martin, A; Martin, S; Spain, A; Toutain, F; Barois, I; Schaefer, R. A hierarchical model of decomposition in terrestrial ecosystems: application to soils of the humid tropics. *Biotropica,* 1993, 25: 130-150.

[89] Scheu, S. Cellulose and lignin decomposition in soils from different ecosystems on limestone as affected by earthworm processing. *Pedobiologia,* 1993, 37: 167-177.

[90] Lee, KE. Earthworms: *Their Ecology and Relationships with Soils and Landuse.* Academic Press, Australia 1985.

[91] Edwards, CA. *Earthworm Ecology*, 2nd edn. CRC Press, Boca Raton, FL, 2004.

[92] Edwards, CA; Bohlen, PJ. *Biology and Ecology of Earthworms*. Chapman and Hall, London, 1996.

[93] Fragoso, C; Lavelle, P; Blanchart, E; Senapati, BK; Jimenetz, JJ; Martinez, MA; Decae¨ns, T; Tondoh, J. Earthworm communities of tropical agroecosystems: *origin, structure and influence of management practices.* In: Lavelle, P; Brussaard, L; Hendrix, P. (Eds.), Earthworm Management in Tropical Agroecosystems. CABI Publishing, Oxon, 1999, pp. 27-55.

[94] Jones, JA. Termites, soil fertility and carbon cycling in dry tropical Africa: a hypothesis. *J. Trop. Ecol.* 1990, 6: 291-305.

[95] Korb, J; Linsenmair, KE. The causes of spatial patterning of the mounds of a fungus-cultivating termite: results from nearest-neighbour analysis and ecological studies. *Oecologia,* 2001, 127: 324-333.

[96] Eggleton, P; Bignell, DE; Hauser, S; Dibog, L; Norgrove, L; Madong, B. Termite diversity across an anthropogenic disturbance gradient in the humid forest zone of West Africa. *Agriculture, Ecosystems and Environment*, 2002, 90: 189-202.

[97] Petal, J; Kusinska, A. Fractional composition of organic matter in the soil of anthills and of the environment of meadows. *Pedobiologia,* 1994, 38: 493-501.

[98] Dean, WRJ; Milton, SJ; Klotz, S. The role of ant nest-mounds in maintaining small scale patchiness in dry grasslands. *Biodiv. Conserv.* 1997, 6: 1293-1307.

[99] Folgarait, PJ. Ant biodiversity and its relationship to ecosystem functioning: a review. *Biodiv. Conserv*. 1998, 7: 1221-1244.

[100] Dauber, J; Wolters, JDV. Microbial activity and functionaldiversity in the mounds of three different ant species. *Soil Biol. Biochem*. 2000, 32, 93-99.

[101] Kuhnelt, W; Walker, N; Butcher, JW; Laughlin, C. Soil Biology: With Special Reference to Animal Kingdom. *Faber and Faber*. 1976, London.

[102] Borror, DJ; De Long DM; Tripleshorn, CA. An introduction to the study of insects. 5th Edition. Saunders College Publishing, Philadelphia, Pennsylvania, 1981.

[103] Hruska, AJ. Weedy groundcover increases damage to cassava by white grubs in Costa Rica. *Trop. Agric. (Trinidad)* 1987, 64: 212–216.

[104] Dangerfield, JM; Milner, AE. Millepede faecal pellet production in selected natural and managed habitats of Southern Africa: implications for litter dynamics. *Biotropica,* 1996, 28: 113-120.

[105] Lavelle, P; Spain, AV. Soil Ecology. Kluwer Academic Publishers, Dordrecht, 2001.

[106] Schulze, ED; Mooney, HA. Ecosystem function of biodiversity: a summary. In: Schulze E.D. Mooney, H.A. (Eds.), *Biodiversity and Ecosystem function*. Springer-Verlag, Berlin, 1994, pp. 497-510.

[107] Solbrig, OT. Biodiversity: an introduction. In: Solbrig, O.T., Emden, H.M., Oordt, P.G.W.J. (Eds.), *Biodiversity and global change*. CAB International and International Union of Biological Sciences (IUBS), Paris, France, 1994, pp. 13-20.

[108] McNaughton, SJ; Osterheld, M; Frank, DA; Williams, KJ. Ecosystem level patterns of primary productivity and herbivory in terrestrial habitats. *Nature*, 1989, 341, 142–144.

[109] Wardle, DA; Yeates, GW; Barker, GM; Bonner, KI. The influence of plant litter diversity on decomposer abundance and diversity. *Soil Biology and Biochemistry,* 2006, 38: 1052-1062.

[110] Coleman, DC; Crossley Jr., DA. *Fundamentals of Soil Ecology*. Academic Press, San Diego, 1996.

[111] Gonzalez, G; Seastedt, TR. Soil fauna and plant litter decomposition in tropical and subalpine forests. *Ecology*, 2001, 82: 955-964.

[112] Adl, S.M. *The Ecology of Soil Decomposition*. CABI Publishing, Wallingford, UK. 2003.

[113] Pandey, RR; Sharma, G; Tripathi, SK; Singh, AK. Litterfall, litter decomposition and nutrient dynamics in a subtropical natural oak forest and managed plantation in northeastern India. *Forest Ecology and Management*, 2007, 240: 96-104.

[114] Lavelle, P. Earthworm activities and the soil system. *Biology and Fertility of Soils*, 1988, 6: 237-251.

[115] Tian, G; Brussaard, L; Kang, BT. Breakdown of plant residues with contrasting chemical compositions: effect of earthworms and millipedes. *Soil Biology and Biochemistry,* 1995, 27: 277-280.

[116] Syers, JK; Sharpley, AN; Keeney, DR. Cycling of nitrogen by surface-casting earthworms in a pasture ecosystem. *Soil Biology and Biochemistry*, 1979, 11: 181-185.

[117] Mulongoy, K; Bedoret, A. Properties of wormcasts and surface soils under various plant covers in the humid tropics. *Soil Biology and Biochemistry*, 1989, 21: 197-203.

[118] Liu, ZG; Zou, ZM. Exotic earthworms accelerate plant litter decomposition in a Puerto Rican pasture and a wet forest. *Journal of Applied Ecology*, 2002, 12: 1406-1417.

[119] Lee, KE; Wood, TG. Termites and Soils. Academic Press, London and New York, 1971, 251 pp.

[120] Wood, TG. Termites and the soil environment. *Biology and Fertility of Soils*, 1988, 6: 228-236.

[121] Joo, SJ; Yim, MH; Nakane, K. Contribution of microarthropods to the decomposition of needle litter in a Japanese cedar (Cryptomeria japonica D. Don) plantation. *Forest Ecology and Management,* 2006, 234: 192-198.

[122] Schaefer, M. Secondary production and decomposition. In: Rohrig, E., Ulrich B., (Eds.), Temperate deciduous forest. Ecosystems of the world 7. Elsevier, Amsterdam, 1991, pp. 175-218.

[123] Kajak, A. Effects of epigeic macroarthropods on grass litter decomposition in mown meadow. *Agriculture, Ecosystems and Environment*, 1997, 64: 53-63.

[124] Lawrence, KL; Wise, DH. Unexpected indirect effect of spiders on the rate of litter disappearance in a deciduous forest. *Pedobiologia,* 2004, 48: 149-157.

[125] Naeem, S; Thompson, LJ; Lawler, SP; Lawton, JH; Woodfin, RM. Declining biodiversity can alter the performance of ecosystem. *Nature,* 1994, 368: 734-737.

[126] Matson, PA; Parton, WJ; Power, AG; Shift, MJ. Agricultural intensification and ecosystem properties. *Science,* 1997, 277: 504-509.

[127] Sinha, B; Bhadauria, T; Ramakrishnan, PS; Saxena, KG; Maikhuri, RK. Impact of landscape modification on earthworm diversity and abundance in the Hariyali sacred landscape, Garhwal Himalaya. *Pedobiologia,* 2003, 47: 357-370.

[128] Beare, MH; Reddy, VM; Tian, G; Srivastava, SC. Agricultural intensification, soil biodiversity and agro-ecosystem function in the tropics: the role of decomposer biota. *Applied Soil Ecology*, 1997, 6(1): 87-108.

In: Soil Nutrients
Editor: Mohammad Miransari

ISBN: 978-1-61324-785-3

Chapter 4

SOIL MICRONUTRIENTS: ROLE IN PLANT GROWTH AND INTEGRATED NUTRIENT MANAGEMENT SYSTEMS

A. A. Deshmukh[1], N. N. Nawani*[1], M. B. Khetmalas,[1] and P. N. Razdan[2]

[1]Dr. D. Y. Patil Biotechnology and Bioinformatics Institute, Tathawade, Pune, India

[2]Dr. D. Y. Patil Vidyapeeth, Pune, India

ABSTRACT

Soil represents a dynamic niche, which is evident from the diversity and variation it exhibits with topographical location, physico-chemical parameters, vegetation and climatic conditions. Soil management and resource conservation is one of the prime focus areas of agriculture due to the fact that soil fertility has a direct implication on food quality. A healthy soil would ensure adequate retention and release of water and nutrients and promote root growth. Soils provide nutrients to plants, which are categorized as basic, macro and micronutrients. The inorganic matter of the soil i.e., trace minerals present in it are crucial for plant growth. The availability of these nutrients in soil is dependent on several biotic and abiotic factors and interactions, for e.g., fluctuations in pH of soil vary the availability of plant nutrients. Today, the major cause of concern is increasing food demand versus declining soil health, declining fertilizer response, nutrient mining and emergence of multi-nutrient deficiencies in soil. Besides, soil pollution due to recalcitrant xenobiotics, pesticides or chemical fertilizers poses further ecological and environmental threats. Several reports suggest the use of microorganisms for bioremediation and improving the nutrient quality of soil. Soil microorganisms play a crucial role in soil nutrient restoration and balance the nutrient level for adequate plant growth. This review gives an introspective view of the status of soil ecology with respect to micronutrients and envisages the role of Integrated Nutrient Management (INM) in maintaining soil health.

---

* E-mail: neelunawani@yahoo.com.

Keywords: bioremediation, integrated nutrient management, nutrient deficiency

INTRODUCTION

Soil is one of the most important natural resources and has a direct implication to human existence by virtue of its ability to endure a large diversity of biota. The nutritive quality of food directly depends on the soil health and fertility. In order to have a healthy life, the human race must protect and conserve this exceptionally useful resource. This review discusses the present status of soil, remedies to reverse the distortion, nutrient deficiencies and possible use of integrated nutrient management in restoring the soil quality.

SOIL STATUS

Anthropocene was the term proposed by Nobel Prize winning atmospheric chemist, Paul Crutzen. As suggested by him, the human civilization is entering a new geological period that is human-made and has evolved from a relatively stable and nurturing environment. Proposals are under consideration to incorporate 'Anthropocene' in the geological time scale. The Anthropocene age clearly remarks the impact of human activities on Earth and its resources (Zalasiewicz et al. 2008). Global soil change forms a major component of this newly proposed geological era, a study of which will be fundamental in strategizing control over elements contributing to rapid change in the global ecosystems (Richter 2007). As other ecological and environmental parameters, soil also has undergone changes in its topology, characteristics and composition. Some of these changes have occurred due to natural phenomena and others due to forcing variables resulting from ecological constraints and human intervention. The second half of twentieth century was remarked by Green Revolution where a quarter of land was converted to cultivated systems and the soil was extensively exposed to a variety of chemical fertilizers and nutrients. The Green Revolution was followed by Industrial Revolution which further led to intensive deforestation enforcing further changes in the ecosystems. The result of uncontrolled exposure of soil to chemicals is the reduced organic content which will have a drastic impact on the agricultural productivity in future. There have been several alternatives to meet the food demand like improved irrigation practices, conventional breeding and the still debatable acceptance of transgenic crops. All these however will not be effective if the rate of soil change due to exposure to the chemical compounds is not controlled which is possible by replenishing the soil with organic matter to revert it close to its original quality. Figure 1 shows the major biomes of the world and Figure 2 depicts the land quality.

The millennium ecosystem assessment monitored ten categories of ecosystems including the forests, cultivated, dryland, coastal, marine, urban, polar, freshwater islands and mountains over a period of time in order to study the interrelation of these ecosystems. Some noticeable transitions are recorded in the terrestrial forests and cultivated land which have led to ecological changes. The changes in ecosystem have gradually allowed either selection or extinction of species. Today several plant and animal species are threatened which transcends the indimency of conservation methods adopted so far. The shifts experienced by ecosystems have a direct or indirect implication on soil quality and consequently on crop productivity. The biomes which have experienced transitions and hard hit alterations include the marine, freshwater and forest ecosystems. All the ecosystems are interrelated and have a definite impact on agricultural productivity. Figure 3 shows the tension zones i.e. biodiversity hot spots which must be protected and conserved and Figure 4 describes the types of the stress in these zones.

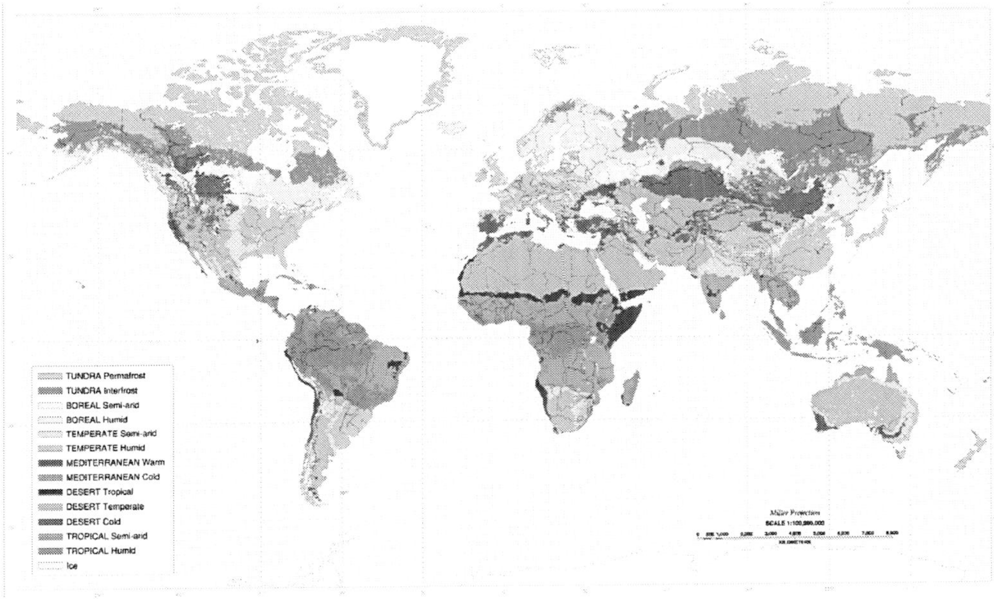

Figure 1. Major biomes of the world.

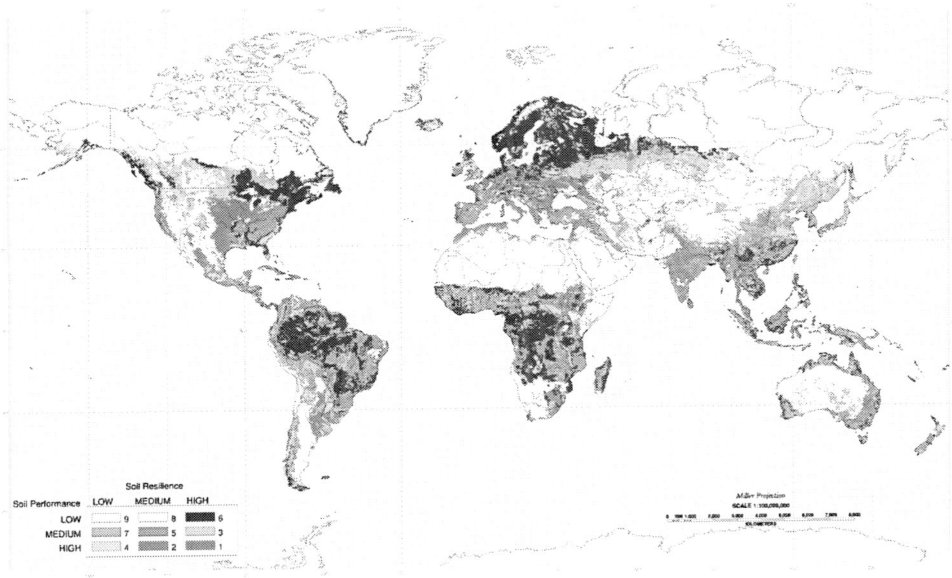

Figure 2. Land quality of different regions over the world.

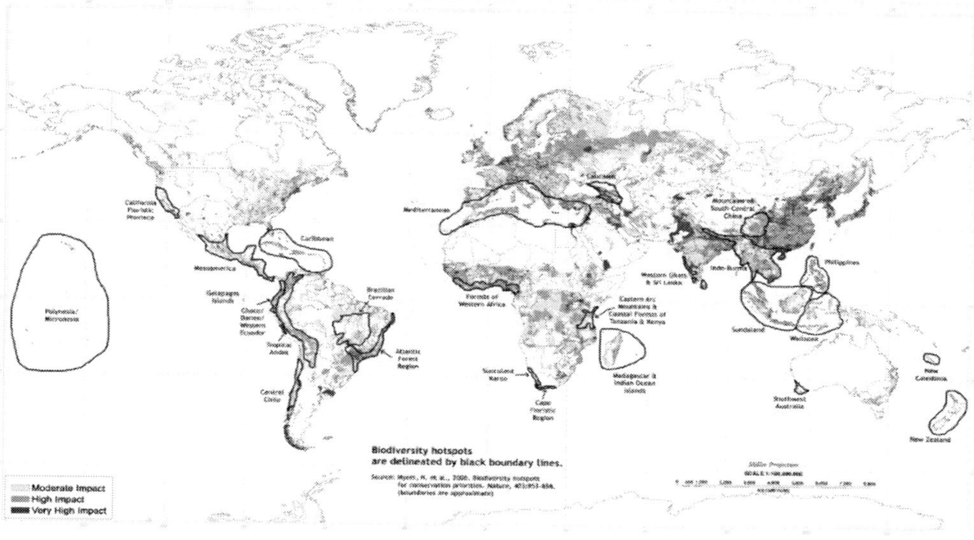

Figure 3. Threatened biodiversity hot spots and tension zones.

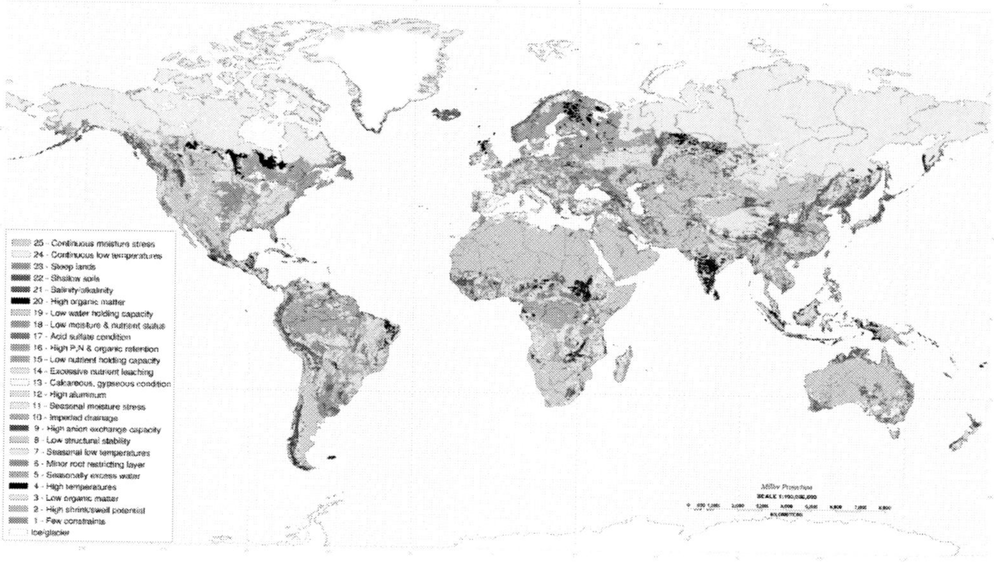

Figure 4. Major types of stresses in tension zones.

As clearly depicted, very small part of land exhibits few constraints whereas most of the regions exhibit some or other constraints which will have a direct implication on the agricultural productivity. The current scenario explains the global diversity and offers possible explanation for global cropping patterns. With gradual shifts in the biomes, the cropping patterns are expected to change and with stresses more and more land is expected to lose its integrity and quality to support any agricultural practices. Table 1 suggests the improved cropping patterns which would help in restoring the quality of cultivable lands to a certain extent (Vincent and Wen 1995).

Table 1. Types of cultivation for improving crop productivity

Type of cropping	Sub-class	Properties
Annual field cropping	Shifting cultivation	Alternate cropping and land clearing
	Fallow cultivation	Alternate cropping and fallow (bushes and grasses allowed to grow)
	Ley cultivation	Long period of arable crops followed by long period of grass or legume cultivation
	Arable cultivation	Only agricultural crops are grown
	Rice cultivation	Annual cropping of paddy with or without controlled water supply
	Irrigated cultivation	Annual field cropping (rain or artificial water supply)
Perrenial field cropping	Non-irrigated	Crops harvested after a long duration, e.g. banana, pineapple, coffee
	Irrigated	Environmental benefits and reduced farming cost
Tree and shrub cropping	Non-irrigated tree crop cultivation	Oaks, mesquites
	Irrigated tree crop cultivation	Rubber, coconut
	Non-irrigated shrub crop cultivation	Legume, shrubs
	Irrigated shrub crop cultivation	Tea

REMEDIAL MEASURES

Remediation is the process used to remove or neutralize the contaminants or decrease them to permissible limits for environment protection and safety. Bioremediation is a pollution control technology that uses biological systems to catalyze the degradation or transformation of various toxic chemicals to less toxic forms. Microbes decompose hazardous substances to their basic, non-toxic elements they particularly play a major role in degradation of xenobiotics with their evolved mechanisms. However, before any remediation is attempted, it is necessary to identify the seriousness of hazard and ways to remedy the situation. The major potential damaging contaminants of soil are oil, petroleum, pesticides, inorganic fertilizer residues besides other propellants and explosives. The characterization of soil for application of remedial measures involves identification of its physical, chemical, thermal, electrical and ignition sensitivity properties. Soil sampling is usually necessary to determine location and condition of the contamination in the ground. Soil remediation can be accomplished by destroying or decomposing the pollutants *in situ*. In case of contamination of soil with explosives either *in situ* applications or removal of the contaminant from soil followed by its destruction.

The effect of a bioremediation process on the geological area should always be assessed. The method of choice should offer a best combination of practicability and effectiveness. The major factors involved are technical, time and site constraints and timescale of effectiveness. One of the natural remedial measures is biodegradation where organic substances decompose by the metabolic and enzymatic processes of microorganisms and other scavengers. The organic substances, such as animal waste, decomposing carcass and plant matter provide energy and nutrients to the microbes. As it occurs at all levels of the food chain, the microbes produce byproducts that return other nutrients or gases to their atmosphere for use by plants or other organisms. By isolating the microorganisms necessary to detoxify or degrade chemicals and compounds found in soil, ecologists can minimize the damage to the ecosystem by providing the best possible environment for these microbes to proliferate. Microorganisms have evolved to occupy ecological niches containing high concentration of pollutants like pesticides and heavy metals. Hence microorganisms are exploited in bioremediation and biomonitoring strategies for environments containing toxic pollutants. Using genetic engineering, scientists can enhance the ability of bacteria to metabolize waste products contaminating soil ecosystem.

Soils contaminated with oil are restored by bioaugmentation where oil degrading microbes are introduced in the soil system and optimization of conditions for maximum removal of the hydrocarbons. Addition of surfactants of biological origin can enhance the release of hydrocarbons which tend to absorb to the organic matrices due to their hydrophobicity (Zhou and Zhu 2007). Although a large number of oil cleaning mechanisms have been studied, there are very few technologies developed and practically implemented (Leahy and Colwell 1990). Removal of pesticides from the environment is attempted with methanogenic sludge (Baczynski and Pleissner 2010). *Serratia marcescens* is reported to degrade diazinon, chlorpyrifos, coumaphos and parathion (Abo-Amer 2011). Metals from industries or other potential sources also affect the soil quality and agricultural productivity. Maximum load of metal ions which enters the soil is a result of industrial activities and mining. Biosorption which utilizes dead biomass like algae, fungi or moss to sequester heavy metals is often a good remedial practice. Microbial cells can accumulate heavy metals by physicochemical and biological processes. Adsorption to dead cells, extracellular polysaccharides, capsules and slime layers and walls of algae, fungi and yeast are used for biosorption. Bioaccumulation is growth dependent process and biosorption involves mechanisms like ion exchange, chelation and complexation. To develop an efficient biosorbent and its reuse by subsequent desorption processes, the knowledge of the mechanisms of metal binding is very important. Metal binding by biosorption method occurs as a two step process, where the first step involves stoichiometric interaction between the metal and the reactive chemical groups in the cell wall, and the second step is an inorganic deposition of increased amount of metal. Since the cell wall has a variety of polysaccharides and proteins it offers a number of active sites capable of binding metal ions. The variation in cell wall composition of microbes leads to differences in the type and amount of metal ion binding to them. The anionic nature of bacterial surfaces enables them to bind metal cations through electrostatic interactions. Because of their thickness and anionic character which is given by peptidoglycan, teichoic acid and teichuronic acids; the cell walls of gram positive bacteria display high capacity for metal binding.

Bacillus species are omnipresent and offer prime sites for metal binding. Reports indicate that peptidoglycan is responsible for metal binding by *Bacillus subtilis* where as in *Bacillus licheniformis* metal binding is predominantly due to teichuronic acid. Besides *Bacillus* sp., *Pseudomonas* sp. and *Zooglea ramigera* are also reported for their biosorption attributes. In *E. coli*, the phosphoryl groups of LPS and phospholipids are demonstrated to be most probable sites for binding of metal cations (Langley and Beveridge 1999). Bacterial metallothioneins, which are low molecular mass cysteine rich metal binding protein, are known to bind divalent cations such as Ca^{2+}, Cu^{2+}, and $Zn^{2+.}$ The de novo designed metal binding peptides offer array of motifs with enhanced bisorption capacity. This can be achieved by creating metal-binding motif consisting of $(Glu\text{-}Cys)_nGly$. These peptides emulate the structure of phytochelatin, a metal-chelating molecule that plays a major role in metal detoxification. Fungi and actinomycetes offer the advantage of having a high percentage of cell wall material, which exhibits excellent binding properties. The genera *Rhizopus, Aspergillus* and *Streptovertcillum* show excellent metal biosorption potential. *Rhizopus arrhizus* has been used for accumulation of uranium and thorium (Tobin et al. 1984). Similarly studies on uranium uptake by *Aspergillus niger* and *Saccharomyces cerevisiae* have indicated involvement of carboxylic group for biosorption (Wang and Chen 2006). *Sargassum natans* and *Ascophyllum nodosum* have high biosorption capacities for various metals. *Chlorella* species and cyanobacteria also display biosorption properties. In Pheophycean members, algin is present which contributes significantly to metal binding. It is suggested that the amino, carboxyl group, nitrogen and oxygen moieties could participate in formation of coordinated band with metal ion. Sulphate covalent bonding between divalent cation and algal cell wall proteins is also reported. The potential metal binding groups in algal cell wall are carboxylates, amines, imidazoles, phosphates, sulfhydryls, sulfates and hydroxyls. Of these the amino and carboxyl groups are available for coordination bonding with metal ions such as lead (II), copper (II) and chromium. Various parameters affect the process of bioabsorption for e.g. effect of change in pH on metal biosorption indicates an increase in cation uptake with increase in pH. Hence the extent of metal ion removal depends on the physical, chemical and biological characteristics of the soil and parameters that govern the growth features of the biosorbing agent (Das et al. 2008).

Biotic (fungal biomass) and abiotic agents like sugarcane bagasse, sawdust, rice straw and charcoal are successful in removing reactive dyes. Fungi like *Penicillium commune*, *P. freii* and *P. allii* removed direct violet dye by 96, 64 and 65% respectively after two hours of incubation. Of the abiotic components, rice straw was more efficient in dye removal than bagasse or sawdust. Rice straw was effective in removing 72% of direct violet dye within 24 hours. These studies indicate the possibility of reversal of soil pollution (El-Rahim 2006). Other fungi like *Phanerochaete chrysosporium* are used for degradation of a number of toxic xenobiotics such as aromatic hydrocarbons (Benzo alpha pyrene, phenanthrene, pyrene), chlorinated organics (alkyl halides, chloroanilines, DDT, pentachlorophenols, trichlorophenols, polychlorinated biphenyls, trichlorophenoxyacetic acid) and nitrogen aromatics (2, 4 Dinitrotoluene, 2,4,6- trinitrotoluene) (Cameron et al. 2000).

Phytoremediation is another strategy for sustaining the soil nutrients and removing the toxic materials from soil. Some plants can absorb and store toxins; other plants can break down nasty chemicals and turn them into less harmful substances. Plants grown for phytoremediation can also prevent toxins from moving from a contaminated site into other areas.

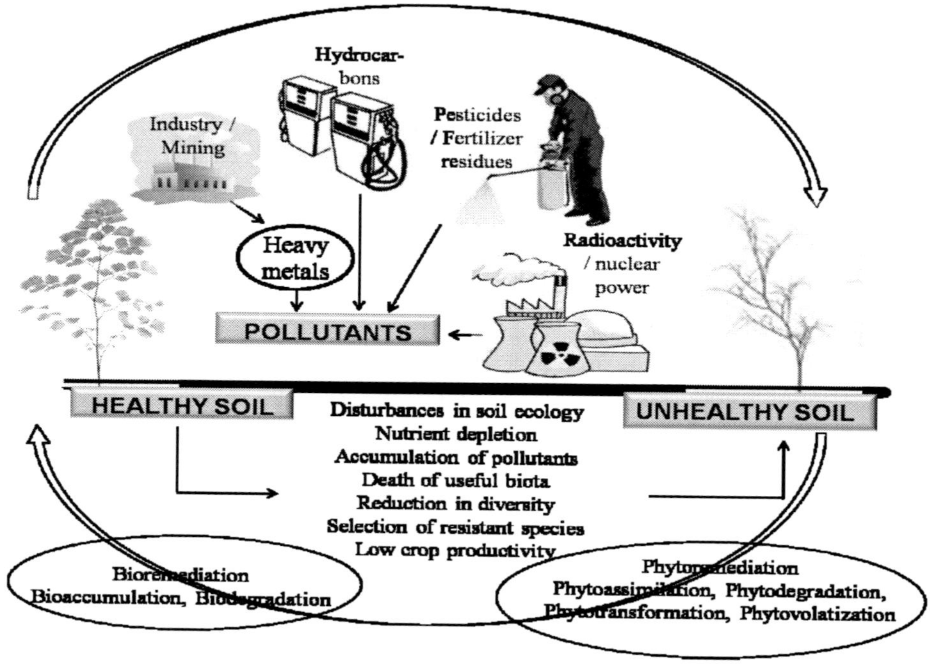

Figure 5. Soil pollution and remedial measures.

The remediation usually employed in heavy metal contaminated soils is conversion of contaminated land to no-food crop cultivating land which would include plantation of flowers and ornamentals. Alternatively plantation of species of plants called hyper accumulators with high affinity for heavy metals can be done. However, the biomass of hyper accumulator species cannot be further reused as green manure or mulch as this practice could return the heavy metal to soil. Hence such plant species are dried and burnt and the ash is disposed off safely. Plants which have been used for accumulation of heavy metals are mustard (Salt et al. 1997), sunflower and tobacco (Boonyapookana et al. 2005), willow (Kuzovkina et al. 2004) and several others. Some ornamental plants like *Melastoma affine* **remove aluminum** from the soil hence carry an aesthetic value and are ecologically significant (Asio et al. 2009). Most members of the Brassica family mustard greens, collards, broccoli etc. are used for removing lead from soil (Lim et al. 2004). Water Hyssop (Bacopa monnieri) removes lead, mercury, cadmium and chromium from bogs and wetlands, and makes a ground cover for muddy shores.

Other related phytoremedial measures include phytoextraction or phytoaccumulation where translocation of metal contaminants from the roots to different parts of plant occurs (Khellaf and Zerdaoui 2009). However, the plant should no longer be used as food crop. Rhizofiltration is used to remediate polluted water bodies where the remedial plant is initially acclimatized to the polluted environment. The pollutants are either adsorbed onto root surface or directly by the root (Verma et al. 2006). Phytostabilisation is the immobilisation of contaminated pool of metals where the contaminants are absorbed and accumulated by roots, or only held in rhizosphere prevents their spread in the food chain. White Lupin has been used for the stabilization of cadmium (Vázquez et al. 2006). Phytodegradation is another remedial measure where pesticides or other organic contaminants in the soil are metabolized or transformed by the plant (Schwitzguebel et al. 2006). Figure 5 gives an overview of the soil pollution and the remedial measures that can restore the soil to its original form. Thus all these mechanisms aid in the replenishment of soil health so that it can be used for cultivation of food crops.

NUTRIENT DEFICIENCIES

A soil is referred to as nutrient deficient when it cannot provide essential nutrients for plant growth. However this is a relative term as it would depend on the type of plant intended for cultivation. In general if the nutrients are not available in sufficient quantity to meet the requirements of a growing plant then the condition is that of nutrient deficiency. Contrary to this when nutrient is in excess such that it harms the plant or interferes with normal growth of plant, the condition is referred to as toxicity. Plants often show characteristic symptoms in nutrient deficient conditions. The primary symptoms that indicate nutrient deficiencies are stunted growth, chlorosis and necrosis. When root and stem elongation, photosynthesis and other metabolic activities are hampered the plant shows slow growth leading to stunting, thus nutrients which directly affect these parameters would stunt plant growth in conditions of poor availability. Chlorosis occurs when deficiency of nutrients necessary for photosynthesis exists. This causes the plant or plant parts to discolor to light green to yellow or appear as spots (Bennett 1994). Necrosis causes the parts of the plant to brown and eventually die. Deficiency of nitrogen reduces crop quality and yield (Jones 1998). Generally nitrogen deficiency in cereals results in grains with low protein content. Phosphorus is required for metabolic activities (ATP) in a plant and its deficiency results in delayed maturation of plant and discoloration and curling of leaves (Das and Sen 1981). Soils high in $CaCO_3$ lead to phosphorus deficiency in plants due to the precipitation of available phosphorus. Potassium and magnesium act as a cofactor for several enzymes involved in metabolism and photosynthesis. Their deficiency reduces the growth rate and causes chlorosis and necrosis. Boron is required for proper cell constitution and its deficiency causes distortion of the leaves and other plant parts (Shorrocks 1997; Mengel and Kirkby 2001). Iron deficiency also causes chlorosis and slow growth as iron is necessary for photosynthesis. Zinc is essential in growth initiation and seems to play a role in the production of growth hormones. Its deficiency causes chlorosis and can cause severe stunting and death of the plant (Cakmak et al. 1999, Cakmak 2005). Copper is required in photosynthesis and protein synthesis and its deficiency not only causes discoloration but also decreases resistance to diseases (Solberg et al. 1999). Nickel is required for growth and seed germination and enhances nitrogen fixation in legumes. The deficiency causes necrosis and poor germination and growth (Gerendas et al. 1999). Nickel plays a major role in plant development as Ni is a component of urease which is present in a wide range of plant species.

Thus primarily the nutrients are not only essential for plant growth but also for the appropriate growth and health of the consumer. Several studies indicate the effect of nutrient deficiency on plant growth. Yu and Rengel (1998) investigated the effect of copper, zinc or manganese deficiency on the growth and activity of superoxide dismutase in seedlings of lupins which contribute to animal feed (*Lupinus angustifolius L*). Since oxidative stress is a major response in case of injury, under nutrient deficient conditions the plants responsiveness to injury reduces thereby reducing the recovery rate (Cakmak and Marschner 1998). Plants subjected to iron deficiency stress display a range of deleterious effects like inhibition of photosynthesis process, other metabolic disturbances and possible onset of oxidative stress. It is also observed that iron stress induces excessive free radical formation due to chlorophyll oxidation (Susin et al. 1993). Eskew et al. (1983) reported that nickel deficient soybean (*Glycine max.* L) accumulate toxic level of urea in the leaf tip. It is also indicated that nickel may have a role in phytoalexin synthesis and plant disease resistance (Brown et al. 1987). Patric et al. (1987) reported a study on grain from barley (*Hordeum vulgare* L) showed lesser germination rates when grown in culture solution with negligible nickel concentration. Plants grown without nickel additions to their growth media produced grain with very low percent germination. Nickel deficiency also affected grain viability indicating the requirement for nickel in embryo metabolism during germination. These deficiencies have to be removed for better crop productivity and quality which is possible to a greater extent using integrated management technologies.

Integrated Nutrient Management (INM)

As clearly indicated above, the major concern today is the growing demand for quality food product which directly poses a demand on high quality output of food crops, hence sustainable agriculture practices with respect to maintenance of soil fertility and stabilized crop production must be considered. Use of organic manures to improve the physical, chemical and biological properties of the soil is often the choice for any agronomist. In addition to replenishing nutrients, crop rotation can help conserving and improving moisture, nutrient holding capacity of the soil and consequently increasing crop productivity along with maintaining the quality of crop produced. Crop rotation can also be used to eradicate the problem of pest associated to a particular crop plant, since many pests and microbes attack only one type of plant, a yearly change of crop offers a form of protection.

For any life form to grow the basic nutrients required are carbon and nitrogen. Plants require some nutrients in large quantities and others in small quantities. The nutrients required by plants in large quantities are called macro or major nutrients like nitrogen, phosphorus, sulfur, calcium, potassium and magnesium where nitrogen, phosphorus and potassium are also known as primary nutrients. The elements like calcium, magnesium and sulphur are known as secondary nutrients. Besides these, some nutrients are required by plants in very small quantities, these are called minor nutrients or trace elements, for e.g. manganese, cobalt, molybdenum, copper, iron, zinc, boron and chlorine. These nutrients are essential for plant growth as they play an important role in nutrient availability and metabolism, several enzymatic reactions, photosynthesis and growth hormone production. The availability of micronutrients in soil is influenced by several factors like pH, redox potential, organic matter content, microbial diversity responsible for harnessing the trace nutrients and individual soil constituents.

It is very essential for agronomist to access the index of soils acidity on which the bioavailibity of trace nutrients depends. Figure 6 shows neutral pH to be the best range for growth of most of the crops. The role of prevailing microflora and pH of the soil, in some instances, determines the availibity of nutrients. For example, plants take up nitrogen in the ammonium (NH_4^+) or nitrate (NO_3^-) form. Near neutral pH (pH 7.0), the microbial conversion of NH_4^+ to nitrate (nitrification) is rapid. In acid soils (pH < 6.0), nitrification is slow, hence plants with the ability to use NH_4^+ have an advantage. This situation is especially considered by an agronomist in determing the 'nitrogen' content or 'N' nutrition of legumes. It is observed that, the soil with acidic pH will not support the activity of *Rhizobium,* the bacterium responsible for nitrogen fixation in association with legumes. The pool available for soluble phosphorus uptake by plant is comparatively low and is highly pH dependent. Phosphorous has a tendency to form a wide range of stable compounds in soil. Thus, it is least mobile nutrient in the soil. The uptake of phosphorus in the rhizosphere depends on the plant species, nutritional status of the soil and soil conditions. Phoshate-solubilizing microorganisms (PSM) could play an important role in supplying phosphate to plants in a sustainable manner which is partly dependent on the pH of the soil for a favorable growth of the PSM. They are widely applied in agronomic practices to maintain soil fertility and increase crop quality.

In general the amount of trace elements can be higher in soils with higher organic content although this need not necessarily indicate their availability to plants. Soils with higher organic content seem to fix more elements and also promote exchangeable Mn and soluble Zn (Holmgren et al. 1993). Microbes influence the content of trace elements and also make them available for uptake by plants. This involves direct or indirect transformations where the ions are released from the organic matter by decomposition and are made available for plant growth. Alterations in pH due to biochemical reactions mediated by microbes indirectly affect the usable mineral content of soil (Welkie and Miller 1993). Microbes also mediate oxidation of elements leading to a change in their valency which affects their solubility and availability in soil. Table 2 gives an account of the micronutrient deficiency recorded in Indian soils.

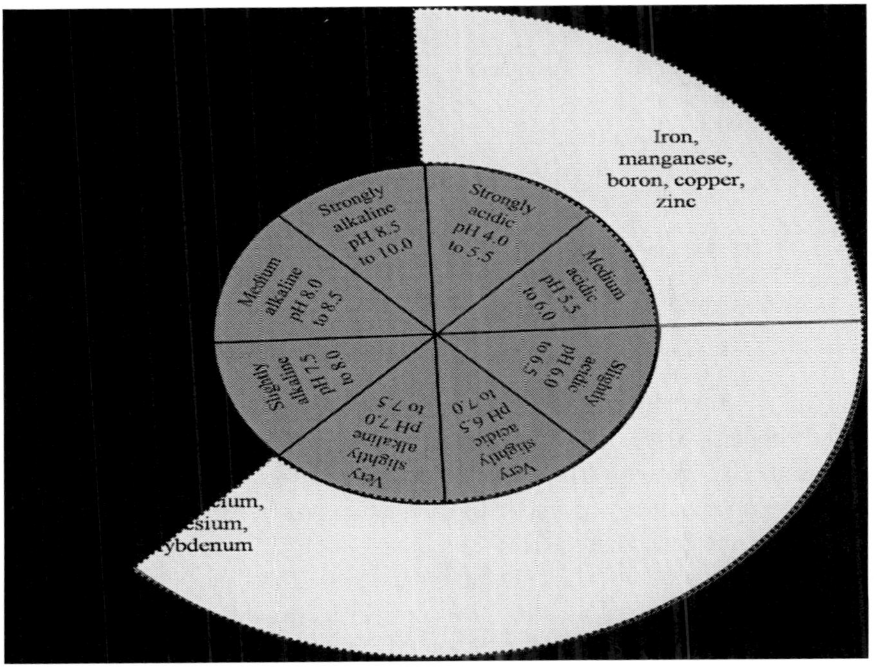

Figure 6. pH range suitable for most crops and availability of nutrients.

Table 2. Micronutrient status of soil of representative regions of India (Singh 2008)

State	Zinc deficient region	Copper deficient region	Iron deficient region	Manganese deficient region
Punjab	Bathinda	Nil	Bathinda, Sangrur, Faridkot	Firozpur, Faridkot
Haryana	Bhiwani, Rohtak, Sonipat, Panipat, Jhajjar, Gurgaon, Faridabad	Kaithal	Hisar	Nil
Uttaranchal	Nil	Nil	Haridwar	Nil
Uttar Pradesh	Saharanpur, Shahjahanpur, Bareilly, Fatehpur, Jhansi, Sitapur and others	Saharanpur, Shahjahanpur, Bareilly, Budaun	Saharanpur, Shahjahanpur, Bareilly, Bijnor, Mohoba, Hamirpur, Sitapur	Saharanpur, Shahjahanpur
Madhya Pradesh	Jabalpur, Panna, Satna, Gwalior, Balaghat, Rajgarh and others	Hoshangabad	Hoshangabad	Nil
Bihar	Gopalganj, Muzzafarpur, Siwan, Rohtas, Jamul, Nawada and Bhabhua	Nil	Madhubani	Pumia
Andhra Pradesh	Nizamabad, Guntur, Krishna, Anantapur, Kumool, Nalgonda, Nellore and Chittoor	Nil	Nil	Nil
Orissa	Kalhandi, Boudh, Sonepur and Balangir	Kalhandi	Mayurbhanj	Nil
Gujarat	Bhavnagar and Amreli	Nil	Surendranagar	Pattan, Kachchh, Gandhinagar and Panchmahal
Tamilnadu	Karur, Thanjavur, Thiruvannamalal, Pudukkottai and others	Nilgiris and Dindigul	Periyar, Dindigul, Salem, Thiruvan and others	Dindigul, Coimbatore, Dharmapuri, Teni and North Arcot

As seen, most of the Indian soils appear deficient in zinc and deficiency in iron, manganese and copper is lesser than that of zinc. Zinc deficiency results in stunted growth of plant and a rosette-like growth at the shoot tips is seen for fruits. Crops grown in zinc-deficient soils have poor root development, low yield and brown spots (Graham et al. 1987).

Integrated nutrient Management (INM) is a breakout option from giving an agronomist a shift from the prevailing traditional cultivation methods. INM can be a useful corrective measure for nutrient deficiencies in soil. INM aims in increasing the sustainable quality of soil so that it can harbor food of higher quality and be environmentally safe at the same time. To put it together INM includes the appropriate usage of organic, inorganic and biological resources to sustain optimum yields to produce maximum yield, improve soil characteristics and provide crop nutrition package which are technically sound practically feasible and environmentally safe. INM recognizes the interrelationship among plant nutrients and the integrated decision making process maximizing the efficiency of nutrient uptake. Biofertilizers constitute a major fraction in INM as they are ecofriendly, cost effective and rich source for plant nutrients. The solubilization and mobilization of nutrients by phosphate solubilizing bacteria and vesicular arbuscular mycorrhiza is helpful in increasing the soil fertility. Nutrient mining is increasing at an alarming rate which might lead to a catastrophe leading to environmental degradation. High yielding profitable crops remove considerable quantities of nutrient from the soil. More than 60% of agricultural land has organic matter content much below the critical level and the rate of depletion of organic matter is still alarming. Hence INM is a very suitable strategy for maintaining sustainable agricultural practices to increase crop production and maintain soil fertility.

Several examples demonstrate the effectiveness of INM in not only enhancing the macro- and micronutrient levels in soil but in general improve soil health holistically. The semiarid conditions of Rajasthan, India soil had less available nutrient content and organic content and hence could not sustain high crop yields. Rao (2003) carried out a study for two years at an agronomy farm in Rajasthan with INM involving combinational use of chemicals, chemical fertilizers and organic sources. The studies were carried out for groundnut and mustard. Their study revealed the availability of nitrogen, phosphorous, sulphur and calcium after groundnut harvest. Thus possible nutrient mining of these nutrients could be understood. The studies for groundnut harvest revealed that, the available nitrogen content, phosphorous content, sulphur and calcium content in the post harvest soil after the application of farmyard manure at 10 t ha^{-1} were increased and thus significant over the control. Similar results were observed for mustard harvest, the available nitrogen and phosphorous content after application of farmyard manure at 10 t ha^{-1} were significantly increased over the control. A comparative study for farmyard manure and poultry manure on available nutrient status indicated that application of poultry manure 5 t ha^{-1} also brought a significant increase in post harvest available nitrogen, phosphorous, sulphur and calcium. Similarly poultry manure application reported an increase in post harvest available nitrogen and phosphorous for mustard harvest. The study also indicated that application of 60 kg P_2O_2 ha^{-1}, as compared to 20 kg P_2O_2 ha^{-1} improved significantly the post harvest available nitrogen, phosphorus, sulphur and calcium content for groundnut harvest and nitrogen and phosphorus content for mustard harvest. This analysis is particularly important in selection of chemical fertilizer for next successive cropping season.

Further studies were carried out by Rao (2003) using gypsum on the above mentioned seasonal crops. Gypsum when applied either in single or in split dose at 250 kg ha^{-1} had a significant increasing effect on available sulphur content in the soil. Thus this study emphasizes the integrated use of organic manure, phosphorous and gypsum on groundnut and mustard in rotation. The proportion of component added was varied and its effect was determined in terms of soil fertility. The availability of phosphorous supplement encouraged various interactions in the rhizosphere of legume crop and organic manure (farmyard and poultry manure) had a favorable effect in maintaining the bioavailability of nutrients from the native pool of nutrients. The organic manure not only has a role in enriching the bioavailibity of soil micronutrients but it also diminishes the possibility of soil erosion, since the organic manure serves as rich nutrient source to soil microflora as well as to the earthworms. Hence the application of farm manure plus organic manure enriches the formation soil colloids.

The inclusion of legumes in a cropping system improves the structure and productivity of soil and increases the plant growth owing to provision of nitrogen and other growth promoting factors. Application of organic manure increases efficiency of nitrogen fixation and facilitates nutrient release. Thomas and Lal (2004) studied the integrated use of biofertilizers, chemical fertilizers and organic manure on yield of green gram in black gram-wheat-green gram system on sandy loam alluvial soil. Their studies indicated an increase in crop yield by a combinatorial use of fertilizer, organic manure forms i.e. farm compost and poultry manure or farm compost and vermicompost and biofertilizers and/or organic sprays. The positive effect of green gram cropping under INM system on the physio-chemical properties of soil was evident from this study. The different combinations of fertilizers (nitrogen and phosphorous) with foliar and soil applications of organic manures enhanced plant dry weight and nodulation pattern. A combination of phosphate solubilizing bacteria plus 2% foliar application of 33% cow's urine maintained a significant effective dry weight. Similarly results of high yield were observed using the combination of phosphate solubilizing bacteria and *Rhizobium*. The nodulation count in experimental plants was high after application of low dose of chemical fertilizers. However, nodule count was highest after treatment with combination of phosphate solubilizing bacteria plus *Rhizobium*. The trend of higher nodulation pattern with the usage of combinatorial method of organic manure proves the enhancement of synergistic effect in soil interactions. The seed yield of black gram under study also increased by the above discussed treatment as compared to the control.

Mahanta and Borah (1998) have reported the role of poultry manure and its effect on increasing black gram yield. The post harvest status for available nutrient pool increased with organic carbon, phosphorous and potassium being higher when soils were treated with farm compost and poultry manure. The integrated use of fertilizers and organic sources improved the organic carbon status of soil. Studies on rice cropping system in Tamilnadu, India revealed increase in rice yield due to use of green manure and full dose of recommended nitrogen, potassium and phosphorous. Furthermore, decomposition of green manure solubilises nitrogen, phosphorus and potassium as well as other micronutrients in soil thereby improving the plant quality. It is reported that an external application of nitrogen and phosphorus reduces drain of native form of the elements (Subramani et al. 2008). Application of green manure with leguminous crops helps in reducing nutrient mining from soil and also maintains the nutrient balance.

Studies on semiarid regions of Gujarat, India during late kharif season demonstrated that higher yield attributes, greater nutrient uptake and protein content was achieved by integrated use of inorganic fertilizer, biofertilizer and organic manure (Patel et al. 2010). Another successful example of an economic plant growth and INM is given by studies on maize by Gaur et al. (2004). The major problem faced by this agro economically important crop is poor soil fertility because of low water holding capacity, nutrient depletion and soil erosion which is associated with poor rain fed dry land areas. Gaur et al. (2004) presented the role of INM along with other recommended cultural practices in their study. They considered the effect of farm yard manure, applied nitrogen, applied phosphorous through chemical fertilizer plus seed treatment with *Azotobacter* and phosphate solubilizing bacteria. The study revealed that the grain yield of maize was increased with *in situ* moisture conservation practices and organic and inorganic fertilizers in rainfed maize cropping system.

Thus INM helps in achieving the dynamic equilibrium within soil components by addition or removal of macro or micronutrients in a specific proportion from a particular patch of soil. INM helps to harness sustainable production of economically important crop plants in a way that not only avoids nutrient mining but also improves crop quality.

Biodynamic Approach

Biodynamics is considered to be oldest organized alternative agriculture movement in the world. This movement has a very ancient origin, it started in 1924 by Rudolf Steiner, the founder of anthroposophy. Just like organic farming, biodynamic farming uses no synthetic chemical fertilizers and pesticides. Instead soil nutrient balance is maintained by compost and animal and green manure addition, controlling pests naturally, rotating crops and diversifying crops and livestock. Rudolf Steiner's concept of a healthy agriculture took into account not only crop rotations, sound stocking rates, and organic manuring, but also cosmic factors, namely the influence of the moon and planets. This makes biodynamic farming differ in certain aspects from organic farming. There are other major difference between organic farming and biodynamic farming, biodynamic farming uses only eight specific preparations to the soil and the biodynamic preparations are not used in organic farming. The biodynamic preparations aim in enhancing soil and crop quality and stimulate the composting process.

In order to achieve a healthier soil and healthier plant, Biodynamic farming presents an array of agricultural practices which are organic in nature, sustainable and ecofriendly. These practices are more labor intensive and hence emphasize more hands-on presence in the farmyard. The existing modern agricultural practices are in a spiral of problematic chemical treatment, so biodynamic approach provides a favorable effect on agroeconomic status on a global scale. Since biodynamic practices are organic, they will contribute to reducing demand for synthetic or artificial fertilizers, pesticides, herbicides etc. Biodynamic is a worldwide movement for the renewal of agriculture based on an understanding of the spiritual forces at work in nature and in human social life (Reganold et al. 1993).

Table 3. Biodynamic preparations

Preparation Name	Gross Content	Application method
500	Cow manure fermented in a cow horn	Soil spray
501	Silica fermented in a cow horn	Crop spray
502	Flower heads from *Achillea millefolium* fermented in stag bladder	Manure or compost piles
503	Flower heads from German chamomile (*Matricaria recutita*) fermented in a cow intestine	Manure or compost piles
504	Stinging nettle (*Urtica dioica*) fermented in the soil	Manure or compost piles
505	Oak bark (*Quercus alba*) fermented in the skull of a domestic animal	Manure or compost piles
506	Flower heads of dandelion (*Taraxacum officinale*) fermented in a cow's mesentery	Manure or compost piles
507	Juice pressed from valerian flowers (*Valeriana officinalis*)	Manure or compost piles

Biodynamic agriculture doesn't rely on any integrated formulation rather harmonizes factors controlling the natural living processes. It necessitates an introspection of uniqueness of each landscape and understanding of its ecology to establish a blended approach of spirituality and science. Soil improvement is thus achieved by humus or organic content management through composting or fermentation. It also relies on the determination of adequate sowing time, growth cycle and harvest time based on the favorable positions of celestial objects. The biodynamic preparations which are used by several crop growers to enhance productivity are enlisted in Table 3.

All these preparations stimulate composting or humus formation in soil and enable a slow release of the fermented components in soil. The science part of biodynamics thus allows the amplification of organic content of soil, however no clear explanation is offered for the spiritual component or the effect of celestial objects on improving soil quality. The biodynamic preparations facilitate impulses in the natural world. Rather than trying to control nature, they seem to work with and harmonize what wants to emerge.

Concluding Remarks

Considering the above facts, our future food security depends on development and implementation of appropriate nutrient management strategies. The prevailing scenario maneuvers policy makers to develop a framework to address nutrient deficiencies and toxicity issues for soil restoration. Use of site-specific nutrient management systems will be more effective in developing sustainable agricultural practices and providing effective nutrition to meet the global food demand.

Acknowledgments

The authors acknowledge the US Department of Agriculture, Natural Resources Conservation Service, Soil Survey Division, World Soil Resources, Washington, D.C. for granting permission to reproduce the global maps (Figure 1 to 4) in this chapter.

References

Abo-Amer A. (2011) Biodegradation of Diazinon by *Serratia marcescens* DI101 and its Use in Bioremediation of Contaminated Environment. *Journal of Microbiology and Biotechnology*, 21, 71-80.

Asio, V. B., Jahn, R., Perez, F. O., Ian, A. N. and Sergio, M. A. Jr. (2009) A review of soil degradation in the Philippines. *Annals of Tropical Research*, 31, 69-94.

Baczynski, T. P. and Pleissner, D. (2010) Bioremediation of chlorinated pesticide-contaminated soil using anaerobic sludges and surfactant addition. *Journal of Environmental Science and Health Part B*, 45, 82-8.

Bennett, W.F. (ed). Nutrient Deficiencies and Toxicities in Crop Plants. St. Paul, MN; APS Press; 1993; pp 202.

Boonyapookana, B., Parkpian, P., Techapinyawat, S., Delaune, R. D., Jugsujinda, A. (2005) Phytoaccumulation of lead by sunflower (*Helianthus annus*), tobacco (*Nicotiana tabacum*) and Vetiver (*Vetiveria zizanioides*). *Journal of Environmental Science and Health*, 40, 117-137.

Brown, P. H., Welch, R. M. and Cary, E. E. (1987) Nickel: A micronutrient essential for higher plants. *Plant Physiology*, 85, 801-803.

Cakmak, I. and Marschner, H. (1988) Enhanced superoxide radical production in roots of zinc-deficient plants. *Journal of Experimental Botany*, 39, 1449-1460.

Cakmak, I. (2005) The role of potassium in alleviating detrimental effects of abiotic stresses in plants. *Journal of Plant Nutrition and Soil Science*, 168, 521-530.

Cakmak, I., Kalayci, M., Ekizc, H., Braund, H. J., Kilince, Y. and Yilmazc A. (1999) Zinc deficiency as a practical problem in plant and human nutrition in Turkey: A NATO-science for stability project. *Field Crops Research*, 60, 175-188.

Cameron, M. D., Timofeevski, S. and Aust, S. D. (2000) Enzymology of Phanerochaete chrysosporium with respect to the degradation of recalcitrant compounds and xenobiotics. *Applied Microbiology and Biotechnology*, 54, 751-758.

Das, B. K. and Sen, S. P. (1981) Effect of nitrogen, phosphorus and potassium deficiency on the uptake and mobilization of ions in Bengal gram (Cicer arietinum). *Journal of Biosciences*, 3, 249-258.

Das, N., Vimala, R. and Karthika, P. (2008) Biosorption of heavy metals- an overview. *Indian Journal of Biotechnology*, 7, 159-169.

El-Rahim, W. M. A. (2006) Assessment of textile dye remediation using biotic and abiotic agents. *Journal of Basic Microbiology*, 46, 318-28.

Eskew, D. L., Welch, R. M. and Cary, E. E. (1983) Nickel: An essential micronutrient for legumes and possibly all higher plants. *Science*, 222, 621-623.

Gaur B.L. and Kumawat S.K. (2004) Evaluating integrated nutrient management and in situ moisture conservation in rainfed maize (Zea mays L.) production system. *Indian Journal of Dryland Agricultural Research and Development,* 19, 91-93.

Gerendas, J., Polacco, J.C., Freyermuth, S.K. and Sattelmacher, B. (1999) Significance of nickel for plant growth and metabolism. *Journal of Plant Nutrition and Soil Science*, 162, 241-256.

Graham, R. D., Welch, R. M., Grunes, D. L., Cary, E. E. and Norvell, W. A. (1987) Effect of zinc deficiency on the accumulation of boron and other mineral nutrients in barley. *Soil Science Society of America Journal*, 51, 652-657.

Holmgren, G.G.S., Meyer, M.W., Chaney, R.L., Daniels, R.B. (1993) Cadmium, lead, zinc, copper, and nickel in agricultural soils of the USA. *Journal of Environmental Quality*, 22, 335-348.

Jones, Jr., J.B. Plant Nutrition Manual. Boca Raton, FL; CRC Press; 1998; pp 149.

Khellaf, N. and Zerdaoui, M. (2009) Phytoaccumulation of zinc by the aquatic plant, Lemna gibba L. *Bioresource Technology*, 100, 6137-6140.

Kuzovkina, Y. A., Knee, M., Quigley, M. F. (2004) Cadmium and copper uptake and translocation in five Willow (Salix L.) species. *International Journal of Phytoremediation*, 6, 269-287.

Langley, S. and Beveridge, T. J. (1999) Effect of O-Side-Chain-Lipopolysaccharide Chemistry on Metal Binding. *Applied and Environmental Microbiology*, 65, 489-498.

Leahy, J. G. and Colwell, R. R. (1990) Microbial Degradation of Hydrocarbons in the Environment. *Microbiological Reviews*, 54, 305-315.

Lim, M. J., Salido, A. L. and Butche, D. J. (2004) Phytoremediation of lead using Indian mustard (Brassica juncea) with EDTA and electrodics. *Microchemical Journal*, 76, 3-9.

Mahanta, B. and Borah, A. (1998) Effect of organic amendments for the management of Meloidogyne incognita on green gram and black gram. *Journal of Agricultural Science Society of North-East India,* 11, 73-76.

Mengel, K. and E.A. Kirkby. *Principles of Plant Nutrition*. Netherlands; Kluwer Academic Publishers; 2001; pp 849.

Patel, C. S., Patel, J. B., Suthar, J. V. and Patel, P. M. (2010) Effect of integrated nutrient management on clusterbean (Cyamopsis tetragonoloba (L.) Taub) seed production cv. Pusa Navbahar. *International Journal of Agricultural Sciences*. 6, 206-208.

Patrick, H. B., Ross, M. W. and Earle, E. C. (1987) Nickel: A micronutrient essential for higher plants. *Plant Physiology*, 85, 801-803.

Rao, S. S. (2003) Nutrient balance and economics of integrated nutrient management in groundnut (Arachis hypogaea L.) - mustard (Brassica juncea L.). *Madras Agricultural Journal,* 90, 465-471.

Reganold, J. P., Palmer, A. S., Lockhart, J. C. and Macgregor, A. N. (1993) Soil quality and financial performance of biodynamic and conventional farms in New Zealand. *Science,* 260, 344–349.

Richter, D. (2007) Humanity's transformation of Earth's soil: Pedology's New Frontier. *Soil Science,* 172, 957-967.

Salt, D.E., Pickering, I. J., Prince, R. C., Gleba, D., Dushenkov, S., Smith, R.D. and Raskin, I. (1997) Metal accumulation by aquacultured seedlings of Indian Mustard. *Environmental Science and Technology*, 31, 1636-1644.

Schwitzguebel, J. P., Meyer, J. and Kidd, P. (2006) Pesticides removal using plants: phytodegradation versus phytostimulation. *Phytoremediation Rhizoremediation*, 179-198.

Shorrocks, V. M. (1997) The occurrence and correction of boron defiency. *Plant Soil,* 193, 121-148.

Singh, M.V. (2008) Micronutrient fertility mapping for Indian Soil, Technical Bulletin. AICRP Micronutrients, *IISS, Bhopal.* Vol. 7, 1-60.

Subramani, T., Durairaj, S. N. and Pandian, B. J. (2008) Effect of nutrient management on grain yield and nutrient uptake of advance kar rice based cropping system in Tambaraparani command area. *Madras Agricultural Journal*, 95, 328-332.

Susin, S., Abian, J., Sanchez–Baeza, F., Peleato, M. L., Abadia, A., Gelpi, E. and Abadia, J. (1993) Riboflavin 3- and 5- sulfate, two novel flavins accumulating in the roots of iron-deficient sugar beet (Beta vulgaris). *Journal of Biological Chemistry*, 268, 20958–65.

Thomas, A. and Lal, R. B. (2004) Performance of blackgram (Vigna Mungo L.) under integrated nutrient management (INM) in a legume based cropping system for inceptisols of NEPZ. *Indian Journal of Dryland Agricultural Research and Development*, 19, 81-87.

Tobin, J.M., Cooper, D. G. and Neufeld, R. J. (1984) Uptake of metal ions by *Rhizopus arrhizus* biomass. *Applied and Environmental Microbiology*, 47, 821-824.

Vázquez, S., Agha, R., Granado, A., Sarro, M. J., Esteban, E., Peñalosa, J. M. and Carpena, R. O. (2006) Use of White Lupin Plant for Phytostabilization of Cd and As Polluted Acid Soil. *Water Air and Soil Pollution*, 177, 349-365.

Verma, P., George, K. V., Singh, H. V., Singh, S. K., Juwarkar A. and Singh R. N. (2006) Modeling rhizofiltration: heavy-metal uptake by plant roots. *Environmental Modeling and Assessment,* 11, 387-394.

Vincent, E. and Wen, T. (eds). *Global and national soils and terrain digital databases* (SOTER). ISRIC, Wageningen; 1995; pp 137.

Wang, J. and Chen, C. (2006) Biosorption of heavy metals by Saccharomyces cerevisiae: A review. *Biotechnology Advances* 24, 427-451.

Welkie, G.W. and Miller, G. W. *Plant iron uptake physiology by non-siderophore systems.* In: LL Barton and BC Hemming eds. Iron Chelation in Plants and Soil Microorganisms. Academic Press: San Deigo; 1993; 345-369.

Yu, Q. and Rengel, Z. (1999) Microdeficiency influences plant growth and activities of superoxide dismutase in narrow leafed Lupins. *Annals of Botany* 83, 175-182.

Zalasiewicz, J., Williams, M., Smith, A., Barry, T. L., Coe, A. L., et al. (2008) Are we now living in the Anthropocene? *Geological Society of America Today*, 18, 4–8.

Zhou, W. and Zhu, L. (2007) Enhanced desorption of phenanthrene from contaminated soil using anionic/nonionic mixed surfactant. *Environmental Pollution*, 147, 350-357.

In: Soil Nutrients
Editor: Mohammad Miransari
ISBN: 978-1-61324-785-3

Chapter 5

ROLE OF MICROBIAL, ENZYMATIC AND PEDOFAUNA ACTIVITIES IN MAKING GOOD THE ADAPTATION OF *TRACHYPOGON* TO PHOSPHORUS DEPLETED ENVIRONMENTS

D. López-Hernández[1]*, M. Toro[1] and J.C. López-Gutiérrez[2]

[1]Universidad Central de Venezuela. Instituto de Zoología y Ecología Tropical, Laboratorio de Estudios Ambientales, Caracas, Venezuela

[2]Ecosystem Science and Management Program University of Northern British Columbia, Canada

ABSTRACT

Savanna ecosystems have a low primary productivity, strong seasonality, and acid soils with low phosphorus (P) content. Among P fractions, organic P (Po) comprises around 50% of the total savanna soil P and is plant-available only after mineralization. The following is a study of the phosphorus dynamics in Trachypogon savannas that takes into account the soil-plant system and the role of microbial, enzymatic and pedofauna activities in making good the adaptation of *Trachypogon spp*. to phosphorus depleted environments. The information here presented summarizes several studies conducted in an Experimental Station located in Venezuelan Central Llanos (Northeastern Venezuela) known as Estación Experimental La Iguana (EELI). We study, at ELLI, arbuscular mycorrhizae (AM), microbial and phosphatase activity in the rhizosphere of *Trachypogon plumosus* Ness. in three acid savanna soils differing in taxonomic order and P content during the dry, rainy and transitional seasons. Microbial number and activity, P mineralization and AM dynamics were quantified in rhizospheric samples of *T. plumosus*. We hypothesize that some specific mechanisms must be operating in the plant rhizosphere to allow plant growth in savannas. Thus, AM, soil microorganisms and APA (Acid phosphatase activity) in the rhizosphere are considered as crucial mechanisms for P availability and plant P-uptake. Our interest focused on the biochemical processes

* E-mail: danilo.lopez@ciens.ucv.ve.

occurring in the rhizosphere of native plant species (*Trachypogon plumosus*) of P-deficient soils. Our data showed that the three soil orders, although differing in their P-sorption values, have low fertility. AM was found to be high for a grass species. APA increased in the rainy season in all cases, but, DHA and microbial counts decreased suggesting that soil microorganisms do not mediate the increase in APA. AM colonization, seasonal changes in microbial activity and in APA seem to be important processes for P availability and *T. plumosus* P-uptake in savanna ecosystems. A more detailed study on the Po transformation rates using Zou et al. method in relation to APA, AM status, DHA (Dehydrogenaseactivity), bacterial and fungal plate counts in the rhizosphere of the native *Trachypogon plumosus* indicated that less available Po fractions (moderately labile, moderately and highly resistant) seem to be involved in short term P-cycling transformations as they significantly varied with season. During the rainy season plant P content (576-1160 mg P kg^{-1} dry weight) and APA (44-200 mg PNP kg^{-1} dry soil) were higher, while microbial number and activity (DHA) were lower. The higher P availability in the Entisol resulted in a better plant nutritional status and inhibited APA. *T. plumosus* seems to be highly dependent on AM symbiosis (45-71% AM colonized root length, 0.6-8 AM spores g^{-1} dry soil), specially during the rainy season. Po mineralization processes, mediated by biological associations in the rhizosphere, are crucial for understanding seasonal P-cycling and fertility in Trachypogon savanna soils. We also found that organic matter accumulation in termite's mounds affects P dynamics in Trachypogon savannas by influencing P sorption-desorption processes. Moreover, termitaria and earthworm biogenic structure constitute nutritional hotspots appropriate for savanna plant establishment. A comparison of adsorption and available P values in mounds of *Nasutitermes ephratae* (a termite species common in Trachypogon savannas) and adjacent soils indicated that the higher P content in the mounds of this grass-feeder termite is most likely due to the incorporation of faeces and grass debris in their nest material. The important reduction on P-sorption capacities (increase in E values) were, in turn, strongly related to the accumulation of organic C in their nests.

INTRODUCTION

Most of Venezuelan savanna ecosystems are located in dystrophic and well drained soils dominated by grasses such as *Trachypogon plumosus* Ness., locally known as Trachypogon savannas (Ramia, 1967). In those savannas other shrub and tree like species, such as *Curatella americana* Linn., *Mauritia flexuosa* Linn. and *Bowdichia virgiloides* H.B. and K. can be found, however, they do not form a continuous canopy (Sarmiento, 1984). Trachypogon is a genus characterized by its low productivity, digestibility and palatability values that seems to be well adapted to acid and nutrient-depleted (especially in nitrogen and phosphorus) soils, (López-Gutiérrez et al., 2004a,b; López-Hernández et al., 2005).

Despite the predominance of nutrient poor soils and seasonality of precipitations, savannas are key productive areas in Venezuela, Colombia and Brazil, where agricultural activities have been established and represent a substantial income for their economies. For this reason, most of the research on savannas has emphasized nutritional aspects to characterize their ecological features and to assess the impact of different land uses (López-Hernández et al., 2005). In tropical savannas, phosphorus has been recognized as one of the most limiting nutrients for plant production because of soil acidity and the high phosphorus adsorption induced by the reactivity of phosphate with iron an aluminum sesquioxides (López-Hernández, 1977). Although, considerable amounts of soluble fertilizer have been added to some savannas soils as a management practice to quench phosphorus adsorption, most of the added fertilizer (over 90%) has been irreversible fixed, so from an economical point of view P-fertilizer addition in savannas is highly inefficient in general. Some assays have shown an increase in plant production and P content in native grasses in well drained savannas under fertilization (San José and García-Miragaya, 1981).

Another management tool strongly associated with natural savannas is the regular burning of vegetation. Although, fire frequency can be very variable (every 1 to 5 years) according to the type of savannas and the human uses, its influence on carbon and nutrient cycling can indeed be remarkable (Chacón et al., 1992; Hernández-Valencia and López-Hernández, 2002; Bustamante et al., 2006). Fire is used to manage rangeland and protected areas maintaining a savanna ecosystem by preventing bush encroachment and selecting for fire tolerant species. In Venezuelan Trachypogon savannas, fire is a fundamental vegetation management tool, used to eliminate grass material considered inedible by cattle, and also to stimulate regrowth of perennial grasses. When fire is suppressed, vegetation develops from open savannas to densely wooded ecosystems (Cerrado type), at least, in the most humid savannas (San José and Fariñas, 1983; Devineau et al., 1984; Sawadogo et al., 2005). Apart from tree densification, fire suppression induces, in turn, an increase in total soil carbon, nitrogen and other nutrients availability. Changes in soil fauna populations, soil enzymatic activities and microbial biomass also occur as soon as a savanna area is protected from fire (López-Hernández et al., 2005, 2008). When a savanna stand burns, the rapid combustion of the organic matter together with the ash dispersion and volatilization of the elements with gaseous cycles (e.g. N and S) induce a decline in the effective input of organic matter and nutrients to the soil (Frost and Robertson, 1985; Hernández-Valencia and López-Hernández, 2002). Thus, N and P cycling are especially sensitive to frequent fires resulting in a loss of N and P in the system.

The cumulative effect of small but persistent losses of nutrients caused by fire and agricultural activities may decrease production and, in turn, cause land degradation (Cobo et al., 2009). However, the existence of recurrent fires is a natural condition and a drastic nutrient depletion by fire, particularly on N and P is incompatible with the existence of savannas *per se*. Consequently, savanna vegetation evolved under frequent fires, which implies a positive or near to steady-state nitrogen and phosphorus budget (López-Hernández et al., 2011). Most of the research in Trachypogon savannas has been devoted to the study of the primary productivity in different sites in the Orinoco Llanos and the fate or losses of nutrients during fires in local studies. Less information exists on the mechanisms of adaptation of those savannas to the minuscule amounts of available phosphorus and potentially available nitrogen present in soil. The nitrogen problem in savannas has been treated in recent reviews (Bustamante et al. 2006; López-Hernández et al., 2011), but comparatively lower information exits on synthesis concerning the mechanisms or strategies that plants living in those ecosystems use to cope with the deficient levels of P.

The following is a study of the phosphorus dynamics in Trachypogon savannas that takes into account the soil-plant system and the role of microbial, enzymatic and pedofauna activities in making good the adaptation of Trachypogon to phosphorus depleted environments. The information here presented summarizes several studies conducted in an Experimental Station located in Venezuelan central Llanos known as Estación Experimental La Iguana (EELI). We first consider aspects on the general characteristics and primary productivity of Trachypogon savannas.

General Characteristic of the Soils of Trachypogon Savannas

Trachypogon savannas are generally associated to acid soils with high P fixation capacity and consequently low P availability, low organic matter, nitrogen and exchangeable base contents (∑exchangeable bases < 5 cmol kg^{-1} soil), but high aluminum saturation due to its acidity (Table 1). The dystrophic nature of soils is a consequence of parent materials with low nutrient contents and/or pedogenetic processes which favors the progressive loss of nutrients. Strong weathering associated to high temperature and also relatively high precipitation also results in the accumulation of resistant primary minerals (e.g. quartz), sesquioxides and clays with low exchange activities (López-Hernández and Hernández-Valencia, 2009). On the other hand, soil water regime is seasonal with periods of excessive wet and eventually anaerobic conditions, alternated with periods where the soil becomes dry and below the permanent wilting point. In those areas where groundwater fluctuation or impeded surface water drainage occurs, it is possible to find a lateritic layer or concretions which evidence the accumulation of sesquioxides due to the redox environment (López-Hernández and Hernández-Valencia, 2009). According to the Soil Taxonomy classification system proposed by United States Department of Agriculture (USDA, 1978), the soils of Trachypogon savannas are identified mainly as oxisols, ultisols and entisols or ferralsols, plinthosols, lixisols and arenosols in the World Reference Base for Soil Resource.

General soil characteristics of three soils in Estación Experimental La Iguana are presented in Table 1.

Table 1. General characterization of savanna soils in Estación Experimental La Iguana. After López-Gutiérrez et al. 2004b

SOIL ORDER	Entisol	Vertisol	Ultisol
pH	5.4	5.7	5.4
Sand (%)	94.4	38.8	73.2
Silt (%)	1.6	29.2	13.2
Clay (%)	4.0	32.0	13.6
Texture	Sand	Clay loam	Sandy loam
P-Total (mg kg^{-1})	90.1	137.3	102.1
P-$NaHCO_3$ (mg kg^{-1})	9.2	1.6	2.6
N (%)	0.08	0.11	0.11
C (%)	0.54	5.1	2.36
Fe (mg kg^{-1})	25.4	75.6	109.6
Al (meq 100 g^{-1})	0.85	0.20	0.70
H (meq 100 g^{-1})	0.20	0.15	0.20
Ca (meq 100 g^{-1})	1.25	3.08	1.88
Mg (meq 100 g^{-1})	0.54	7.90	2.41
Na (meq 100 g^{-1})	0.01	0.31	0.09
K (meq 100 g^{-1})	0.13	0.39	0.22
CEC	2.42	15.66	8.28

Vegetation of Trachypogon Savannas

The Trachypogon savannas are dominated by grass species of the genus *Trachypogon*, particularly *Trachypogon plumosus* and *T. vestitus*. Other associated species are the grasses *Axonopus canescens, A. anceps, Andropogon selloanus, Leptocoryphium lanatum, Paspalum carinatum, Sporobolus indicus, S. cubensis* and several species of the genus *Aristida*; legumes of the genera *Mimosa, Cassia, Desmodium, Eriosema, Galactia, Indigofera, Phaseolus, Stylosanthes, Tephrosia* and *Zornia*, and sedges of the genera *Rhynchospora* and *Bulbostylis* (López-Hernández and Hernández-Valencia, 2009). Tree presence varies from absent in open savannas to dense in savanna parklands where tree crowns may be close together with minimal overlapping. Trees can also be found scattered or in groups locally called "matas". In Trachypogon savannas pyrophyte species are abundant such as *Byrsonima crassifolia, Curatella americana* and *Bodwichia virgilioides* and in a lesser proportion *Copaifera officinalis, Casearia silvestris, Genipa americana* and *Cochlospermum vitifolium*; however, in all those cases the herbaceous layer plays an important role defining structural and functional attributes of the plant community. Plants of Trachypogon savannas are adapted to soils with low nutrient content and consequently they have low nutrient demands, low nutrient contents in their tissues and a low organic matter production.

Primary Productivity in Trachypogon Savannas

Since the preliminary work done by Blydenstein in 1962 at Estación Biológica de los Llanos (EBLL), Calabozo, Venezuela many studies have evaluated the primary productivity of Trachypogon savannas. Above ground net primary production (APP) in Trachypogon savannas located in different regions of Venezuelan llanos ranges from 0.56 t ha^{-1} yr^{-1} in Cabruta in the Central southern llanos with an annual precipitation about 1400 mm to 6.35 t $ha^{-1}yr^{-1}$ in the dry savannas of the eastern llanos in Jusepín, Estado Anzoátegui (Hernández-Valencia and López-Hernández 2002; López-Hernández et al., 2011). Below-ground primary production (BPP) was also found to vary widely within savannas ranging from 1.12 t ha^{-1} yr^{-1} to 9.05 t $ha^{-1}yr^{-1}$. Overall, the total primary production (TPP) in the studied savannas ranges from to 1.97 to 11.97 t ha^{-1} yr^{-1}. These values are below those presented by Grace et al. (2006) for different savannas around the world (14.4 t ha^{-1} yr^{-1}) and those presented by López-Hernández et al. (2011) for savannas of Central Africa, confirming that total primary productivity in Venezuelan Trachypogon savannas is among the lowest in the world (Sarmiento, 1984). However, the precise relationship between precipitation and net primary production over large areas is difficult to assess because of the variability in the proportion of trees, shrubs and herbaceous components as well as the differences in soil chemical and physical characteristics (Grace et al., 2006).

Fertilization experiments in Trachypogon savannas have resulted in higher aboveground biomass production with N or P alone, although more significant effects were obtained when both fertilizers were added together (Medina et al. 1978; San José and García-Miragaya, 1981). In spite of the increase in biomass production, P concentration in leaves remained higher in fertilized plants than in controls, while N concentrations in fertilized plants slightly surpassed that of control plants, indicating that at the end of the growing season translocation of P from old to young leaves is relatively smaller than that of N (Medina et al. 1978).

Study Site

The studies were carried out at Estación Experimental La Iguana, located in Guárico State in Northeastern Venezuela (8° 25' N and 65° 24' W). La Iguana Experimental Station is under the influence area of the Orinoco River watershed and corresponds to representative savannas of the Venezuelan Central Plains. These savannas are dominated by *T. plumosus* (Poaceae) with the presence of isolated trees and shrubs such as *Curatella americana* (Dellineaceae), *Copernicia tectorum* (Palmae), *Byrsonima crassifolia* (Malpigheaceae) and *Bowdichia virgilioides* (Papiloneaceae). The climate is markedly tropical isothermic, with a mean annual precipitation of 1342 mm, most of which falls during the rainy season (May to August), and a mean annual temperature of 27.9°C. Since Estación Experimental La Iguana simultaneously has edaphic substrates different in age and genesis (Ponce et al., 1994), we selected soils from three different orders Entisol (Ustosix Quarsipzamments), Vertisol (Entic Chromusterts) and Ultisol (Aquic Paleustults) where *Trachypogon plumosus* was the dominant species.

Role of Microbial and Enzymatic Activities

Arbuscular Mycorrhiza and Enzymatic Activities in the Rhizosphere of Trachypogon Plumosus Ness. Their Role in Plant P-Uptake

Although Po is not readily available for plant uptake, it can become available after mineralization to inorganic forms (Ascencio, 1997; Kellogg et al., 2006). Acid phosphatase activity (APA) allows the use of less readily available P forms hydrolyzing Po to Pi (Clarholm, 1993). Most of the Po mineralizing activity occurs at the rhizosphere where phosphatases released from plant roots (Helal and Dressler, 1989), soil microorganisms (Asmar et al., 1995) and earthworms (Satchell and Martin, 1984; Lavelle et al., 1994; Caipo, 2010) are present. Arbuscular mycorrhizae (AM) contribute to plant growth and nutrient uptake, especially P (Smith and Gianinazzi-Pearson, 1988). Contribution can be a direct effect through an exploration of a larger volume of soil and/or a hyphal access to soil P pools not available to plants (Koide, 1991; Li et al., 1991; Jayachandran et al., 1992) and AM mycelia may have P mineralizing activity (Jayachandran et al., 1992; Tarafdar and Marshner, 1994). APA (Fox and Comerford, 1992; Clarholm, 1993) as well as AM colonization (Tarafdar and Marshner, 1994; Joner and Jakobsen, 1995) can be induced by P deficiency. Furthermore, rhizospheric microorganisms and AM can synergistically contribute to plant P nutrition (Toro et al., 1996; Singh and Kapoor, 1998).

We hypothesize that some specific mechanisms must be operating in the plant rhizosphere to allow plant growth in savannas. Thus, AM, soil microorganisms and APA in the rhizosphere are considered as crucial mechanisms for P availability and plant P-uptake. Our interest focused on the biochemical processes occurring in the rhizosphere of native plant species (*Trachypogon plumosus*) of P-deficient soils. We studied microbial activity, as measured by viable plate counts and dehydrogenase activity (DHA), mineralization activity (APA) and AM dynamics in the rhizosphere of *T. plumosus* growing on three seasonal savanna soils differing in phosphorus content and soil order.

Soil Chemical and Textural Analyses: Phosphorus Forms

Soils were slightly acid and as expected they had a very low Pt content (Table 1). P-$NaHCO_3$, which is considered to be plant available (Bowman and Cole, 1978), is extremely low in the Ultisol and Vertisol soils. This result supports findings for other seasonal tropical ecosystems where P is one of the most limiting macronutrients (Roy and Singh, 1995; López-Gutiérrez et al. 2004a).

In general, available P is higher as texture becomes finer (O'Halloran et al., 1987; López-Hernández and Niño, 1993; Zubillaga and Giuffré, 1999), our results showed exactly the opposite, P-$NaHCO_3$ was lower as texture became finer, being extremely low in the case of the Vertisol and Ultisol. P adsorption in these soils was higher as silt and clay content increased (Figure 1); similar results were presented by López-Hernández (1977). The higher P adsorption in those soils appears more related to the Fe content than to the exchangeable aluminum (Table 1). The Vertisol in this study differs from others in the world because of the acid pH instead of the basic pH of those derived from calcareous material which are very fertile soils (USDA, 1978), on the contrary the available-P in the Entisol is high (Table 1) indicating than the P-sorbing surfaces in this soil are much reduced as corroborated with the P-sorption isotherms (Figure 1).

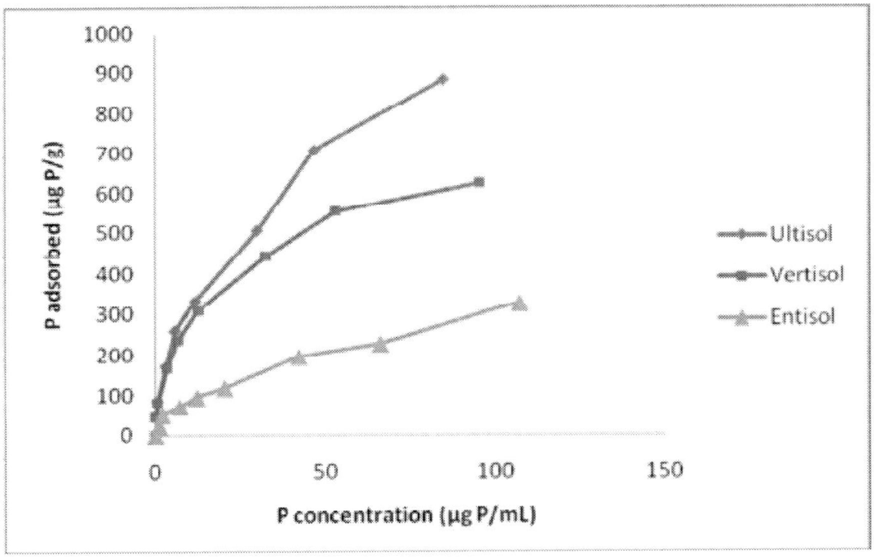

Figure 1. Phosphate adsorption isotherms in selected soils of Estación Experimental La Iguana. Adapted from Flores and Ramos (2008).

Enzymatic Results and Viable Plate Counts

APA significantly increased during the rainy season in all soil orders. This increase, however, was proportionally higher for the Entisol (Figure 2). These results suggest that P mineralization, as expressed by APA, increases during the rainy season, when plant nutrient uptake and nutrient leaching are higher (Roy and Singh, 1995).

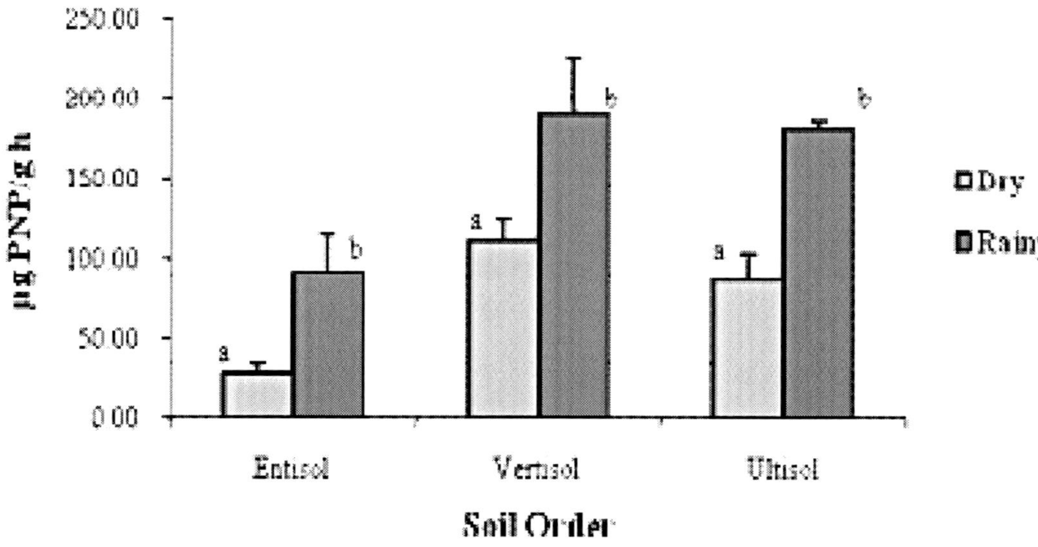

Figure 2. Seasonal changes in acid phosphatase activity in the rhizosphere of *T. plumosus* in three acid savanna soils (means + SD, n=3). Values followed by the same letter do not differ significantly. Duncan´s test (P=0.05). After López-Gutiérrez et al. 2004a.

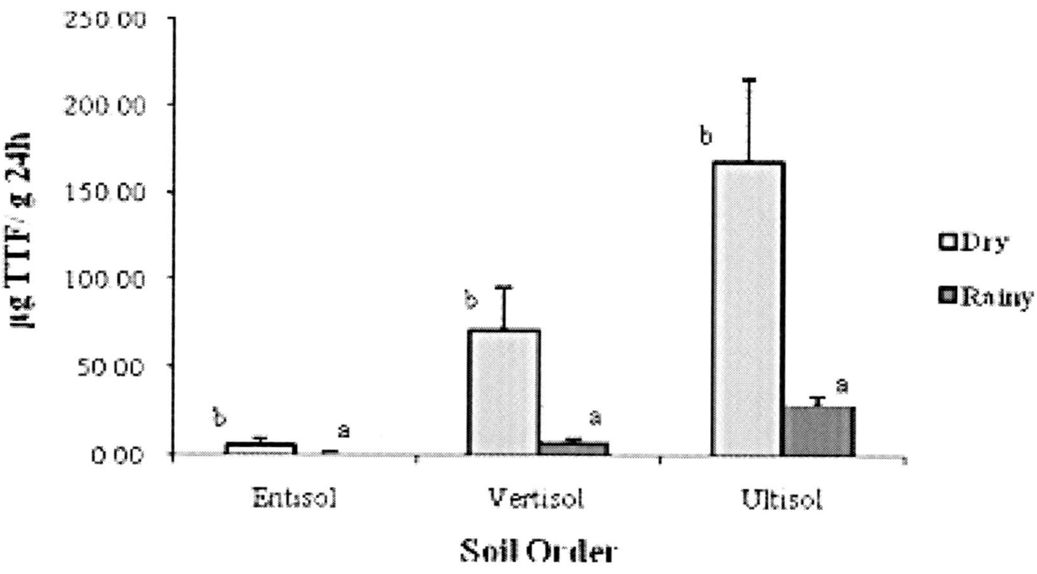

Figure 3. Seasonal changes in dehydrogenase activity in the rhizosphere of *T. plumosus* in three acid savanna soils (means + SD, n=3). Values followed by the same letter do not differ significantly. Duncan´s test (P=0.05). After López-Gutiérrez et al. 2004a.

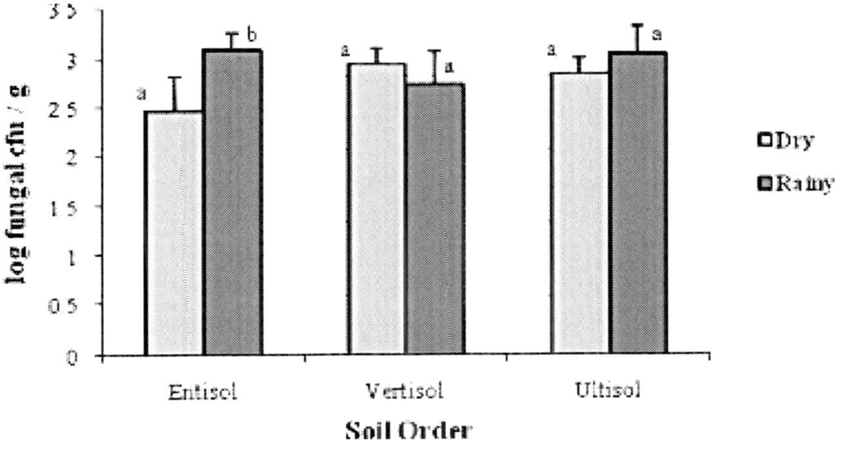

Figure 4. Seasonal changes in viable fungal plate counts in the rhizosphere of *T. plumosus* in three acid savanna soils (means + SD, n=3). Values followed by the same letter do not differ significantly. Duncan´s test (P=0.05). After López-Gutiérrez et al. 2004a.

Dehydrogenases are not expected to be free in soil, therefore, DHA expresses living cell activity and that is why it is an indicator of biological activity (Frankenberger and Dick, 1983). DHA significantly decreased during the rainy season in all soil orders (Figure 3). Even though several soil factors such as soil texture affect DHA, this enzymatic activity can be used to detect soil seasonal changes (Cooper and Warman, 1997). Furthermore, DHA has been shown to correlate with changes in the microbial biomass due to long-term soil amendment (Goyal et al., 1993). In our study DHA (Figure 3) appeared to be related with organic matter content, which, in turn control biological activity. DHA in the Entisol which was poor in soil carbon (0.54%, Table 1) was very low while DHA values increased in the Vertisol and Ultisol which had higher soil carbon contents (5.1 and 2.36%, respectively).

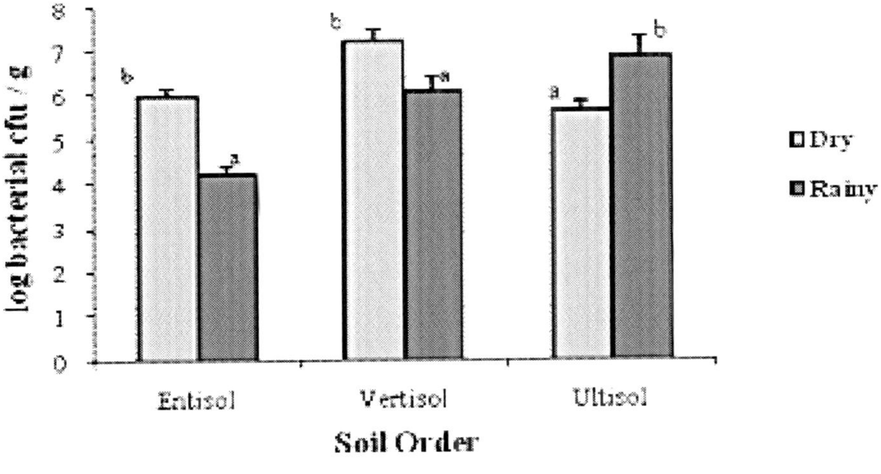

Figure 5. Seasonal changes in viable bacterial plate counts in the rhizosphere of T. plumosus in three acid savanna soils (means + SD, n=3). Values followed by the same letter do not differ significantly. Duncan´s test (P=0.05). After López-Gutiérrez et al. 2004a.

In an attempt to relate DHA with another indicator of soil microbial activity, we quantified viable soil fungi and bacteria. Viable plate counts are not expected to show the same pattern as enzyme activities because selective media recovers only a proportion of soil microorganisms and might include those in dormant phase (Frankenberger and Dick, 1983; Riis et al., 1998). The results showed that viable fungal counts (Figure 4) were significantly different between seasons only for the Entisol, being higher during the rainy season.

Bacterial counts (Figure 5) significantly increased during the rainy season for the Ultisol, but decreased for the Entisol and the Vertisol during the rainy season. The latter result agrees with the microbial activity, as expressed by DHA. A higher bacterial count and DHA during the dry season support the idea that in seasonal ecosystems microbial grazers and plants are not as active during the dry season as soil microorganisms are (Singh et al., 1989). In fact, Ingham et al. (1986) showed how in a seasonally dry ecosystem protozoa, fungivorous microarthropods and bacterial feeding nematode populations decreased while bacterial populations increased during the dry season. In savannas (Singh et al., 1991), as well as in other tropical ecosystems affected by low fertility and strong seasonality (Singh et al., 1989; Campo et al., 1998), soil microorganisms, besides being adapted to dry conditions, play an important role in the seasonal P-cycling by immobilizing P during the dry season and releasing P during the rainy season. The importance of soil microorganisms in these studies had been evaluated in terms of the content of nutrients in the microbial biomass. Our results confirmed these earlier findings taking into account viable counts and activity of soil microorganisms (DHA), instead of the nutrient content present in the living cells or microbial biomass. Furthermore, microorganisms in the rhizosphere may contribute to P nutrition through the synthesis and release of phosphatases when P is not available (Speir and Cowling, 1991; Clarholm, 1993). The decrease in DHA and bacterial counts during the rainy season, however, suggests that the increase in APA during that season is not mediated by rhizospheric microorganisms, i.e., soil extracellular phosphatases may have been already present in the soil or have plant root origin.

Mycorrhizal Status

AM colonization, expressed as percent of colonized root length (%CRL), was rather high in *T. plumosus* (Figure 6); considering that herbaceous plants with rapidly growing fine roots and abundant root hairs are less responsive to AM colonization (St. John, 1980).

According to Baylis's hypothesis (1975), grasses with fine roots, abundant root hairs and rapid growth are less sensitive to AM colonization (St. John, 1980). Some authors (St. John et al., 1983; St. John and Uhl, 1983) observed CRL of 54.7% for *Panicum pilosum* Sw. and 29.8% for *Andropogon bicornis* Linn., in savannas of the Venezuelan Amazonia. The high %CRL in our results suggests that AM may be a crucial mechanism for *T. plumosus* P-uptake in savanna ecosystems. AM colonization of *T. plumosus*, was higher in the Entisol and Vertisol during the dry season, but not in the Ultisol (Figure 6). Several environmental factors, seasonality being one of them, may cause differences in %CRL. Such variations may depend upon the specific structural, developmental and physiological characteristics of the plant–AM combination (Smith and Read, 1997). Furthermore, the ability of the mycorrhiza to absorb nutrients is linked to the development of external mycelium and the proportion of the fungus which is metabolically active (Hepper, 1981) which might not be reflected on the %CRL (Tisserant et al., 1993).

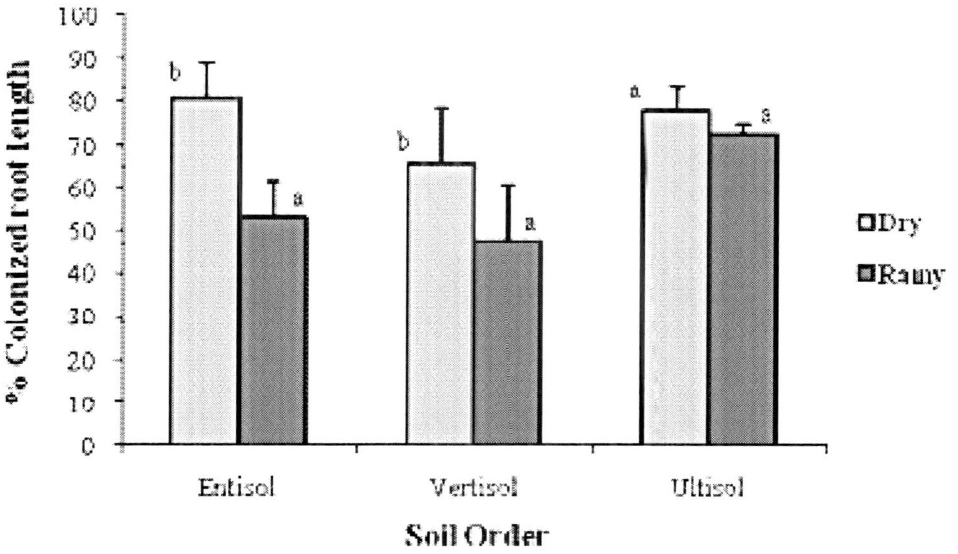

Figure 6. Seasonal changes in AM colonization of *T. plumosus* in three acid savanna soils (means + SD, n=3). Values followed by the same letter do not differ significantly. Duncan's test (P=0.05). After López-Gutiérrez et al. 2004a.

On the other hand, Nelsen and Safir (1982) showed how drought stress, even in fertile soils, causes high levels of AM colonization because low moisture levels can reduce nutrient diffusion plant availability. Accordingly, our results showed that, at least for the Entisol and the Vertisol, AM colonization (Figure 6) was higher during the dry season, when P mineralization, as expressed by APA (Figure 2) was lower and microbial immobilization (Figs. 3 and 5) was higher. Thus, AM colonization was higher when P was less available for plant uptake.

Mineralization Processes in the Rhizosphere of Trachypogon as Mechanisms for Plant P Uptake in Native Savanna Soils

Microbial P can represent up to 6% of the total P in savanna ecosystems (Hernández-Valencia and López-Hernández, 1999) which indicates the importance of soil microorganisms in catalyzing P transformation in these soils. Furthermore, symbiotic arbuscular mycorrhiza (AM), can contribute greatly to plant P nutrition and be induced by P deficiency and the presence of organic matter in soil (St. John et al., 1983; McArthur and Knowles; 1993, Tarafdar and Marschner, 1994; Joner and Jakobsen; 1995, López-Gutiérrez et al. 2004 a,b). AM contribution may be indirect due to AM hyphae exploring a larger soil volume and inducing physiological changes that favor the establishment of P solubilizing and mineralizing microorganisms in the mycorrhizosphere (Linderman, 1988; Dighton, 1991; Kim et al., 1998; Singh and Kapoor, 1998). Alternatively, it may be direct as a consequence of the production of extracellular phosphatases and the access to distant P sources otherwise not available to the host plant (Li et al., 1991; Jayachandran et al., 1992; Joner and Johansen, 2000).

Therefore, if Po mineralization is essential for plant P nutrition in savanna ecosystems, there should be seasonal changes in soil Po fractions, Po mineralization, microbial activity, phosphatase activity and mycorrhizal status in the rhizosphere induced by P deficiency. We further studied these aspects in three soils (Entisol, Vertisol and Ultisol) from the EELI taking into account dry, rainy and transitional seasons as well. The soils differed in genesis, texture and total soil P content, however, they all had acid pH, low P fertility (Ponce et al., 1994) and *T. plumosus* was the plant dominant species.

The aim of the work was to study mineralization processes in the rhizosphere of *T. plumosus* as mechanisms for plant P uptake in native savanna soils. Due to the importance of biological associations in the rhizosphere for plant mineral nutrition, we focused on seasonal changes in Po fractions, P transformation rates, enzymatic activities, microorganisms and AM symbiosis.

Seasonal Changes in Po Fractions

The labile Pi fraction (Pi-lab) was lower at the end of the rainy season (EndRain) for the Entisol and higher for the Vertisol. In all seasons Pi-lab was much lower in the Vertisol and the Ultisol than in the Entisol. Labile Po (Po-lab) did not change with season in all three soils, while moderately labile Po (Po-mola) decreased in all three soils during EndRain. Moderately resistant Po (Po-mores) was lower during the dry season for the Entisol and the Ultisol and highly resistant Po (Po-hires) was also lower during this season in the Entisol (Table 2). According to the fractionation method of Bowman and Cole (1978), Po can represent a large fraction (33 to 82%) of total soil P (Table 2) indicating its importance (Dalal, 1977; Oberson et al., 1993), especially in highly weathered soils (e.g, savanna), where P may limit primary productivity (San José and García-Miragaya; 1981; Tiessen et al., 1984; Roy and Singh, 1995). Differences in P availability and soil P fractions in these soils may be a result of their differences in P sorption capacities (Figure1). Seasonal changes in the different fractions (Table 2) suggest that Po-mola is probably being mineralized as P demand increases during the rainy season; this coincides with a decrease of Pi-lab in the Entisol. In a seasonal Colombian savanna, Oberson et al. (1999) found a higher Po content during the dry season, when decomposition processes are decreased. The changes observed in Po-forms may be a result of a residual effect of the higher P demand during the rainy season.

Tiessen et al. (1984) assert that available P is related to the different Po fractions present in weathered soils, while Oberson et al. (1993) postulate that residual Po can be a reserve that is hydrolyzed when soil Pi content is low. Presumably less active Po fractions have been shown to react to changes in land use, fertilization and seasonality (Guggenberger et al., 1996; Maroko et al., 1999; Daroub et al., 2001) which, along with our results, support the idea that such fractions are important when assessing fertility in such soils because they are related to short term transformations in P-cycling.

Table 2. Phosphorus in various soil fractions according to Bowman and Cole (1978) during the different seasons in three savanna soils in Estación Experimental La Iguana (mg P per kg of dry soil). After López-Gutiérrez et al. 2004b

Soil	Season	Pt	Pi-lab	Po-lab	Po-mola	Po-mores	Po-hires	Po-t
Entisol	Dry	90.09 a	9.32 b	15.38 a	31.25 b	4.15 a	1.63 a	52.42 a
		(6.73)	(2.99)	(2.78)	(3.39)	(1.80)	(1.36)	(4.48)
	BegRain	86.63 a	8.17 b	25.02 a	14.41 a	9.59 ab	7.45 b	56.47 a
		(11.22)	(2.16)	(10.61)	(5.27)	(4.49)	(1.80)	(8.00)
	Rainy	76.62 a	8.72 b	15.08 a	26.57 b	13.81 b	7.57 b	63.02 a
		(16.88)	(1.93)	(7.03)	(4.27)	(7.12)	(4.28)	(5.57)
	EndRain	77.49 a	5.66 a	21.20 a	14.82 a	13.91 b	9.24 b	59.16 a
		(17.50)	(2.17)	(11.06)	(6.81)	(10.56)	(7.72)	(14.09)
Vertisol	Dry	137.25 a	1.62 a	11.49 a	16.86 b	9.85 a	7.38 a	45.59 a
		(25.49)	(0.24)	(1.83)	(3.01)	(4.72)	(5.98)	(5.47)
	BegRain	118.25 a	1.49 a	11.48 a	17.98 b	10.28 a	14.90 a	54.64 a
		(9.66)	(0.68)	(4.23)	(1.16)	(6.39)	(5.34)	(5.04)
	Rainy	116.56 a	1.17 a	10.73 a	15.64 b	15.92 a	11.10 a	53.40 a
		(10.62)	(0.52)	(3.10)	(1.45)	(3.74)	(6.63)	(11.70)
	EndRain	106.01 a	2.52 b	8.18 a	10.89 a	16.37 a	10.74 a	46.18 a
		(26.28)	(0.95)	(2.94)	(3.70)	(9.56)	(6.01)	(10.54)
Ultisol	Dry	102.01 a	2.60 a	11.89 a	18.50 b	6.54 a	8.25 a	45.17 a
		(10.38)	(0.84)	(1.05)	(6.60)	(0 66)	(4.88)	(10.65)
	BegRain	83.66 a	2.71 a	12.79 a	15.77 ab	8.77 a	13.81 a	51.13 a
		(13.47)	(1.19)	(6.26)	(4.15)	(3.10)	(4.30)	(11.44)
	Rainy	88.71 a	2.25 a	9.24 a	17.57 b	10.38 ab	9.90 a	47.08 a
		(11.32)	(0.62)	(2.89)	(4.01)	(6.14)	(6.58)	(9.66)
	EndRain	90.14 a	2.16 a	9.82 a	11.15 a	14.34 b	8.21 a	43.52 a
		(26.04)	(1.80)	(3.81)	(3.03)	(4.29)	(7.53)	(6.23)

BegRain: Beginning of rainy. EndRain: End of rainy. Pt: Total P. Pi-lab: Labile inorganic P. Po-lab: Labile organic P. Po-mola: Moderately labile organic P. Po-mores: Moderately resistant organic P. Po-hires: Highly resistant organic P. Po-t: Total organic P.

Means (SD in parenthesis) followed by the same letter do not differ significantly according to LSD test (n=6; $P<0.05$). Different seasons are compared in each soil order.

P in Plants and Acid Phosphatase Activity (APA)

P content in *T. plumosus* (P-plant) growing in the Entisol was not significantly different among seasons and doubled the P-plant for the Vertisol and the Ultisol (Figure 7A). In the rainy season P-plant was significantly higher for the Vertisol and the Ultisol. Acid phosphomonoesterase activity (APA) was also higher during the humid seasons in the Entisol and the Vertisol, while it was not significantly different among seasons in the Ultisol (Figure 7B).

The higher P-plant in the Entisol (Figure 7A) is a consequence of a higher P availability and lower P-adsorption values (Figure 1 and Table 2) to plants. The increase in P-plant during the rainy season in the Vertisol and the Ultisol concurs with the period of higher P biotic demand which usually results in a higher P content in plant aerial biomass (Medina et al., 1978).

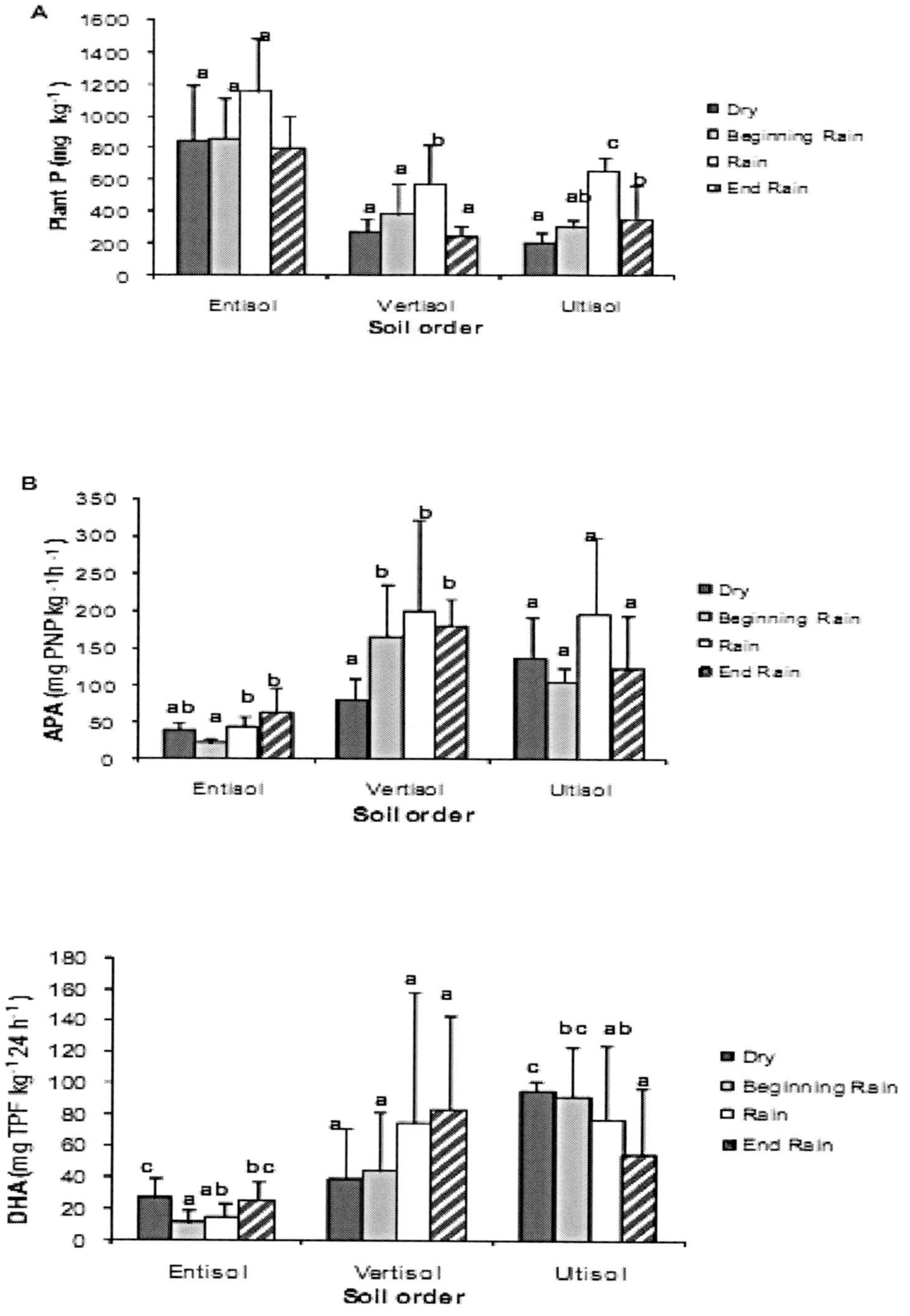

Figure 7. Mean of plant P content (A) rhizosphere acid phosphomonoesterase activity (B) and dehydrogenase activity (C) of *T. plumosus* during different seasons in Estaciόn Experimental La Iguana. Values followed by the same letter do not differ significantly according to LSD test. Error bars represent SD. Different seasons are compared in each soil order. After López-Gutiérrez et al. 2004b.

Soil Microbial Activity and Number

Microbial activity, expressed as dehydrogenase activity (DHA), was higher during the dry season for the Entisol and the Ultisol but not significantly different for the Vertisol (Figure 7C). Bacterial numbers were significantly higher during the beginning of the rainy season (BegRain), while fungal numbers were higher during the rainy and EndRain seasons in the Entisol. Fungal numbers were higher during the dry and the BegRain seasons, while bacterial numbers did not differ among seasons in the Vertisol. Finally, bacterial numbers were higher during EndRain, while fungal numbers were not significantly different among seasons in the Ultisol (Table 3).

Acid phosphomoesterase activity (APA), increased in the Entisol and the Vertisol during the rainy season when nutrient lixiviation and plant uptake are higher (Roy and Singh, 1995) (Figure 7B). In general, the high Pi-lab in the Entisol may be inhibiting APA (Ascencio, 1997), regardless of season (Table 2). Enzymatic activity in the rhizosphere may be higher during the growing season (Sinsabaugh et al., 2003). In fact, Antibus and Linkins III (1992) observed how seasonal changes in Po mineralization and APA were related to the physiological activity of the vegetation in a pine forest. Furthermore, Grierson and Adams (2000) found seasonal variation in APA in an eucalypt forest and emphasized the importance of rigorous sampling throughout the different seasons because infrequent measurements could lead to erroneous interpretations in ecosystem nutrient cycling. This idea is also supported by the findings of Oberson et al. (1993, 1996, 1999) of seasonal fluctuations in residual Po and APA in agricultural systems.

We found seasonal changes in different Po fractions (Table 2), but they did not seem to be directly related to seasonal changes in APA in all three soils (Figure 7B). In the case of the Vertisol, the increase in P-plant (Figure 7A) coincided with the increase in APA and Pi-lab, and the decrease in Po-mola during the EndRain season. No particular trends were apparent for the Entisol and the Ultisol.

Table 3. Number of bacteria and fungi during the different seasons in three savanna soils in Estación Experimental La Iguana (log cfu per g dry soil). After López-Gutiérrez et al. 2004b

	Entisol		Vertisol		Ultisol	
Season	Bacteria	Fungi	Bacteria	Fungi	Bacteria	Fungi
Dry	6.38 a (0.81)	4.29 a (0.18)	5.65 a (0.12)	4.79 a (0.39)	5.69 a (0.22)	4.81 a (0.35)
BegRain	6.63 b (1.16)	4.27 a (0.24)	6.68 a (1.30)	4.84 b (0.20)	6.46 a (0.94)	4.65 a (0.18)
Rainy	6.47 a (0.65)	4.93 b (0.51)	5.73 a (0.62)	4.40 a (0.74)	5.55 a (0.29)	4.67 a (0.50)
EndRain	5.99 a (0.10)	4.83 b (0.12)	6.87 a (1.09)	4.64 b (0.19)	6.87 b (1.15)	4.70 a (0.16)

BegRain: Beginning of rainy. EndRain: End of rainy.

Means (SD in parenthesis) followed by the same letter do not differ significantly according to LSD test ($n=6$; $P<0.05$). Different seasons are compared in each soil order.

The lack of correlation between seasonal changes in P-fractions and APA may be explained by the fact that, even though APA may be induced by P deficiency, the importance of this activity in the transformation of Po to plant available Pi also depends on the susceptibility to hydrolysis of the different compounds in the Po fractions and the different catalytic features of soil phosphatases (Fox and Comerford, 1992; Rao et al., 2000; Yadav and Tarafdar, 2001).

Soil Microbial Activity and Number

DHA decreased during the rainy season (Figure 7C) in agreement with reports of lower microbial activity during the rainy season in tropical ecosystems affected by seasonality (Campo et al., 1998). Our results corroborated previous findings (Hernández-Hernández and López-Hernández, 2002) but consider soil microbial activity (DHA) instead of the nutrient content in the live cells of the microbial biomass (López-Gutiérrez et al., 2004a).

Rhizosphere microorganisms can contribute also to P nutrition through the synthesis and liberation of phosphatases when P is not available (Speir and Cowling, 1991; Clarholm, 1993). Nonetheless, the decrease in DHA during the rainy season (Figure 7C) suggests that the increase in APA (Figure 7B) during this season is not mediated by rhizosphere microorganims; as we previously asserted extracellular phosphatases may have been already present or have plant root origins.

Viable fungal and bacterial counts (Table 3) do not follow the same pattern as the DHA probably because DHA is an indicator of microbial activity instead of number. However, it is worth noting that only fungal numbers are higher during the rainy season in the Entisol, all other changes correspond to the transitional seasons (Table 3). This confirms previous view that, in seasonally marked tropical ecosystems, the rainy season is not the most favorable period for soil microbial growth (Singh et al., 1989; Campo et al., 1998; López-Gutiérrez et al., 2004a). Furthermore, changes in microbial number in the transitional seasons may be related to nutrient fluxes to the soil solution with the onset of the first rains (Campo et al., 1998).

P Transformation Rates

Seasonal changes in P transformation rates according with the method of Zou et al. (1992) were only observed in net P solubilization (Pns) in the Entisol (data not shown) being significantly higher during the beginning of the rain season (BegRain) and in the rainy seasons. With the exception of Pns, which was higher in the Entisol than in the other soil orders, the Ultisol showed the highest P transformation rates, gross P mineralization (Pgm) and microbial P immobilization (Pmi). Solution P and labile organic P during incubation were lower in the Vertisol, while the Entisol accumulated the highest amount soil solution P (Figure 8).

Although the transformation rates (Figure 8) Pmi and Pgm are not significantly affected by seasonality in any of the soils, the incubation method allows the study of short term P cycling and demonstrates that P is limiting (Lodge et al., 1994). This was shown by the fact that Pmi > Pgm in all three soils and that microorganisms were competing for solution P (Zou et al., 1995). Furthermore, the incubation process allows the comparison of the P transformation rates with intermittently measured parameters, such as bacterial and fungal counts, DHA and APA. For instance, microbial activity was, in general, higher in the Ultisol (Figure 7C) where Pmi was also higher, however differences in microbial activity between the other two soils were not reflected in the corresponding Pmi. Similarly, Pgm and APA were higher in the Ultisol.

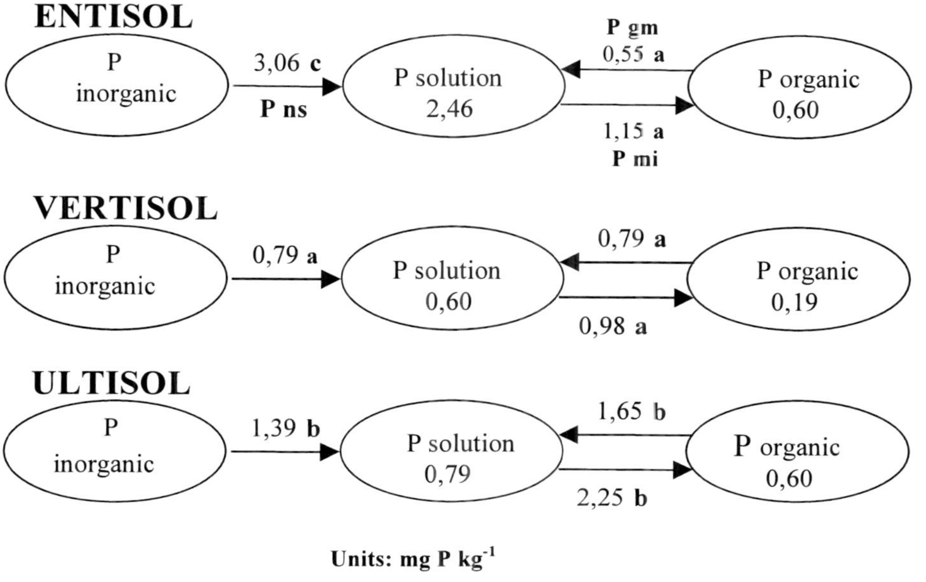

Figure 8. Mean of net solubilization (Pns), gross mineralization (Pgm) and microbial immobilization (Pmi) rates in the rhizosphere of *T. plumosus* during the different seasons in Estación Experimental La Iguana. Numbers on arrows correspond to transformation rates. Numbers in ovals represent net changes in P fractions. Values followed by the same letter do not differ significantly according to LSD test. Different rates are compared between soils. After López-Gutiérrez et al. 2004b.

Moreover, we observed that in the Entisol, where Pi-lab was higher (Table 2), the Pns was also higher and the Pgm and Pmi were lower (Figure 8). This agrees with the higher nutritional status of *T. plumosus* and the lower APA and DHA (Figures 7A-C) and a probable inhibition of Po mineralization induced by the Pi-lab. In the Vertisol, Pgm did not correspond with the high APA, but the Pmi agreed with the lower DHA.

Mycorrhizal Status of T. Plumosus

The AM infective potential, as determined by the Most Probable Number (MPN) method, was not significantly different among soils according to the most probable number method (Table 4). AM colonization, expressed as percentage of AM colonized root length (%CRL) in *T.plumosus*, was higher during the rainy season in Entisol and the Vertisol but not significantly different among seasons in the Ultisol (Figure 9A). Viable AM spore number was relatively high in these soils (Figure 9B) since 3.6 spores g^{-1} dry soil is considered to be a high density (McGee et al., 1997). AM spore number followed a similar pattern than AM colonization in the Entisol and the Vertisol, and decreased during BegRain in the Ultisol (Figure 9B).

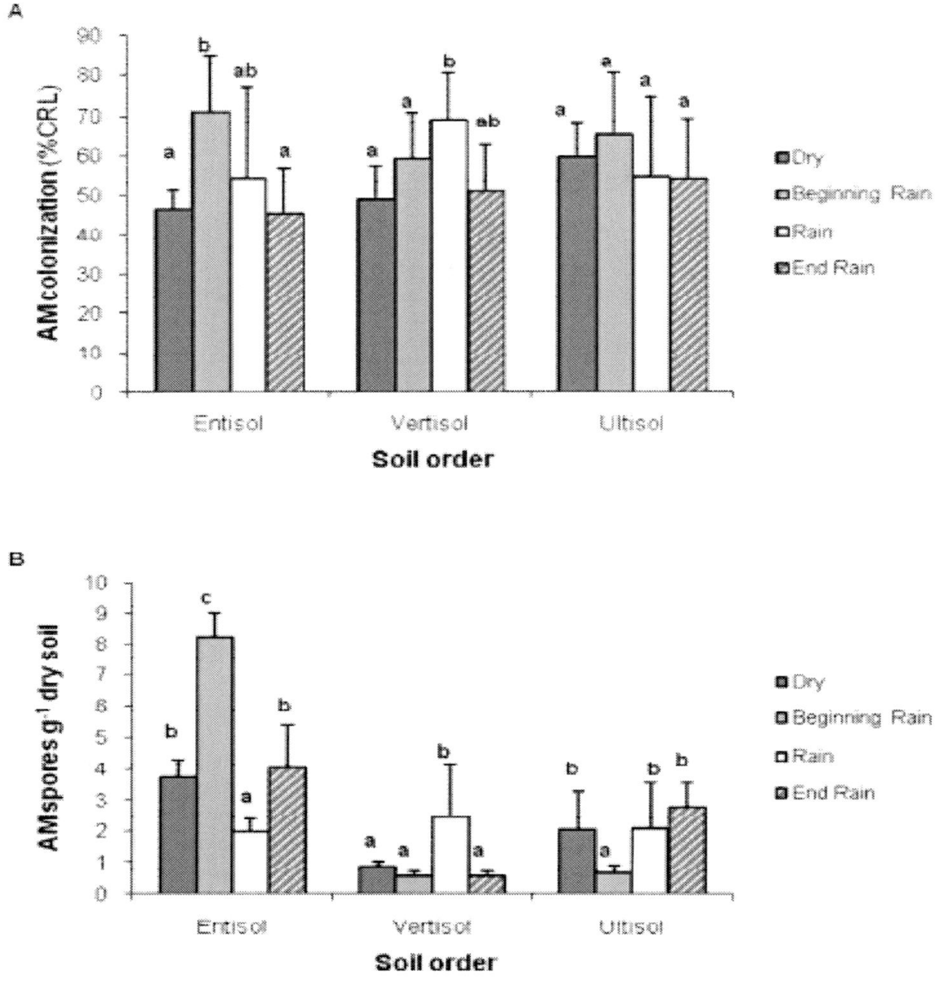

Figure 9. Mean of the percentage of AM colonized root length (A) and viable AM spore number (B) in the rhizosphere of *T. plumosus* during different seasons in Estación Experimental La Iguana. Values followed by the same letter do not differ significantly according to LSD test (n=6, P<0.05). Error bars represent SD. Different seasons are compared in each soil order. After López-Gutiérrez et al. 2004b.

Table 4. Number of AM infective propagules in three savanna soils in Estación Experimental La Iguana (Ω in 100 g dry soil). After López-Gutiérrez et al. 2004b

Soil order	Number of AM infective propagules	Confidence limits (95%)
Entisol	435.4	100.2 – 456
Vertisol	435.4	100.2 – 456
Ultisol	189.5	44.4-201.84

Whereas AM infective potential was not significantly different among soils (Table 4), the values were 2 to 3 times higher than those reported for Colombian natural savannas (Toro and Sieverding, 1986), however, a well-managed cassava (*Manihot esculenta* Crantz.) crop may have values six-fold higher than natural savannas (Sieverding, 1991). Proper agricultural management of savanna soil may increase the soil AM potential, increasing the efficiency of plant AM symbiosis for P-uptake (Dodd et al., 1990).The large confidence limits of the MPN method (Porter, 1979) and/or the rupture of hyphal network caused by the dilution and mixing of the soil (Jasper et al., 1993), may explain why differences among soils were not encountered.

The higher AM colonization (Figure 9A), as well as the increase in AM spore number during the rainy season (Figure 9B), concurred with the higher plant P demand (Figure 7A). Nevertheless, seasonality, intrinsic characteristics of the plant-fungus combination and also soil properties can alter these parameters (Smith and Read, 1997). Thus, it would be uncertain to adjudicate such changes to seasonality alone.

T. plumosus showed a particularly high mycorrhizal status for a grass in terms of AM infective potential spore density and root colonization.The fact that colonization was, in general, high and increased during the rainy season (Figure 9A) when P demand by plants also increased (Figure 8A), suggests the importance of AM symbiosis as a mechanism for P uptake of *T. plumosus* in savanna ecosystems. This is emphasized when considering that one of the main strategies of grass species for P uptake in acid soils is a preferential distribution of resources to the roots (Rao et al., 1996) and *T. plumosus* posses a dense and fibrous radical system that allows an intensive exploration of the soil substrate (Medina and Bilbao, 1991).

AM colonization (Graham et al., 1981) and APA are expected to be inducible by Pi deficiency, thus, the increase in both parameters during the rainy season (Figures 9A and 7B) should agree with decreases in Pi-lab. We did not find significant decreases in Pi-lab accompanying the increases in AM colonization and APA, albeit such increases coincided with the period of higher P demand and uptake by the vegetation (Figure 7A). In fact, the mechanisms regulating AM enzyme expression are not fully understood (García-Garrido et al., 2002). Moreover, in the case of the Entisol C and N content were lower (Table 1) resulting in a lower microbial activity (Figure 7C) and, together with the higher Pi-lab content, may have caused the lower APA and Pgm (Figure 8). Regardless of the presence or not of seasonal changes in the different mycorrhizal parameters studied, the sustained high level of AM colonization throughout the different seasons indicate that *T. plumosus* requires AM symbiosis to achieve an adequate P nutrition in savanna ecosystems.

ROLE OF PEDOFAUNA ACTIVITY

Role of Pedofauna in the Generation of Biogenic "Hotspots" in Trachypogon Savannas

In savanna ecosystems termites and earthworms are among the most conspicuous components of soil macrofauna and play an important role on the decomposition processes of organic matter and nutrient cycling (Lavelle et al. 1994; López-Hernández 2001; Chapuis-Lardy et al., 2011).

Invertebrates belonging to macrofauna are key species of the soil functioning. They participate in litter decomposition, mix organic and mineral matter, create and maintain soil structure by digging burrows and modifying aggregation, regulate microbial diversity and activity (Lavelle et al., 1994, 2006). The gut and biogenic structures (burrows, casts, nests) of earthworms and termites are specific habitats where soil microbial activities are either stimulated or attenuated (Martin, 1991). Freshly egested earthworm casts, i.e. the by-products of gut passage, are generally characterized by an intense mineralization of organic matter and the release of nutrients for plants (Lavelle et al. 1994; López-Hernández et al. 1993). Conversely, the mineralization in old casts is reduced which allows for carbon and nutrient storage in soil, in the long term (Martin, 1991; Lavelle et al. 1994).

Thus, biogenic structures produce by earthworms (casts), ants and termites have been considered as "hotspots" for nutrients and places to potentially start a successional development from savannas to a "microforest".

Phosphorus Contents in Casts of Earthworms and Adjacent Soils

Earthworm activity creates a series of geochemical and biological effects that are especially important for the cycling of P in soils (Chapuis-Lardy et al., 2011). The availability of P is strongly influenced by the physical adsorption or fixation of P in soil (see, Figure1 and Table 1). López-Hernández et al. (1993), using isotopic techniques, reported higher P availability (phosphorus in solution) and E_1(P isotopically exchanged in 1min) in fresh casts of *P. corethrurus* for two soils, one from the savanna of Lamto, Ivory Coast, with a moderate to low P-sorption capacity and the other from Laguna Verde, Veracruz, Mexico, with a high P-sorption capacity. These changes in P availability were ascribed to (i) the greater pH of the gut content (Barois and Lavelle 1986) (ii) changes in sorption complexes induced by competition for sorbing sites between orthophosphates and carboxyl groups of a mucus glycoprotein produced by the earthworm in its gut (López-Hernández et al., 1993) and (iii) an increase in microbial activity during digestion (López-Hernández et al., 1993).

Casts of earthworms usually contain larger amounts of organic P (e.g., for *P. corethrurus*, Chapuis-Lardy et al., 1998), probably derived from a selective ingestion of soil particles (López-Hernández et al., 1993; Chapuis-Lardy et al., 1998). Several studies have reported increased enzymatic activity and the stimulation of microbial activity in earthworm casts (Sharpley and Syers 1976; López-Hernández et al., 1993; Brossard et al., 1996; Le Bayon and Binet 2006; Caipo, 2010). Devliegher and Verstraete (1996) showed that the increase of inorganic phosphorus in soils in the presence of earthworms was due to the production of alkaline phosphatases by earthworms (Satchell and Martin, 1984; Park et al., 1990) and a stimulation of microbial (acid) phosphatase production by earthworm activity (Satchell and Martin 1984). Devliegher and Verstraete (1996) concluded that the increase of Po mineralization and Pi availability in presence of *Lumbricus terrestris* is more a gut-associated than a cast-associated process.

Phosphorus Fractionation in Earthworm Casts and Associated Soils in Trachypogon Savannas

In a recent P-fractionation study of earthworm casts and adjacent soils, in Trachypogon savannas of the Orinoco Basin, Caipo (2010) reported very low total phosphorus values (65 mg kg $^{-1}$) for the Quartzipsamment analyzed. However, a small, yet important, increase in total P was found in the casts (74 mg kg $^{-1}$, Table 5). The small increment in the casts is distributed in the inorganic resin, bicarbonate, NaOH and HCl forms. There was no difference between fresh and old casts. However, the P content in the subsoil (10-30 cm) was much lower than in the associated casts. Other parameters related to the P-cycling (e.g. APA) were also higher in the fresh casts with respect to the adjacent soils (Caipo, 2010). The higher P availability observed in Trachypogon savannas in the presence of earthworms may be linked to a rapid turnover of an increased organic P content.

Table 5. P fractionation (mg kg^{-1}) of soil and earthworm casts in Trachypogon savannas of the Orinoco Basin (After Caipo, 2010). Values followed by the same letter do not differ significantly according to Duncan test ($P<0.05$)

FRACTION	Soil		Cast	
Inorganic P	0-10 cm	10-30 cm	Fresh	Old
Resin	0.15	0.19	0.58	0.67
Water soluble	0.40	0.23	0.00	0.00
$NaHCO_3$	0.00	0.00	0.00	0.00
$NaHCO_3$ fumigated	1.40b	0.39b	3.24a	3.27a
NaOH sonicated	1.14	0.90	1.26	0.42
NaOH without sonication	1.46	0.49	3.93	3.09
HCl	0.70b	0.65bc	1.17a	1.14a
Total Pi	5.25ac	2.85b	10.18a	8.59a
Organic P				
$NaHCO_3$	2.06a	1.66ab	1.55b	1.57b
$NaHCO_3$ fumigated	2.98	1.85	5.06	4.42
NaOH sonicated	6.72a	3.30b	7.67a	8.70a
NaOH without sonication	16.13	12.09	16.58	12.78
Total Po	27.89	18.90	30.86	27.47
Residual P	31.65	22.99	32.99	36.94
Total P	64.79	44.74	74.03	73.00

The Role of Feeding and Construction Habits on the P Content of Termite Mounds

Termites play an important role in the transformation processes of organic compounds in savannas and tropical forest ecosystems, as a consequence of: (i) their significant biomass and diversity within those ecosystems, (ii) their consumption rates and (iii) their effects on the physical, chemical and biochemical characteristics of the adjacent soils (Lee and Wood 1971; López-Hernández and Febres 1984; Lavelle et al. 1994; López-Hernández, 2001; Brossard et al., 2007).

Termite nests are abundant in the savanna soils of Africa, Australia and South America, and their establishment is strongly related to the soil physical properties. Literature for epigeous termites is abundant, but less information exists for subterranean termites which are less conspicuous, although their role as ecosystem engineers is well recognised. The modification of soil materials by termites (epigeous and hypogeous nesters) is partly a consequence of the input of organic compounds (of salivary or faecal origin) to mound construction (Lee and Wood 1971). Therefore, organic compounds, including N and P-compounds, are, in general, more abundant in termite nests than in the surrounding soils. Mounds act as a sink for some elements and protect those with gaseous phases (N and S) against seasonal savanna fires (López-Hernández and Febres, 1984). A large body of literature discussing the accumulation of nutrients (C, N, P and available cations) in termite mounds and adjacent soils is available (Lee and Wood, 1971; López-Hernández and Febres, 1984; Okello-Oloya et al., 1985; Lavelle et al., 1994; López-Hernández, 2001; Brossard et al., 2007). Since, evidence suggests that nutrients are accumulated in mounds from the nearby areas; therefore the abundance of termite populations could play an important role in controlling nutrient cycling in tropical savannas, in which available nutrients can often be a limiting factor for plant growth and development.

P Content of Termite Mounds in Relation to Adjacent Soils

Data on the P content of termite mounds in relation to adjacent soils are sometimes contradictory (Lee and Wood, 1971; Wood et al., 1983; López-Hernández et al., 1989a,b). Lee and Wood (1971), in an extensive study of mounds and other structures of a variety of Australian termites, found little difference in P concentration between termite structures and the soil from which they were built. However, Okello-Oloya et al. (1985) for Australian *Amitermes* mounds, and Nutting et al. (1987) for subterranean North American desert termites, found that extractable P was significantly greater in mounds than adjacent soils. In terms of P content of termite mounds, Wood et al. (1983) suggested, based on the role of the termites' feeding habits, that available P appears to be more abundant in mounds of soil-feeding termites than in the mounds of plant debris-feeding termites. However, López-Hernández et al. (1989 a,b) and López-Hernández (2001) reported, for neo-tropical termites in the EELI (Table 6), significantly higher levels of both total and available P in termite mounds of *Nasutitermes ephratae*, a common plant debris-feeding termite from Trachypogon savannas of the Orinoco Llanos compared with the adjacent soil. In a comprehensive P-fractionation study (Table 6), all P levels were significantly higher in the *Nasutitermes* mounds than in the adjacent soils.

As expected, resin-P is almost negligible in savanna soils, whereas mound values are as high as those for less weathered fertile mollisols. A similar result was found for inorganic (Pi), organic (Po) and water soluble P. So far, it seems that termite feeding and nest-building habits have a strong influence on P availability parameter in mound, nest and gallery structures and those "hotspots" can influence P-cycling in Trachypogon savannas.

Table 6. P-fractionation (µg/g) in soil and mounds of *N. ephratae* living in Trachypogon savannas (adapted from López-Hernández 2001)

	Mound	Soil
Resin	12.50± 0.56	1.92± 0.01
Pi water soluble	4.59± 0.98	0.72± 0.21
Po water soluble	19.47± 1.57	11.88± 3.75
NaOH-Pi	87.52± 0.11	12.21± 3.46
NaOH-Po	56.93± 0.10	18.50± 0.64
HCl-P	14.80± 0.96	1.85± 0.56
P-Residual	288.2	167.9
Total P	484.01	214.98
Microbial P	26.68± 7.46	7.26± 2.53

P-Sorption of Termite Mounds in Relation to Adjacent Soils

A study of on different termite species from Central Africa (Ivory Coast) and Trachypogon savannas of South America indicates that P dynamics (availability and sorption-desorption processes) differ within termite species according with their feeding and construction habits (López-Hernández et al., 2006). While the plant-debris-feeding species of termites (*N. ephratae* and *T. geminatus*) collect and accumulate pieces of grass in their mounds (López-Hernández et al., 2006; Brossard et al., 2007), *Cubitermes* species (humivorous) appear to feed on organic matter mixed with soil particles mostly from the topsoil. *Macrotermes*, while foraging on plant litter above the ground, gets most of its nest material from deeper layers. Subterranean fungus-growing termites (e.g. *A. cavithorax* and *M. toumodiensis*) also forage mostly on the soil surface and get their nest materials from deeper layers. Such variation in feeding and construction habit implies that these species will influence soil P status differently.

The comparison of adsorption and available P values in mounds of termite species with different feeding and construction habits in soil of South America and Central Africa savannas indicated that the generally higher P content in the mounds of grass-feeding and soil-humus-feeding termites, compared with adjacent soils, is most likely due to the incorporation of faeces in their nest material. The important reduction on P-sorption capacities (increase in E values) are in turn strongly related to the accumulation of organic C in their nests. Phosphorus-sorption processes are strongly reduced in organic matter-enriched mounds compared with adjacent soils, as demonstrated by the competition of phosphates with organic molecules for adsorption sites under laboratory studies (López-Hernández et al., 1986). For *Trinervitermes geminatus* and *Nasutitermes ephratae*, the abundant amount of grasses accumulated in the nests (*Loudetia simplex* and *Trachypogon sp.*, respectively), can account for the reported high available P values (Table 7) as well as the high exchangeable potassium levels of these mounds in comparison with adjacent soils (López-Hernández and Febres, 1984).

Table 7. P-Exchangeable values (E) and water soluble P (Cp, μgP ml^{-1}) in nests and adjacent soils *for Macrotermes bellicosus, Trinervitermes geminatus, Cubitermes severus, Nasutitermes ephratae, Ancistrotermes cavithorax,* and *Microtermes toumodensis* (adapted from López-Hernández et al. 2006)

Termite species	Mound		Soil	
	E(1 min)	*Cp*	*E(1 min)*	*Cp*
Macrotermes bellicosus	0.29	0.002	0.34	0.021
Trinervitermes geminatus	3.11	0.221	0.19	0.010
Cubitermes severus	0.33	0.016	0.13	0.006
Nasutitermes ephratae	1.88	0.124	0.46	0.006
Ancistrotermes cavithorax	0.33	0.006	0.20	0.008
Microtermes toumodensis	0.45	0.009	0.24	0.009

Macrotermes mounds are built from subsoil materials (Pomeroy, 1983; López-Hernández and Febres, 1984) and contain finer soil particles, which are selected for their construction. *Macrotermes* species forage mostly from the organic C-deficient subsoil and thus their mounds, built with clay-enriched materials, have very high P-sorbing capacities and consequently a lower P availability than the adjacent topsoil (Table 7). For the subterranean species (*A. cavithorax* and *M. toumodiensis*, Table 7), there is more P sorption in nest structures than in associated soils. The low water-soluble P found in the subsoil reflects strong P sorption in this soil layer, probably due to the presence of the high P-sorbing clay-enriched materials.

Phosphatase Activities of Termite Mounds In Relation to Adjacent Soils

Since the strong accumulation of inorganic P in the mounds of several termite species, a comparison of APA in termite nests with those of adjacent soils have been done (López-Hernández et al., 1989b and more recently by Roose-Amelang et al., 2005). APA in the soil and mounds of *Nasutitermes ephratae* in EELI ranged from 0.56 to 2.32 μmol PNP g $^{-1}$ h^{-1}. Comparison of APA for both mound and adjacent soil showed no significant differences (López-Hernández et al., 1989b). Increases in enzyme activities and microbial biomass are expected to occur in mounds as a result of the increase in their organic-matter content. However, APA is repressible because the activity of the enzyme in plants and microbial cells are determined by the cell P requirement. Therefore enzyme activity in the *Nasutitermes ephratae* mounds might have been inhibited as a consequence of the relatively high available P found in those structures. Contrary to the results of López-Hernández et al. 1989b, Roose-Amelang et al. (2005) found that soil-feeding termites affect soil phosphatase activity, with lower activities in mature nests than in surrounding soils. They also found that APA tended to decrease according with ageing from fresh to old to mature nests. The greater activity expected in fresh nests could be attributed not only to organic matter but to organic matter quality. The higher levels of inorganic P in termite nests presented in Table 6 may result from a higher phosphatase activity in the fresh materials as have been observed in fresh in other biogenic structures as earthworm casts (Sharpley and Syers, 1976; López-Hernández et al,. 1993; Zhang et al., 2000). Roose-Amelang et al. (2005) concluded that the released inorganic P could have a retroactive effect inhibiting APA as reported by López-Hernández et al. (1989b).

Origin of the Phosphate in the Mounds

The enrichment of P on the mound may have two main sources: 1) a biological origin from salivary and faeces used in the mound building, 2) a mineral origin from an accumulation of clay within the mound. Wood et al. (1983) made a differentiation between plant-debris feeding termites which employ the selection of certain particle sizes such clay from subsoil (e.g. African *Macrotermes spp*. López-Hernández et al., 2006) or coarse sand particles (South American and African *Nasutitermes*) cementing soil particles with saliva alone or with saliva and faeces (Lee and Wood 1971; Wood et al., 1983) and soil-feeding termites that mainly employ ingestion of soil, primarily organic-rich fractions, resulting in an intimate mixing of faecal soil and soil carried in the mouth (Wood et al., 1983). Recently, Camacaro (2006) confirmed previous reports from López-Hernández et al. (1989a,b) on phosphorus accumulation in mounds with respect to adjacent soils for plant-debris feeding termites in Venezuelan Trachypogon savannas, emphasizing that the origin of the P in mounds of the savannas analyzed, is essentially from biological activity since the termites built their nests from coarse materials with a scarce amount of fine particles within the termitaria.

CONCLUSIONS

Mycorrhizosphere Influence

Linderman (1988) coined the term mycorrhizosphere and hypothesized that rhizobacteria and AM fungi function in concert to promote plant growth. In fact, specific systems of plant, AM and rhizobacteria have been described (Andrade et al., 1997; Toro et al., 1997). In the present study, on the other hand, a natural system is described in which P mineralization, rhizobacterial community and AM colonization seem to be working in asynchrony to overcome the P limitations set by seasonality. In savannas, as well as in tropical ecosystems affected by seasonality, asynchrony in such processes in the mycorrhizosphere might be essential for plant P-uptake.

The fact that colonization is, in general, high and increases during the rainy season (Figure 9A) when P demand by plants also increases (Figure 7A), suggests the importance of AM symbiosis as a mechanism for P uptake of *T. plumosus* in savanna ecosystems.

In acid savanna soils the Po fractions considered as less available may be involved in seasonal P cycling. Therefore, they need to be considered when studying nutrient cycling and soil fertility. Po mineralization, as well as AM colonization, seems to be important for *T. plumosus* P uptake, especially during the rainy season when P demand increases. Soil microorganisms do not seem to mediate increases in mineralization during the rainy season;

in fact, they are more active during the dry season. Future studies considering the high spatial variability in these soils, may help to further elucidate the mechanisms behind Po mineralization in savanna ecosystems.

Drilosphere and Termitosphere Influence

Earthworms markedly change the biogeochemical status of P (availability, Po pools, enzymatic activities) in gut and in hot spots of the soil, as casts and burrow-linings (drilosphere effects). However, their impact on P dynamics and availability in the soil depends on the close relationships between the properties of soil, the organic P source, the specific burrowing behavior and food preferences of the worms.

The materials used by termites to build their nest structures significantly impact the physico-chemical characteristics related to P-sorption and P availability parameters from that of the adjacent soils. These differences can be analyzed as a consequence of the termites' feeding behaviour and materials utilized in nest building.

The generally higher P content in the mounds of grass-feeding and soil-humus-feeding termites, compared with adjacent soils, is most likely due to the incorporation of plant debris and faeces in the nests. The important reduction on P-sorption capacities are also strongly related to the accumulation of organic C in their nests.

The results presented confirm the overall view the role of feeding and construction habits on the P content of termite mounds when compared with surrounding soils. The estimates provide some useful insight into the relative importance of the biological activity of subterranean and epigeous termites on P status in savannas.

Acknowledgments

We thank Fondo Nacional de Ciencia y Tecnología (Proyecto S1-2000000649) and Consejo de Desarrollo Científico y Humanístico-UCV (Proyectos 03.31.4109.98 and PI 03-00-6415-2006) for partially financing the study. Finally, we thank D. Hernández, P. Duque and I. Bazó for their technical assistance in the laboratory.

References

Andrade, G., Mihara, K.L., Linderman, R.G. and Bethlenfalvay, G.J. (1997). Bacteria from rhizosphere and hyphosphere soils of different arbuscular-mycorrhizal fungi. *Plant Soil,* 192, 71-79.

Antibus, R.K.and Linkins III, A.E. (1992). Effects of liming a red pine forest floor on mycorrhizal numbers and mycorrhizal and soil acid phosphatase activities. *Soil Biology and Biochemistry,* 24, 479-487.

Ascencio, J. (1997). Root secreted acid phosphatase kinetics as a physiological marker for phosphorus deficiency. *Journal of Plant Nutrition*, 20, 9-26.

Asmar, F., Singh, T., Nielsen, G. and Nielsen, N.E. (1995). Barley genotypes differ in activity of soluble extracellular phosphatase and depletion of organic phosphorus in the rhizosphere soil. *Plant and Soil*, 172, 117-122.

Barois, I. and Lavelle, P. (1986). Changes in respiration rate and some physicochemical properties of a tropical soil during transit through *Pontoscolex corethrurus* (Glossoscolecidae, Oligochaeta). *Soil Biol. Biochem.*, 18, 539-541.

Baylis, G.T.S. (1975). *The magnolioid mycorrhiza and mycotrophy in root systems derived from it.* In: Sanders, F.E., Mosse, B., Tinker, P.B. (Ed.) Endomycorrhizas. Academic Press, New York and London, pp. 373-389.

Blydenstein, J. (1962). La sabana de Trachypogon del Alto Llano. *Boletín Sociedad Venezolana Ciencias Naturales,* 23, 139-206.

Bowman, R.A. and Cole, C.V. (1978). An exploratory method for fractionations of organic phosphorus from grassland soils. *Soil Science*, 125, 95-101.

Brossard, M., Lavelle P. and Laurent, J.Y. (1996). Digestion of a vertisol by the endogeic earthworm *Polypheretima elongate*, Megascolecidae, increases soil phosphate extractability. *Eur. J. Soil Biol.*, 32,107-111.

Brossard, M., López-Hernández, D., Lepage, M. and Leprun, J.C. (2007). Nutrient storage in soils and nests of mound-building *Trinervitermes* termites in Central Burkina Faso: consequences for soil fertility. *Biology Fertil. Soils*, 43, 437-447.

Bustamante, M.M.C., Medina, E., Asner, G.P., Nardoto, G.B. and García-Montiel, D.C. (2006). Nitrogen cycling in tropical and temperate savannas. *Biogeochem.,* 79, 209-237.

Caipo, S. (2010). *Fraccionamiento del fósforo y actividades enzimáticas en coprolitos de lombrices de tierra en sabanas del Orinoco Venezolano*. Trabajo Especial de Grado. Escuela de Química. Universidad Central de Venezuela.

Camacaro, J. (2006). *Rol del género Cortaritermes en los cambios que sufren los constituyentes inorgánicos retenidos en sus estructuras biogénicas.* Trabajo Especial de Grado. Escuela de Biología. Universidad Central de Venezuela.

Campo, J., Jaramillo, V.J. and Maass, J.M. (1998). Pulses of soil phosphorus availability in a Mexican tropical dry forest, effects of seasonality and level of wetting. *Oecologia*, 115, 167-172.

Casida, L.E., Klein, D.A. and Santoro, T. (1964). Soil dehydrogenase activity. *Soil Science,* 98, 371-376.

Chacón, P., Lamotte, M. and López-Hernández, D. (1992). Dynamique de la matière organique de la strate herbacée dans une savane à Trachypogon du Venezuela, Comptes Rendus de l'Academie des Sciences, Paris, t. 315, série III, p 209-212.

Chapuis-Lardy, L., Brossard M., Lavelle P. and Schouller E (1998). Phosphorus transformations in a Ferralsol through ingestion by *Pontoscolex corethrurus,* a geophagous earthworm. *Eur. J. Soil Biol.,* 34,61–67.

Chapuis-Lardy L., Le Bayon R.C., Brossard M., Lopez-Hernandez D., Blanchart E. (2011). Role of soil macrofauna in P cycling. /In/"Phosphorus in Action -- Biological Processes in Soil Phosphorus Cycling", Bünemann E.K., Oberson A. and Frossard E. (Eds). Springer Soil. Biology Series 26, Springer, NY, USA, pp. 199-213. DOI 10.1007/978-3-642-15271-9_8.

Clarholm, M. (1993). Microbial biomass P, labile P, and acid phosphatase activity in the humus layer of a spruce forest, after repeated additions of fertilizers. *Biol. Fert. Soils*, 16, 287-292.

Cobo, J.G. Dercon, G. and Cadisch, G. (2009). Nutrient balances in African land use systems across different spatial scales. A review of approaches challenges and progress. Agric. Ecosyst. Environ. (2009), doi:10.1016/j.agee.2009.11.006

Cooper, J.M. and Warman, P.R. (1997). Effects of three fertility amendments on soil dehydrogenase activity, organic C and pH. *Canadian Journal of Soil Science*, 77, 281-283.

Dalal, R.C. (1977). Soil organic phosphorus. *Advances in Agronomy*, 29, 83-117.

Daroub, S.H., Ellis, B.G. and Robertson, G.P. (2001). Effect of cropping and low-chemical input systems on soil phosphorus fractions. *Soil Science*, 166, 281-291.

Devineau J.L., Lecordier C. and Vuattoux R. (1984). Évolution de la diversité spécifique du peuplement ligneux dans une succession préforestière de colonisation d'une savane protégée des feux (Lamto, Côte d'Ivoire). *Candollea*, 39,103- 134.

Devliegher, W and Verstraete, W. (1996). *Lumbricus terrestris* in a soil core experiment: effects of nutrient-enrichment processes (NEP) ad gut-associated processes (GAP) on the availability of plant nutrients and heavy metals. *Soil Biol. Biochem*., 28,489-496.

Dighton, J. (1991). Acquisition of nutrients from organic resources by mycorrhizal autotrophic plants. *Experientia*, 47, 362-369.

Dodd, J.C., Arias, I., Koomen, I. and Hayman, D.S. (1990). The management of populations of vesicular arbuscular mycorrhizal fungi in acid-infertile soils of a savanna ecosystem. *Plant Soil*,122, 229-240.

Flores, E. and Ramos, F. (2008). Adsorción de P en suelos ácidos de sabanas de pedogénesis contrastantes. Trabajo Especial de Grado. Departamento de Química. Instituto Universitario de Tecnología Dr. Federico Rivero Palacio.

Fox, T.R. and Comerford, N.B. (1992). Rhizosphere phosphatase activity and phosphatase hydrolyzable organic phosphorus in two forested spodosols. *Soil Biology and Biochemistry,* 24, 579-583.

Frankenberger, Jr., W.T. and Dick, W.A. (1983). Relationship between enzyme activities and microbial growth and activity indices in soil. *Soil Science Society of America Journal*, 47, 945-951.

Frost, P.G.H. and Robertson, F. (1985). The ecological effects of fire in savannas. In: Walker B.H. (ed). Determinants of tropical savannas. IVBS # 3. IRL Press. pp 93-140.

García-Garrido, J.M., Ocampo, J.A. and García-Romera, I. (2002). Enzymes in the arbuscular-mycorrhizal symbiosis. In: Burns, R.G., Dick, R.P. (Eds.), Enzymes in the Environment. Marcel Dekker, New York, NY, pp. 125-152.

Goyal, S., Mishra, M.M., Dhankar, S.S., Kappor, K.K. and Batra, R. (1993). Microbial biomass turnover and enzyme activities following the application of farmyard manure to field soils with and without previous long-term applications. *Biol. Fert. Soils*,15, 60-64.

Grace J, San José JJ, Meir P, Miranda H and Montes R (2006). Productivity and carbon fluxes of tropical savannas. *Journal Biogeography*, 33, 387-400.

Graham, J.H., Leonard, R.T. and Menge, J.A. (1981). Membrane-mediated decrease in root exudation responsible for phosphorus inhibition of vesicular-arbuscular mycorrhiza formation. *Plant Physiology*, 68, 548-552.

Grierson, P.F and Adams, M.A., (2000). Plant species affect acid phosphatase, ergosterol and microbial P in a Jarrah (*Eucalyptus margitana* Donn ex Sm.) forest in south-western Australia. *Soil Biology and Biochemistry*, 32,1817-1827.

Guggenberger G., Haumaier, L, Thomas, R. J. and Zech, W. (1996). Assessing the organic phosphorus status of an Oxisol under tropical pastures following native savanna using 31P NMR spectroscopy. *Biol. Fert. Soil*, 23, 332-339.

Helal, H. and Dressler, A. (1989). Mobilization and turnover of soil phosphorus in the rhizosphere. *Z. Pflanzenernahr. Bondenk*, 152, 175-180.

Hepper, C.M. (1981). Techniques for studying the infection of plants by vesicular arbuscular fungi under axenic conditions. *New Phytol.*, 88, 641-647.

Hernández-Hernández, R.M. and López-Hernández, D. (2002). Microbial biomass, mineral nitrogen and carbon content in savanna soil aggregates under conventional and no-tillage. *Soil Biol. Biochem.*, 34, 1563-1570.

Hernández-Valencia, I. and López-Hernández, D. (1999). Allocation of phosphorus in a tropical savanna. *Chemosphere*, 39, 199-207.

Hernández-Valencia, I. and López-Hernández, D. (2002). Pérdida de nutrimentos por quema de vegetación en una sabana de Trachypogon. *Revista de Biología Tropical*, 50, 1013-1019.

Ingham, E.R., Trofymow, J.A., Ames, R.N., Hunt, H.W., Morley, C.R., Moore, J.C. and Coleman, D.C. (1986). Trophic interactions and nitrogen cycling in a semi-arid grassland soil. I Seasonal dynamics of natural populations, their interactions and effects on nitrogen cycling. *Journal of Applied Ecology*, 23, 597-614.

Jasper, D.A., Abbott, L.K. and Robson, A.D. (1993). The survival of infective hyphae of vesicular-arbuscular mycorrhizal fungi in dry soil: An interaction with sporulation. *New Phytologist,* 124, 473-479.

Jayachandran, K., Schwab, A.P. and Hetrick, B.A. (1992). Mineralization of organic phosphorus by vesicular-arbuscular mycorrhizal fungi. *Soil Biology and Biochemistry*, 24, 897-903.

Joner, E.J. and Jakobsen, I. (1995). Growth and extracellular phosphatase activity of arbuscular mycorrhizal hyphae as influenced by soil organic matter. *Soil Biology and Biochemistry*, 27, 1153-1159.

Joner, E.J. and Johansen, A. (2000). Phosphatase activity of external hyphae of two arbuscular mycorrhizal fungi. *Mycological Research*, 104, 81-86.

Kellogg, L, Bridgham, S. and López-Hernández, D. (2006). Organic phosphorus mineralization. A comparison of isotopic and non-isotopic methods. *Soil Science American Journal,* 70, 1349-1358.

Kim, K.Y., Jordan, D. and McDonald, G.A. (1998). Effect of phosphate-solubilizing bacteria and vesicular-arbuscular mycorrhizae on tomato growth and soil microbial activity. *Biol. Fertil. Soil,* 26, 79-87.

Koide, R.T. (1991). Nutrient supply, nutrient demand and plant response to mycorrhizal infection. *New Phytol.,* 117, 365-386.

Lavelle, P., Dangerfield, M., Fragoso., C., Eschenbrenner, V., López-Hernández, D., Pashanashi, and Brussard, L. (1994). *The relationship between soil macrofauna and tropical soil fertility.* In: The Biology Management of Tropical Soil Fertility. Woomer, O.L., Swift M.J. (eds). TSBF: Wiley-Sayce. pp 137-169.

Lavelle, P, Decaëns, T., Aubert, M., Barot, S., Blouin, M., Bureau., F., Margerie, P., Mora P and Rossi, J.P. (2006). Soil invertebrates and ecosystem services. *Eur. J. Soil Biol.,* 42:S3-S15.

Le Bayon, R.C. and Binet, F. (2006). Earthworms change the distribution and availability of phosphorous in organic substrates. *Soil Biol. Biochem*. 38,235-246 doi:10.1016/j.soilbio.2005.05.013.

Lee, K.E. and Wood, T.G. (1971). Physical and chemical effects of Australian termites and their pedological significance. *Pedobiologia*, 11, 376-409.

Li, X.L., George, E. and Marshner, H. (1991). Extension of the phosphorus depletion zone en VA-mycorrhizal white clover in a calcareous soil. *Plant and Soil*, 136, 41-48.

Linderman, R.G. (1988). Mycorrhizal interactions with rhizosphere microflora: The mycorrhizosphere effect. *Phytopathology,* 78, 366-371.

Lodge, J., McDowell, W.H. and McSwiney, C.P. (1994). The importance of nutrient pulses in tropical forests. *Trends in Ecology and Evolution*, 9, 384-387.

López-Gutiérrez, J.C., Toro, M. and López-Hernández, D. (2004a). Arbuscular mycorrhiza and enzymatic activities in the rhizosphere of *Trachypogon plumosus* Ness in three acid savanna soils. *Agriculture Ecosystems and Environment,* 103, 405-411.

López-Gutiérrez, J.C., Toro, Marcia and D. López-Hernández. (2004b). Seasonality of organic phosphorus mineralization in the rhizosphere of the native savanna grass, *Trachypogon plumosus*. *Soil Biol. Biochem*., 36, 1675-1684.

López-Hernández D. (1977). La Química del Fósforo en Suelos Ácidos. 123p. Casa Editora: Ediciones de la Biblioteca. Universidad Central de Venezuela. Caracas.

López-Hernández, D. (2001). Nutrient dynamics (C, N and P) in termite mounds of *Nasutitermes ephratae* from savannas of the Orinoco llanos (Venezuela). *Soil Biol. Biochem.,* 33, 747-753.

López-Hernández, D. and Febres, A. (1984). Changements chimiques et granulométriques produits dans des sols de Côte d'Ivoire par la présence de trois espèces de termites. *Rev. Ecol. Biol. Sol.,* 21,477-489.

López-Hernández, D. and Niño, M. (1993). Phosphorus mineralization during laboratory incubation in soils derived from different textured parent material. *Geoderma.* 56, 527-537.

López-Hernández, D. and Hernández-Valencia, I. (2009). *Nutritional aspects in Trachypogon savannas as related to nitrogen and phosphorus cycling.* In: International Commission on Tropical Biology and Natural Resources, [eds. Kleber Del Claro et al.]. in Encyclopedia of Life Support Systems (EOLSS), Developed under the Auspices of the UNESCO, Eolss Publishers, Oxford ,UK.

López-Hernández, D., Siegert, G. and Rodríguez, J.V. (1986). Competitive adsorption of phosphate with malate and oxalate by tropical soils. *Soil Sci. Soc. Amer. J.,* 50, 1460-1462.

López-Hernández, D., Hernández-Hernández, R.M. and Brossard, M., (2005). Historia del uso reciente de tierras de las sabanas de América del Sur. *Interciencia* 30: 623-630.

López-Hernández, D., Hernández-Valencia, I. and Güerere, I., (2008). Cambios en parámetros físicos, químicos y biológicos en una sabana protegida de quema y pastoreo durante veinticinco años. *Bioagro*, 20,151.

López-Hernández, D., Brossard, M. and Fournier, A. (2011). Savanna biomass production, N biogeochemistry, and cycling: A comparison between Western Africa (Ivory Coast and Burkina Faso) and the Venezuelan Llanos. Recent Res. Devel. Soil Sci., 3: 1-34. Research Signpost Kerala, India.

López-Hernández, D., Niño, M., Fardeau, J.C., Nannipieri, P. and Chacón, P. (1989a). Phosphorus accumulation in savanna termite mound in Venezuela. *J Soil Sci.*, 40,635-640.

López-Hernández, D., Niño, M., Nannipieri, P. and Fardeau, J.C. (1989b). Phosphatase activity in *Nasutitermes ephratae* termite nest. *Biol. Fert. Soils*, 7,134-137.

López-Hernández, D., Lavelle, P., Fardeau, J.C. and Niño, M. (1993). Phosphorous transformations in two P-sorption contrasting tropical soils during transit through *Pontoscolex corethrurus* (Glossoscolecidae, Oligochaeta). *Soil Biol. Biochem.* 25,789–792.

López-Hernández, D., Brossard, M., Fardeau, J.C. and Lepage, M. (2006). Effect of different termite feeding groups on P sorption and P availability in African and South American savannas. *Biol. Fert. Soils*, 42, 207-214.

Maroko, J.B., Buresh, R.J. and Smithson, P.C., (1999). Soil Phosphorus fractions in unfertilized fallow-maize systems on two tropical soils. *Soil Science Society of America Journal* 63, 320-326.

Martin, A. (1991). Short- and long-term effects of the endogeic earthworm *Millsonia anomala* (Omodeo) (Megascolecidae, Oligochaeta) of tropical savannas, on soil organic matter. *Biol. Fert. Soils* 11:234-238.

McArthur, D.A.J.and Knowles, N.R., (1993). Influence of vesicular-arbuscular mycorrhizal fungi on the response of potato to phosphorus deficiency. *Plant Physiology* 101, 147-160.

McGee, P.A., Pattinson G.S., Heath R.A., Newman C.A., Allen, S.J. (1997). Survival of propagules of arbuscular mycorrhizal fungi in soils in eastern Australia used to grow cotton. *New Phytologist* 135: 773–780.

Medina, E.and Bilbao, B. (1991). Significance of nutrient relations and symbiosis for the competitive interaction between grasses and legumes in tropical savannas. In: Esser, G., Overdieck, D. (Ed.), Modern Ecology: Basic and applied aspects. Elsevier Science Publ. B. V., Amsterdam, pp. 295-319.

Medina, E., Mendoza, A. and Montes, R., (1978). Nutrient balance and organic matter production in the Trachypogon savannas of Venezuela. *Tropical Agriculture* 55, 243-253.

Nelsen, C.E., Safir, G.R. (1982). Increased drought resistance in onion plants by micorrhizal infection. *Planta.* 154, 407-411.

Nutting W.L., Haverty M.I. and Lafage J.P. (1987). Physical and chemical alteration by two subterranean termite species in Sonoran Desert grassland. *J. Arid. Environ.* 12, 233-239.

Oberson, A., Fardeau, J.C., Besson, J.M., Sticher, H., (1993). Soil phosphorus dynamics in cropping systems according to conventional and biological agricultural methods. *Biology and Fertility of Soils* 16, 111-117.

Oberson, A., Besson, J.M., Maire, N. and Sticher, H. (1996). Microbial processes in soil organic phosphorus transformations in conventional and biological cropping systems. *Biology and Fertility of Soils* 21, 138-148.

Oberson, A., Friesen, D.K., Tiessen, H., Morel, C.and Stahel, W. (1999). Phosphorus status and cycling in native savanna and improved pastures on acid low-P Colombian Oxisol. *Nutrient Cycling in Agroecosystems* 55, 77-88.

O'Halloran, I.P., Stewart, J.W.B.and Kachanoski, R.G. (1987). Influence of texture and management practices on the forms and distribution of soil phosphorus. *Can. J. Soil Sci.* 67, 147-163.

Okello-Oloya T., Spain A.V.and John R.D. (1985). Selected chemical characteristics of the mounds of two species of *Amitermes* (lsoptera, Termitinae) and their adjacent surface soil from Northeaestern Australia. *Rev. Ecol. Biol. Sol.* 22:291-311.

Park S.C., Smith T.J.and Bisesi M.S. (1990). Hydrolysis of bis (4-nitro phenyl) phosphate by the earthworm *Lumbricus terrestris*. *Soil Biol. Biochem*. 22:729-730.

Pomeroy, D.E. (1977). The distribution and abundance of large termite mounds in Uganda. *J. Appl. Ecol.* 14, 465-75.

Ponce, M.E., González, V., Brandín, J. and Ponce, M. (1994). Análisis de la vegetación asociada a una toposecuencia en los Llanos Centro-Orientales de Venezuela. *Ecotropicos* 7, 11-22.

Porter, W.M. (1979). The most probable number method for enumerating infective propagules of vesicular arbuscular mycorrhizal fungi in soil. *Australian Journal of Soil Research* 17, 515-519.

Ramia, M. (1967). Tipos de sabanas en los llanos de Venezuela. *Boletín de la Sociedad Venezolana de Ciencias Naturales* 27, 264-288.

Rao, I.M., Borrero, V., Ricaurte, J., García, R. and Ayarza, M.A. (1996). Adaptive attributes of tropical forage species to acid soils. II. Differences in shoot and root growth responses to varying phosphorus supply and soil type. *Journal of Plant Nutrition* 19, 323-352.

Rao, M.A., Violante, A., Gianfreda, L. (2000). Interaction of acid phosphatase with clays, organic molecules and organo-mineral complexes: kinetics and stability. *Soil Biology and Biochemistry* 32, 1007-1014.

Riis, V., Lorbeer, H. and Babel, W. (1998). Extraction of microorganisms from soil: Evaluation of the efficiency by counting methods and activity measurements. *Soil Biology and Biochemistry* 30, 1573-1581.

Roose-Amsaleg C., Mora P. and Harry, M. (2005). Physical, chemical and phosphatase activities characteristics in soil-feeding termite nests and tropical rainforest soils. *Soil Biol. Biochem.* 37:1910-1917.

Roy, S. and Singh, J.S. (1995). Seasonal and spatial dynamics of plant-available N and P pools N-mineralization in relation to fine roots in a dry tropical forest habitat. *Soil Biology and Biochemistry* 27, 33-40.

San José, J.J. and García-Miragaya, J. (1981). Factores ecológicos operacionales en la producción de materia orgánica de las sabanas de Trachypogon. *Boletín de la Sociedad Venezolana de Ciencias Naturales* 36, 347-374.

San José, J.J. and Fariñas, M. (1983). Changes in tree density and species composition in protected Trachypogon savanna, *Venezuela. Ecol.* 64: 447-453.

Sarmiento, G. (1984). The ecology of neotropical savannas. Harvard University Press, Cambridge, pp.

Satchell, J.E.and Martin, K. (1984). Phosphatase activity in earthworm faeces. *Soil Biology and Biochemistry* 16, 191-194.

Sawadogo, L., Tiveau, D. and Nygård, R. (2005). Influence of selective tree cutting, livestock and prescribed fire on herbaceous biomass in the savannah woodlands of Burkina Faso, West Africa. *Agric. Ecosystems Environ*. 105: 335-345.

Sharpley A., Syers, J. (1976). Potential role of earthworms casts for the phosphorus enrichment of run-off water. *Soil Biol. Biochem*. 8:341-346.

Sieverding, E. (1991). Vesicular-Arbuscular mycorrhiza management in tropical agrosystems. GTZ, Germany, 371 pp.

Singh, S. and Kapoor, K.K. (1998). Effects of inoculations of phosphate-solubilizing microorganisms and an arbuscular mycorrhizal fungus on mungbean grown under natural soil conditions. *Mycorrhiza* 7, 249-253.

Singh, J.S., Raghubanshi, A.S., Singh, R.S.and Srivastava, S.C. (1989). Microbial biomass acts as a source of plant nutrients in dry tropical forest and savanna. *Nature,* 338, 499-500.

Singh, R.S., Srivastava, S.C., Raghubanshi, A.S., Singh, J.S. and Singh, S.P. (1991). Microbial C, N, and P in dry tropical savanna: Effects of burning and grazing. *Journal of Applied Ecology* 28, 869-878.

Sinsabaugh, R.L., Saiya-Cork, K., Long, T., Osgood, M.P., Neher, D.A., Zak, D.R. and Norby, R.J. (2003). Soil microbial activity in a Liquidambar plantation unresponsive to CO_2-driven increases in primary production. *Applied Soil Ecology* 24, 263-271.

Smith, S.E. and Gianinazzi-Pearson, V. (1988). Physiological interactions between symbionts in vesicular-arbuscular mycorrhizal plants. *Ann. Rev. Plant Physiol. Plant Mol. Biol.* 39, 221-244.

Smith, S.E. and Read, D.J. (1997). Mycorrhizal symbiosis. 2nd ed. Academic Press, Inc, Cambridge, 605 pp.

Speir, T.W. and Cowling, J.C. (1991). Phosphatase activities of pasture plants and soils: Relationship with plant productivity and soil P fertility indices. *Biology and Fertility of Soils* 12, 189-194.

St. John, T.V. (1980). Root size, root hairs and mycorrhizal infection: a re-examination of Baylis´s hypothesis with tropical tress. *New Phytologist* 84, 483-487.

St. John, T.V.and Uhl, C. (1983). Mycorrhizae in the rain forest of San Carlos de Río Negro, Venezuela. *Acta Científica Venezolana* 34, 233-237.

St. John, T.V., Coleman, D.C., Reid, C.P.P. (1983). Association of vesicular arbuscular mycorrhizal hyphae with soil organic particles. *Ecology* 64, 957-959.

Tarafdar, J.C. and Marshner, H. (1994). Phosphatase activity in the rhizosphere of VA mycorrhizal wheat supplied with inorganic and organic phosphorus. *Soil Biology and Biochemistry* 26, 387-395.

Tiessen, H., Stewart, J.B. and Cole, C.V. (1984). Pathways of phosphorus transformations in soil of different pedogenesis. *Soil Science Society of America Journal* 48, 853-858.

Tisserant, B., Gianinazzi-Pearson, V., Gianinazzi, S. and Gollotte, A. (1993). In planta histochemical staining of fungal alkaline phosphatase activity for analysis of efficient arbuscular mycorrhizal infections. *Mycol. Res*. 97, 245-250.

Toro, M., Azcón, R. and Barea, J.M. (1997). Improvement of Arbuscular Mycorrhizal development by inoculation with Phosphate Solubilizing Rhizobacteria to benefit rock phosphate bioavailability (32P) and nutrient cycling. *Appl. Environ. Microbiol.* 63, 4408-4412.

Toro, M., Azcón, R. and Herrera, R. (1996). Effects of yield and nutrition of mycorrhizal and nodulated *Pueraria phaseoloides* exerted by P-solubilizing rhizobacteria. *Biol. Fertil. Soilss.* 21, 23-29.

Toro, T.S. and Sieverding, E. (1986). Evaluación cuantitativa y cualitativa de hongos formadores de micorriza vesículo arbuscular en la región de Mondomo, Colombia. *Suelos Ecuatoriales*, 16, 122-129.

United States Department of Agriculture (USDA). (1978). Soil Taxonomy. A basic system of soil classification for making and interpreting soil surveys. *Agriculture Handbook No.* 436. India Offset Press, New Delhi, pp. 375-382.

Wood, T.G., Johnson, R.A. and Anderson, J. M. (1983). Modification of soils in Nigerian savanna by soil-feeding *Cubitermes*, *Termitidae. Soil Biol. Biochem*., 15,575-579.

Yadav, R.S. and Tarafdar, J.C. (2001). Influence of organic and inorganic phosphorus supply on the maximum secretion of acid phosphatase by plants. *Biology and Fertility of Soils*, 34, 140-143.

Zhang, B.G., Li, G.T., Shen, T.S., Wang, J.K. and Sun, Z. (2000). Changes in microbial biomass C, N and P and enzyme activities in soil incubated with the earthworms *Metaphire guillelmi* or *Eisenia fetida. Soil Biol. Biochem.,* 32,2055-2062.

Zou, X., Binkley, D. and Doxtater, K.G. (1992). A new method for estimating gross phosphorus mineralization and immobilization rates in soils. *Plant and Soil,* 147, 243-250.

Zou, X., Binkley, D. and Caldwell, A. (1995). Effects of dinitrogen-fixing trees on phosphorus biogeochemical cycling in contrasting forests. *Soil Science Society of America Journal,* 59, 1452-1458.

Zubillaga, M.S. and Giuffré, L. (1999). Soil phosphorus mobilization in different taxonomic orders. *Journal of Plant Nutrition and Soil Science*, 162, 201-205.

In: Soil Nutrients
Editor: Mohammad Miransari
ISBN: 978-1-61324-785-3

Chapter 6

SOLUBILIZATION OF MINERAL AND ROCKS: MICROORGANISMS ACTING TO IMPROVE SOIL FERTILITY

Maria Leonor Lopes-Assad, Márcia Maria Rosa-Magri and Sandra Regina Ceccato-Antonini*
Centro de Ciências Agrárias, Universidade Federal de São Carlos,
Campus de Araras, SP, Brazil

ABSTRACT

In this chapter the authors will consider the importance of rocks for plant nutrition and the role of the microorganisms in this process. The mechanisms involved with the nutrient release from different types of rocks and minerals as phosphate rocks, micas, feldspars and others will also be considered. The microbial solubilization of minerals for increasing nutrient availability to crops is the focus of this chapter, as a challenging approach to the use of rocks for agricultural soils.

1. INTRODUCTION

Nowadays disparities in social, economic and political development have been found in the world in spite of impressive social progress since the 70´s of the last century in many countries (Estes, 2010). Rapid population growth along with the urban migration has compromised the environment resulting in deforestation, exhausted soil and animal resources, recurrent floods and other disasters, both natural and man-made, including the social consequences of depleted minerals and other natural resources (Estes, 2010).

In this context, agriculture is essential for its social and economic role regarding the impact that it may cause. In tropical regions where many of the world´s poorest countries are

* Centro de Ciências Agrárias, Universidade Federal de São Carlos, Campus de Araras – Via Anhanguera, km 174 – P.O. Box 153 – 13600-970 – Araras – SP – Brazil. E-mail: antonini@cca.ufscar.br.

located, soil is generally highly weathered. The intense action of climate and organisms promotes chemical reactions and high loss of soluble elements essential for plant mineral nutrition.

The essential nutrients to plant life occur in various concentrations in aluminosilicates and ferromagnesian silicates, the major rock-forming minerals, and in other accessory minerals. They vary in structure and composition, and may be a primary source of many nutrients required for plant growth (Harley and Gilkes, 2000).

But the processes of the releasing of nutrients contained in minerals are relatively slow or very slow, and do not meet the demands of most crops. In addition, intensive farming systems on fertile soils require replacement of chemical elements to ensure the rapid uptake by plants. Thus, cultivated soils generally require the addition of nutrients, as fertilizers, to guarantee ecologically sustainable productivity.

Global fertilizer consumption follows the economy and particularly the performance of agriculture, which is also affected by unfavorable weather. Therefore, after a sharp drop in 2008/2009 brought on by the financial and economic downturn, world fertilizer consumption began to recover in 2009/2010. Indeed, data from the International Fertilizer Industry Association - IFA (Table 1) show that the global fertilizer consumption of nitrogenous (N), phosphate (P_2O_5) and potash fertilizers (K_2O) increased 4.1%, 11.5% and 0.3%, respectively, compared with the estimations for 2009/2010 and 2008/2009. Moreover, the rebound in global demand for fertilizers and low stocks in early 2010 caused a strong increase in sales, and the fertilizer industry has operated at 82% of installed capacity, compared with 74% in 2009 (Heffer and Prud'homme, 2010).

In order to achieve greater environmental and agronomic sustainability, several alternatives have been adopted mainly by small farmers, who cannot always afford to pay the same prices for fertilizers as their counterparts on big-industrial farms. An alternative is the use of a natural rock or stonemeal, which is sometimes combined with an industrial fertilizer.

The stonemeal practice may be unfeasible because the time required for the mineral alteration contained in rocks, and the consequent availability for plants, is very large and incompatible with the needs of the culture.

Table 1. Global fertilizer consumption (Mt nutrients) of nitrogenous (N), phosphate (P_2O_5) and potash fertilizers (K_2O) from 2007 to 2012, measured, estimated and forecasted by INA - International Fertilizer Industry Association

Year	N	P_2O_5	K_2O	Total
2007/2008	100.5	38.4	28.9	167.9
2008/2009	98.5	33.7	23.5	155.6
2009/2010 (e)	102.6	37.5	23.5	163.7
Change	*+4.1%*	*+11.5%*	*+0.3%*	*+5.2%*
2010/2011 (f)	104.2	39.8	27.4	171.4
Change	*+1.6%*	*+6.0%*	*+16.3%*	*+4.7%*
2011/2012 (f)	106.9	41.7	29.4	177.9
Change	*+2.6%*	*+4.7%*	*+7.2%*	*+3.8%*

(e) estimated; (f) forecasted

Source: Heffer and Prud'homme (2010).

Phosphate-rich rocks when crushed and applied directly to the soil rely on compounds produced by plant roots (exudates) to extract the phosphorus that plants need (Manning, 2008). In this regard, the soil microorganisms also play an important role in the release of minerals from the rocks. As already estimated, in fertile gardens or organic soil, there will be higher number of organisms than the total number of human beings that have ever been on the earth: 10^{12} bacteria, 10^4 protozoa, 10^4 nematodes, 25 km of fungi and countless other species (Young and Crawford, 2004). Due to the size of most of these microorganisms, their surface coverage amounts in total to only $10^{-5}-10^{-6}$ % of the total soil surface area (Hinsinger et al., 2009). The community of microorganisms within soil is very complex and particularly important with regard to the services it provides for agriculture, waste management, and water quality, as well as the natural and semi-natural environment.

Considering the importance of the presence of rocks for plant nutrition and the role of the microorganisms in this process, this chapter will deal with the microbial solubilization of minerals for increasing nutrient availability to crops.

2. Stonemeal: An Old Practice for Current Problems

Efforts to utilize rock powder directly into the soil have been attempted for years. This technique, known as stonemeal or soil remineralization, is defined as an agricultural practice of incorporating powdered rocks and/or minerals to the soil. Liming and natural phosphate are examples of this practice (Leonardos et al., 2000).

The use of ground rocks for the soil fertilization was initially proposed by Julius Hensel, in 1894, when his book "Bread from stones" was published. Hensel stated that healthy food, coming from a healthy soil results in healthy and nourished people, and can even eliminate the hunger in the world (Hensel, 1894).

Rocks are defined as solid materials, naturally formed, constituted by minerals and/or amorphous substances like volcanic glass, gels and natural coal. One of the main classifications of rocks is based on the way they are formed. Igneous, for example, result from the magma cooling; metamorphic rocks result from the transformation of mineral materials, consolidated or not by the temperature increase and/or pressure, whether or not the fusion point is reached; and sedimentary rocks are formed from the compactation and/or cementation of fragments produced by weathering (Madureira Filho et al., 2000).

A rock may consist of only a mineral (monomineralic rock) or by various minerals (plurimineralic rock). The examples of the former are quartzite and marble; the latter are granite and basalt. Indeed, even the monomineralic rocks show some diversity in the type of mineral. The quartz (SiO_2), for example, is a mineral species, however assorted colours are found (purple, yellow, uncolour, green and so on).

Many mineral classifications can be found. One of the most utilized has grouped the mineral species known as mineral classes based on the anion or anionic radical dominant in the chemical formulation of the mineral. Examples are the carbonates, phosphates, oxides, silicates, nitrates, borates and so on. The silicates are predominant in the Earth´s crust, comprising roughly 97% of its volume (Madureira Filho et al., 2000). Many rocks from the crust are predominantly composed of silicates and called silicate rocks.

Silicate rocks can be igneous, metamorphic or sedimentary rocks. A variety of them contain minerals, such as micas and feldspars, which have in their composition important elements for plant nutrition. The mineral solubilization to produce industrial fertilizers may occur by chemical processes using an acid application, but the cost is high. The use of stonemeal has as principle the weathering of rocks with release of major species like as Ca^{2+}, Na^{+}, Mg^{2+}, K^{+} or minor species as Ni^{2+}, CO^{2+}, Cr^{3+} and others, and correction of pH (Formoso, 2006). The rocks have different compositions (Table 2), and for this reason, the effect of them in soil fertility will depend on the type of the rock to be utilized.

Many rocks are utilized as fertilizer or natural corrective, since they are appropriately managed. The choice of rock and conditions for stonemeal depends on various factors (Formoso, 2006) like the rock chemical composition, the distance between the source and the area where it will be applied and the amount of organic matter in the soil.

According to Martins et al. (2008), studies about the use of rocks as nutrient sources evidentiated an alternative to be evaluated in the search for options replacing soluble sources of potassium. The rocks may provide long-term nutrient sources for crops, promoting the increase of cation exchange capacity of the soils, due to the formation of new clay minerals during the process of fertilizer alteration (Melamed et al., 2009). Preliminary results also pointed to the effects of water retention in sandy soils (Costa et al., 2010).

The use of soil remineralization by rock powder is also a viable alternative in economic and ecological contexts because its processing is inexpensive, involving the grinding of the rock. The gradual release of nutrients decreases the loss by lixiviation and favors a long-term action of the product (Melamed et al., 2009). Moreover, some rocks that can be used as stonemeal are mining tailings and their use can diminish the environmental impacts caused by mineral exploration.

Table 2. Some types of rocks and main nutrients that can be provided (Adapted from Lapido-Loureiro and Ribeiro, 2009)

Rock	Elements
	Macronutrients
Ultrabasic rocks	Mg, Ca
Basalt / Gabbro	Mg, Ca
Carbonatite	Ca, Mg, P e K[(a)]
Syenites and nepheline syenites	K
Granites and gneiss	K
Gypsum and industrials rejects	S, Ca
Calcitic and dolomitic limestone	Ca, Mg
	Micronutrients
Ultramafic igneous rock	Cr, Co, Ni, Fe, Mn
Basalt	Cu, Zn, Mn, Fe, Co, Ni
Granite	Mo, Cu, Zn, Cl, Si[(b)]
Black schist	Cu, Zn, Ni, Mo, Fe, B
Red schist / sandstone	Mo, Cu

[(a)] Mainly the phlogopite-rich carbonatites; [(b)] Although it is not considered a micronutrient, its importance for the plant mineral nutrition is increasingly recognized.

Stonemeal has been proposed as a slow release fertilizer for highly weathered soils and leaching environments where soluble fertilizers may be easily removed (Leonardos et al., 1987; Coroneos et al., 1996; Gillman et al., 2002). Indeed, Anda et al. (2009), using various rates of finely-grained basalt (<250 μm) in an Oxisol from Indonesia with cocoa (*Theobroma cacao* L.) in a greenhouse for 15 months, observed a continuous increase of soil pH along with high application rates of basalt. A significant increase of Ca, Mg, K and Na, with concomitant suppression of toxic elements (Al and Mn), suggested that basalt is a promising material to be used for restoring the chemical properties of highly weathered soils.

Similarly, Ribeiro et al. (2010), evaluating the effect of three potassium-rich silicate rocks (alkaline ultramaphic, pyroclastic breccia and phlogopite) in a greenhouse experiment, verified that K was released in greater amounts to the soil from ultramafic alkaline and pyroclastic breccia and considered them promising for agricultural use as fertilizer .

However, some studies have pointed out little promising results. Escosteguy and Klamt (1998) have incubated during 30, 150 and 300 days a microcrystalline and an olivine-basalt in 0, 5, 10, 25, 50 e 100 t ha^{-1} rates, in two Brazilian soils (a Dark Red Latosol and a Red Yellow Podzolic soil). They observed that the release of the contents of K, Ca and Mg to the soil varied with incubation period, but not with rock type, with the exception of the concentrations of Ca and P, which were higher for the olivine-basalt. But the small release of nutrients from the rocks indicated that these materials can not be used as the major source of plant nutrients. Bolland and Baker (2000), using granite powder in sandy soils from Western Australia, have also concluded that the dust granite had no value as a fertilizer.

Most reactions involved with the mineral dissolution in soil are dependent on pH, redox (for Fe and Mn minerals), soil solution composition and temperature (Harley and Gilkes, 2000). Organic acids released by roots and rhizosphere microbiota may affect the dissolution rates of silicate minerals directly through three mechanisms. First, by affecting the dissolution rate far from equilibrium; second, by affecting the saturation state; and third, by affecting the speciation of aluminum (and possibly other elements) in the solution (Drever and Stillings, 1997). The role of organic acids in the mineral solubilization will be discussed in further section of this chapter.

Although many geochemical studies of weathering are based on the equilibrium concept, the soil environment is diverse and dynamic so that no uniform equilibrium in the condition of the soil solution exists (Harley and Gilkes, 2000). Biological activity due to plants and microorganisms strongly influences the chemical and physical conditions of the soil, where most of the biodiversity on earth exists. The rhizosphere probably represents the most dynamic habitat on earth; and certainly is the most important zone in terms of defining the quality and quantity of the terrestrial food resource for humans (Hinsinger et al., 2009).

The application of silicate rocks, as it can be concluded, restores a fraction of minerals to the soil, which in turn may be a stock of mineral nutrients. However, the full use of the minerals will only be possible with high soil microbial activity. This enables the fast and effective mineral release through the production of organic acids, chelating compounds and ions by the microorganisms (Nahas et al., 1990; Borges et al., 2007), which in general, are the main products of microbial secondary metabolism. Therefore, powdered rock fertilizers are likely to be most effective if soils and their plant/microbiological communities can provide an adequate source of acidity for dissolution and be a sink for dissolved ions (Harley and Gilkes, 2000).

3. Mineral Solubilization Processes by Microorganisms

3.1. Microorganisms and Weathering: A Geochemistry View

Soil is a natural product resulting from the weathering of rocks and minerals, and can have different properties according to the origin of the parental material, climate relief and various kinds of living and dead organisms, as Dokuchaev (1889), commonly regarded as the "father of Pedology", and soil science forerunners as Jenny (1941) and Duchaufour (1960) have already shown. Several further studies confirmed that most weathering processes involve activities of microbes, earthworms, termites and the roots of higher plants and their associated microbes by a range of physical, chemical and biological processes (Young and Crawford, 2004; Verboom and Pate, 2006a; Lambers et al., 2009), with complex interactions, even in extreme environments.

Notions that the soil biota affects soil morphology and processes were first recorded by Charles Darwin in a book published in London in 1881 (entitled *The formation of vegetable mould, through the action of worms, with observations on their habits*) describing his observations of the prodigious mixing of plant and mineral matter in soil by earthworms.

The soils probably represent some of the most complex and poorly studied habitats of terrestrial ecosystems (Decaëns, 2010). The biological communities inhabiting soils are among the most diversified, comprising a wide range of life-forms and functions (Lavelle et al., 1997; Young and Crawford, 2004), involving a large number of ecological processes and providing key ecosystem services for human populations (Lavelle et al., 2006).

The organisms living in soils play therefore an important role and life on this planet is sustained by a small volume of soil surrounding roots and influenced by living roots, called the rhizosphere (Hinsinger et al., 2009). This micro-environment converges much of earth's biodiversity, perhaps being the most dynamic habitat on earth (Jones and Hinsinger, 2008).

Microorganisms can live on different conditions. Cowan and Ah Tow (2004) showed that the presence of microorganisms in Antarctic continent indicates that life is also present in this extreme environment. Verboom and Pate (2006b), studying soil profiles and rooting morphologies from semiarid ecosystems of the southwest section of Western Australia, presented important evidence that soil formation is a process broadly linked to the geography and physiology of higher plants and their microbial associates.

Microorganisms use several elements in their metabolism. Reith et al. (2009) cited that C, H, O, N, P, S, Mg, Na, Ca, Fe, B, Cr, Co, Cu, Fe, Mn, Mo, Ni, Se, W, V and Zn are used by cell components and As, Se, Te, Sb, Hg and U are directly oxidized or reduced by prokaryotes in a variety of catabolic reactions in order to gain energy. To use an element, organisms convert it from an oxidation state to another, playing an essential role in the recycling of elements.

The capacity of microorganisms in dissolving inorganic material was observed by Muntz in 1890, with the start of studies on the microbial action in the decomposition and weathering of rocks and on the biogeochemical cycles of the elements. After the concept of rhizosphere (a zone of high biological activity around the roots) was launched by Hiltner in 1904, Sackett established the function of microorganisms in the phosphorus solubilization in soil four years later (Moreira and Siqueira, 2006a). From these studies, many others concerning microbial action on mineral dissolution were carried out with promising results alternatively to soluble fertilizers.

Several microorganisms, as fungi, including mycorrizhae, bacteria, archaea and algae, are involved on the breakdown of rocks. Microorganisms control the natural cycling of C, S, P, and in special the cycling of N, because it is not available from silicate weathering. In the soil environment, microorganisms are responsible, directly or indirectly, for biochemical processes, transforming minerals and rocks and transferring nutrients and energy. Some of these biochemical processes are summarized by Moreira and Siqueira (2006b) as organic matter decomposition, organic compound mineralization, inorganic N- and S- transformations (ammonification, nitrification, denitrification, S-oxydation and the reduction of SO_4^{-2}), production of various metabolites such as hormones, antibiotics, siderophores and organic acids and degradation of xenobiotics.

The transformation of ground rock in soils involves two basic processes: the physical and mechanical disaggregation without chemical modification of minerals, and chemical transformations by modifying minerals and promoting the ion release into the soil solution. The principal agent for these basic processes is water. Reactions such as hydrolysis, dissolution, complexing reactions, hydration and oxidation are important for mineral transformations. The hydrolysis is considered the most important rock alteration processes (Duchaufour, 1977) and the aqueous solution attack is usually loaded with ion H^+, from the alterations of silicates and dissociated organic acids.

The minerals have different degrees of resistance to weathering, in general associated with the crystallization order: the higher the temperature of crystallization, the less stable is the mineral, offering a lower resistance to weathering. Indeed quartz, for example, as the last mineral to crystallize at temperatures around 500°C, is the most resistant mineral, while the olivine, which crystallizes at about 1400°C, is considered the most susceptible to alteration (Toledo et al., 2000). Among the micas, the biotite, rich in iron, alters rapidly by losing iron and potassium, while muscovite, rich in potassium, is very resistant. Among the feldspars, the lower the silica content, the faster the alteration process (Duchaufour, 1977).

Based on the knowledge of mineralogical and geochemical processes, some agronomists have confirmed the potential of silicate rock powder as fertilizers and/or soil conditioners. However the low solubility of this rock powder is the limiting factor. Rhizosphere processes and other biological activity may further enhance mineral dissolution. The physico-chemical conditions that predominate in the rhizosphere are very important to understand the role that the microorganisms play on soil nutrient availability. The oxygen concentration in this zone is very low due to the high oxygen demand for the breakdown of carbonaceous compounds and the highly dense microbial populations. Consequently, the CO_2 concentration is high, creating an anaerobic environment, which favor reduction reactions (Vega, 2007).

Additionally, the rhizosphere pH is generally lower than the bulk soil in 1-2 units and the main mechanisms which contribute to this effect are the production of CO_2 by respiration processes; pump of H^+ in nutrient uptake by plant and microbes; organic matter decomposition; N_2 fixation by the symbiotic bacteria; and the release of organic acids by roots and microbes (Marschner, 1995). For some authors, the production of organic acids is the major mechanism involved in the solubilization of rock phosphate, although other mechanisms might be involved (Kim et al., 1997).

3.2. Role of the Organic Acids and other Metabolites

Ground rock has been proposed as fertilizer for soil fertilization (Gillman, 1980; Hinsinger et al., 1996; Leonardos et al., 2000) although the nutrient release can be very slow and incompatible with crop demand. Thus, considering the role of microorganisms on rocks and sediment weathering and the multiple functions of soil biota, it is possible to accelerate available nutrient processes through the usage of soil microorganisms.

Many microorganisms found in soil are able to solubilize ''unavailable'' forms of silicate minerals, such as micas and feldspars, by excreting organic acids, which either directly dissolve rock potassium or chelate silicon ions to bring the K and Si into solution (Groudev, 1987; Friedrich et al., 1991; Sheng et al., 2008). The capacity of roots to reduce the concentration of ions in the soil solution to very low levels and organic acids produced by soil organisms may accelerate the rock weathering, increasing the efficiency of rocks applied in powder form to provide nutrients to plants (Harley and Gilkes, 2000).

Other metabolites involved in the phosphate rock solubilization are phosphatase enzymes, nitric and sulphuric acid produced by bacteria, carbonic acid formed as the result of plant decomposition (Vega, 2007) and siderophores, low molecular weight compounds synthesized under iron-deficient conditions (Ghaderi et al., 2008).

In vitro studies have shown that P-solubilizing microorganisms can simultaneously synthesize and release pathogen-suppressing metabolites, as siderophores, phytohormones and lytic enzymes (Vassilev et al., 2006), which can also help in the solubilization process.

Gluconic, oxalic, citric, lactic, tartaric and aspartic acids are some of the organic acids produced by microbial metabolism, mostly by oxidative respiration or by fermentation of organic carbon sources (Kim et al., 1997; Di Simine et al., 1998; Madigan et al., 2004). These low-molecular-weight organic acids in the soil solution are also derived from plant roots, dead plant material and atmospheric inputs (Smith, 1976; Curl and Truelove, 1986; Ryan et al., 2001).

Bolan et al. (1994), using a range of organic acids (acetic, formic, lactic, malic, tartaric, oxalic, and citric) in soil pre-incubated with monocalcium phosphate and North Carolina phosphate rock, verified that the agronomic effectiveness of both fertilizers increased in the presence of organic acids. The increase was greater with phosphate rock than with the monocalcium phosphate. They also showed that the adsorption of P by soils follows the order tricarboxylic acid > dicarboxylic acid > monocarboxylic acid.

The release of K mineralization occurs mainly by the action of organic and inorganic acids produced by biological activity (Wallander and Wickman, 1999; Harley and Gilkes, 2000; Yuan et al., 2004), by the release of H^+ or as complexants. On the other hand, the removal of soluble K by microbial assimilation promotes the release of mineral K by increasing the concentration gradient during the hydrolysis reaction of minerals (Harley and Gilkes, 2000).

Phosphate-solubilizing soil and rhizosphere microorganisms have been distinguished by their relative abilities to dissolve phosphate minerals in pure culture (Sperber, 1958; Gaur et al., 1973; Nahas et al., 1990) and in association with plant roots (Perez et al., 2007). This activity was frequently attributed to organic acids and chelating metabolites (Agnihotri, 1970; Asea et al., 1988; Nahas et al., 1990; Jones, 1998; Ryan et al., 2001).

However, Hinsinger et al. (2003) emphasized that a major process that contributes to the root-induced pH changes in the rhizosphere is the release of charges carried by H^+ or OH^- to compensate for an unbalanced cation–anion uptake at the soil–root interface. The authors considered that, in addition to the ions taken up by the plant, all the ions crossing the plasma membrane of root cells (e.g., organic anions exuded by plant roots) should be taken into account, since they all need to be balanced by an exchange of charges, i.e., a release of either H^+ or OH^-. The root exudation and respiration can contribute to some extent to the rhizosphere pH decrease as the result of a build-up of the CO_2 concentration, forming carbonic acid in the rhizosphere that may dissociate in neutral to alkaline soils. Ultimately, plant roots and associated microorganisms can also alter rhizosphere pH via redox-coupled reactions and soil pH itself; pH buffering capacity also have a dramatic influence on root-mediated pH changes (Hinsinger et al., 2003).

Indeed, Illmer and Schinner (1992) have concluded that the production of organic acids is not the main mechanism involved in phosphate solubilization, because they only find three solubilizing acids produced by bacteria among 24 tested. The authors considered the possibility of the acidification of the environment through the release of H^+ protons in a process of cation exchange or from the respiratory chain. Marschner (1995) had pointed out that the acidification should be a consequence of respiration and decomposition of organic matter due to the presence of H_2CO_3.

Besides roots, many soil microbes can produce organic acids and thus contribute to rhizosphere acidification as documented for ectomycorrhizal and saprophytic or pathogenic fungi massively producing oxalic acid (Dutton and Evans, 1996; Wallander, 2000; Casarin et al., 2004). Hinsinger et al. (2009) also reported large changes in pH as a consequence of the microbial-mediated oxidation of nitrogen or sulphur.

The organic acids can decrease or can act as a complexing or a chelating agent, forming compounds with calcium, iron and aluminum (Sperber, 1958; Whitelaw, 2000). Vega (2007) explained that when P-solubilizing rhizosphere bacteria are inoculated in neutral and alkaline soils, the acid production decreases the rhizosphere pH, favoring the solubility of calcium phosphates and apatites. In acid soils, the organic acids propitiate the formation of complexes with Fe and Al ions, which in turn facilitates the dissolution of these minerals. Organic acids can also compete with phosphates for fixation sites on the surface of soil colloids, as demonstrated by He and Zhu (1998).

Many works support that organic acids have a role in nutrient acquisition and plant growth by mobilization of scarcely soluble nutrients (e.g. P and K) and detoxification of potentially harmful elements like Al (Hue et al., 1986; Jones and Darrah, 1994; Marschner, 1995; Jones, 1998; Gerke et al., 2000). They are also assumed to be involved in soil pedogenesis, pathogen attraction and alleviation of anaerobic stress in plant roots (Curl and Truelove, 1986; Jones, 1998; Lundstrom et al., 2000).

In a interesting study about the mechanism that controls the dissolution rate of kaolinite in the presence of oxalate under far-from equilibrium conditions, Cama and Ganor (2006) considered that the focus of discussions about the effects of organic acids on minerals weathering has been on three major questions: 1) Is the dissolution rate faster in the presence of organic acid than in the presence of non-organic acid under the same pH? 2) To what extent do organic acids affect mineral dissolution rates under natural conditions? 3) What is the mechanism responsible for the enhancement of dissolution rate in the presence of organic acids? The authors verified that the kaolinite dissolution rate was catalyzed by oxalic acid and they affirmed that for the first question the answer is positive for oxalic acid, but the other two questions are still the subject of scientific debates.

Seshadri et al. (2004), studying the mineral phosphate solubilization by ten *Aspergillus niger* strains using different phosphate sources (tricalcium, dicalcium, ferric and aluminum phosphates and bone meal), showed that all the *A. niger* strains possess a mineral phosphate solubilizing potential but the results vary according to the phosphate source used.

Magdziak et al. (2011), studying the effect of two different Ca/Mg ratios (4:1 and 1:10) and trace element ions (Cd, Cu, Pb and Zn) in solution on the efficiency of low-molecular-weight organic acid formation in *Salix viminalis* rhizosphere, concluded that the organic acids can influence the mobility of metal ions in the rhizosphere.

Akter and Akaji (2005) conducted experiments in a controlled glasshouse during a 42 day-period using basalt rock particles with 1.00 – 2.00 μm in size. They observed the positive effect of soybean on the release of Si, Ca, Mg, Mn and Al from the rock particles. Thus, they concluded that weathering may be caused partially by the absorption of nutrient elements directly through the interface of fine roots and rock particles, and it is most likely associated with alterations of the local rhizosphere conditions surrounding the roots.

Sandnes et al. (2005) studied low-molecular-weight organic acids in root exudates and soil solutions of Norway spruce and Silver Birch grown in rhizoboxes, sterile microcosms, and in the field. The rhizosphere of birch was found to contain more organic acids at higher concentrations than spruce. Growth conditions were important for the composition and the amount of organic acids.

Sheng et al. (2008) isolated a silicate mineral-solubilizing bacterial strain Q12 from the surfaces of weathered feldspar inoculated with three silicate minerals (feldspar, muscovite, and biotite) to investigate potassium and silicon mobilization. They showed that the strain grows better on biotite than on feldspar and muscovite, because the former was the best potassium source for the growth of that strain. The solubilization of potassium and silicon from the silicate minerals by mostly organic acids such as gluconic and acetic acids, results in the solubilization of feldspar. The gluconic acid seemed to be the most active agent for the solubilization of the three silicate minerals studied.

Gluconic acid was also released by a strain of *Stenotrophomonas maltophilia*, isolated from phosphate mines, resulting in the solubilization of tricalcium phosphate (Xiao et al., 2009b), and by *Pseudomonas fluorescens* from zinc phosphate (Di Simine et al., 1998).

Ghaderi et al. (2008) verified that *Pseudomonas* strains with significantly higher siderophore production and phosphatase activity had the least effect on P release from iron hydroxide, while two other isolates with higher H^+ production were the most efficient solubilizers. Nevertheless, using the fungi *Penicillium expansum, Mucor ramosissimus* and *Candida krissii*, the alkaline phosphatases in the culture medium were also helpful for rock phosphate solubilization, along with citric, oxalic and gluconic acids also present in the culture media (Xiao et al., 2009a).

3.3. Microbial Action in Different Minerals and Rocks

Bacteria, actinomycetes, yeasts and molds have been used in mineral solubilization with promising results (Al-Falih, 2005; Sheng, 2005; Sugumaran and Janarthanam, 2007; Hamdali et al., 2008a, b; Lian et al., 2008; Xiao et al., 2008; Hamdali et al., 2010). Microorganisms are able to affect the chemical environment of the minerals, accelerating the soil weathering by the inorganic and organic acids they produce, CO_2 generation, introduction of organic chelates and reactions of ionic change, as seen in the sections above. The usage of insoluble fertilizers, such as rock powders, which are less expensive and sustainable sources for soil nutrition, is then viable due to the microbial biosolubilization. A list of contributions to this subject in the last two decades is presented in Table 3.

The solubilization ability depends on several factors such as the nature of the nitrogen and carbon sources and the type of mineral or rock. With insoluble phosphates, for example, the microorganisms are better able to solubilize calcium phosphate than iron phosphate or aluminum (Silva Filho and Vidor, 2000; Gyaneshwar et al., 2002).

Xiao et al. (2008) have optimized the cultural conditions for rock phosphate solubilization by using two molds and a yeast and verified that the content of P gradually decreased with the enlargement of the particle size with all the isolates; the optimum initial pH was lower (5.5) for the yeast and near neutrality (7.0-7.5) for the molds; and different C:N ratios (20:1, 30:1 and 40:1, for *Candida krissii*, *Mucor ramosissimus* and *Penicillium expansum*, respectively) to achieve the highest P solubilization. Other physical factors like temperature and oxygen (Lopes-Assad et al., 2010) can significantly influence the amount of soluble nutrients provided by the microorganisms.

Maltose as carbon source and ammonium nitrate as nitrogen source rendered maximal P solubilization by using the bacterium *Stenotrophora maltophilia* (Xiao et al., 2009b). The solubilization of zinc phosphate by *Pseudomonas fluorescens* was only possible when glucose was provided as carbon source (Di Simine et al., 1998).

In further sections, the microbial action on different groups of rocks or minerals will be discussed in view of soil fertility and also in relation to vegetal growth promotion, through the facilitated availability of nutrients for plant absorption.

3.3.1. Phosphate and Phosphatic Rocks

Phosphorus is the tenth most common element in nature. Currently, there are 607 known minerals with phosphorus in their chemical composition (Webmineral, 2011), of which only the group of apatite has economic importance (Lapido-Loureiro and Melamed, 2009). Apatite is a phosphate mineral that can occur in igneous rocks, mainly carbonatites, and sedimentary rocks. In igneous rocks, the most common varieties of apatite are the fluorapatite and hydroxyapatite, and in sedimentary rocks, carbonate-apatite and carbonate-fluorapatite. Many apatites are rich in rare earth elements and they generally have low solubility (Lapido-Loureiro and Melamed, 2009).

Phosphorus deficiencies are limiting crop production in agricultural soils worldwide. Available sources of local raw phosphate rock (PR) are being recognized for their use in soil fertility improvement. The bioavailability of phosphorus is essential for the enhanced efficiency of PRs and can be increased by acid treatments. The utilization of microbes, which produce organic acid, notably *Aspergillus niger*, is a sustainable method instead of using strong inorganic acids, but the fungal acid production is determined by the mineral content of the growth media.

Schneider et al. (2010) compared the phosphorus mobilization efficiency of *A. niger* cultivations with three phosphate rock (one highly reactive sedimentary phosphate rock from North Carolina, USA, and two less reactive igneous phosphate rocks from Catalão, Brazil, and Kapuskasing, Canada). The authors observed that citric and oxalic acid contributed to the phosphate rock solubilization and would benefit farmers in developing countries where conventional fertilizers are not adequately accessible.

Table 3. A list of contributions to the understanding of microbial solubilization of minerals and rocks in the last two decades

Microorganism	Type of mineral/rock	Aims	Reference
Candida tropicalis, Geotrichum capitatum, Geotrichum candidum, Rhodotorula minuta and *Rhodotorula rubra*	Calcium phosphate	The role of selected soil yeasts on the solubilization of insoluble P	Al-Falih (2005)
Glomus mosseae and yeasts	Phosphates	Interaction between mycorrhizal fungi and yeast for P solubilization	Alonso et al. (2008)
Penicillium	Inorganic phosphate	P solubilization in solution culture and soil	Asea et al. (1998)
P-solubilizing microorganisms	Tilemsi phosphate	Effect of rock-solubilizing microorganisms on P uptake and yield of wheat	Babana and Antoun (2006)
Bacteria	K-feldspar	Effect of silicate-dissolving bacteria on tomato yield	Badr (2006)
Bacteria	Biotite and phlogopite	Biotic dissolution kinetics and processes under aerobic conditions	Balland et al. (2010)
Suillus tomentosus	Biotite	Hyphal attachment to biotite surfaces and nutrient release	Balogh-Brunstad et al. (2008)
Pseudomonas cepacia	Kaolinite and montmorillonite	P solubilization mediated by *P. cepacia*	Bar-Yosef et al. (1999)
Bacillus mucilaginosus	Waste mica	Potassium release by K-solubilizing microorganism	Basak and Biswas (2009)
Bacillus mucilaginosus and *Azotobacter chroococcum*	Waste mica	Growth promotion and nutrient uptake by *Sorghum vulgare* Pers.	Basak and Biswas (2010)
Serratia marcescens	Phosphates	Phosphate solubilization	Ben Farhat et al. (2009)
Glomus mosseae	Biotite	Role of endomycorrhizae on biotite weathering, plant growth and K uptake	Berthelin and Leyval (1982)
Thiobacillus ferrooxidans	Phlogopite	Effects of microbe on weathering	Bigham et al. (2001)
Aspergillus awamori	Phosphate rock and waste mica	Changes in nutrient contents	Biswas et al. (2009)
Aspergillus niger	Phosphorite	Effect on plant growth (barley)	Bojinova et al. (1999)
Community of cyanobacteria, diatoms and heterotrophic bacteria	Different types of silica	Effect of microorganisms and biofilms on biologically induced breakdown	Brehm at al. (2005)
Pseudomonas fluorescens	Inorganic phosphate	Relation between bacterial phylogeny and P solubilization	Browne et al. (2009)
Fungi	Rocks in general	The role of fungi in geomicrobiological processes	Burford et al. (2003)
Ectomycorrhizal fungi	Hydroxyapatite and calcium carbonate	Effect of the ectomycorrhizal fungi on the P bioavailability and mobilization	Casarin et al. (2004)
Aspergillus niger	Phosphate	Effect on phosphate solubilization	Chuang et al. (2007)
Pseudomonas fluorescens	Zinc phosphate	Characterization of a P-solubilizing bacterium isolated from a forest soil	Di Simine et al. (1998)
Microbial communities on termite mound	Phosphate rock	Effect on *Acacia seyal* growth	Duponnois et al. (2005)
Bacillus	Phosphate	Effect of P solubilization on sunflower (*Helianthus annuus*) growth	Ekin (2010)
Pseudomonas fluorescens and others isolates from oil palm tree *(Elaeis guineensis)* rhizosphere	Phosphate	Evaluation of solubilization process	Fankem et al. (2006)

Microorganism	Type of mineral/rock	Aims	Reference
Pseudomonas	Phosphate from variable charge mineral	Ability of three bacterial strains to release P from iron (III) hydroxide	Ghaderi et al. (2008)
P-solubilizing microorganisms	Phosphate	Critical analysis of the role of P-solubilizing microorganisms on phosphate solubilization	Gyaneshwar et al. (2002)
Micromonospora aurantiaca and *Streptomyces griseus-related strains*	Rock phosphate	Solubilization tests and plant growth promotion of wheat	Hamdali et al. (2008a)
Actinomycetes	Rock phosphate	Screening for plant growth-promoting activities by P-solubilizing microorganisms	Hamdali et al. (2008b)
Actinomycetales, Streptomyces lividans and *Streptomyces griseus*	Phosphate rock	Physiological studies and comparative analysis of P solubilization abilities of Actinomycetales	Hamdali et al. (2010)
P- and K-solubilizing bacteria	Rocks in general	Effect of P and K solubilization on eggplant	Han and Lee (2005)
P- and K-solubilizing bacteria	Rocks in general	Effect of P and K solubilization on pepper and cucumber	Han et al. (2006)
Bacillus subtilis, Shewanella putrefaciens, Streptomyces acisdiscabies and *Schizophyllum commune*	Biotite	Effects on the release of Al, Fe, K, Mg, Mn, Si, and Ti	Hopf et al. (2009)
P- and K-solubilizing bacteria strains	Phosphate rock	Dissolved mineral phosphate and potassium	Hu et al. (2006)
Arbuscular mycorrhizal fungi and organic phosphorus mineralizing bacteria	Phosphate rock	Long-term (18-year) fertilizer experiment	Hu et al. (2009)
Microorganisms from forest soils	Inorganic phosphates	Solubilization tests	Illmer and Schinner (1992)
Fungi	Rock phosphate	Evaluation of fungal-rhizobial biofilms in P solubilization	Jayasinghearachchi and Seneviratne (2006)
Achromobacter xylosoxidans	Insoluble phosphate	Evaluation of P-solubilizing and plant growth promotion potential	Jha and Kumar (2009)
Trichoderma spp.	Tricalcium phosphate	Evaluation of P-solubilizing potential	Kapri and Tewari (2010)
Enterobacter agglomerans and cloned *Escherichia coli*	Hydroxyapatite	The solubilization of phosphate by a known P-solubilizing and a genetically modified bacteria	Kim et al. (1997)
Azotobacter chroococcum	Rock phosphate	Effect of phosphate solubilization by analogue-resistant mutants of bacteria on sorghum	Kumar et al. (1999)
Azotobacter chroococcum	Fertilizer	Establishment of P-solubilizing strains in rhizosphere and plant growth promotion in wheat	Kumar et al. (2001)
Rhizospheric microorganisms	Assorted minerals	A review on the existing literature about interactions among plants, microorganisms and soils and the role of rhizosphere in the weathering of rocks and minerals	Lambers et al. (2009)
Laccaria laccata, Agrobacterium radiobacter and beech roots	Assorted minerals	Microbial and plant interaction for mineral mobilization	Leyval and Berthelin (1989)

Table 3 (Continued)

Microorganism	Type of mineral/rock	Aims	Reference
Rhizospheric microorganisms	Mica	Influence of symbiotic and non-symbiotic microorganisms on rhizospheric weathering of mica	Leyval and Berthelin (1991)
Aspergillus fumigatus	Potassic minerals	K solubilization by a thermophilic fungus	Lian et al. (2008)
Aspergillus niger	Alkaline ultramafic	Small-scale batch fermentations for K release	Lopes-Assad et al. (2010)
Aspergillus niger	Phosphates	Phosphate solubilization with fungus inoculated in vinasse	Nahas and Assis (1992)
Aspergillus niger	Fluoroapatite	P solubilization in vinasse medium	Nahas et al. (1990)
Yeasts	Rock phosphate	Characterization of the solubilizing capacity of 25 yeast isolates	Narsian et al. (2010)
Pisolithus sp. and *Pseudomonas sp.*	Rock phosphate	Influence of *Pisolithus* fungi on rock phosphate solubilization by *Pseudomonas*	Ouahmane et al. (2009)
Burkholderia, Serratia, Ralstonia and Pantoea	Tricalcium phosphate	Evaluation of P solubilization by bacteria	Perez et al. (2007)
Agricultural soil fungi	Inorganic phosphates	Evaluation of P solubilization by soil fungi	Pradhan and Shukla (2005)
Bacteria, *Aspergillus flavus, A. niger* and *Penicillium canescens*	Phosphates	Phosphate solubilization by organic acids	Rashid et al. (2004)
Aspergillus niger and *A. tubingensis*	Rock phosphates	Influence of fungi isolated from rhizosphere on solubilization of rock phosphate	Reddy et al. (2002)
Soil microorganisms	Phosphates	A review of prospects for using soil microorganisms to improve P acquisition by plants	Richardson (2001)
Torulaspora globosa	Alkaline ultramafic	K solubilization by a yeast	Rosa et al. (2009)
Trichoderma spp	Tricalcium phosphate	Solubilization tests with nine fungal isolates compared to *Bacillus megaterium* and the influence on yield and growth of chickpea	Rudresh et al. (2005)
Aspergillus niger	Rock phosphate	Effect of organic acids from *A. niger* on rock phosphate solubilization	Schneider et al. (2010)
Aspergillus niger	Tricalcium, dicalcium, ferric and aluminum phosphates and bone meal	Solubilization of phosphate by ten fungal strains using different P sources	Seshadri et al. (2004)
Bacillus edaphicus	K minerals	Effect on K uptake and growth promotion of cotton and rape	Sheng (2005)
Bacillus edaphicus	Feldspar and illite	K-mobilization by *B. edaphicus* and mutants	Sheng and He (2006)
Bacillus globisporus	Muscovite, feldspar and biotite	Investigation of K and silicon mobilization by a strain isolated from the surfaces of weathered feldspar	Sheng et al. (2008)
Bacillus, Pseudomonas, Enterobacteriaceae, Penicillium, Aspergillus and *Paecilomyces*	Phosphates (Ca, Al or Fe)	Phosphate solubilization	Silva Filho and Vidor (2000)
Fungus and bacteria	Phosphates	Phosphate solubilization	Silva Filho et al. (2002)
Aspergillus and *Enterobacteriaceae*	Aluminum phosphate and apatite	Phosphate solubilization	Souchie et al. (2005)
Aspergillus and *Penicillium*	Rock phosphate	Influence of environmental factors and salinity on P solubilization	Srividya et al. (2009)

Microorganism	Type of mineral/rock	Aims	Reference
Acidithiobacillus	Phosphate and potash rocks	Chemical attributes of Argisoil after biofertilizer application	Stamford et al. (2009)
Bacillus cereus	Micas in kaolin and quartz sands	Extraction of Al and Fe from micas	Styriakova et al. (2003)
Bacillus cereus and *Acidithiobacillus ferrooxidans*	Phlogopite	Weathering of phlogopite	Styriakova et al. (2004)
Bacteria	Microcline, orthoclase and muscovite mica	K solubilization and effect on plant growth	Sugumaran and Janarthanan (2007)
P-solubilizing bacteria	Tricalcium phosphate	Potentiality of P-solubilizing bacteria as plant growth promoters in peanut	Taurian et al. (2010)
P-solubilizing organisms	Rock phosphate	An overview of biotechnological techniques, mainly based on solubilization processes involving agroindustrial residues	Vassilev and Vassileva (2003)
P-solubilizing microorganisms	Rock phosphate	Potentiality of P-solubilizing microorganisms for the simultaneous synthesis and release of pathogen-supressing metabolites	Vassilev et al. (2006)
Yarrowia lipolytica	Phosphate rock and calcium carbonate	Effect of the mode of biocatalyst application on the solubilization	Vassileva et al. (2000)
P-solubilizing microorganisms	Insoluble phosphates	A review dealing with the wide range of metabolic abilities of P-solubilizing microorganisms in different environments	Vassileva et al. (2010)
Rhizosphere bacteria and mycorrhizal fungi	Phosphates	A review on the benefits of rhizosphere bacteria and mycorrhizal fungi on plant nutrition	Vega (2007)
Torulaspora globosa	Tricalcium phosphate	Evaluation of a P-solubilizing yeast	Vora and Shelat (1998)
Penicillium	Phosphate	P solubilization associated with wheat roots	Wakelin et al. (2004)
Mycorrhizal fungi	Biotite and microcline	Effect of K-rocks in mycorrhizal and non-mycorrhizal *Pinus* seedlings	Wallander and Wickman (1999)
Aspergillus niger and *Penicillium oxalicum*	Rock phosphate	Effect of lanthanum on solubilization	Wang et al. (2004)
Aspergillus niger and *Penicillium oxalicum*	Rock phosphate	The dynamic solubilization of two concentrations of rock phosphate in a liquid culture with different dosages of glucose	Wang et al. (2005)
Burkholderia cepacia	Phosphate	The effectiveness of the film bags with the bacterium to improve the absorbability of a fertilizer for plants and to promote the plant growth	Wu (2010)
Candida krissii, Penicillium expansum and *Mucor ramosissimus*	Rock phosphate	The optimization of solubilization by fungi isolated from phosphate mines	Xiao et al. (2008)
Candida krissii, Penicillium expansum and *Mucor ramosissimus*	Rock phosphate	Characterization of three P-solubilizing fungi and their effect on wheat seedling growth	Xiao et al. (2009a)
Stenotrophomonas maltophilia	Rock phosphate	Characterization of a P-solubilizing strain isolated from phosphate mines	Xiao et al. (2009b)
Cenococcum geophilum and two isolates of *Pisolithus*	Vermiculite and phlogopite	Evaluation of the growth, proton and oxalate efflux, K absorption and mineral depletion	Yuan et al. (2004)

Besides *Aspergillus,* the genus *Penicillium* and bacteria such as *Bacillus*, *Pseudomonas* and *Enterobacter* are effective P-solubilizing microorganisms (Whitelaw, 2000; Vassilev and Vassileva, 2003).

Wakelin et al. (2004) verified that the phosphate solubilization by *Penicillium* spp. was shown to occur both internally as well as externally to the roots of wheat plants cultivated in soil in Southern Australia. Among the 47 isolates evaluated for the solubilization of phosphate rock from Idaho in a liquid culture medium, the species *Penicillium bilaiae* (RS7B-SD1) was the most effective, mobilizing 101.7 mg/L of soluble P after 7 days. Other isolates with promising results included the species *Penicillium simplicissimum* (58.8 mg/L of soluble P), and five strains of *Penicillium griseofulvum* (varying from 47.6 to 56.1 mg/L of soluble P). Isolates of *Penicillium* sp. and *Aspergillus* sp from agricultural soils have solubilized a higher amount of soluble P (275 μg/mL and 480 μg/mL of soluble P) after three days of incubation (Pradhan and Shukla, 2005).

Jayasinghearachchi and Seneviratne (2006) have evaluated the biofilm formation by the fungi *Penicillium* spp., *Aspergillus* spp. and *Pleurotus ostreatus* associated with the nitrogen-fixing bacteria *Bradyrhizobium elkanii* in the solubilization of phosphate rock. The authors have verified a significantly higher release of soluble P when the association was used instead of the pure fungal cultures. The nitrogen provided by *B. elkanii* to the fungal nutrition and consequently resulting in the production of organic acid explains this beneficial effect of co-inoculation.

In the context of multifunctional activities of the microorganisms, the co-inoculation of different microbial groups is also an effective technique of solubilization and plant growth promotion. Babana and Antoun (2006) have evaluated the inoculation of solubilizing microorganisms isolated from the Mali soils, a commercial isolate of the mycorrhizal fungi *Glomus intraradices* in wheat, in single or mixed cultures with *Aspergillus awamori* (C1), *Penicillium chrysogenum* (C13) and an isolate of *Pseudomonas* sp. (BR2), in the presence or absence of the mycorrhizal fungus (Gi). In the 45-day wheat plants, the higher extension of root colonization by Gi (62%) was observed in the wheat fertilized with phosphate rock powder and inoculated with Gi and BR2. The inoculation of the wheat with the consortium Gi, BR2 and C1 in the treatments with fertilization promoted an increase in the grain production (1.71 mg/g of dry mass compared to 1.19 mg/g in the non-inoculated control). In the treatments without fertilization or inoculation, the fungi Gi colonized only 5.5% of the root extension in wheat, 45 days after planting.

Hamdali et al. (2008a) have found satisfactory results in the P released by the actinobacteria *Micromonospora aurantiaca* and *Streptomyces griseus*, isolated from the extraction mines in Morocco, from rock powders, as well as a stimulatory effect in wheat growth (an increase of 50% and 80% in the weight of aerial and root parts, respectively). An inhibition of 14% in the growth of *Pythium ultimum*, the causative agent of damping off, was also observed with these species of actinobacteria.

A strain of *Pseudomonas cepacia* was very effective to produce gluconic and 2-ketogluconic acids from glucose in a bioreactor containing rock phosphates (Bar-Yosef et al., 1999). The protons resulted from the dissociation of acids reacted with the rock phosphate, releasing phosphate ions which were then precipitated with Ca to form soluble fertilizers (super phosphates).

Taurian et al. (2010) selected 37 epiphytic isolates and 73 endophytic isolates from a collection of 133 bacteria isolated from rhizosphere, phyllosphere and plant tissues from peanuts in order to evaluate the phosphate-solubilizing activity. These isolates were further tested for other plant growth-promoting attributes. The inoculation effects of some of the isolates were evaluated on peanut growth. A high number of isolates were able to produce siderophore, especially the endophytes of the root nodules. Antibiosis was evaluated against the phytopathogen fungus *Sclerotinia minor* and *S. sclerotiorum*. The inhibitory effects of endophytes from nodules were the highest on fungal growth, while the minority of isolates was able to produce auxin-like molecules. The enhancing effects of these bacteria on the seed germination of peanut seedlings was variable with the isolate J49 from the genus *Pantoea*, being the most promising one because of the production of auxin-like molecules in peanut in inoculation experiments. The authors concluded that peanut soils in the province of Cordoba contain bacteria with important plant growth promoting properties, which represent a potential source of new strains that could be used as biological inoculants in agriculture.

Yeasts have been successfully utilized as phosphate solubilizing-organisms, especially the species *Candida tropicalis*, *Geotrichum capitatum*, *Geotrichum candidum*, *Rhodotorula minuta* and *Rhodotorula rubra* in calcium phosphate (Al-Falih, 2005). The author have observed that the release of soluble phosphate occurred along with a significant pH decrease, with the species *G. capitatum* showing the best result (45 µg/mL of soluble P).

Alonso et al. (2008) isolated yeasts from the genera *Rhodotorula, Rhodosporidium* and *Cryptococcus* from the spore surface of *Glomus mosseae*, a mycorrhizal fungus, and observed that these yeasts were able to solubilize phosphate *in vitro*. This fact indicated an association between the solubilizing yeasts and the mycorrhizal fungi capable of helping plant with nutrient absorption.

Among 25 yeast isolates, Narsian et al. (2010) verified that 6 yeasts belonging to the genera *Saccharomyces, Hansenula, Kloeckera, Rhodotorula* and *Debaryomyces* exhibited the highest solubilization results from tricalcium phosphate and low-grade rock phosphates. Although acidic pH of the spent media was observed in all cases, the authors could not establish a correlation between final acidity and the amount of phosphate released.

The application of P-solubilizing organisms is undoubtedly a successful practice to increase P uptake and plant growth. However, very little information is available concerning the ecological consequences of the inoculation of P-solubilizing microorganisms. Wang et al. (2007) considered that complex interactions between plant species and exotic P-solubilizing organisms affected the structure of the bacterial community in the rhizosphere, but the plant species was more important in determining the soil microbial structure than the exotic microorganisms introduced. The success, therefore, depends on the interaction plant-microbe and should be analyzed in a case-by-case basis.

3.3.2. Micas and Feldspars

Micas are a group of silicate minerals that inhabit plates or sheets (phyllosilicate minerals). They consist of hydrated aluminum silicate with cations such as Mg^{2+}, Fe^{3+}, K^{+}, Li^{+}; the main mineral species are biotite, muscovite and phlogopite. Their most notable characteristic is the highly defined cleavage. At lower levels, mica is a very common constituent of crustal rocks and may occur in igneous, metamorphic and sedimentary rocks. Phlogopite is iron-poor, but biotite and muscovite are iron-rich micas. Illite is a hydrate quite similar to muscovite, with slightly more silicon, magnesium, iron, water and less potassium. In general, micas have a low solubility. A great number of studies have demonstrated that microorganisms can accelerate elemental release from silicate minerals in aerobic conditions (Balland et al., 2010).

Feldspars are a very important family of silicate minerals. Along with quartz, they are the most common constituents of granitic rocks and granitites, forming a large part of the continental crust. In addition to the high frequency of this group, they are widely distributed, being absent only in some ultramafic rocks and rare alkaline rocks (Madureira Filho et al., 2000).

Berthelin and Leyval (1982) have cultivated maize under axenic conditions in laboratory devices inoculated with symbiotic (*Glomus mosseae*) and/or non symbiotic microflora in a K^+-deficient medium, with biotite as the K^+ source. They observed that *G. mosseae* stimulated growth and K uptake; non-symbiotic microflora also increased plant growth but promoted much more biotite weathering and K uptake. The endomycorrhizae, and more particularly the non-symbiotic microflora, also increased Ca and Mg absorption by plants.

Leyval and Berthelin (1989) have inoculated *Fagus silvatica* L. with the ectomycorrhizal fungus *Laccaria laccata* and a phosphate-solubilizing rhizobacteria *Agrobacterium radiobacter* along with rock phosphate and phlogopite as the only sources of P, Fe, Mg and Al during two years. They observed that beech dry matter and P, Mg, Fe and K uptake were higher for plants inoculated either with *A. radiobacter* or with *L. laccata*; however, those elements were not higher for the plants inoculated with both microorganisms. They also verified that the mineral element mobilization from rock phosphate and from phlogopite was also higher when plants were inoculated with the bacteria or with ectomycorrhizal fungi and concluded that the absence of significant effects of dual inoculation suggested competition between microorganisms.

Wallander and Wickman (1999) reported the use of K from biotite and microcline by ectomycorrhizae in *Pinus*. Yuan et al. (2004) reported the use of vermiculite and phlogopite by ectomycorrhizal fungi with *Eucalyptus*.

Sheng (2005) has evaluated the action of *Bacillus edaphicus* in the K solubilization from illite and in the plant growth promotion with cotton plants. The author observed an increase of 30% in the K immobilized by the plant in treatments with bacterial inoculation along with a significant increase in the N and P plant contents. The plant growth was 24% higher compared to the uninoculated plants, indicating the effect of plant growth promotion by the microorganism.

Badr (2006) utilized the silicate-solubilizing bacterium *Bacillus cereus* as a biofertilizer in combination with feldspar (11.25% K_2O), rice straw and chicken manure in tomato cultivation. The bacterial inoculation increased the potassium available in the compost (16.7 g/kg - 44 % of the total K) when compared to the uninoculated treatment, in addition to the superior response of tomato plants to the inoculated compost when added to sandy soils with low K content.

The potassic solubilization from minerals such as microcline, orthoclase and muscovite mica by the bacteria *Bacillus mucilaginosus* was studied by Sugumaran and Janarthanam (2007), who observed an increase of 125% in the plant growth of the inoculated peanut.

The direct contact cell-mineral was essential for the solubilization of potassic minerals by the mold *Aspergillus fumigatus*. When the mold was separated from the mineral by a membrane of porosity 0.22 μm, the solubilization was lower (Lian et al., 2008).

Balogh-Brunstad et al. (2008) verified the weathering function of *Suillus tomentosus* in liquid cultures with biotite micas incubated at room temperature and suggested that under the conditions of the experiment, the ectomycorrhizal fungus *S. tomentosus* indirectly weathers biotite surfaces by lowering the solution pH and generating complex forming ligands through the production of low molecular weight organic acids and fungal respiration, regardless the attachment and associated dissolution features.

Nishanth and Biswas (2008) evaluated the efficient action of *A. awamori* in releasing phosphorus and potassium from composts of rock phosphate and waste mica on the yield and nutrient uptake by wheat (*Triticum aestivum*). The results showed increase in the soluble P and K, significantly higher biomass yield and uptake of P and K if compared to the composts without the inoculant.

Hopf et al. (2009) studied the microbial assisted dissolution of biotite using *Bacillus subtilis*, *Shewanella putrefaciens*, *Streptomyces acisdiscabies* and *Schizophyllum commune* incubated for 35 days in batch reactors at slightly alkaline pH conditions (ca. pH 9.5), using an experiment without a biological component as control treatment. They observed that biotite dissolves incongruently with a preferential release of the interlayer cations and then by the octahedral and subsequently the tetrahedral site. They calculated K release rates as well as Mg/Si and K/Si element ratios and observed a preferential dissolution of these elements from biotite in the presence of microorganisms, indicating a significant effect of microorganisms on biotite dissolution.

Basak and Biswas (2009) studied the dynamics of K release from waste mica inoculated with potassium-solubilizing microorganism (*Bacillus mucilaginosus*) and evaluated its effectiveness as potassic-fertilizer using sudan grass (*Sorghum vulgare* Pers.) var Sudanensis as test crop grown under two Alfisols. They observed that mica significantly enhanced the biomass yield, K recovery (%) and uptake by sudan grass compared to control (without K). These parameters increased further when mica was inoculated with the bacterial strain in both soils compared to uninoculated mica.

In another assay, Basak and Biswas (2010), also using a waste mica, verified the effect of co-inoculation of potassium-solubilizing (*Bacillus mucilaginosus*) and nitrogen-fixing (*Azotobacter chroococcum* A-41) bacteria on growth promotion and nutrient uptake by sudan grass (*Sorghum vulgare* Pers.) in a Typic Haplustalf. They observed a significantly higher biomass accumulation and nutrient acquisition in all treatments with mica and/or bacterial strain, when compared to control, and that the co-inoculation of waste mica with both bacteria resulted in higher biomass production and nutrient acquisition. The authors concluded that the co-inoculation of potassium-solubilizing and nitrogen-fixing bacteria to waste mica could be a promising and alternative option for utilizing waste mica as K fertilizer to crops and maintaining greater nutrients availability in soil.

3.3.3. Other Minerals and Rocks

Besides insoluble phosphates, micas and feldspars, ferromagnesian minerals and rocks as granite, basalt, phonolite and alkaline ultramafic could be employed in soil fertilization combined with solubilizing microorganisms.

Some silicate rocks, when grinded and utilizing a process similar to liming, have showed significant K release in laboratory tests, in incubated soils and in controlled cultivation (Castro et al., 2006; Oliveira et al., 2006). In some cases (other than K), the rocks provide other nutrients and an alkalinizing effect was observed (Hinsinger et al., 1996; Bolland and Baker, 2000; Gillman et al., 2001, 2002; Machado et al., 2005; Resende et al., 2005; Barbosa

Filho et al., 2006; Moreira et al., 2006; Anda et al., 2009), probably due to the function of silicate ion, which can act as a proton receptor, just as the carbonate from lime does.

The co-inoculation of the P-solubilizing *Bacillus megaterium* and the K-solubilizing *Bacillus mucilaginosus* significantly increased the availability of P and K in the soil along with the absorption of N, P and K by cucumber and red pepper plants. It also brought an increase in plant growth promotion (Han et al., 2006). The same observation was taken by Han and Lee (2005) using eggplant, in which the authors verified an increase in 14% and 21% in K of the aerial part and root, respectively, along with a dry mass 30% higher in the inoculated plants.

Silva et al. (2005) studied the properties of granite powders, a granite manufacturing waste product, to assess their potential use as amendments and fertilizers on acid soils. Two types of powders were evaluated: one produced during granite cutting with a diamond-edged disc saw, comprised only rock powder; and the other produced during cutting with a multi-blade bandsaw, with calcium hydroxide and metal filings applied during the cutting procedure. The acid neutralizing capacity (ANC) of the granite powders was determined in short-term (2–3 h) and medium-term (1–30 d) experiments. The powders showed a buffering capacity at around pH 8, which was related to the rapid dissolution of basic cations, and another buffering effect at pH < 4.5, related to the dissolution of Fe and Al. The ANC measured in the short-term experiments, to a final pH of 4.5, varied between 5 and 61 cmol $H^{+}kg^{-1}$ of powder. The ANC in pH 4.5 obtained in the medium-term experiments was much higher than that obtained in the short-term experiments, reaching a maximum ANC value of 200 cmol $H^{+}kg^{-1}$ powder. There was no great difference in the neutralizing capacity measured at between 1 and 30 d. The most common elements in acid solutions obtained at the end of medium-term experiments were Mg and Ca for disc saw powders, whereas Ca and Fe (at pH < 5) were the elements with the highest solubility in the bandsaw powders. The rapid release of these cations suggested the possible effective use of the granite powders as a source of nutrients to be added to acid soils.

Rosa et al. (2009) evaluated the action of the yeast *Torulaspora globosa* (strain 1S112), isolated from the sugar cane rhizosphere, in the solubilization of alkaline ultramafic rock powder, observing the production of organic acids by the yeast and the release of soluble potassium. Moreover, this strain had an effective action in the biological control of phytopathogens and characteristics of a plant growth promoter (Rosa et al., 2010). The same yeast was evaluated for phosphate solubilization with promising results (Vora and Shelat, 1998).

Lopes-Assad et al. (2010) have evaluated the action of two strains of *A. niger* (CCT911 and CCT4355) in the solubilization of alkaline ultramafic rock powder, obtaining as much as 60% of K release after 35 days of incubation, in 125-mL flasks with agitation. There were no better results in the solubilization in higher fermentation scales (500, 1000 and 2000 mL-flasks), probably due to the lower medium oxygenation, indicating this is a very important factor to be taken into account for the effective production of secondary compounds responsible for mineral solubilization.

4. Final Considerations

An ever-increasing amount of phosphorus and potassium, as essential elements for plant growth, is required to meet agricultural needs. The practice of incorporating powdered rocks to the soil is not new, especially considering that potential resources exist in many countries in the form of *in natura* rocks, mineral wastes or mining tailings.

The natural fertility of soils is directly connected to the availability and release of chemical elements from the rocks to the soil. However the long periods needed for the alteration of the minerals contained in the rocks and consequently their availability to plant nutrition became this alternative quite unfeasible. The utilization of the rock powder singly will not guarantee the sustainability of the crop production and should be adopted in association with other agricultural techniques.

The effectiveness of some medium-to-low reactive rocks needs to be enhanced especially by biological processes. The role of the microorganisms in the process of mineral alteration of the rocks is crucial for nutrient availability. Bacteria and fungi from the rhizosphere are able to produce a variety of substances, like the organic acids which play a role in the nutrient acquisition and plant growth by the mobilization of scarcely soluble nutrients (e.g. P and K) and detoxification of potentially harmful elements like Al. The microorganisms can affect the chemical environment and accelerate soil weathering also by the generation of carbonic gas, introduction of chelates and ionic exchange reactions.

Satisfactory and beneficial effects of microbial inoculation combined with rocks have been found for a great variety of crops. Plant growth promotion is also acquired due to the solubilizing organisms, bringing about greater nutrient availability. A well-nourished plant is more resistant to pathogens and pests, resulting in healthy plants and consequently a diminishing usage of industrial fertilizers and pesticides.

The development of specific technologies both for *in vitro* production of inoculants and field trials should be stimulated and their economics and applicability must be assessed on a case-by-case basis. More information regarding the ecological consequences of the inoculation of solubilizing organisms to crops is required, but studies have revealed that plant species seem to be more important in determining the microbial community structure than the introduction of exotic microorganisms.

References

Agnihotri, V.P. (1970). Solubilization of insoluble phosphates by some soil fungi isolated from nursery seedbeds. *Canadian Journal of Microbiology*, 16, 877–880.

Akter, M.; Akagi, T. (2005). Effect of fine root contact on plant-induced weathering of basalt. *Soil Science and Plant Nutrition*, 51, 861-871.

Al-Falih, A.M. (2005). Phosphate solubilization *in vitro* by some soil yeasts. *Qatar University Science Journal,* 25, 119-125.

Alonso, L.M.; Kleiner, D.; Ortega, E. (2008). Spores of the mycorrhizal fungus *Glomus mosseae* host yeasts that solubilize phosphate and accumulate polyphosphates. *Mycorrhiza,* 18, 197-204.

Anda, M.; Shamshuddin, J.; Fauziah, C.I.; Omar, S.R.S. (2009). Dissolution of ground basalt and its effect on oxisol chemical properties and cocoa growth. *Soil Science,* 174, 264-271.

Asea, P.E.A.; Kucey, R.M.N.; Stewart, J.W.B. (1988). Inorganic phosphate solubilization by two *Penicillium* species in solution culture and soil. *Soil Biology and Biochemistry*, 20, 459-464.

Babana, A.H.; Antoun, H. (2006). Effect of Tilemsi phosphate rock-solubilizing microorganisms on phosphorus uptake and yield of field-grown wheat (*Triticum aestivum* L.) in Mali. *Plant and Soil,* 287, 51–58.

Badr, M.A. (2006). Efficiency of K-feldspar combined with organic materials and silicate dissolving bacteria on tomato yield. *Journal of Applied Science Research,* 2, 12, 1191-1198.

Balland, C.; Poszwa, A.; Leyval, C.; Mustin, C. (2010). Dissolution rates of phyllosilicates as a function of bacterial metabolic diversity. *Geochimica et Cosmochimica Acta,* 74, 5478-5493.

Balogh-Brunstad, Z.; Keller, C.K.; Dickinson, J.T.; Stevens, F.; Li, C.Y.; Bormann, B.T. (2008) Biotite weathering and nutrient uptake by ectomycorrhizal fungus, *Suillus tomentosus*, in liquid-culture experiments. *Geochimica et Cosmochimica Acta,* 72, 2601-2618.

Bar-Yosef, B.; Rogers, R.D.; Wolfram, J.H.; Richaman, E. (1999). *Pseudomonas cepacia*-mediated rock phosphate solubilization in kaolinite and montmorillonite suspensions. *Soil Science Society of America Journal*, 63, 6, 1703-1708.

Barbosa Filho, M.P.; Fageria, N.K.; Santos, D.F.; Couto, P.A. (2006). Aplicação de rochas silicáticas como fontes alternativas de potássio para a cultura do arroz de terras altas. *Espaço e Geografia*, 9, 63-84.

Basak, B.B.; Biswas, D.R. (2009). Influence of potassium-solubilizing microorganism (*Bacillus mucilaginosus*) and waste mica on potassium uptake dynamics by sudan grass (*Sorghum vulgare* Pers.) grown under two Alfisols. *Plant and Soil,* 317, 235-255.

Basak, B.B.; Biswas, D.R. (2010). Co-inoculation of potassium-solubilizing and nitrogen-fixing bacteria on solubilization of waste mica and their effect on growth promotion and nutrient acquisition by a forage crop. *Biology and Fertility of Soils,* 46, 641-648.

Ben Farhat, M.; Farhat, A.; Bejar, W.; Kammoun, R.; Bouchaala, K.; Fourati, A.; Antoun, H.; Bejar, S.; Chouayekh, H. (2009). Characterization of the mineral phosphate-solubilizing activity of *Serratia marcescens* CTM 50650 isolated from the phosphate mine of Gafsa. *Archives of Microbiology,* 191, 815-824.

Berthelin, J.; Leyval, C. (1982). Ability of symbiotic and non-symbiotic rhizospheric microflora of maize (*Zea mays*) to weather micas and to promote plant-growth and plant nutrition. *Plant and Soil,* 68, 369-377.

Bigham, J.M.; Bhatti, T.M.; Vuorinen, A.; Tuovinen, O.H. (2001). Dissolution and structural alteration of phlogopite mediated by proton attack and bacterial oxidation of ferrous iron. *Hydrometallurgy,* 59, 301-309.

Biswas, D.R.; Narayanasamy, G.; Datta, S.C.; Singh, G.; Begum, M.; Maiti, D.; Mishra, A.; Basak, B.B. (2009). Changes in nutrient status during preparation of enriched organomineral fertilizers using rice straw, low-grade rock phosphate, waste mica, and phosphate-solubilizing microorganism. *Communications in Soil Science and Plant Analysis,* 40, 2285-2307.

Bojinova, D.; Velkova, R.; Karev, K. (1999). Agrochemical action of organic mineral fertilizers produced through biotreatment of natural phosphate. *Agrochimica,* 43, 118-125.

Bolan, N.S.; Naidu, R.; Mahimairaja, S.; Baskaran, S. (1994). Influence of low-molecular-weight organic-acids on the solubilization of phosphates. *Biology and Fertility of Soils*, 18, 311-319.

Bolland, M.D.A.; Baker, M.J. (2000). Powdered granite is not an effective fertilizer for clover and wheat in sandy soils from Western Australia. *Nutrient Cycling in Agroecosystems*, 56, 59-68.

Borges, K.B.; Borges, W.S.; Pupo, M.T.; Bonato, P.S. (2007). Endophytic fungi as models for the stereoselective biotransformation of thioridazine. *Applied Microbiology and Biotechnology,* 77, 669–674.

Brehm, U.; Gorbushina, A.; Mottershead, D. (2005). The role of microorganisms and biofilms in the breakdown and dissolution of quartz and glass. *Palaeogeography, Palaeoclimatology, Palaeoecology,* 219, 117-129.

Browne, P.; Rice, O.; Miller, S.H.; Burke, J.; Dowling, D.N.; Morrissey, J.P.; O'Gara, F. (2009). Superior inorganic phosphate solubilization is linked to phylogeny within the *Pseudomonas fluorescens* complex. *Applied Soil Ecology,* 43, 131-138.

Burford, E.P.; Fomina, M.; Gadd, G.M. (2003). Fungal involvement in bioweathering and biotransformation of rocks and minerals. *Mineralogical Magazine*, 67, 1127-1155.

Cama, J.; Ganor, J. (2006) The effects of organic acids on the dissolution of silicate minerals: A case study of oxalate catalysis of kaolinite dissolution. *Geochimica et Cosmochimica Acta ,* 70, 2191-2209.

Casarin, V.; Plassard, C.; Hinsinger, P.; Arvieu, J.C. (2004). Quantification of ectomycorrhizal fungal effects on the bioavailability and mobilization of soil P in the rhizosphere of *Pinus pinaster*. *New Phytologist*, 163, 177–185.

Castro, C.; Oliveira, F.A.; Moreira, A.; Salinet, L.H.; Veronesi, C.O. (2006). Rochas brasileiras como fonte alternativa de potássio para a cultura do girassol. *Espaço e Geografia,* 9, 2, 179-193.

Chuang, C.C.; Kuo, Y.L.; Chao, C.C.; Chao, W.L. (2007). Solubilization of inorganic phosphates and plant growth promotion by *Aspergillus niger*. *Biology and Fertility of Soils,* 43, 575-584.

Coroneos, C.; Hinsinger, P.; Gilkes, R.J. (1996). Granite powder as a source of potassium for plants: A glasshouse bioassay comparing two pasture species. *Fertilizer Research*, 45, 143-152.

Costa, L.G.; Lopes, O.M.M.; Peres, J.G.; Lopes-Assad, M.L. (2010). Efeito de pó de rocha na retenção de água no solo. Proceedings of the 9th Congreso Latinoamericano y del Caribe de Ingeniería Agricola, 34th Congresso Brasileiro de Engenharia Agrícola. Vitória-ES, Brazil.

Cowan, D.A.; Ah Tow, L. (2004). Endangered Antarctic environments. *Annual Review of Microbiology,* 58, 649–90.

Curl, E.A.; Truelove, B. (1986). *The rhizosphere*. Berlin: Springer-Verlag, 1986.

Decaens, T. (2010). Macroecological patterns in soil communities. *Global Ecology and Biogeography,* 19, 287-302.

Di Simine, C. D.; Sayer, J. A.; Gadd., G. M. (1998). Solubilization of zinc phosphate by a strain of *Pseudomonas fluorescens* isolated from a forest soil. *Biology and Fertility of Soils,* 28, 1, 87-94.

Dokuchaev, V.V. (1889). *The zones of Russia*. Moscow: Akademy Nauk.

Drever, J.I.; Stillings, L.L. (1997). The role of organic acids in mineral weathering. *Colloids and Surfaces A-Physicochemical and Engineering Aspects*, 120, 167-181.

Duchaufour, P. (1960). *Précis de pédologie*. Paris: Masson, 438p.

Duchaufour, P. (1977). Pédogenése et classification. In: Duchaufour, P.; Souchier, B. ed. *Pedologie*. Paris : Masson, v.1, 478p.

Duponnois, R.; Paugy, M.; Thioulouse, J.; Masse, D.; Lepage, M. (2005). Functional diversity of soil microbial community, rock phosphate dissolution and growth of *Acacia seyal* as influenced by grass-, litter- and soil-feeding termite nest structure amendments. *Geoderma,* 124, 349-361.

Dutton, M.V.; Evans, C.S. (1996). Oxalate production by fungi: its role in pathogenicity and ecology in the soil environment. *Canadian Journal of Microbiology*, 42, 881–895.

Ekin, Z. (2010). Performance of phosphate-solubilizing bacteria for improving growth and yield of sunflower (*Helianthus annuus* L.) in the presence of phosphorus fertilizer. *African Journal of Biotechnology*, 9, 3794-3800.

Escosteguy, P.; Klamt, E. (1998). Basalto moído como fonte de nutrientes. *Revista Brasileira de Ciência do Solo*, 22, 11-20.

Estes, R.J. (2010). The world social situation: development challenges at the outset of a new century. *Social Indicators Research*, 98, 363–402

Fankem, H.; Nwaga, D.; Deubel, A.; Dieng, L.; Merbach, W.; Etoa, F.X. (2006). Occurrence and functioning of phosphate-solubilizing microorganisms from oil palm tree (*Elaeis guineensis*) rhizosphere in Cameroon. *African Journal of Biotechnology*, 5, 2450-2460.

Formoso, M.L.L. (2006). Some topics on geochemistry of weathering: a review. *Anais da Academia Brasileira de Ciências*, 78, 809-820.

Friedrich, S.; Platonova, N.P.; Karavaiko, G.I.; Stichel, E.; Glombitza, F. (1991). Chemical and microbiological solubilization of silicates. *Acta Biotechnologica*, 11, 187–196.

Gaur, A.C.; Madan M.; Ostwal K.P. (1973). Solubilization of phosphatic compounds by native microflora of rock phosphates. *Indian Journal of Experimental Biology*, 11, 427-429.

Gerke J.; Roemer, W.; Beissner, L. (2000). The quantitative effect of chemical phosphate mobilization by carboxylate anions on P uptake by a single root. II. The importance of soil and plant parameters for uptake of mobilized P. *Journal of Plant Nutrition and Soil Science,* 163, 213–219.

Ghaderi, A.; Aliasgharzad, N.; Oustan, S.; Olsson, P.A. (2008). Efficiency of three *Pseudomonas* isolates in releasing phosphate from an artificial variable-charge mineral (iron III hydroxide). *Soil and Environment*, 27, 1, 71-76.

Gillman, G.P. (1980) The effect of crushed basalt scoria on the cation exchange properties of a highly weathered soil. *Soil Science Society of America Journal*, 44, 465–468.

Gillman, G.P.; Burkett, D.C.; Coventry, R.J. (2001). A laboratory study of application of basalt dust to highly weathered soils: Effect on soil cation chemistry. *Australian Journal of Soil Research*, 39, 799-811.

Gillman, G.P.; Burkett, D.C.; Coventry, R.J. (2002). Amending highly weathered soils with finely ground basalt rock. *Applied Geochemistry*, 17, 987-1001.

Groudev, S.N. (1987). Use of heterotrophic microorganisms in mineral biotechnology. *Acta Biotechnologica*, 7, 299–306.

Gyaneshwar, P.; Kumar, G.N.; Parekh, L.J.; Poole, P.S. (2002). Role of soil microorganisms in improving P nutrition of plants. *Plant and Soil*, 245, 83-93.

Hamdali, H.; Hafidi, M.; Virolle, M.J.; Ouhdouch, Y. (2008a). Growth promotion and protection against damping-off of wheat by two rock phosphate-solubilizing actinomycetes in a P-deficient soil under greenhouse conditions. *Applied Soil Ecology*, 40, 510-517.

Hamdali, H.; Hafidi, M.; Virolle, M.J.; Ouhdouch, Y. (2008b). Rock phosphate-solubilizing actinomycetes: screening for plant growth-promoting activities. *World Journal of Microbiology and Biotechnology*, 24, 2565–2575.

Hamdali, H.; Smirnov, A.; Esnault, C.; Ouhdouch, Y.; Virolle, M.J. (2010). Physiological studies and comparative analysis of rock phosphate solubilization abilities of Actinomycetales originating from Moroccan phosphate mines and of *Streptomyces lividans*. *Applied Soil Ecology*, 44, 24-31.

Han, H.S.; Lee, K.D. (2005). Phosphate and potassium-solubilizing bacteria effect on mineral uptake, soil availability and growth of eggplant. *Research Journal of Agriculture and Biological Science*, 1, 176-180.

Han, H.S.; Supanjani, K.D.; Lee, K.D. (2006). Effect of co-inoculation with phosphate and potassium-solubilizing bacteria on mineral uptake and growth of pepper and cucumber. *Plant and Soil Environment*, 52, 130-136.

Harley, A.D.; Gilkes, R.J. (2000). Factors influencing the release of plant nutrient elements from silicate rock powders: a geochemical overview. *Nutrient Cycling in Agroecosystems*, 56, 11-36.

He, Z.L.; Zhu, J. (1998). Microbial utilization and transformation of phosphate adsorbed by variable charge minerals. *Soil Biology and Biochemistry*, 30, 7, 917-923.

Heffer, P.; Prud'homme, M. (2010). Short-term fertilizer outlook 2010-2011. 36th IFA Enlarged Council Meeting, New Delhi, India, 2-4 December. 8p. (cited 2011 January 30). Available from: http://www.fertilizer.org/ifa/Home-Page/LIBRARY/Publication-database.html/Short-Term-Fertilizer-Outlook-2010-2011.html.

Hensel, J. (1894). Bread from Stones: a new and rational system of land fertilization and physical regeneration. Scanned by Soil and Health Library, July 1998. (cited 2011 January 15). Available from: http://www.soilandhealth.org/01aglibrary/ 010173. hensel.pdf.

Hinsinger, P.; Bengough, A.G.; Vetterlein, D.; Young, I.M. (2009). Rhizosphere: biophysics, biogeochemistry and ecological relevance. *Plant and Soil*, 321, 117-152.

Hinsinger, P.; Bolland, M.D.A.; Gilkes, R.J. (1996). Silicate rock powder: Effect on selected chemical properties of a range of soils from Western Australia and on plant growth as assessed in a glasshouse experiment. *Fertilizer Research*, 45, 69-79.

Hinsinger P.; Plassard, C.; Tang, C.; Jaillard, B. (2003). Origins of root-mediated pH changes in the rhizosphere and their environment constraints – A review. *Plant Soil*, 248, 43-59.

Hopf, J.; Langenhorst, F.; Pollok, K.; Merten, D.; Kothe, E. (2009). Influence of microorganisms on biotite dissolution: an experimental approach. *Geochemistry*, 69, 45-56.

Hu, J.L.; Lin, X.G.; Wang, J.H.; Chu, H.Y.; Yin, R.; Zhang, J.B. (2009). Population size and specific potential of P-mineralizing and -solubilizing bacteria under long-term P-deficiency fertilization in a sandy loam soil. *Pedobiologia*, 53, 49-58.

Hu, X.F.; Chen, J.S.; Guo, J.F. (2006). Two phosphate- and potassium-solubilizing bacteria isolated from Tianmu Mountain, Zhejiang, China. *World Journal of Microbiology and Biotechnology*, 22, 983-990.

Hue, N. V.; Craddock, G.R.; Adams, F. (1986). Effect of organic acids on aluminum toxicity in subsoils. *Soil Science Society of America Journal*, 50, 28–34.

Illmer, P.; Schinner, F. (1992). Solubilization of inorganic phosphates by microorganisms isolated from forest soils. *Soil Biology and Biochemistry*, 24, 4, 389-395.

Jayasinghearachchi, H.S.; Seneviratne, G. (2006). Fungal solubilization of rock phosphate is enhanced by forming fungal-rhizobial biofilms. *Soil Biology and Biochemistry*, 38, 2, 405-408.

Jenny, H. (1941). *Factors of soil formation*. New York, McGraw-Hill, 281 p.

Jha, P.; Kumar, A. (2009). Characterization of novel plant growth promoting endophytic bacterium *Achromobacter xylosoxidans* from wheat plant. *Microbial Ecology*, 58, 179-188.

Jones, D.L. (1998). Organic acids in the rhizosphere—a critical review. *Plant Soil*, 205, 25–44.

Jones, D.L.; Darrah, R. (1994). Role of root derived organic acids in mobilization of nutrients from the rhizosphere. *Plant Soil*, 166, 247-257.

Jones, D.L.; Hinsinger, P. (2008). The rhizosphere: complex by design. *Plant and Soil*, 312, 1-6.

Kapri, A.; Tewari, L. (2010). Phosphate solubilization potential and phosphatase activity of rhizospheric *Trichoderma* spp. *Brazilian Journal of Microbiology*, 41, 787-795.

Kim, K.Y.; McDonald, G.A.; Jordan, D. (1997). Solubilization of hydroxyapatite by *Enterobacter agglomerans* and cloned *Escherichia coli* in culture medium. *Biology and Fertility of Soils*, 24, 4, 347-352.

Kumar, V.; Behl, R.K.; Narula, N. (2001). Establishment of phosphate-solubilizing strains of *Azotobacter chroococcum* in the rhizosphere and their effect on wheat cultivars under green house conditions. *Microbiological Research*, 156, 87-93.

Kumar, V.; Punia, S.S.; Lakshminarayana, K.; Narula, N. (1999). Effect of phosphate solubilizing analogue-resistant mutants of *Azotobacter chroococcum* on sorghum. *Indian Journal of Agricultural Sciences*, 69, 198-200.

Lambers, H.; Mougel, C.; Jaillard, B.; Hinsinger, P. (2009). Plant-microbe-soil interactions in the rhizosphere: an evolutionary perspective. *Plant and Soil*, 321, 83-115.

Lapido-Loureiro, F.E.; Melamed, R. (2009). O fósforo na agroindústria brasileira. In: Lapido-Loureiro, F.E.; Melamed, R.; Figueiredo Neto, F. (Ed.). *Fertilizantes: agroindústria e sustentabilidade*. CETEM/MCT, pp.257-304.

Lapido-Loureiro, F.E.; Ribeiro, R.C.C. (2009). Fertilização natural: rochagem, agricultura orgânica e plantio direto. Breve síntese conceitual. In: Lapido-Loureiro, F.E.; Melamed, R.; Figueiredo Neto, F. (Ed.). *Fertilizantes: agroindústria e sustentabilidade*. CETEM/MCT, pp.149-173.

Lavelle, P.; Bignell, D.; Lepage, M.; Wolters, V.; Roger, P.; Ineson, P.; Heal, O.W.; Dhillion, S. (1997). Soil function in a changing world: the role of invertebrate ecosystem engineers. *European Journal of Soil Biology*, 33, 159-193.

Lavelle, P.; Decaens, T.; Aubert, M.; Barot, S.; Blouin, M.; Bureau, F.; Margerie, P.; Mora, P.; Rossi, J.P. (2006). Soil invertebrates and ecosystem services. *European Journal of Soil Biology,* 42, 3-15.

Leonardos, O.H.; Fyfe, W.S.; Kronberg, B.I. (1987). The use of ground rocks in laterite systems - an improvement to the use of conventional soluble fertilizers. *Chemical Geology,* 60, 361-370.

Leonardos, O.H.; Theodoro, S.H.; Lopes-Assad, M.L.L. (2000). Remineralization for sustainable agriculture: a tropical perspective from a Brazilian viewpoint. *Nutrient Cycling in Agroecosystems,* 56, 3-9.

Leyval, C.; Berthelin, J. (1989). Interactions between *Laccaria laccata, Agrobacterium radiobacter* and beech roots - influence on P, K, Mg, and Fe mobilization from minerals and plant-growth. *Plant and Soil,* 117, 103-110.

Leyval, C.; Berthelin, J. (1991). Weathering of a mica by roots and rhizospheric microorganisms of pine. *Soil Science Society of America Journal,* 55, 1009-1016.

Lian, B.; Wang, B.; Pan, M.; Liu, C.; Teng, H.H. (2008). Microbial release of potassium from K-bearing minerals by thermophilic fungus Aspergillus fumigatus. *Geochimica et Cosmochimica Acta,* 72, 87-98.

Lopes-Assad, M.L.; Avansini, S.H.; Rosa, M.M.; Carvalho, J.R.P.; Ceccato-Antonini, S.R. (2010). The solubilization of potassium-bearing rock powder by *Aspergillus niger* in small-scale batch fermentations. *Canadian Journal of Microbiology,* 56, 7, 598-605.

Lundström, U.S.; van Breemen, N.; Bain, D. (2000). The podzolization process: a review. *Geoderma,* 94, 91-107.

Machado, C.T.T.; Resende, A.V.; Martins, E.S.; Sobrinho, D.A.S.; Nascimento, M.T.; Faleiro, A.S.G.; Linhares, N.W.; Souza, A.L.; Corazza, E.J. (2005). Potencial de rochas silicáticas no fornecimento de potássio para culturas anuais: II. Fertilidade do solo e suprimento de outros nutrientes. Proceedings of the 30th Congresso Brasileiro de Ciência do Solo. Recife-PE, Brazil. (cd-rom).

Madigan, M.T.; Martinko, J.M.; Parker, J. (2004). *Microbiologia de Brock.* 10th ed., Pearson Prentice Hall, São Paulo, 579p.

Madureira Filho, J.B.; Atencio, D.; McReath, I. (2000). Minerais e rochas: constituintes da Terra sólida. In: Teixeira, W.; Toledo, C.M.; Fairchild, T.R.; Taioli, F. (Ed.). *Decifrando a Terra.* Oficina de Textos, pp.28-42.

Magdziak, Z.; Kozlowska, M.; Kaczmarek, Z.; Mleczek, M.; Chadzinikolau, T.; Drzewiecka, K.; Golinski, P. (2011). Influence of Ca/Mg ration phytoextraction properties of *Salix viminalis.* II. Secretion of low molecular weight organic acids to the rhizosphere. *Ecotoxicology and Environmental Safety*, 74, 33–40.

Manning, D.A.C. (2008). Phosphate minerals, environmental pollution and sustainable agriculture. *Elements*, 4, 2, 105-108.

Marschner, H. (1995). *Mineral nutrition of higher plants.* 2nd ed., Academic Press, London, 889 p.

Martins, E.S.; Gouveia, C.G.; Resende, A.R.; Matos, M.S.F. (2008). Agrominerais–rochas silicáticas como fontes minerais alternativas de potássio para a agricultura. In: Luz, A.B.; Lins, F.A.F. (Ed.). *Rochas e minerais industriais.* CETEM/MCT, pp.205-209.

Melamed, R.; Gaspar, J.C.; Miekeley, N. (2009). Pó-de-rocha como fertilizante alternativo para sistemas de produção sustentáveis em solos tropicais. In: F. E. Lapido-Loureiro,

F.E.; Melamed, R.; Figueiredo Neto, F. (Ed.) *Fertilizantes: agroindústria e sustentabilidade*. CETEM/MCT, pp.385-395.

Moreira, A.; Castro, C. de; Oliveira, F. A. de; Salinet, L. H.; Sfredo, G. J. (2006). Efeito residual de rochas brasileiras como fertilizantes e corretivos da acidez do solo. *Espaço e Geografia*, 9, 2, 163-177.

Moreira, F.M.S.; Siqueira, J.O. (2006a). Histórico, evolução e tendências. In: Moreira, F.M.S.; Siqueira, J.O. (Ed.). *Microbiologia e bioquímica do solo*. Editora UFLA, pp.1-15.

Moreira, F.M.S.; Siqueira, J.O. (2006b). Transformações bioquímicas e ciclos dos elementos no solo. In: Moreira, F.M.S.; Siqueira, J.O. (Ed.). *Microbiologia e bioquímica do solo*. Editora UFLA, pp.313-405.

Nahas, E.; Assis, L.C. (1992). Solubilização de fosfatos de rochas por *Aspergillus niger* em diferentes tipos de vinhaça. *Pesquisa Agropecuária Brasileira*, 27, 325-331.

Nahas, E.; Banzatto, D.A.; Assis, L.C. (1990). Fluorapatite solubilization by *Aspergillus niger* in vinasse medium. *Soil Biology and Biochemistry*, 22, 8, 1097-1101.

Narsian, V.; Abu, A.; Samaha, S.M.; Patel, H.H. (2010). Rock phosphate dissolution by specific yeast. *Indian Journal of Microbiology*, 50, 57-62.

Nishanth, D.; Biswas, D.R. (2008). Kinetics of phosphorus and potassium release from rock phosphate and waste mica enriched compost and their effect on yield and nutrient uptake by wheat (*Triticum aestivum*). *Bioresource Technology*, 99, 3342-3353.

Oliveira, F. A.; Castro, C.; Moreira, A.; Rosa, J.L. (2006). Eficiência da adubação residual com rochas brasileiras para a cultura da soja. *Espaço e Geografia*, 9, 2, 231-246.

Ouahmane, L.; Revel, J.C.; Hafidi, M.; Thioulouse, J.; Prin, Y.; Galiana, A.; Dreyfus, B.; Duponnois, R. (2009). Responses of *Pinus halepensis* growth, soil microbial catabolic functions and phosphate-solubilizing bacteria after rock phosphate amendment and ectomycorrhizal inoculation. *Plant and Soil*, 320, 169-179.

Perez, E.; Sulbáran, M.; Ball, M.B.; Yarzábal, L.A. (2007). Isolation and characterization of mineral phosphate-solubilizing bacteria naturally colonizing a limonitic crust in the south-eastern Venezuelan region. *Soil Biology and Biochemistry*, 39, 2905–2914.

Pradhan, N.; Shukla, L.B. (2005). Solubilization of inorganic phosphates by fungi isolated from agriculture soil. *African Journal of Biotechnology*, 5, 850-854.

Rashid, M.; Khalil, S.; Ayub, N.; Alam, S.; Latif, F. (2004). Organic acid production and phosphate solubilization by phosphate solubilization microorganisms (PSM) under *in vitro* conditions. *Pakistan Journal of Biological Sciences,* 7, 187-196.

Reddy, M.S.; Kumar, S.; Babita, K. (2002). Biosolubilization of poorly soluble rock phosphates by *Aspergillus tubingensis* and *Aspergillus niger*. *Bioresource Technology,* 84, 187-189.

Reith, F.; Dürr, M.; Welch, S.; Rogers, S.L. (2009). Geomicrobiology of the regolith. In: Scott, K.M.; Pain, C.F. (Ed.). *Regolith Science*. CSIRO and Springer, pp.127-157.

Resende, A.V.; Machado, C.T.T.; Martins, E.S.; Nascimento, M.T.; Sobrinho, D.A.S.; Faleiro, A.S.G.; Linhares, N.W.; Souza, A.L.; Corazza, E.J. (2005). Potencial de rochas silicáticas no fornecimento de potássio para culturas anuais: I. Respostas da soja e do milheto. Proceedings of the 30th Congresso Brasileiro de Ciência do Solo. Recife-PE, Brazil. (cd-rom).

Ribeiro, L.D.; Santos, A.R.; Souza, L.F.D.; Souza, J.S. (2010). Potassium-rich silicate rocks as plant nutrient sources. *Revista Brasileira de Ciência do Solo*, 34, 891-897.

Richardson, A.E. (2001). Prospects for using soil microorganisms to improve the acquisition of phosphorus by plants. *Australian Journal of Plant Physiology*, 28, 897-906.

Rosa, M.M.; Avansini, S.; Lopes-Assad, M.L.; Ceccato-Antonini, S.R. (2009). Biossolubilização de pó de rocha para liberação de potássio pela levedura *Torulaspora globosa*. Proceedings of the 17th Simpósio Nacional de Bioprocessos, Natal-RN, Brazil. (cd-rom).

Rosa, M.M.; Tauk-Tornisielo, S.M.; Rampazzo, P.E.; Ceccato-Antonini, S.R. (2010). Evaluation of the biological control by the yeast *Torulaspora globosa* against *Colletotrichum sublineolum* in sorghum. *World Journal of Microbiology and Biotechnology*, 26, 1491-1502.

Rudresh, D.L.; Shivaprakash, M.K.; Prasad, R.D. (2005). Tricalcium phosphate solubilization abilities of *Trichoderma* spp. in relation to P uptake and growth and yield parameters of chickpea (*Cicer arietinum* L.). *Canadian Journal of Microbiology*, 51, 3, 217-222.

Ryan, P.R.; Delhaize, E.; Jones, D.L. (2001). Function and mechanism of organic anion exudation from plant roots. *Annual Review of Plant Physiology and Plant Molecular Biology*, 52, 527-560.

Sandnes, A.; Eldhuset, T.D.; Wollebaek, G. (2005). Organic acids in root exudates and soil solution of Norway spruce and silver birch. *Soil Biology and Biochemistry*, 37:259-269.

Schneider, K.D.; van Straaten, P.; de Orduna, R.M.; Glasauer, S.; Trevors, J.; Fallow, D.; Smith, P.S. (2010). Comparing phosphorus mobilization strategies using *Aspergillus niger* for the mineral dissolution of three phosphate rocks. *Journal of Applied Microbiology*, 108, 366-374.

Seshadri, S.; Ignacimuthu, S.; Lakshminarasimhan, C. (2004). Effect of nitrogen and carbon sources on the inorganic phosphate solubilization by different *Aspergillus niger* strains. *Chemical Engineering Communications*, 191, 1043-1052.

Sheng, X.F. (2005). Growth promotion and increased potassium uptake of cotton and rape by a potassium-releasing strain of *Bacillus edaphicus*. *Soil Biology and Biochemistry*, 37, 1918-1922.

Sheng, X.F.; He, L.Y. (2006). Solubilization of potassium-bearing minerals by a wildtype strain of *Bacillus edaphicus* and its mutants and increased potassium uptake by wheat. *Canadian Journal of Microbiology*, 52, 66-72.

Sheng, X.F.; Zhao, F.; He, L.Y.; Qiu, G.; Chen, L. (2008). Isolation and characterization of silicate mineral solubilizing *Bacillus globisporus* Q12 from the surfaces of weathered feldspar. *Canadian Journal of Microbiology*, 54, 1064–1068.

Silva, M.T.B.; Hermo, B.S.; García-Rodeja, E.; Freire N.V. (2005). Reutilization of granite powder as an amendment and fertilizer for acid soils. *Chemosphere*, 61, 993–1002.

Silva Filho, G.N.; Narloch, C.; Scharf, R. (2002). Solubilização de fosfatos naturais por microrganismos isolados de cultivos de *Pinus* e *Eucalyptus* de Santa Catarina. *Pesquisa Agropecuária Brasileira*, 37, 847-854.

Silva Filho, G.N.; Vidor, C. (2000). Solubilização de fosfatos por microrganismos na presença de fontes de carbono. *Revista Brasileira de Ciência do Solo*, 24, 2, 311-329.

Silva Filho, G.N.; Vidor, C. (2001). Phosphate solubilizing activity of microorganisms in the presence of nitrogen, iron, calcium and potassium. *Pesquisa Agropecuária Brasileira*, 36, 1495-1508.

Smith, W.H. (1976). Character and significance of forest tree root exudates. *Ecology*, 57, 324–331.

Souchie, R.A.; Barea, J.M.; Saggin Júnior, O.J.; Silva, E.M.R. (2005). Solubilização de fosfatos em meios sólidos e líquidos por bactérias e fungos do solo. *Pesquisa Agropecuária Brasileira*, 40, 1149-1152.

Sperber, J. I. (1958). The incidence of apatite-solubilizing organisms in the rhizosphere and soil. *Australian Journal of Agricultural Research,* 9, 6, 778-781.

Srividya, S.; Soumya, S.; Pooja, K. (2009). Influence of environmental factors and salinity on phosphate solubilization by a newly isolated *Aspergillus niger* F7 from agricultural soil. *African Journal of Biotechnology,* 8, 9, 1864-1870.

Stamford, N.P.; Moura, P.M.; Lira, M.A.; Santos, C.; Duenhas, L.H.; Gava, C.A.T. (2009). Chemical attributes of an Argisoil of the Vale do São Francisco after melon growth with phosphate and potash rocks biofertilizers. *Horticultura Brasileira,* 27, 447-452.

Styriakova, I.; Bhatti, T.M.; Bigham, J.M.; Styriak, I.; Vuorinen, A.; Tuovinen, O.H. (2004). Weathering of phlogopite by *Bacillus cereus* and *Acidithiobacillus ferrooxidans*. *Canadian Journal of Microbiology*, 50, 3, 213-219.

Styriakova, I.; Styriak, I.; Nandakumar, M.P.; Mattiasson, B. (2003). Bacterial destruction of mica during bioleaching of kaolin and quartz sands by *Bacillus cereus*. *World Journal of Microbiology and Biotechnology,* 19, 6, 583-390.

Sugumaran, P.; Janarthanam, B. (2007). Solubilization of potassium-containing minerals by bacteria and their effect on plant growth. *World Journal of Agricultural Sciences,* 3, 350-355.

Taurian, T., Anzuay, M.S., Angelini, J.G., Tonelli, M.L., Luduena, L., Pena, D., Ibanez, F. and Fabra, A. (2010). Phosphate-solubilizing peanut associated bacteria: screening for plant growth-promoting activities. *Plant and Soil,* 329, 421-431.

Toledo, M.C.M.; Oliveira, S.M.B.; Melfi, A.J. (2000). Intemperismo e formação do solo. In: Teixeira, W.; Toledo, M.C.M.; Fairchild, T.R.; Taioli, F. (Ed.). *Decifrando a Terra.* Oficina de Textos, pp.140-166.

Vassilev, N.; Vassileva, M. (2003). Biotechnological solubilization of rock phosphate on media containing agro-industrial wastes. *Journal of Applied Microbiology and Biotechnology,* 61, 435–440.

Vassilev, N.; Vassileva, M.; Nikolaeva, I. (2006). Simultaneous P-solubilizing and biocontrol activity of microorganisms: potentials and future trends. *Applied Microbiology and Biotechnology,* 71, 137-144.

Vassileva, M.; Azcon, R.; Barea, J.M.; Vassilev, N. (2000). Rock phosphate solubilization by free and encapsulated cells of *Yarrowia lipolytica*. *Process Biochemistry*, 35, 7, 693-697.

Vassileva, M.; Serrano, M.; Bravo, V.; Jurado, E.; Nikolaeva, I.; Martos, V.; Vassilev, N. (2010). Multifunctional properties of phosphate-solubilizing microorganisms grown on agro-industrial wastes in fermentation and soil conditions. *Applied Microbiology and Biotechnology,* 85, 1287-1299.

Vega, N.W.O. (2007). A review on beneficial effects of rhizosphere bacteria on soil nutrient availability and plant nutrient uptake. *Revista Facultad Nacional de Agronomia Medellin*, 60, 1, 3621-3643.

Verboom, W.H.; Pate, J.S. (2006a). Bioengineering of soil profiles in semiarid ecosystems: the 'phytotarium' concept. A review. *Plant and Soil,* 289, 71-102.

Verboom, W.H.; Pate, J.S. (2006b). Evidence of active biotic influences in pedogenetic processes. Case studies from semiarid ecosystems of south-west Western Australia. *Plant and Soil,* 289, 103-121.

Vora, M.S.; Shelat, H.N. (1998). *Torulospora globosa*: a unique strain solubilizing tricalcium phosphate. *Indian Journal of Agricultural Science,* 68, 630-631.

Wakelin, S. A.; Warren, R.A.; Harvey, P.R.; Ryder, M.H. (2004). Phosphate solubilization by *Penicillium* spp. closely associated with wheat roots. *Biology and Fertility of Soils,* 40, 36-43.

Wallander, H. (2000). Uptake of P from apatite by *Pinus sylvestris* seedlings colonized by different ectomycorrhizal fungi. *Plant and Soil,* 218, 249-256.

Wallander, H.; Wickman, T. (1999). Biotite and microcline as a K source in mycorrhizal and non-mycorrhizal *Pinus sylvestris* seedlings. *Mycorrhiza,* 9, 25–32.

Wang, G.H.; Jin, J.; Xu, M.N.; Pan, X.W.; Tang, C. (2007). Inoculation with phosphate-solubilizing fungi diversifies the bacterial community in rhizospheres of maize and soybean. *Pedosphere,* 17, 2, 191-199.

Wang, G.H.; Zhou, D.; Yang, Q.; Jin, J.; Zhao, Y. (2004). Effect of lanthanum on solubilization of rock phosphate in liquid culture by *Aspergillus niger* and *Penicillium oxalicum*. *Chinese Journal of Rare Earth,* 229, 177-180.

Wang, G.H.; Zhou, D.; Yang, Q.; Jin, J.; Liu, X.B. (2005). Solubilization of rock phosphate in liquid culture by fungal isolates from rhizosphere soil. *Pedosphere,* 15, 4, 532-538.

Webmineral. (2011). Mineral Species containing Phosphorus (P). (cited 2011 January 30). Available from: http://webmineral.com/chem/Chem-names-P.shtml.

Whitelaw, M.A. (2000). Growth promotion of plants inoculated with phosphate solubilizing fungi. *Advances in Agronomy,* 69, 99–151.

Wu, C.S. (2010). Promoting fertilizer use via controlled release of a bacteria-encapsulated film bag. *Journal of Agricultural and Food Chemistry,* 58, 10, 6300-6305.

Xiao, C.Q.; Chi, R.A.; He, H.; Qiu, G.Z.; Wang, D.Z.; Zhang, W.X. (2009a). Isolation of phosphate-solubilizing fungi from phosphate mines and their effect on wheat seedling growth. *Applied Biochemistry and Biotechnology,* 159, 2, 330-342.

Xiao, C.Q.; Chi, R.A.; He, H.; Zhang, W.X. (2009b) Characterization of tricalcium phosphate solubilization by *Stenotrophomonas maltophilia* YC isolated from phosphate mines. *Journal of Central South University of Technology,* 16, 4, 581-587.

Xiao, C.Q.; Chi, R.A.; Huang, X.H.; Zhang, W.X.; Qiu, G.Z.; Wang, D.Z. (2008). Optimization for rock phosphate solubilization by phosphate-solubilizing fungi isolated from phosphate mines. *Ecological Engineering,* 33, 187-193.

Young, I.M.; Crawford, J.W. (2004). Interactions and self-organization in the soil-microbe complex. *Science,* 304, 1634-1637.

Yuan, L.; Huang, J.; Li, X.; Christie, P. (2004). Biological mobilization of potassium from clay minerals by ectomycorrhizal fungi and eucalypt seedling roots. *Plant and Soil,* 262, 1-2, 351-361.

In: Soil Nutrients
Editor: Mohammad Miransari
ISBN: 978-1-61324-785-3

Chapter 7

Role of Microorganisms in Nitrogen Cycling in Soils

Egamberdieva Dilfuza *

Department of Biotechnology, Faculty of Biology and Soil Sciences,
National University of Uzbekistan, 100174 Tashkent, Uzbekistan

Abstract

The agricultural crops are dependent on inorganic nitrogen for growth, and the amounts of plant-available ammonium- or nitrate-nitrogen are limited in the soil. Microorganisms play an important role in improving soil fertility and involved in all aspects of N cycling, including N_2 fixation, nitrification, denitrification and ammonification. They decompose plant residues, soil organic matter and release inorganic nutrients that can then be taken up by plants. The environmental factors have direct and indirect effects on soil nutrient turnover processes, including N cycling. Thus changes in N dynamics in soils are closely connected with altering in microbial activities. The knowledge of the factors influencing soil microorganisms involved in N cycling is fundamental for sustainable environmental management. This chapter presents an overview of the role of microorganisms in N cycling and also elucidates whether biotic and abiotic factors alter their population and activities.

Keywords: N cycle, nitrogen fixation, ammonification, nitrification, denitrification

Introduction

Nitrogen (N) is an extremely important element to all forms of life, the main component of most organic compounds in plants including amino and nucleic acids, enzymes, chlorophyll, and proteins (Atlas and Bartha, 1987; Stark and Richards, 2008).

* E-mail: egamberdieva@yahoo.com.

In the soil nitrogen interacts with the soil particles, water, microorganisms and plants Zak et al., 2000). The inorganic nitrogen is most important element in plant growth and the amounts of plant-available forms ammonium-(NH_4) and nitrate (NO_3) are limited in the soil. Since major part of nitrogen occurs in organic form and must be mineralized to inorganic nitrogen before it can be taken up by plants (Rosswal, 1976; Blackburn and Henriksen, 1983; Bock, et al., 1989; Schimel and Bennett, 2004; Boyle et al., 2008). According Phillips et al. (2006) and Dijkstra and Cheng (2008) environmental change will have direct and indirect effects on plant nutrient uptake, soil nutrient turnover processes, including N and C cycling, ammonification, nitrification and denitrification. About 50% of N applied as fertilizer is utilised by plants and the remainder is temporarily retained in roots, immobilized into soil organic matter, leached or lost to the atmosphere (Dowdell, 1982).

Stark and Richards (2008) suggested that in the soil system, N fluxes can be divided into inputs (atmospheric N deposition, biological N-fixation, incorporation of crop residues) and outputs (adsorption of ammonium to the soil matrix, nitrate leaching, ammonia volatilisation, emission of gaseous N compounds by nitrification and denitrification processes, immobilisation). Microorganisms play an important role in enhancing soil physical structure, improving soil fertility and they involved in all aspects of N cycling, including N_2 fixation (N_2 to NH_4+), nitrification (NH_4+to NO_3-), denitrification (NO_3- to N), and ammonification, the release of ammonia (NH3) through breakdown of organic matter and immobilization (Blackburn and Henriksen, 1983; Dueri et al., 2007). Thus they make the nutrients more or less available to plants (Lynch and Whipps, 1991; Coleman and Crossley, 1996; Schimel and Bennett 2004). The fungi including mycorrhizae, may also influence N turnover in soils due to their ability producing extracellular enzymes that promote the depolymerisation of N compounds (Jones et al., 2005). According Beier et al., (2004) elevated temperature and moisture stimulate microbial activity and changing relative N losses from soil systems by stimulating mineralisation, or by promoting N uptake by plants as well as microbial N immobilisation.

Changes in N dynamics in soils are closely connected with altering in microbial activities involved in N cycle by biotic and abiotic factors, Therefore it is very important to determine the factors controlling the microbial processes of N cycling in the soil. This chapter presents an overview of the role of microorganisms in N cycling and also elucidates whether biotic and abiotic factors alter their population and activities.

NITROGEN FIXATION

In agricultural production and natural ecosystems N cycling relies on biological N fixation by diazotrophic and free living N fixing bacteria (Orr et al., 2011). Biological nitrogen fixation is an important process because it provides ammonia (NH_3) for plants and animals and it is carried out by free-living bacteria and by bacteria which form mutualistic associations with plants (Burns and Hardy, 1975; Franche et al., 2009). Some of these species live symbiotically, in association with leguminous or with non-leguminous plants. Non-symbiotic nitrogen fixation can contribute significant plant-available nitrogen in N-deficient soils (Boddey et al., 1995; Bhattacharjee et al, 2008). Free-living nitrogen fixers represent a range of bacteria including rhizosphere bacteria and endophytic bacteria (Mirza et al., 2000).

N_2 fixation is recorded in at least 26 genera of bacteria (Burns and Hardy, 1975; Elmerich and Newton, 2007) and some of the species presented in Table 1.

The genus *Azotobacter* (Shabaev et al., 1991; Kennedy et al., 1997), *Azospirillium,* (Li and Castellano, 1987), *Alcaligenes, Beijerinckia, Campylobacter, Derxia*, (Rennie, 1980), *Bacillus* (Li et al., 1992), *Pseudomonas*, (Rozycki et al., 1999), *Burkholderia* (Vandamme et al., 2002), *Klebsiella, Pantoea* (Balandreau, 1983; Elmerich et al., 1992) are being capable of fixing N. Nitrogenase activity of microorganisms associated with ectomycorrhizal fungi *Laccaria laccata, Hebeloma crustuliniforme, Rhizopogon vinicolor* and *Thelephora* sp. (Li and Hung, 1987; Li and Castellano, 1987). Several free living bacteria such as *B. megaterium, B. halodurans, Cellulomonas* sp., *Micobacterium lacticum, Pantoea agglomerans* and *Azospirillum* sp. were found in pea rhizosphere (Table 2). However their nitrogen fixing capabilities were very low.

Table 1. Examples of bacterial species, which are able to fix atmospheric N_2

Bacterial strain	References
Azotobacter chroococcum	Mrkovacki and Milic, 2001; Shabaev et al., 1991; Kennedy et al., 1997
Pantoea agglomerans	Feng et al., 2006
Klebsiella spp. *Klebsiella pneumoniae.*	Balandreau, 1983; Wright and Weaver, 1981
Cellulomonas sp.	Egamberdiyeva and Hoflich, 2002
Bacillus azotoformans *B. mycoides* *B. cereus* *B. thuringiensis* *B. licheniformis* *B. megaterium*	Li et al., 1992; Rozycki et al.,1999
Mycobacterium flavum	Biggins and Postgate, 1969
Azospirillum spp. *A.brasilense*	Li and Castellano, 1987; Riggs et al., 2001
Burkholderia gladioli *B. vietnamiensis,* *B.kururiensis,* *B. tuberum* *B. phynatum*	Rozycki et al.,1999; Estrada-De Los Santos et al., 2001; Vandamme et al., 2002; Baldani et al., 2000
Pseudomonas sp. *P. glathei* *P. fluorescens*	Rozycki et al., 1999; Shaharoona et al., 2006
Xanthomonas oryzae	Rozycki et al., 1999
Agrobacterium tumefaciens	
Serratia marcescens	Gyaneshwar et al., 2001
Herbaspirillum seropedicae	Gyaneshwar et al., 2002; Baldani et al., 2000, Riggs et al., 2001

Table 2. Nitrogen fixing activities of bacterial strains isolated from the rhizosphere of pea grown in loamy sand soil

Bacteria	Nitrogenase-activity 10^{-4}nmol C_2H_4/h
Azospirillum sp. D13	678.9
Bacillus megaterium 10/1	11.3
*B. halodurans*10/2	6.76
B. lentus 37/2	18.0
Bacillus sp. 39/2	6.2
B. pumilus 39/1	33.3
B. cohnii 19	120.6
Cellulomonas sp. 21/2	11.9
Microbacterium lacticum 33	10.38
Pseudomonas sp. 22	1.89
P. fluorescens 45	13.3
Pantoea agglomerans PF76/4	57.14

The occurrence of endophytic nitrogen-fixing bacteria such as *Azoarcus* sp, *Gluconacetobacter* and *Herbaspirillum* were also reported (Hurek and Reinhold-Hurek, 2003; Baldani and Baldani, 2005). Many leguminous plant species can enter into a symbiotic relationship with root-nodule bacteria such as *Bradyrhizobium*, *Rhizobium* and *Sinorhizobium* (Doyle and Luckow, 2003).

Nitrogen-fixing organisms are generally active in plant root zone soil and supply N from the atmospheric N_2 that is converted to NH_4 + which requires high energy input, thus root exudates help support the reaction (DeLuca et al., 1996; Elmerich and Newton, 2007). It is reported that C exudation by roots into the surrounding soil and the availability of mineral nutrients in soil are of considerable importance to increasing microbial populations (Murder, 1975). Root exudates, as a source of energy, are important for the ecological functions, especially nitrogen fixation, of heterotrophic soil microflora. Since their nitrogen fixing activity is strongly dependent on favourable moisture conditions, oxygen concentration, and a supply of organic C substrates plants that are capable of releasing exudates exhibited higher nitrogen fixation activity in soil (Berkum and Bohlool, 1980).

Over the years, a number of studies have characterized free living N fixing bacterial microbial populations in soils from various climatic conditions (Hartley and Schlesinger, 2002; Okon and Labandera-Gonzales, 1994). We have found a significantly greater density (28%-61%) of nitrogen-fixing bacteria in spring than in the other seasons averaged across crops and soil depth (Egamberdieva and Kucharova, 2008). The increased moisture content in soil in spring period and optimal temperature related to the higher atmospheric precipitation, and it enhances the development of microflora. Also Zou et al., (2000) found similar changes in microbial population during different seasonal time that in winter the total number of microorganisms decreased. Among the crops, alfalfa was able to support the highest number (50%-80%) of nitrogen-fixing bacteria in all seasons, while fallow soil had the lowest (Figure 1). According DeLuca et al., (1996) high levels of C will allow more optimal conditions for free-living N fixation. Orr et al. (2011) found in their study that crop rotation had a strong effect on free-living N-fixing bacterial population and activity. They observed more active and diverse diazotroph community under barley.

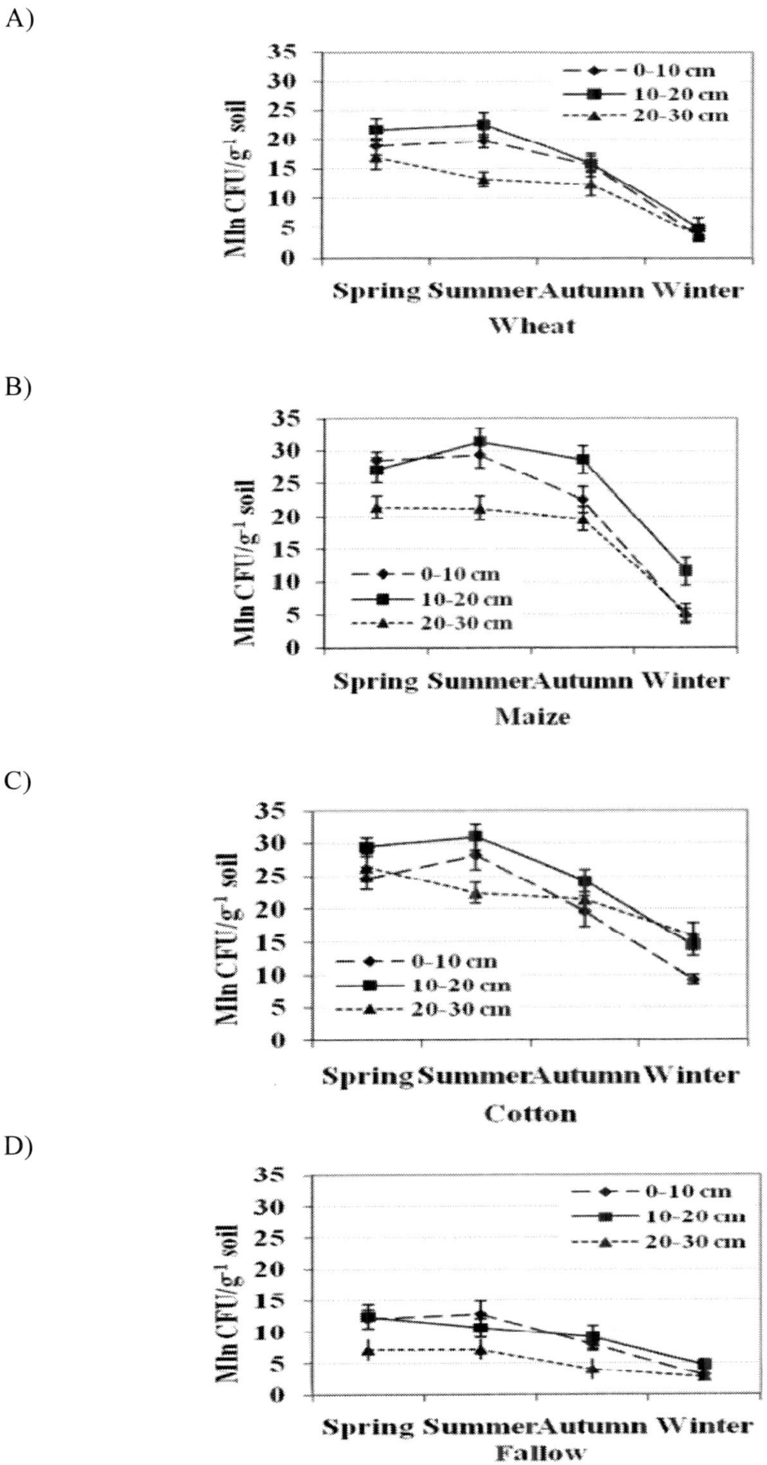

Figure 1. The number (CFU) of nitrogen fixing bacteria under wheat (A), maize (B), cotton (C) and in fallow (D). The samples were taken in spring, summer, autumn and winter (0-30 cm soil depth).

The nitrogen fixing bacteria is higher in nitrogen lower soils to compare to other soil which characterize with higher N content. The arid, saline calcareous soils are characterized with very low nitrogen content, where nitrogen fixing bacteria are dominant (Egamberdieva and Kucharova, 2008).

Bormann et al., (1993) claimed that pine forest ecosystems can be nitrogen self-sufficient due to the non-symbiotic and associative N_2-fixation. Boddey et al., (1995) confirmed that up to 60–80% of sugarcane plant N was derived from biological N fixation. In other studies N fixing *Burkholderia vietnamiensis* strain was found capable of saving 25–30 kgN ha−1 of fertilizer (Nakata, 2002). According Yanni and El-Fattah (1999) the estimated N accumulation by rice plants increased by 11–15 kg ha^{-1} due to *Azotobacter* inoculation. Such studies indicate that biological nitrogen fixation has high inputs of N to crops and it is ideal N source for low input farming systems (DeLuca et al., 1996; Biswas et al., 2000). The application of nitrogenous chemical fertilizer for crop production has increased environment-related concerns. In this context the use of the nitrogen-fixing (NF) bacteria in agricultural practices is very important approach (Franche et al., 2009).

It is well documented that biological nitrogen fixation mediated by nitrogenise enzymes is a process important to the biological activity of soil (Trolldenier, 1977). It requires a high energy input where plants help supply this energy in the form of organic compounds (Atlas and Bartha, 1987). Nitrogenase activity in soil depends on ecological conditions in association with the specific nitrogen fixation capabilities of certain microorganisms and plant genotypes under various climatic conditions (Choudhury et al. 2002). Differences in nitrogenase activity in soils may be best explained by differences in species composition (DeLuca et al., 1996). We also observed differences in nitrogenase activity in soils cultivated with various crops (Egamberdieva and Kucharova, 2008). The significantly greater nitrogenase activity observed in soil under alfalfa during spring is most probably related to greater release of exudates and availability of C substrates, due to alfalfa's extensive rooting system. Lower nitrogenase activity in fallow soil than in cropped soil might be related to fallow soil having a smaller bacterial pool. Alfalfa had a versatile capacity to produce greater root exudates and enrich the soil with nitrogen through nitrogen-fixing activities (Buckley and Schmidt, 2001). A large population of diazotrophic bacteria on plant roots would result in higher rates of nitrogenase activity (Roesch et al., 2008).

AMMONIFICATION

Mineralization of N and immobilization by soil microorganisms are one of the key components of the soil N cycle and are both considered important processes in soils (Janssen, 1996). The organic nitrogenous compounds are decomposed by microbial enzymes to form ammonia (NH_3) and thus the amount of plant available N in soils is increased through those processes. They are also responsible for the digestion of polymers such as cellulose and lignin, which multicellular organisms are unable to utilize (Grayston et al., 1998). The part of the nitrogen that is present in the converted organic material is used in microbial tissue and part is released (mineralized) as inorganic nitrogen (Jenssen, 1996). The immobilization process, results in an increase of organic N in the remaining organic material. The result of this study confirmed that the fraction of mineralized organic N is linearly related to substrate

C/N for equally decomposable materials. According Schimmel and Bennett (2004) N mineralization by soil microbes is the key event in the N cycle making mineral N bio-available, whereas plants only uptake mineral N. The ammonifying microorganisms include species such as *Bacillus*, *Pseudomonas*, *Microbacterium*, *Streptomyces* (Govedarica, 1995; Wirth and Egamberdieva, 2008). The mycorrhizae are also important in acting directly as decomposers by producing the exoenzymes that break down organic polymers in low soil N environments (Schimmel and Bennett, 2004).

Ammonifying bacterial population were predominant groups in the rhizosphere of plants (Bossio et al., 1998). We have observed distinct differences in ammonifying bacterial populations directly associated with different agricultural crops and soil depth, seasonal changes. Marked effects were found to have taken place on the bacterial populations under alfalfa and wheat grown in saline arid soils. Numerous studies have reported on the direct influence of different plant species due releasing different nutrients and organic compounds into the soil on rhizosphere community (Marschner et al., 2004; Bertin et al., 2003). It is known from earlier studies about the adaptation of microbial communities in water stressed soils, whereas their N mineralization activities did not strongly affected by soil condition (Sparling et al., 1990). The highest density of ammonifying bacteria in soil under various crops was observed during summer and the lowest in winter (Egamberdiyeva et al., 2004). The studies of Singh et al. (1989) on temporal dynamics of N mineralization indicated that a net immobilization of N occurs during the summer months when the soil is dry. Microbial N generally accounts for 2% to 6% of the total soil N (Dalal and Mayer, 1987). The nitrogen concentration, lignin, and C/N in plant residue can affect their decomposition rate (Wang et al., 2004). The carbon-to-nitrogen (C:N) ratio is often used as an indication of whether mineralization or immobilization will occur.

According Kumar and Goh (2000) residues with a high N content have a high decomposition rate and residues with high C/N ratio decompose more slowly than those with a low C/N ratio. The temperate ecosystem show a smaller biomass C:N ratio, ranging from 3.8 to 6.1 (Schimel et al., 1985; Powlson et al., 1987), whereas arable soil of Australia has higher biomass C:N ratio of 8.7 to 13.2 (Dalal and Mayer, 1987). We have observed that the N loss was maximal in soil under cotton cultivation (C:N 18.3) and lowest in soil under alfalfa (C:N 9.1) Table 3.

Table 3. C/N ratio of calcareous soil under various crop cultivation

Plant	Soil depth (cm)	Winter	Spring	Summer	Autumn
Cotton	0 - 15	17.8	18.3	16.6	18.3
	15 - 30	15.8	16.5	18.8	19.6
Alfalfa	0 - 15	11.6	9.5	11.1	11.9
	15 - 30	9.7	8.0	9.4	10.6
Maize	0 - 15	11.0	12.0	16.35	16.25
	15 - 30	9.6	10.1	13.46	17.64
Wheat	0 - 15	17.9	14.04	10.49	11.18
	15 - 30	15.6	15.33	11.16	12.62
No plant	0 - 15	16.37	15.90	14.65	16.73
	15 - 30	11.71	12.55	16.36	20.41

The higher C:N ratio occurred in spring and autumn and lowest in winter. Li et al. (2008) also show a seasonal effect on microbial biomass C and N, where an increase in microbial biomass occurred in early spring translating into higher rates of net N mineralization.

It has previously been documented that of crop residues correlates well with N mineralization (Ford et al., 1989). Nutrient availability can place very substantial constraints on the decomposition rate of organic matter by its associated microflora (Rojas, 2005). The gradual increase in C:N ratio suggest that the potential for N immobilization increases (Gurlevik et al., 2004).

NITRIFICATION

Ammonium (NH_4 +) is the most frequently found form of nitrogen in the biosphere and under oxic conditions, ammonia is not stable and is converted to nitrite (NO_2 –) and nitrate (NO_3 –) by nitrifying bacteria in soil and occupies a central position within the nitrogen cycle (Hanaki et al., 1990; Fiencke et al., 2005).

It is readily taken up by plants and moves freely through soil. The end product of nitrification, nitrate can also easily leach into ground, surface water and can cause significant losses of N from soil (Davidson and Hackler., 1994) and may contribute to NO_3 pollution of surface and ground waters. Therefore reducing the rate of conversion from NH_4 + to NO_3 - can help to maintain more nitrogen in the soil for plant use. Treatments of fertilizers in combination with nitrification inhibitors have been suggested as a technique to reduce nitrification rate and NH_3 volatilization (McCarty et al., 1991; Freney et al., 1992).

A nitrification inhibitor may potentially reduce NO_3 losses due to leaching from NH_4-N-libeiating fertilizer materials, including organic nitrogen sources by maintaining nitrogen as NH_4 which is less susceptible to loss from the soil through this route and through NH_3 volatilization (Bremner and McCarty. 1993; Kholdebarin et al., 1998; Subbarao et al., 2006).

Nitrification inhibitors have been successfully employed to reduce nitrate leaching into ground water and nitrous oxide emissions to the atmosphere and N retention in soil (Wolt, 2004; Di and Cameron, 2002). We also observed that the application of fertilizer without nitrification inhibitors had no effect on nitrifying bacterial population. The nitrification inhibitor Potassium oxalate inhibited nitrifiers, which was reflected in the reduction of NO_3 losses through leaching from fertilizer material (Egamberdiyeva et al., 2001). In our study the application of fertilizer increased net nitrification by six times compared to the control. Nitrification inhibitors reduced the rate of nitrification and so increased the thermal time required for NH_4-N depletion and NO_3-N accumulation in soil amended with NH_4 -N-forming materials, compared with fertilizer alone. Some authors suggest the reduction of nitrification after application of nitrification inhibitors (Hyman et al., 1990; Smith and Hadley, 1992; Vallejo et al. 2005). Our results indicate that nitrification inhibitor Potassium Oxalate slows the rate of nitrification and may effectively reduce potential NO_3^- leaching losses. They did not affect negatively on soil biological activities. Similar results observed by other authors where nitrification inhibitors have no effects on microbial biomass, respiration and enzymatic activities (Di and Cameron, 2002; Mahmood et al., 2005).

Other authors also reported that Thiourea inhibited the nitrifyng activity of nitrifiers, which was reflected in the increased availability and efficiency of fertilizer nitrogen to the rice plants and indicated a potentiality as a nitrification inhibitor (Fog, 1988). Amberger (1989) and Zerulla et al. (2001) reported that nitrification inhibitors, such as DCD and DMPP (3,4-dimethylpyrazole phosphate) have been used to increase the efficiency of N supply from fertilizers. According Di and Cameron (2002) nitrification inhibitors can help retain more nutrients, particularly N, in the top layer of the soil, this makes more N available for pasture uptake over a longer period. They also stated that the effectiveness of nitrification inhibitors to reduce NO_3- leaching and N_2O emissions from grazed pastures depends on application time, whereas best period considered shortly after grazing before the urine-N is converted to NO_3-.

Keeney (1983) reported that nitrification rates differed significantly in soils from different climatic regions. In other studies Macduff and White (1985) observed that nitrification is influenced by soil and environmental factors, primarily pH, temperature, moisture and O_2 supply. Nitrification also is assumed to be more sensitive to heavy metal toxicity (Baath, 1989).

In soils the oxidation of ammonia to nitrite is carried out by species *Nitrosomonas* and while the oxidation of nitrite to nitrate is carried out by *Nitrobacter* (Bock et al., 1989). In early studies Riha et al. (1986) observed that heterotrophic bacteria were found to be better competitors for the available ammonium than the nitrifying bacteria. Verhagen et al. (1995) reported that *N. europaea* looses the competition for ammonium from the heterotrophic species *A. globiformis* as well as from roots of *P. lanceolata.* Other authors found low nitrifying bacteria and nitrate in unfertilized soil under grasslands and forests (Spieck and Bock, 2005).

Tortoso and Hutchinson (1990) reported that in a North American temperate soil, numbers of nitrifying bacteria were between $2.8 x 10^4$ and $2.3 x 10^6$ CFU/g soil, respectively. We also observed similar number of nitrifying bacteria in soil differ with crop cultivation (Table 4).

Table 4. The number of nitrifying bacteria under different crops in sierozem soil (10^4 CFU/g soil.(The surface (0-15 cm) and sub-surface (15-30 cm) soil depths)

Plant	Soil depth (cm)	Winter	Spring	Summer	Autumn
Cotton	0 - 15	2.7	4.2	5.3	4.8
	15 - 30	3.8	5.5	5.8	4.4
Alfalfa	0 - 15	1.2	3.5	4.0	2.1
	15 - 30	2.5	4.1	5.3	3.0
Maize	0 - 15	1.6	2.5	2.9	1.9
	15 - 30	2.0	3.1	3.7	2.1
Wheat	0 - 15	4.1	5.2	5.0	4.2
	15 - 30	5.0	5.6	6.1	5.1
No plan	0 - 15	0.1	0.4	0.5	0.3
	15 - 30	0.3	0.5	0.6	0.4

Plant species had strong influence on nitrifying bacterial population in soil and marked effects were found to have taken place on the nitrifyers populations under cotton, alfalfa, and wheat. The highest density of nitrifying bacteria (6.2 x10^4 CFU/g soil) was observed during summer and the lowest in winter in soil under cotton cultivation (Table 4).

Chao et al. (1993) showed higher numbers of nitrifying bacteria in tropical soil ranged between 1.1x10^8 and 1.9x10^7 CFU/g soil. According Laanbroek and Gerards (1991) any stimulation of nitrifying activity is undesirable because of possible losses of nitrogen from the soil and it could lead to an accumulation of its toxic nitrite, in the soil. The stimulation of nitrification process by low amounts of various pollutants has already been observed (Andersch and Anderson, 1991; De Boer and Kowalchuk, 2001; Sverdrup et al., 2007).

DENITRIFICATION

Biological nitrate elimination can be achieved through nitrification and denitrification. The denitrification is involved in the destruction of stratospheric ozone, global warming, and the loss of ammonia fertilizer to the atmosphere as N_2 (Atlas and Bartha. 1987; Shelton et al., 2000). Denitrification is also important in depleting soil fertility and reducing NO_3- pollution of groundwater and various bacterial species such as *Aerobacter, Klebsiella* and *E.coli, Pseudomonas* are involved in the process (Keeney and Nelson, 1982). Cheneby et al. (2000) investigated denitrifying bacteria in three agricultural soils and found a good correlation between the number and diversity of denitrifiers and soil type. The various factors affect on denitrification by bacteria such as levels of NO_3 - NO_2, amount and type of decomposable organic C supply, soil aeration status, pH and temperature (Ottow and Fabig, 1985; Qian et al., 1997).

In other works DeCatanzaro and Beauchamp (1985) concluded that denitrifier activity was related to the amount of water-soluble C made available in plant residues. Paul and Beauchamp (1989) observed that the denitrification rates in various soils were highly correlated with the water-soluble C concentration in the soil, where addition of organic C to soil can slow down denitrification by increasing the immobilization of N in microbial cells.

Root exudates have long been recognised as a major potential source of energy for many microorganisms in soil. They are considerably stimulated in the vicinity of the roots, as a consequence of the release by roots of a range of C-compounds (Lynch and Whipps, 1991; Hinsinger et al., 2006). Several studies reported that plants have a little effect on the ability of soils under root system the denitrify NO_3- (Haider et al. 1987; MeCarty and Bremner, 1993). In earlier studies Haller and Stolp (1985) found an increase of denitrifying activity in the rhizosphere. MeCarty and Bremner (1993) observed in their study that the crop plants did not significantly increase the amount of leachable organic C in surface soil. Thus any water-soluble portion of rhizodeposited C is rapidly immobilized or mineralized by microorganisms in the rhizosphere and has little influence on the ability of subsoils to denitrify.

Lower competition between plants and denitrifying bacteria were found during early spring and fall when plant uptake is minimal, and greater NO3 availability in anaerobic condition, lead to increase denitrification (Zak and Grigal, 1991). Groffinan and Yiedje (1989) suggested that plant uptake was an important mechanism regulating seasonal patterns of denitrification. In other studies Groffman et al., (1992) reported that over 80% of the annual denitrification activity occurred during high soil wetness, early spring and late fall (Liang et al., 1991; Rover et al. 1998). Similar observation is reported by Paul and Beauchamp (1989), where more irrigation raised more NO_3 by mineralization and thus the total denitrification have been increased. In flooded soils commonly used for rice farming, submergence of soils results in the depletion of O_2, creating conditions favourable for denitrification (Aulakh et al., 2000). They also report that substantial amounts of fertilizer N are lost via denitrification in semiarid subtropical soils under nearly saturated and flooded conditions that are prevalent in wetland rice-growing areas.

In conclusion, soil microorganisms have a great influence on many organic matter oxidation, hydrolysis and degradation, and these in turn reflected in the natural cycles of nitrogen and other elements. Those processes are sensitive to any factor occurring in soil ecosystems, including environmental conditions, moisture content, temperature, input of chemicals, soil type and crop management strategies. Therefore, knowledge of the factors influencing soil microorganisms involved in N cycling is fundamental for sustainable environmental management. More research is needed to understand the population dynamics of microorganisms involved in N cycling and how they respond to changes in soil management practice and ecological factors.

REFERENCES

Amberger, A. (1989). Research on dicyandiamide as a nitrification inhibitor and future outlook. *Commun. Soil science plant analysis*, 20, 1933-1955.

Andersch, I. and Anderson, J.P.E. (1991). Influence of pesticides on nitrogen transformations in soil. *Environmental Toxicology and Chemistry*, 30, 153-158.

Aulakh, M.S, Khera, T.S. and Doran, J.W. (2000). Mineralization and denitrification in upland, nearly-saturated and flooded subtropical soil. I. Effect of nitrate and ammoniacal nitrogen. *Biology and Fertility of Soils,* 31, 162–167.

Atlas, RM. and Bartha R. (1987). *Microbial Ecology*: Fundamentals and Applications. 2nd Menlo-Park, CA. The Benjamin/Cummings Publishing Company, p. 310-317.

Balandreau, J. (1983). Microbiology of the association. *Canadian Journal of Microbiology,* 29, 851–859.

Baldani, J.I. and Baldani, V.L.D. (2005). History of biological nitrogen fixation research in graminaceous plants: special emphasis on the Brazilian experience. *Ann. Brazil Acad. Sci.,* 77, 549–579.

Bhattacharjee, R.B., Singh, A. and Mukhopadhyay, S.N. (2008). Use of nitrogen-fixing bacteria as biofertiliser for non-legumes: prospects and challenges. *Applied Microbiology and Biotechnology,* 80, 199–209.

Baath, E. (1989). Effects of heavy metals in soil on microbial processes and popniations, a review. *Water, Air and Soil Pollution,* 47, 335-379.

Baldani, V.L.D., Baldani, J.I. and Dobereiner, J. (2000). Inoculation of rice plants with the endophytic diazotrophs *Herbaspirillum seropedicae* and *Burkholderia* spp. *Biol. Fertil. Soils,* 30: 485-491.

Beier, C., Schmidt, I.K. and Kristensen, H.L. (2004). Effects of Climate and Ecosystem Disturbances on Biogeochemical Cycling in a Semi-Natural Terrestrial Ecosystem. *Water, Air, and Soil Pollution, Focus,* 4, 191-206.

Berkum, P., Van. and Bohlool, B.B. (1980). Evaluation of nitrogen fixation by bacteria in association with roots of tropical grasses. *Microbiology Rev.*, 44(3): 491-517.

Bertin, C., Yang, X., Weston, L.A. (2003). The role of root exudates and allelochemicals in the rhizosphere. *Plant Soil* 256: 67-83.

Biggins, D.R. and Postgate, J.R. (1969). Nitrogen fixation by cultures and cell- free extracts of. *Mycobacteriumflavum 301 General Microbiology*, 56, 181-193.

Biswas, J.C., Ladha, J.K. and Dazzo, F.B. (2000). Rhizobia inoculation improves nutrient uptake and growth of lowland rice. *Soil Science Society of America*, 64, 1644–1650.

Blackburn, T.H. and Henriksen, K. (1983). Nitrogen cycling in different types of sediments from Danish waters. *Limnology and Oceanography*, 28, 477-493.

Boddey, R.M., Oliveira, O.C.D., Urquiaga, S., Reis, V.M., Olivares, F.L.D., Baldani, V.L.D. and Döbereiner, J. (1995). Biological nitrogen fixation associated with sugarcane and rice: contributions and prospects for improvement. *Plant and Soil,* 174,195–209.

Bock, E., Koops, H.P. and Harms, H. (1989). *Nitrifying bacteria.* P, 81-96.

Boyle, S.A., Yarwood, R.R., Bottomley, P.J. and Myrold, D.D. (2008). Bacterial and fungal contributions to soil nitrogen cycling under Douglas fir and red alder at two sites in Oregon. *Soil Biology and Biochemistry,* 40, 443-451.

Bormann, B.T., Bormann, F.H., Bowden, W.B., Pierce, RS, Hamburg, S.P., Wang, D., Snyder, M.C., Li, C.Y. and Ingersoll, R.C. (1993). Rapid N2 fixation in pines, alder, and locust evidence from the sandbox ecosystem study. *Ecology* 74, 583-598.

Bossio, D.A, Scow, K.M. and Gunapala, K.J. (1998). Determinants of soil microbial communities: Effects of agricultural management, season, and soil type on phospholipid fatty acid profiles. *Microbial Ecology,* 36, 1-12.

Bremner, J.M. and McCarty, G.W. (1993). Inhibition of nitrification in soil by allelochemicals derived from plants and plant residues. *Soil Biochemistry,* 8, 181-218.

Buckley, D.H. and Schmidt, T.M. (2001). The structure of microbial communities in soil and the lasting impact of cultivation. *Microbial Ecology*, 42, 11-21.

Burns, R.C. and Hardy, R.W.F. (1975). *Nitrogen Fixation in Bacteria and Higher Plants*. New York: Springer-Verlag.

Chao, W.L, Gan, K.D. and Chao, C.C. (1993). Nitrification and nitrifying potential of tropical and subtropical soils. *Biology and Fertility of Soils*,15,87-90.

Choudhury, A.T.M., Zaman, S.K. and Bhuiyan, N.I. (2002). *Sesbania: a potential nitrogen source for sustainable rice production.* In: Kennedy IR, Choudhury ATMA (eds) Biofertilisers in action. Rural Industries Research and Development Corporation, Canberra, pp 94–101.

Coleman, D.C. and Crossley, D.A. (1996). *Fundamentals of Soil Ecology.* Academic Press, San Diego

Cheneby, D., Philippot, L., Hartmann, A. and Germon. J.C. (2000).16S rDNA analysis for the characterization of denitrifying bacteria isolates from three agricultural soils. *FEMS Microbiology Ecology,* 34, 121–128.

Dalal, R.C., and Mayer, R.J. (1987). Long-term trends in fertility of soils under continuous cultivation and cereal cropping in southern Queensland. VII. Dynamics of nitrogen mineralisation potentials and microbial biomass. *Aust. J. Soil Res*. 25, 461-72.

Davidson, E.A. and Hackler, J.L. (1994). Soil heterogeneity can mask the effects of ammonium availability on nitrification. *Soil Biology and Biochemistry*, 26, 1449-1453.

DeLuca, T.H., Drinkwater, L.E., Wiefling, B.A. and DeNicola, D.M. (1996). Free-living nitrogen-fixing bacteria in temperate cropping systems: influence of nitrogen source. *Biology and Fertility of Soils,* 23, 140–144.

De Boer, W., Kowalchuk, G.A. (2001): Nitrification in acid soils: micro-organisms and mechanisms. *Soil Biol. Biochem.* 33: 853-866.

DeCatanzaro, J.B. and Beauchamp, E.G. (1985). The effect of some carbon substrates on denitrification rates and carbon utilization in soil. *Biology and Fertility of Soils,* 1, 183-187.

Di, H.J. and Cameron, K.C. (2002). Nitrate leaching in temperate agroecosystems: Sources, factors, and mitigating strategies. *Nutr. Cycling Agroecosys,* 46, 237-256.

Dijkstra, F. and Cheng W. (2008). Increased soil moisture content increases plant N uptake and the abundance of 15N in plant biomass. *Plant and Soil,* 302, 263-271.

Doyle, J.J. and Luckow, M.A. (2003). The rest of the iceberg. Legume diversity and evolution in a phylogenetic context. *Plant Physiol.* 1331:900–910.

Dowdell, RJ. (1982). Fate of nitrogen applied to agricultural crops with particular reference to denitrification. *Philosophical Transactions of the Royal Society of London*, 296, 363–373.

Dueri, S., Calanca, P.L. and Fuhrer, J. (2007). Climate change affects farm nitrogen loss - A Swiss case study with a dynamic farm model. *Agricultural Systems,* 93, 191-214.

Egamberdiyeva, D., Mamiev, M. and Poberejskaya, S. (2001). The influence of mineral fertilizer combined with a nitrification inhibitor on microbial populations and activities in calcareous Uzbekistanian soil under cotton cultivation. *The Scientific World Journal,* 1(S2), 108-113.

Egamberdiyeva, D., Juraeva, D., Haitov, B., Gafurova, L. (2004). Microbial and biochemical changes induced by rotation and tillage in a calcareous soil under melon, tomato, wheat and cotton production. In. *Proceeding book of 26th Annual Conservation Tillage Conference for Sustainable Agriculture,* Auburn, USA, 224-228.

Egamberdiyeva, D. and Hoflich, G. (2002). Root colonization and growth promotion of winter wheat and pea by Cellulomonas spp. at different temperatures. *Plant Growth Regulation,* 38, 219-224.

Elmerich, C., Zimmer, W. and Vieille, C. (1992). *Associative nitrogenfixing bacteria.* In: Stacey G, Burris RH, Evans HJ (eds) Biological nitrogen fixation. Chapman and Hall Inc, New York, pp 212–258.

Elmerich, C. and Newton W.E. (2007). *Associative and Endophytic Nitrogen-fixing Bacteria and Cyanobacterial Associations*.S pringer, The Netherlands.

Estrada-De Los Santos, Bustillo-Cristales, R. and Cabellero-Mellado, J. (2001). *Burkholderia*, a genus rich in plant associated nitrogen fixers with wide environmental and geographic distribution. *Applied and Environmental Microbiology*, 67, 2790-2798.

Feng, Y, Shen, D. and Song, W. (2006). Rice endophyte *Pantoea agglomerans* YS19 promotes host plant growth and affects allocations of host photosynthates. *Applied Microbiology,* 100, 938-945.

Fiencke, C., Spieck, E. and Bock, E. 2005. Nitrifying Bacteria. In: Werner, D. and W. E. Newton (eds.), *Nitrogen Fixation in Agriculture, Forestry, Ecology, and the Environment,* 255-276.

Franche, C., Lindström, K. and Elmerich, C. (2009). Nitrogen-fixing bacteria associated with leguminous and non-leguminous plants. *Plant and Soil,* 321, 35–59.

Ford, P.B., Hargrove, W.L. and Neely, C.L. (1989). "Crop residue decomposition under conventional and no-tillage," Irt Anronomv Abstracts. (Agronomy Society of America: Madison, WI). p. 279.

Fog, K. (1988). The effect of added nitrogen on rhe rate of decompositioonf organism atter. *Biology Rev.,* 63, 433-462.

Govedarica, M., Jarak, M., Milosevich, N., Marinkovich, B., Hadzik, V., Ubavic, M. 1995. Number and enzymatic activity of microorganisms in soil under maize grown after sugar beet. *Savremena-poljopriveda.* 43, 133-138.

Grayston, S.J., Wang, S., Campbell, C.D. and Edwards, A.C. (1998) Selective influence of plant species and microbial diversity in the rhizosphere. *Soil Biology and Biochemistry* 30, 369–378.

Groffman, P.M. and Yiedje, J.M. (1989). Denitrification in north temperate forest soils: Spatial and temporal patterns at the landscape and seasonal scales. *Soil Biology and Biochemistry,* 21, 613-620.

Gyaneshwar, P., James, E.K., Mathan, N., Teddy, P.M., Reinhold-Hurek, B. and Ladha, J.K. (2001). Endophytic colonization of rice by a diazotrophic strain of Serratia marcescens. *Bacteriology,* 183, 2634-2645.

Gyaneshwar, P., James, E.K., Reddy, P.M. and Ladha J.K. (2002). Herbaspirillum colonization increases growth and nitrogen accumulation in aluminium-tolerant rice varieties. *New Phytologist*, 154, 131–145.

Groffman, P.M., Gold, A. and Simmons, R.C. (1992). Nitrate dynamics in riparian forests: Microbial studies. *Journal of Environmental Quality*, 21:666-671.

Gurlevik, N., Kelting, D.N. and Allen, H.L. (2004). Nitrogen mineralization following vegetation control and fertilization in a 14-yearold Loblolly pine plantation. *Soil Science Society of America*, 68, 272–281.

Haller, T, Stolp, H. (1985). Quantitative estimation of root exudation of maize plants. *Plant and Soil,* 86, 207-216.

Hanaki, K., Wantawin, C. and Ohgaki, S. (1990). Effects of heterotrophs on nitrification in a suspended-growth reactor. *Water Res.,* 24 (3), 289-296.

Haider, K., Mosiem, A.R. and Heinemeyer, O. (1987). The effect of growing plants on denitrification at high soil nitrate concentrations. *Soil Sci. Soc. Am. J.* 51:97-102.

Hartley, A.E. and Schlesinger, W.H. (2002). Potential environmental controls on nitrogenase activity in biological crusts of the northern Chihuahuan Desert. *Envir.*, 5, 293-304.

Hyman, M.R, Kim, C.Y. and Arp, D.J. (1990). Inhibition of ammoniam onooxygenasine Nittosomonaes unopaeb,cy carbon disulfide. *Bacteriology,* 172, 4775- 4782.

Hinsinger, P., Plassard, C. and Jaillard, B. (2006). Rhizosphere: A new frontier for soil biogeochemistry. *Geochemical Exploration*, 88, 210-213.

Hurek, T. and Reinhold-Hurek, B. (2003). Azoarcus sp strain BH72 as a model for nitrogen-fixing grass endophytes. J Biotechnol 106:169–178 doi:10.1016/j.jbiotec.2003.07.010.

Janssen, B.H. (1996). Nitrogen mineralization in relation to C:N ratio and decomposability of organic materials, *Plant and Soil,* 181, 39-45.

Jones, D.L., Healey, J.R., Willett, V.B., Farrar, J.F. and Hodge, A. (2005). Dissolved organic nitrogen uptake by plants – an important N uptake pathway. *Soil Biology and Biochemistry*, 37, 413–423.

Kholdebarin, B., Mozafar, A. and Frossard, E. (1998). Differential inhibition of nitrification by three metabolic inhibitors. *Plant Nutrition*, 21, 929-936.

Kumar, K. and Goh, K.M. (2000). Crop residues and management practices: effects on soil quality, soil nitrogen dynamics, crop yield, and nitrogen recovery. *Advances in Anatomic Pathology*, 68:197–319.

Keeney, D.R. (1983). *Factor affecting the persistence and bioactivity of nitrification inhibitors.* In: Nitrification inhibitors–potentials and limitations, (Eds.): J.J. Meisinger, G.W. Randall and M.L. Vitosh, ASA Special Publication No. 38, American Society of Agronomy, Madison, pp. 33-46.

Keeney, D. R. and Nelson, D.W. (1982). *Nitrogen-inorganic forms*, In A. L Page and D. R. Keeney (eds.) Methods of Soil Analysis Part 2. Chemical and Microbiological Properties of Soil Analysis. Agronomy Monograph No. 9 ASA, Madison. pp 648-654.

Kennedy I.R., Pereq-Gerk L.L., Wood C., Deaker R., Gilchrist K., Katupitiya S. (1997). Biological nitrogen fixation in non-leguminous field crops: Facilitating the evolution of an effective association between *Azospirillum* and wheat. *Plant Soil*, 194: 65-79.

Laanbroek, H.J. and Gerards, S. (1991). Effects of organic manure on nitrification in arable soils. *Biol. Fert. Soils* 12: 147-153.

Li, J., Zhao, B., Li, X., Jiang, R., So, H. (2008). Effects of long-term combined application of organic and mineral fertilizers on microbial biomass, soil enzyme activities and soil fertility. *Agricultural Sciences in China*, 7 (3), 336-343.

Li, C.Y. and Castellano, M. (1987). *Azospirillum* isolated from within sporocarps of the mycorrhizal fungi *Hebeloma crustuliniforme, Laccaria laccata and Rhizopogon vinicolor. The British Mycological Society*, 88, 563–565.

Li, C.Y., Massicotte, H.B. and Moore L.V.H. (1992). Nitrogen fixing *Bacillus sp.* associated with Douglas-fir tuberculate ectomycorrhizae. *Plant and Soil*, 140, 35-40.

Li, C.Y. and Hung, L.L. (1987). Nitrogen-fixing (acetylenereducing) bacteria associated with ectomycorrhizae. *Plant and Soil*, 98, 425-428.

Liang, B.C., Remillard, M. and MacKenzie, A.F. (1991). Influence of fertilizer, irrigation, and non-growing season precipitation on soil nitrate-nitrogen under corn. *Environmental Quality,* 20, 123–128.

Lynch, J.M., Whipps, J.M. (1991). *Substrate flow in the rhizosphere*. In:Keister DL, Cregan, PB (eds) The rhizosphere and plant growth. Kluwer Academic Publishers, Norwell, Massachusetts, pp 15-24

Macduff, J.H. and White, R.E. (1985) "Net mineralization and nitrification rates in clay soil measured and predicted in permanent grass land from soil temperature and moisture cntent", *Plant and Soil*, 86, 151- 172.

Mahmood, T., R. Ali, K.A. Malik, Z. Aslam and S. Ali. 2005. Seasonal pattern of denitrification under an irrigated wheat-maize system fertilized with urea and farmyard manure in different combinations. *Biol. Fertil. Soils*, 42: 1-9.

Marschner, P., Crowley, D., Yang, C.H. (2004). Development of specific rhizosphere bacterial communities in relation to plant species, nutrition and soil type. *Plant Soil* 261: 199-208.

McCarty, G.W., Bremner, J.M. and Schmidt, E.L. (1991). Effects of phenolic acids on ammonia oxidation by terrestrial autotrophic nitrifying organisms. *FEMS Microbiology Ecology,* 85, 345-450.

MeCarty, G.W. and Bremner, J.M. (1993). Factors affecting the availability of organic carbon for denitrification of nitrate in subsoils. *Biology and Fertility of Soils*, 15, 132-136.

Mirza, M.S., Rasul, G., Mehnaz, S., Ladha, J.K., So, R.B., Ali, S. and Malik, K.A. (2000). *Beneficial effects of inoculated nitrogen-fixing bacteria on rice*. In: Ladha JK, Reddy PM (eds) The quest for nitrogen fixation in rice. International Rice Research Institute, Los Banos, pp. 191–204.

Mrkovacki, N. and Milic, V. (2001). Use of *Azotobacter chroococcum* as potential useful in agricultural application. *Annals of Microbiology*, 51, 145- 158.

Murder, E.G. (1975). *Physiology and ecology of free-living, nitrogenfixing bacteria.* In: Nitrogen fixation by free-living microorganisms.W.D.P. Stewart (ed.). Cambridge University Press,New York.

Nakata, P.A. (2002). The generation of a transposon-mutagenized Burkholderia glumae library to isolate novel mutants. *Plant Science*, 162, 267–271.

Okon, Y. and Labandera-Gonzales, C.A. (1994). Agronomic applications of Azospirillum: an evaluation of 20 years worldwide field inoculation. *Soil Biology and Biochemistry*, 26, 1591–1601.

Orr, C.H., James, A., Leifert, C., Cooper, J.M. and Cummings, S.P. (2011). Diversity and Activity of Free-Living Nitrogen-Fixing Bacteria and Total Bacteria in Organic and Conventionally Managed Soils. *Applied and Environmental Microbiology*, 77, 911-919.

Ottow, J.C.G. and Fabig, W. (1985). Influence of oxygen aeration on denitrification and redox level in different bacterial batch cultures. In: Planetary Ecology. D.E. Caldwell, J.A. Brierley and C.L. Brierley (eds). Van Nostrand Reinhold, New York. P. 427–440.

Paul, J.W. and Beauchamp, E.G. (1989). Denitrification and fermentation in plant-residue-amended soil. *Biology and Fertility of Soils,* 7, 303-309.

Phillips, D.A., Fox, T.C. and Six, J. (2006). Root exudation (net efflux of amino acids) may increase rhizodeposition under elevated CO2. *Global Change Biology,* 12, 561-567.

Powlson, D.S., Brookes, P.C. and Christensen, B.T. (1987). Measurement of soil microbial biomass provides an early indication of changes in total soil organic matter due to straw incorporation. *Soil Biology and Biochemistry*, 19, 159-164.

Qian, J.H., Doran, J.W., Weier, K.L., Mosier, A.R., Peterson, T.A. and Power J.M. (1997). Soil denitrification and nitrous oxide losses under corn irrigated with high-nitrate groundwater. *Environmental Quality*, 26, 348–360.

Rennie, R.J, (1980). N-isotope dilution as a measure of dinitrogen fixation by Azospirillure brasilense. 15 associated with maize. *Canadian Journal of Botany*, 58, 21-24.

Riha, S.J., Campbell, G.S. and Wolfe, J. (1986). A model of competition for ammonium among heterotrophs, nitrifiers and roots. *Soil Science Society of America*, 50, 1463-1466.

Riggs, P.J., Chelius, M.K., Iniguez, A.L., Kaeppler, S.M. and Triplett, E.W. (2001). Enhanced maize productivity by inoculation with diazotrophic bacteria. *Plant Physiology,* 28, 829-836.

Roesch, L.F.W., Camargo, F.A.O., Bento, F.M. and Triplett, E.W. (2008). Biodiversity of diazotrophs within the soil, root and stem of field grown maize. *Plant and Soil*, 302, 91–104.

Rojas, J.C. (2005). Factors influencing responses of loblolly pine stands to fertilization. PhD dissertation, North Carolina State Univ., Raleigh, NC. 147 p.

Rosswall, A. (1976). The internal nitrogen cycle between microorganisms, vegetation and soil. In Svensson, BH. and Soderlund, R. (eds.). Nitrogen, Phosphorus and Sulphur – Global Cycles. SCOPE Report 7, *Ecological Bulletins*, (Stockholm), 22, 157-167.

Rover, M., Heinemeyer, O. and Kaiser, E.A. (1998). Microbial induced nitrous oxide emissions from an arable soil during winter. *Soil Biology and Biochemistry,* 14, 1859–1865.

Rózycki, H., Dahm, H., Strzelczyk, E., Li, C.Y. (1999). Diazotrophic bacteria in root-free soil and in the root zone of pine (*Pinus sylvestris* L.) and oak (*Quercus robur* L.). *Applied Soil Ecology* 12: 239–250.

Shabaev, V.P., Smolin, V.Y. and Strekozova, V.I. (1991). The effect of *Azospirillum brasilense* Sp 7 and *Azotobacter chroococcum* on nitrogen balance in soil under cropping with oats (*Avena sativa* L.). *Biology and Fertility of Soils*, 10, 290-292.

Shaharoona, B., Arshad, M., Zahir, Z.A. and Khalid A. (2006). Performance of *Psedomanas* ssp. Containing ACC-deaminase for improving growth and yield of maize (*Zea mays* L.) in the presence of nitrogenous fertilizer. *Soil Biology and Biochemistry*, 38, 2971-2975.

Shelton, R.D., Sadeghi, A.M. and McCarty, G.W. (2000). Effect of soil water content on denitrification during cover crop decomposition. *Soil Sci.* 165:365-371.

Schimel, D.S., Coleman, D.C. and Harton, K.A. (1985). Soil organic matter dynamics in paired rangeland and cropland toposequences in North Dakota. *Geoderma,* 30, 201-214.

Schimmel, J. and Bennett, J. (2004). Nitrogen mineralization: challenges of a changing paradigm. *Ecology* 85, 591-602.

Singh, J.S., Raghubanshi, A.S., Singh, R.S. and Srivastava, S.C. (1989). Microbial biomass acts as a source of plant nutrient in dry tropical forest and savanna. *Nature,* (London), 399, 499-500.

Smith, K. and Hadley, P. (1992). Nitrogen fertilizerv alueo f activateds ewaged erivedp rotein:e ffecto f environmenat ndn itrificationi nhibitoro n NO3-r eleases, oil microbiala ctivifya ndy ield of summerc abbage. *Fertil. Res,* 33, 47-57.

Spieck, E. and Bock, E. (2005). The lithoautotrophic nitrite-oxidising bacteria. In: Garrity G.M., Brenner D.J., Krieg N.R., Staley J.T. (eds.), Bergey´s Manual of Systematic Bacteriology, 2nd Ed., Vol 2, Part A, 149-153. Springer, USA.

Subbarao, G.V., Ito, O., Sahrawat, K.L, Berry, W.L, Nakahara, K., Ishikawa, T., Watanabe, T., Suenaga, K., Rondon M. and Rao, I.M. (2006). Scope and strategies for regulation of nitrification in agricultural systems - challenges and opportunities. *Critical Reviews in Plant Sciences,* 25, 303-335.

Stark, CH. and Richards, KG. (2008). The Continuing Challenge of Agricultural Nitrogen Loss to the Environment in the Context of Global Change and advancing research. *Dynamic Soil, Dynamic Plant*, 2, 1-12.

Sparling, G.P., Feltham, C.W., Reynolds, J., West, A.W., Singleton, P. (1990). Estimation of soil microbial C by a fumigation-extraction method: use on soils of high organic matter content, and a reassessment of the *k*ec-factor. *Soil Biology and Biochemistry* 22, 301-307.

Sverdrup, L.E., Hagen, S.B., Krogh, P.H. and van Gestel, C.A.M. (2007). Benzo(a)pyrene show low toxicity to three species of terrestrial plants, two soil invertebrates and soilnitrifying bacteria. *Ecotoxicol. Environ. Safety,* 66, 362-368.

Tortoso, A.C., Hutchinson, G.L. (1990). Contributions of autotrophic and heterotrophic nitrifiers to soil NO and N2O emissions. *Applied and Environmental Microbiology*, 56, 1799–1805.

Vallejo, A., García-Torres L., Díez, J.A., Arce, A. and López-Fernández, S. (2005). Comparison of N losses (NO3 −, N2O, NO) from surface applied, injected or amended (DCD) pig slurry of an irrigated soil in a Mediterranean climate. *Plant and Soil*, 272, 313-325.

Vandamme, P., Goris, J., Chen, W.M., Vos, de P. and Willems, A. (2002). Burkholderia tuberum sp. nov. and Burkholderia phymatum sp. nov., nodulate the roots of tropical legumes. *Systematic and Applied Microbiology*, 25, 507–512.

Verhagen, E.L.M., Laanbroek, H.J. and Woldendorp, J.W. (1995). Competition for ammonium between plant roots and nitrifying and heterotrophic bacteria and the effects of protozoan grazing. *Plant and Soil*, 170, 241-250.

Wang, W.J., Baldock, J.A., Dalal, R.C. and Moody, P.W. (2004). Decomposition of plant materials in relation to nitrogen availability and biochemistry determined by NMR and wet-chemical analysis. *Soil Biology and Biochemistry*, 36, 2045–2058.

Wirth, S., Egamberdieva D. (2008). *Soil microbial activities affected by soil depth, seasonal change and cultivation type*. In proceeding book of the International scientific conference "The Role of National University in the development of the higher education and science in Uzbekistan". pp. 93-97.

Wolt, J.D. (2004). A meta-evaluation of nitrapyrin agronomic and environmental effectiveness with emphasis on corn production in the Midwestern USA. *Nutr. Cycl. Agroecosyst,* 69, 23-41.

Wright, S.F. and Weaver, R.W. (1981). Enumeration and identification of nitrogen-fixing bacteria from forage grasses. *Appl. Environ. Microbiol*. 42:97-101.

Yanni, Y.G. and El-Fattah, F.K.A. (1999). Towards integrated biofertilization management with free living and associative dinitrogen fixers for enhancing rice performance in the Nile delta. *Symbiosis*, 27, 319–331.

Zak, D.R. and Grigal., D.F. (1991). Nitrogen mineralization, nitrification and denitrification in upland and wetland ecosystems. *Oecologia,* 88, 189-196.

Zak, D.R., Pregitzer, K.S., King, J.S. and Holmes, W.E. (2000). Elevated atmospheric CO2, fine roots and the response of soil microorganisms: a review and hypothesis. *New Phytologist,* 47, 201-222.

Zerulla, W., Barth, T., Dressel, K.E., Locquenghien, K.H. and Wissemeier, A.H. (2001). 3,4-Dimethylpyrazole phosphate (DMPP) – a new nitrification inhibitor for agriculture and horticulture. *Biology and Fertility of Soils,* 34, 79-84.

Zou, L., Chen, Y.L. and Yan, T.Z. (2000). Ecological distribution and seasonal change of soil microorganisms in pure and mixed plantation. *Forestry Research,* 11, 106-108.

In: Soil Nutrients
Editor: Mohammad Miransari

ISBN: 978-1-61324-785-3

Chapter 8

MANGANESE DEFICIENCY AND TOXICITY IN SOILS AND MECHANISMS ADOPTED BY PLANTS IN ORDER TO FACE THESE NUTRITIONAL DISORDERS: A REVIEW

T. Chatzistathis[a]*, I. Therios[a] and D. Alifragis[b]
[a]Laboratory of Pomology, School of Agriculture, Aristotle University, Thessaloniki, Greece
[b]Laboratory of Forest Soils, School of Forestry and Natural Environment, Aristotle University, Thessaloniki, Greece

ABSTRACT

The mean Mn concentration in soils ranges from extremely low levels to 10000 p.p.m., depending on soil type, as well as on climate. The basic reason for the great variation of Mn concentration among different soil types is the differential Mn concentration of different rocks and minerals from which they were originated, through the process of weathering. In soils, manganese is present under six forms: a) soluble Mn (that of soil solution), b) exchangeable Mn, c) organically bound soluble Mn, d) easily reducible Mn, e) the Mn of oxides-hydroxides, and finally, f) insoluble Mn complexed with organic matter. The main factors responsible for increasing or decreasing soil available Mn are: pH, soil moisture, organic matter, soil texture, climatic conditions, soil microorganisms and especially mycorrhiza, ionic form of N fertilization (NO_3 or NH_4), soil disinfection, root exudations, P content of soils e.t.c.

Under the above mentioned factors, Mn deficiency or toxicity disorders may be a reality in soils, and different plant species are 'obliged' to adopt various strategies in order to increase or reduce Mn availability for them, i.e. to increase or reduce Mn uptake. Therefore, in cases of Mn toxicity, some of the strategies that may be adopted by plants in order to reduce Mn uptake, or to alleviate Mn toxicity, are: (i) increase of rhizosphere pH, (ii) enhanced capacity of root system to convert Mn^{2+} into Mn^{4+} (a form that can not

* Corresponding author. Tel.: +30-2310-991391. E-mail address: ChChatzi@in.gr (T. Chatzistathis).

be absorbed by plants), (iii) enhanced Mn compartmentalization in Mn-toxicity tolerant tissues (e.g. root system or other woody tissues) and reduced transport in Mn-toxicity sensitive ones (shoot or leaves), (iv) arbuscular mycorrhiza fungus (AMF) development for the protection of plants from Mn toxicity, (v) genetic, or biochemical defense through the production of specialized enzymes, proteins or aminoacids e.t.c. In contrast to the above, in the cases of Mn deficiency, many plant species are able to acidify their rhizosphere (reduction of Mn^{4+} to Mn^{2+}), to produce organic root exudates in order to facilitate Mn uptake through the formation of Mn-organic complexes, to adjust root morphology in order to be able to absorb more easily Mn^{2+}, to develop symbiosis between soil fungi and root system (e.g. AMF) in order to increase Mn uptake e.t.c. Emergent knowledge gaps are identified and new research aspects are discussed under the most recent scientific data concerning Mn transformations in soils and Mn uptake by plants under Mn toxic and deficient conditions.

INTRODUCTION

Manganese is considered to be an essential nutrient for plant growth and, despite the fact that this is usually found in abundance in most soils, in plants it is found in many cases in very low concentrations, since plant requirements for Mn are low (Therios, 2009). This is why Mn is considered to be a micronutrient, although its importance for plant growth and metabolism is very high: It is contained in three enzymes' complexes (those of photosystem II, MnSOD and acid phosphatases) and it is an activator for a large number of enzymes (about 35), as for example those catalyzing reactions of oxidation-reduction, decarboxylations, hydrolysis e.t.c. (Burnell, 1988). Due to its great importance which Mn plays in photosystem II, photosynthetic rates may be disturbed under Mn deficiency (Okhi, 1985; Campbell and Nable, 1988; Marschner, 1995; Singh et al., 2001), as well as under Mn toxicity conditions (Houtz et al., 1988; Nable et al., 1988; Macfie and Taylor, 1992; Kitao et al., 1997a; Kitao et al., 1997b; Kitao et al., 1999; Subrahmanyan and Rathore, 2000).

The fact that only a small fraction of soil Mn is available to plants, despite the great, in most cases, total soil Mn content may be explained by the low solubility of soil Mn. The solubility of Mn is of significance since its supply to plants depends mainly on the soluble Mn pool in the soil. In well drained soils, the solubility of Mn always increases with the increase of soil acidity. However, the ability of Mn to form anionic complexes and to complex with organic ligands may contribute to increased Mn solubility in the alkaline pH range (Kabata-Pendias, 2001). So, in highly alkaline ($pH>8$), in organic, calcareous, or calciferous soils, in acid-highly leached soils, and/or in soils with fluctuating moisture content Mn deficiency may be observed (Alifragis and Papamichos, 1994). On the other hand, Mn toxicity conditions are common in areas where the crops are overfertilized with $(NH_4)_2SO_4$ and/or Mn containing fertilizers, and in agricultural areas located nearby to highly polluted areas (El-Jaoual and Cox, 1998; Qing-Ren et al., 2002). Furthermore, Mn toxicity may be observed in soils rich in Mn (e.g. close to mines), or/and irrigated with high Mn concentration water (Chatzistathis et al., 2010).

The aim of this review is to present and discuss the most important research findings and aspects, from the oldest, chronologically, until the most recent ones, concerning Mn transformations in soils under different physicochemical conditions, and its uptake by plants under Mn toxicity and deficient conditions. Finally, factors contributing to increased tolerance of plants to Mn toxicity and deficiency, as well as mechanisms adopted by plants to face these nutritional disorders are fully analyzed and discussed.

Mn Content of Different Rocks and Mn Concentrations in Soils

Manganese concentration of the upper part of the earth's crust is on average 650 p.p.m. (Gilkes and McKenzie, 1988). According to Lindsay (1979), the mean Mn concentration of the lithosphere is about 900 p.p.m. The Mn concentration of different rocks varies a lot; generally, the greatest Mn contents are found in basic-eruptive rocks (basalt, gabbro), varying between 1000 and 2000 p.p.m. Manganese contents vary also widely in acid eruptive (granite, rhyolite), metamorphic (schists), as well as in certain sedimentary rocks, varying among 200 and 1200 p.p.m. Average contents are found in limestones (400-600 p.p.m.), while in sands Mn contents are relatively low (20-500 p.p.m.). The great variation of Mn content between the different rocks is responsible for the wide variation of Mn content of soils: from traces (podzols of Poland) to 10000 p.p.m. (unleached alkali soils of Chad) (Aubert and Pinta, 1977). Total Mn content varies also in relation to soil texture: sandy soils are, generally, less rich in manganese, than clayey soils.

Mn Containing Minerals

From the different minerals containing Mn (different oxides-hydroxides, carbonate, siliceous, sulphureous salts), the most important are those of oxides-hydroxides, which have highly adsorptive capacity for heavy metals due to their negative charge. Taylor and McKenzie (1966) refer that Co in many cases is adsorbed by Mn oxides-hydroxides; this is of great importance for agriculture, since Co availability for plants depends on the abundance of Mn oxides-hydroxides in soils. Only a moderate abundance of Mn oxides-hydroxides could be responsible for Co deficiency, in cases that it is applied as a fertilizer. Some of the most important Mn oxides-hydroxides are birnesite, hollandite, pyrolousite, manganite e.t.c. (McKenzie, 1977).

Mn Sources in Soils

Manganese in soils could be present under three different forms: Mn^{2+}, Mn^{3+} and Mn^{4+}, although the last one is rarely present. The basic Mn sources in soils are the different minerals, by which through the processes of weathering, hydrolysis and dilution, Mn^{2+} is 'liberated' in soil solution (a form that can be absorbed by plants).

Manganese Oxidation and Reduction in Soils

In well aerated soils, under neutral, or alkaline, pH conditions, the form of Mn that is usually present is that of Mn^{4+}. In contrast to the above, under anaerobic conditions Mn^{4+} is reduced to Mn^{2+}, which can be absorbed by plants, according to the equation:

$$MnO_2 + 4H^+ + 2e^- \rightarrow Mn^{2+} + 2H_2O$$

This reaction can be also observed in wet soils, which have not, however, reached their saturation point of soil moisture. According to Graven et al. (1965), 72 hours after reaching the saturation point in soil moisture, Mn concentration of *Medicago sativa* L. plants has been increased from 426 p.p.m. to more than 6000 p.p.m. Finally, according to Mandal (1961), Mn oxides in soils cultivated with *Oryza sativa* plants have been dramatically decreased after the saturation of these soils with water.

Mn Forms in Soils

The basic Mn forms in soils are the following: a) the soluble Mn (that of soil solution, i.e. Mn^{2+}), b) the exchangeable Mn (that which is adsorbed in soil colloids and can be substituted by other divalent cations and 'liberated' in soil solution), c) the organically bound soluble Mn (that which can be absorbed by plants as a complex with organic matter), d) the easily reducible Mn, e) the inert and insoluble Mn (that of the lattice of minerals, as well as that contained in Mn oxides-hydroxides), f) the insoluble complexes of Mn with organic matter (Smith and Paterson, 1990). From the above mentioned Mn forms, only the soluble and the exchangeable Mn are the ones that can be directly absorbed by plants.

Factors that Influence the Concentration of Soluble Mn in Soils

The most important factors that influence the concentration of soluble Mn in soils are: pH, soil moisture, organic matter, redox potential, soil texture, climatic conditions, soil microorganisms, kind of N fertilization (NH_4 or NO_3), soil disinfection, root exudations, mycorrhiza, and finally, the content of soils in phosphoric anions. Subsequently, it is given in detail the influence of each of the above mentioned factors in soluble Mn concentration.

a) *pH.* The lower is the pH, the more are increased the soluble Mn forms. For a pH decrease from 9 to 4 the solubility of Mn^{2+} is increased 100 times (Keramidas, 1997).

b) *The soil moisture.* In wet soils Mn^{4+} is converted into Mn^{2+} (a form that can be absorbed by plants). When high soil moisture is combined with high soil acidity, plants may suffer from Mn toxicity (Alifragis and Papamichos, 1994).

c) *The organic matter.* The different metals have the ability to form with organic matter organometallic complexes. Usually, these complexes are insoluble. However, there are cases when Mn-organic matter complexes constitute about 80-95% of the total soluble Mn forms; in addition, their stability is usually increased with the increase of pH. So, the importance of Mn-organic matter complexes is great under alkaline soil conditions. Consequently, as in the case of Fe, alkaline conditions do not necessarily provoke Mn deficiency in soils (Keramidas, 1997).

d) *The redox potential.* According to Grass et al. (1973), the reduction of Mn takes place when the redox potential is lower than 300 mV. Patrick and Turner (1968) found that the transformation of the easily reducible Mn into the exchangeable one begins taking place in redox potential values around 400 mV. There is an interaction between pH and redox potential in Mn solubilization. According to Gotoh and Patrick (1972), in a soil with pH 5, saturated with water, almost all the quantity of the reducible Mn converted into soluble and exchangeable in redox potential values around 500 mV. In contrast to that finding, when pH values were between 6 and 8, the greatest part of the reducible Mn was converted into soluble and exchangeable one in much lower redox potentials (200-300 mV) (Gotoh and Patrick, 1972).

e) *The soil texture.* The soil texture influences Mn leaching, i.e. in sandy soils Mn is leached easier, than in clayey ones (Alifragis and Papamichos, 1994).

f) *The soil structure.* When the soil structure is deteriorated (for example due to the compression from agricultural machines), and especially when this fact is combined with high water holding capacity, Mn reduction from Mn^{4+} to Mn^{2+} is a reality (bad aeration of deeper soil horizons). In this case, Mn concentration in soil solution may reach very toxic levels.

g) *Soil microorganisms.* Manganese oxidation may take place through biological processes (basically through aerobic bacteria). These bacteria are not very sensitive to soil acidity, but they are very active in a pH range between 6 and 7.5 (Smith and Paterson, 1990). According to Douka (1973), bacteria of the genus *Pseudomonas* and *Citrobacter*, found in western Peloponnesus, are capable of oxidizing, through enzymatic or non-enzymatic processes, Mn^{2+} into Mn oxides.

h) *Soil disinfection.* According to Grasmanis and Leeper (1966), and Cotter and Mishra (1968), after soil chemical disinfection Mn reduction takes place in neutral soils. This enhancement of Mn reduction after disinfection may be ascribed to one of the two following reasons (or to both of them): i) to the release of the organically bound Mn, or/and ii) to the death of the microorganisms which are responsible for the oxidation of Mn^{2+} to Mn^{4+} (Boyd, 1971).

i) *The root exudations.* According to Smith and Paterson (1990), who studied the fluctuation of Mn concentration in soil solution in a soil cultivated with *Hordeum vulgare*, a maximum of Mn concentration was found during the early summer period, together with a slight increase of pH. A possible explanation for this phenomenon was a production of low molecular weight organic substances produced by the roots during the period of the maximum growth rate of plants. These organic substances acted as a factor of creation of complexes between them and Mn, increasing its availability to plants (Smith and Paterson, 1990). Bromfield (1958) found that the root exudations of 20 week-old *Avena sativa* plants were able of solubilizing Mn oxides, and that this solubilization capacity was inversely correlated with soil pH (the quantity of Mn solubilized in pH values greater than 6.5 was very small).

j) *Mycorrhiza.* According to Posta et al. (1994), the root exudations of maize plants inoculated with the fungus *Glomus mosseae* had two times lower capacity for Mn solubilization, than the root exudations produced by non-inoculated maize plants.

k) *The content of soils in phosphoric anions.* A great content of soils in phosphoric anions (soils richly fertilized with P) has a consequence in the formation of Mn phosphate insoluble substances, which are not available for plants (Alifragis and Papamichos, 1994).

Mn Uptake by Plants

Manganese is basically absorbed as Mn^{2+}, or through some of its soluble complexes with organic matter. In the second case, its uptake is enhanced by the existence of mycorrhiza (Alifragis and Papamichos, 1994). Manganese uptake is realised through a mechanism of energetic absorption, i.e. with energy consumption. Clarkson (1988) found that the presence of 10μM DNP in a nutrient solution reduced by 80% the absorption of Mn by *Saccharum Officinarum* plants, something which proves that Mn uptake is realized through a mechanism of energetic absorption.

Factors that Influence Mn Uptake by Plants

There are a lot of factors influencing Mn uptake. These factors could be subdivided into two categories: a) soil factors (most of them have been analysed above, through the influence they exert in increasing soluble, i.e. plant available, Mn and they are: pH, soil moisture, organic matter, redox potential, soil texture, soil microorganisms, Mn concentration in soil solution, soil temperature, soil $CaCO_3$ content, soil content in different oxides-hydroxides, the presence of other nutrients in soil that can competitively or synergistically in Mn uptake), b) non-soil factors, such as the kind of N fertilization, root exudations, mycorrhiza, and the plant species.

A. Soil Factors

i) *pH.* Mn uptake is closely related to soil pH, more than that of other micronutrients (Marschner, 1988). As mentioned above, soil pH reduction increases Mn solubility and its uptake by plants.

ii) *Redox potential.* As mentioned above, the decrease of redox potential in lower levels than a critical value is followed by the convertion of easily reducible Mn into soluble and interchangeable one, therefore, into plant available Mn (Patrick and Turner, 1968; Gotoh and Patrick, 1972).

iii) *The organic matter.* The influence of organic matter on soil available Mn is very complicated. So, although organic matter has the ability to form complexes with Mn, as well as with other metals, and to reduce its availability for plants, under alkaline conditions Mn-organic matter complexes constitute about 80-95% of the total soluble

Mn forms and they contribute very decisively in Mn nutrition of higher plants. The uptake of these complexes is positively influenced by mycorrhiza (Alifragis and Papamichos, 1994).

iv) *The activity of microorganisms which are responsible for Mn oxidation.* As already discussed above, the oxidation of Mn^{2+} into Mn oxides can also be done by microorganisms. Such kind of microorganisms may be bacteria of genus *Arthrobacter sp.*, which are particularly active in a range of pH varying between 5.7 and 7.5 (Marschner, 1988). The result of the action of these microorganisms is the decrease of Mn availability for plants and its reduced uptake. The most usual genus of bacteria which are responsible for Mn oxidation are: *Arthrobacter, Bacillus, Pseudomonas* and *Hyphomicrobium.* Other less frequent occuring genus are those of *Corynebacterium, Proteus, Flavobacterium, Enterobacter* and *Citrobacter.* From the fungi responsible for Mn oxidation we should distinguish those of the genus *Streptomyces,* as well as that of *Gaeumannomyces graminis* (Ghiorse, 1988).

v) *Mn concentration in soil solution.* The quantity of Mn which is available for plants is that of soil solution and not the total one. The directly available for plants Mn (that of soil solution) is only a small percentage of the total Mn of soil and it is influenced by many factors, such as pH, redox potential e.t.c. The quantity of the directly available for plants soil Mn is controlled not only by the soil or plants' characteristics, but also by the interactions among root system with that part of soil surrounding it (rhizosphere properties) (Reisenauer, 1988).

vi) *Soil temperature.* Mn uptake is influenced by soil temperature and its' deficiency is more likely to take place under low soil temperatures. Finally, Mn uptake is influenced by soil temperature, more than that of Fe (Marschner, 1988).

vii) *$CaCO_3$ content.* The availability of Mn for plants and, hence, its uptake is negatively influenced by high $CaCO_3$ content due to the increase of pH (decrease of Mn solubility). Chatzistathis and Alifragis (2004), who did twice addition of $CaCO_3$ in a soil with toxic Mn concentrations (nearby to an old mine) managed to reduce from 2.5 to 23 times the concentrations of DTPA extractable Mn in *Populus I-214, Populus alba, Robinia pseudoacacia, Juglans regia* and *Eucalyptus* sp. plantations and from 10 to 19 times the concentration of Mn in the leaves of all plant species, compared to those found before the addition of $CaCO_3$. Apart from the direct influence of $CaCO_3$ on the decrease of Mn solubility, another reason for the reduction of Mn uptake under high $CaCO_3$ conditions can be the competition between Ca and Mn for uptake by plants (El-Jaoual and Cox, 1998).

viii) *Soil texture.* In clayey soils Mn can be bound in the negatively charged surface of clay minerals (Norvell, 1988). In that case, reduced Mn uptake by plants is very likely to happen. On the other hand, in sandy soils, as previously mentioned, Mn is usually leached easier than in other ones, such as sandy-clay-loams, clay-loams, sandy-clays e.t.c.

ix) *Soil content in minerals of Fe, Al, Si e.t.c. oxides-hydroxides.* The most important oxides-hydroxides minerals are those of Fe, Al, Si and Mn. The size of their positive, or negative, charge depends on soil pH (by increasing pH increases the negative charge of these minerals, while by decreasing pH increases the positive charge). So, under neutral or alkaline soil conditions Mn can be adsorbed in the surface of oxides-hydroxides (Norvell, 1988).

x) *The presence of other nutrients which may act synergistically or competitively with Mn.* Many competitive or synergistic effects among Mn and other divalent cations, such as Fe, Ca and Mg, have been observed. The most important is the interaction between Mn and Fe. Clarkson (1988) supports that Mn and Fe act competitively for a place of absorption in root surface. According to El-Jaoual and Cox (1998), the nature of interaction between Mn and Fe differs among plant species and while in most of them is competitive, for some others is neutral, or even synergistic. From our experience, genotypic differences in the nature of interactive effect of Mn-Fe may be observed not only between plant species, but also among cultivars of the same species. Indeed, in four olive cultivars studied, for three of them we found competitive effect between Mn and Fe, while in one of them ('Kothreiki') the effect was synergistic (Chatzistathis, 2008). According to Ghasemi-Fasaei et al. (2002), high Fe supply in soil may reduce Mn uptake from plants and its transport from root to shoot. The ratio Fe/Mn has a great importance in a balanced nutrition of different cotton varieties: Therefore, it was higher for the resistant in Mn deficiency cultivar 'C-307' and lower for the sensitive in Mn deficiency 'C-Sgl 70-517'. According to El-Jaoual and Cox (1998), the increased levels of Ca in soil, or nutrient, solution has a negative effect on Mn uptake by plants. Very often, of great importance is the Ca/Mn ratio. In *Pistacia vera* plants, this ratio for optimum mineral nutrition should be greater than 80 (El-Jaoual and Cox, 1998). According to Quartin et al. (2001), increased Mg concentrations act competitively to Mn uptake by plants. Phosphorus addition in soil may increase Mn uptake and deteriorate the severity of Mn toxicity symptoms (El-Jaoual and Cox, 1998; Sarkar et al., 2004). However, in over P-fertilized soils Mn availability for plants is decreased via the formation of Mn phosphate insoluble substances, which are not available for plants (Alifragis and Papamichos, 1994). Finally, it was found that other metals, such as Cu and Zn, compete with Mn uptake by plants (El-Jaoual and Cox, 1998; Nautiyal and Chatterjee, 2002). In contrast to these findings, Monnet et al. (2001) found that increased Zn concentrations in hydroponic cultures increased also Mn uptake and its concentrations in the leaves of *Lolium perenne* plants.

B. Non Soil Factors

i) *The kind of N fertilization (NO_3 or NH_4).* Plants receiving NH_4 fertilization acidify their rhizosphere and increase the availability of Mn, and therefore, its uptake by plants (Tong et al., 1997), in contrast to those receiving NO_3 fertilization (Grass et al., 1973).

ii) *Root exudations.* As previously mentioned, root exudations are responsible for forming Mn complexes in alkaline soils (Smith and Paterson, 1990), or for Mn reduction (Bromfield, 1958). In both cases, the result is the increase of Mn uptake.

iii) *Mycorrhiza.* Mycorrhiza contributes positively in the uptake of the organic complexes of Mn (Alifragis and Papamichos, 1994). Nevertheless, in inoculated with *Glomus mosseae* maize plants the capacity of root exudations for Mn reduction was two times lower, compared to that of the control (non inoculated with *Glomus mosseae*) ones (Posta et al., 1994). Furthermore, according to Kothari et al. (1991), the proportion of microorganisms which are responsible for Mn uptake was 20-30 times greater in non-inoculated with *Glomus mosseae* maize plants, than in those inoculated with that fungus.

iv) *The plant species and/or cultivar.* Among the different plant species, there are great differences concerning Mn uptake. In a highly polluted area, maize was found to have the highest Mn concentrations in leaves, compared to all other plant species grown there (Qing-Ren et al., 2002). According to Graham (1988), from a great variety of plant species being susceptible to Mn deficiency the most important are those of *Avena sativa*, *Citrus* plants, *Malus domestica*, *Prunus* sp. and several species of Brassicaceae family. Among the plant species with high sensitivity to Mn toxicity it should be mentioned those of *Nicotiana tabacum* (Nable et al., 1988; Houtz et al., 1988; El-Jaoual and Cox, 1998), *Solanum tuberosum* (Sarkar et al., 2004), *Hordeum vulgare* (Quartin et al., 1998; Alam et al., 2001), cotton plants (Sirkar and Amin, 1974; Foy, 1995), *Glycine max* (Wu, 1994), *Persea americana* (Tracy, 1991), *Cyperus papyrus* (Foy et al., 1998). Apart from the differential Mn uptake capacity existing between plant species, genotypic differences may also exist among different cultivars of the same species (Heenan and Carter, 1976; Foy et al., 1981; Wissemeier and Horst, 1991; Wu, 1994; Foy et al., 1995; Khabaz-Saberi et al., 1997; Huang and Graham, 1997; Foy et al., 1998; Gonzalez et al., 1998; Rout et al., 2001; Quartin et al., 2001; Sadana et al., 2002; Fecht-Christoffers et al., 2003; Gherardi and Rengel, 2003; Chatzistathis et al., 2009).

MN TRANSPORT AND DISTRIBUTION WITHIN PLANT TISSUES

As previously mentioned, Mn is usually absorbed by plants in the form of Mn^{2+}. In some cases, Mn chelates (usually complexes with EDTA) may contribute to the mineral nutrition of higher plants. These complexes are absorbed more slowly, than Mn^{2+} (Clarkson, 1988). Under Mn deficient conditions in soil the plants may have extremely low Mn concentrations in their root system and in young leaves, while they may have sufficient Mn concentrations in their mature (old) leaves. This happens because, according to Loneragan (1988), many plant species can accumulate a considerable quantity of Mn in their root system or in their shoot when they are grown under Mn sufficient conditions, and, therefore, when Mn deficiency occurs they are able to remobilize part of this Mn quantity from these tissues to leaves in order to cover their Mn nutritional needs. Chatzistathis et al. (2011), found that leaf Mn concentrations of the olive cultivars 'Kothreiki' and 'FS-17', irrigated for a 130-day experimental period with 50% modified Hoagland's nutrient solution not containing Mn (0 μM Mn), were Mn sufficient (greater than 20 mg/kg d.w.) for the whole experimental period, due possibly to the fact that the initial 'pool' of Mn existed in their root system was enough (95 μg/g. d.w. for 'FS-17' and 102 μg/g. d.w. for 'Kothreiki') to cover the Mn nutritional needs of plants.

In contrast to that, under Mn toxicity conditions, Mn concentrations in the root system of *Mentha spicata*, *Citrus* sp., *Pseudoacacia* sp. *Juglans regia, Populus* sp. and *Eucalyptus* sp. were much greater, than those of the above ground parts (shoot and leaves) (Loneragan, 1988;

Chatzistathis and Alifragis, 2004; Papadakis, 2004; Asrar et al., 2005). Chatzistathis et al. (2009), who studied the uptake and distribution of Mn in two Greek olive cultivars ('Koroneiki' and 'Kothreiki'), grown in three soil types with different physicochemical properties but with sufficient to over-sufficient Mn concentrations, found that in all the three soils both cultivars accumulated greater percentage (%) of the total plant Mn content in their root system (44-56% for 'Koroneiki' and 74-80% for 'Kothreiki'), than in leaves and stems. Furthermore, this high ability of olive plants (especially those of cultivar 'Kothreiki') to accumulate Mn in their root system could be advantageous in soils with very high Mn concentrations and may constitute a detoxification mechanism for olive trees, protecting their above-ground part from Mn toxicity (Chatzistathis et al., 2009). Despite the fact that for many tree species Mn concentrations in root system were much greater, than those of shoots, other researchers found that Mn concentrations in the root system of many annual species were lower than those of leaves (Foy et al., 1995; Schier and McQuattie, 2000; Lidon, 2001b; Candan and Tarhan, 2003). Therefore, it seems possible that apart from Mn availability (deficiency, excess, or toxicity), plant species (annual or perennial/ tree species) play a very important role in Mn distribution among tissues. Finally, according to Horst and Marschner (1978), plants have the ability to accumulate considerable amounts of Mn in their vacuoles under Mn toxicity conditions.

Mn Toxicity Conditions and Symptoms

Manganese toxicity may be observed basically in acid soils (pH<5). Other factors which may be responsible for Mn toxicity is the parent material with high Mn concentrations (volcanic rocks) (Tracy, 1991), the reducing conditions in soils (flooded soils) (El-Jaoual and Cox, 1998), the exaggerate use of fertilizers leading to soil acidification (ammonium sulphate), the soil disinfection of glasshouses, the abuse of fertilizers, fungicides and pesticides having high Mn content, and the proximity of agricultural areas with other, highly polluted, industrial regions (Qing-Ren et al., 2002).

The problem of Mn toxicity may be faced either by adding organic matter in soils (formation of complexes between the organic matter and Mn or/ and adjustment of pH), or by adding $CaCO_3$ in order to increase pH (Aubert and Pinta, 1977; Chatzistathis and Alifragis, 2004). In glasshouse soils which have been disinfected Mn excess can be leached out through a suitable irrigation system (Schubert, 1992). Finally, the abuse of fertilizers, fungicides and pesticides having high Mn content should be avoided, especially in acid soils, and tolerant to Mn toxicity genotypes should be grown in areas where Mn toxicity is a common problem (El-Jaoual and Cox, 1998).

Although among plant species the appearance of Mn toxicity symptoms may be different, basically three categories of Mn toxicity symptoms should be distinguished: a) symptoms of physiologically old and non-active plant tissues (the symptoms of this category appear as brown spots and chlorotic peripheral stains in leaves), b) inhibition of plant growth and leaf curling, c) chlorosis of young leaves. The last category of symptoms is usually known as 'Fe deficiency due to Mn toxicity' and is due to Mn-Fe antagonism (Horst, 1988). Apart from Fe deficiency, under Mn toxicity the uptake and distribution of other also nutrients may be disturbed, such as those of P, Ca, Mg, Zn, Na, K and Cu (El-Jaoual and Cox, 1998; Foy et al.,

1998; McQuattie and Schier, 2000; Lidon, 2000; Lidon, 2001a; Quartin et al., 2001; Lidon, 2002; Nautiyal and Chatterjee, 2002). When exaggerate great amount of Mn is accumulated in the root system of many plant species, its growth may be inhibited. This inhibition of root growth is usually combined with brown speckles in its surface, a phenomenon which is indicative of oxidized phenols' accumulation (Horst, 1988; Marschner, 1995).

Manganese is an activator of a great variety of enzymes (about 35), such as those catalyzing reactions of oxidation-reduction, decarboxyliosis, hydrolysis e.t.c. Therefore, under Mn toxicity conditions may be disturbed: a) the metabolism of carbohydrates and N (Burnell, 1988; Marschner, 1995), b) the reactions of Crebs' cycle (Marschner, 1995), c) photosynthesis and chlorophyll biosynthesis (Houtz et al., 1988; Nable et al., 1988; Macfie and Taylor, 1992; Kitao et al., 1997a; Kitao et al., 1997b; Kitao et al., 1999; Subrahmanyan and Rathore, 2000), d) the biosynthetic path of sikimic acid, e) the activity of the enzymes IAA oxidase (Sirkar and Amin, 1974; Burnell, 1988), polyphenol oxidase (PPO) (Sirkar and Amin, 1974; Chatzistathis et al., 2010), allantoate amylohydrolase (Burnell, 1988; Marschner, 1995; Vadez et al., 2000) e.t.c. Chatzistathis et al. (2010) found that periodic measurements of polyphenol oxidase activity in the leaves of the olive cultivars 'Picual' and 'FS-17' could be used for early detection of Mn toxicity (before the appearance of symptoms). In contrast to that, periodic measurements of the values of the chlorophyll fluorescence ratios F_v/F_m and F_v/F_0 were proved to be non-reliable means to predict the appearance of the visible symptoms of Mn toxicity in olive leaves (cvs. 'Kothreiki' and 'FS-17') (Chatzistathis et al., 2011).

Factors Contributing to the Increase of Plants' Resistance to Mn Toxicity

Apart from the interventions that can be done in soils (soil liming, addition of organic matter e.t.c.) in order to reduce, or avoid, Mn toxicity problems for plants, other interventions may also be done in plants' environment (for glasshouse cultures), such as those of light and temperature. According to El-Jaoual and Cox (1998), the increase of temperature until a critical point increases the resistance of plants to Mn toxicity. This increased resistance of plants to Mn toxicity may be attributed, according to the same authors, to the enhanced plant growth rate. On the other hand, Marsh et al. (1989) found that in *Solanum tuberosum* plants high temperature decreased plant growth rate and increased Mn uptake, so deteriorated the severity of Mn toxicity symptoms. Furthermore, the high light intensity had a negative consequence on the increase of Mn uptake by plants and on the severity of toxicity symptoms (Horiguchi, 1988). In contrast to Horiguchi (1988), Wissemeier and Horst (1992) found that light intensity did not deteriorate Mn toxicity symptoms in *Vigna unguiculata* plants. Another important factor contributing to the reduction of Mn toxicity is the supply of salicylic acid: According to Shi and Zhu (2008), the supply of salicylic acid in *Cucumis sativus* plants, grown under Mn toxicity, has considerably reduced the transport of Mn from root to shoots, the severity of Mn toxicity symptoms, and the production of reactive oxygen species (ROS).

Furthermore, by increasing the supply of various nutrients in soil it is feasible to increase the resistance of vegetative tissues to Mn toxicity. According to El-Jaoual and Cox (1998), and Alam et al. (2001), the beneficial role of Fe in reducing Mn toxicity symptoms in plants may be attributed to a combined mechanism of reduced Mn uptake and Mn accumulation in vegetal tissues. Another very important nutrient that plays a highly important role in reducing the negative effects of Mn toxicity is P. The beneficial role of P in the reduction of Mn toxicity symptoms may be focused on the inactivation of Mn ions inside plant tissues (Sarkar et al., 2004). The role of Si, although it is not very clear, it seems, according to the findings of many researchers, that it contributes to the increased resistance of plant tissues to Mn toxicity and to a more regular distribution of it among them, instead of its accumulation in particular tissues (Marschner, 1995; El-Jaoual and Cox, 1998; Jucker et al., 1999). Rogalla and Romheld (2002) found that in *Cucumis sativus* plants suffered from Mn toxicity and supplied with Si more than 90% of the total Mn was bound in the cell walls, compared to those not supplied with Si, where about 50% of the total Mn was bound in cell walls. Calcium competes with Mn uptake by plants, since both nutrients are divalent cations. Furthermore, Ca contributes to soil pH elevation and reduces Mn availability for plants (Chatzistathis and Alifragis, 2004). Magnesium plays a similar role with that of Ca (competes Mn ions for uptake by plants); however, it is not so effective in reducing Mn toxicity as Ca, since great quantities of Mg are required in order to avoid the appearance of Mn toxicity symptoms. Furthermore, great quantities of Mg supply may disturb the balanced mineral nutrition of higher plants (especially the uptake of Ca) (Davis, 1996). Concerning Mo capacity to reduce Mn toxicity symptoms, it should be focused on its capacity to adjust and control the negative consequences of Mn toxicity on Fe availability for plants ('Fe deficiency due to Mn toxicity') (El-Jaoual and Cox, 1998).

STRATEGIES ADOPTED BY PLANTS TO FACE MN TOXICITY

Many strategies may be adopted by plants in order to face Mn toxicity problems. Some of the most important of these strategies are analyzed below:

a) *Increase of rhizosphere pH.* Some plant species under Mn toxicity are able of increasing their rhizosphere pH in order to reduce the uptake of Mn (conversion of Mn^{2+} into Mn^{4+}, a form that can not absorbed by plants) (El-Jaoual and Cox, 1998).

b) *Enhanced root capacity to convert Mn^{2+} into Mn^{4+}.* It has been found by El-Jaoual and Cox (1998) that the root of rice plants, under anaerobic conditions (reduction of Mn^{4+} to Mn^{2+}), have the ability to oxidize the divalent form of Mn into Mn^{4+}, a non available form for plants.

c) *Enhanced Mn accumulation and compartmentalization in tolerant to Mn-toxicity tissues/organs and reduced Mn transport in sensitive to Mn-toxicity ones.* According to El-Jaoual and Cox (1998), under Mn toxicity conditions, *Rhododendron* spp. plants have the ability to accumulate the excess Mn in the woody parts of those plants, where Mn becomes inactive. Furthermore, it was found that *Helianthus annuus* plants may accumulate great quantities of Mn in its inflorescences, where Mn ions are converted into a metabolically inactive form (Blamey et al., 1986). In some plant species, their resistance to Mn toxicity is mainly attributed to the retention of excess Mn in their root system and to the reduced transport of it to their shoots (El-Jaoual and Cox, 1998; Chatzistathis and Alifragis, 2004; Chatzistathis et al., 2011). Although for most metals it was found that their transfer and accumulation takes place in particular organs (usually in vacuoles), for Mn it does not exist a particular organ-target in which it is usually accumulated, but this depends a lot on plant species (Lidon, 2001b).

d) *Arbuscular mycorrhiza fungus (AMF) development for the protection of plants from Mn toxicity.* According to Nogueira et al. (2002; 2004; 2007), under Mn toxicity mycorrhiza plays a beneficial role by increasing P uptake (Mn ions become physiologically inactive because of the formation of P-Mn complexes). Furthermore, AMF is capable of decreasing Mn uptake, through two basic processes: i) by increasing the action of aerobic bacteria, which are responsible for the oxidation of Mn from Mn^{2+} to Mn^{4+} and, ii) by reducing Mn reduction in the rhizosphere (Nogueira et al., 2002; 2004; 2007).

e) *Genetic or biochemical defense through the production of specialized enzymes, proteins or aminoacids e.t.c.* In some plant species, resistance to Mn toxicity is related to the ability of particular tissues (e.g. shoots) to tolerate excessive/toxic Mn concentrations (El-Jaoual and Cox, 1998). Lidon and Teixeira (2000) found that the rice cultivar 'Safari' had high tolerance to survive under toxic Mn concentrations, since it produced a protein of molecular weight of 36.5 kDa in its chloroplasts, which assembled in its function and role to those of SOD. According to Khan et al. (2000), it is possible that the greater increases of proline concentration in the roots of maize tolerant cultivars, compared to those of the sensitive ones, could be related to a possible tolerance mechanism to Al and Mn toxicity of this amino-acid.

MN DEFICIENCY CONDITIONS AND SYMPTOMS

The soil conditions which may lead to Mn deficiency are generally the following:

a) Deep organic soils formed on calcareous parent materials.
b) Alluvial soils, rich in $CaCO_3$.
c) Badly drained soils with great organic matter content.
d) Very acid, highly leached, soils.
e) Old acid soils on which organic matter and $CaCO_3$ added for many years.
f) Soils with great moisture fluctuations (Alifragis and Papamichos, 1994).

Apart from the soil conditions that may lead to Mn deficiency, from the other, non-soil, factors which may favour Mn deficiency the most important is temperature. In Great Britain, Mn deficiency is more usual in cool regions, than in the warmer ones. This may happen due to the fact that under low temperature conditions root growth is restricted. Later, when soil temperature increases plants are capable of overcoming Mn deficiency, probably because of the greater root system growth (Reuter et al., 1988). Manganese deficiency may be observed in vegetal species with low reducing capacity of the root, with restricted ability of rhizosphere acidification, or/and with low transpiration rates (Reuter et al., 1988).

The basic symptoms of Mn deficiency are appeared in the young leaves, where interveinal chlorosis may be observed, which is in contrast to the green veins of the leaf area. This kind of symptoms may be confused with those of Mg deficiency. When the intensity of symptoms is very severe, necrotic spots in the leaf area may be found (Hannam and Ohki, 1988). Other symptoms of Mn deficiency are changes in the leaf chloroplasts structure (Welkie και Pound, 1958; Papadakis et al., 2007) and in chlorophyll concentration (Papadakis et al., 2007), as well as in the ratio chlorophyll a/b (Anderson and Pyliotis, 1969).

STRATEGIES ADOPTED BY PLANTS TO FACE MN DEFICIENCY

Some of the most important strategies that may be adopted by plants, in order to face Mn deficiency, are the following:

a) *The ability of some vegetal species to acidify their rhizosphere under alkaline conditions* (reduction of Mn^{4+} to Mn^{2+}, which can be absorbed by plants) (Reuter et al., 1988).

b) *Production of organic root exudates in order to facilitate Mn uptake, through the formation of Mn-organic complexes.* The uptake of these complexes is usually favored by mycorrhiza (Alifragis and Papamichos, 1994). According to Maruyama et al. (2005), who made a comparison of iron availability in leaves of barley and rice, the difference in the Fe acquisition ability between these two species was affected by the differential mugineic acid secretion. A similar possible mechanism might be responsible, according to Chatzistathis et al. (2009), for the differential Mn uptake between the olive cultivars 'Koroneiki' and 'Kothreiki'.

c) *The modification of root morphology in order to be able to absorb more easily* Mn^{2+}. According to Rengel (2001), who made a review on genotypic differences in micronutrient use efficiency of many crops, micronutrient-efficient genotypes were capable of increasing the available soil micronutrient pools through changing chemical and microbiological properties of the rhizosphere, as well as by growing thinner and longer roots and by having more efficient uptake and transport mechanisms. In our experiments, we found that the differential root system morphology of the olive cultivar 'Kothreiki' (this was richly-branched and with much greater root hair development and density, compared to that of the cultivars 'Koroneiki' and 'Chondrolia Chalkidikis') was probably responsible for the significantly greater Mn uptake (Chatzistathis, unpublished data).

d) *The development of AMF in order to increase Mn uptake.* Generally, AMF contributes to the uptake of soluble organic complexes of Mn under alkaline soil conditions (Alifragis and Papamichos, 1994).

Furthermore, under Mn deficiency conditions, particular emphasis should be given in the production of genotypes with increased Mn uptake ability and use efficiency (Reuter et al., 1988; Huang and Graham, 1997; Khabaz-Saberi et al., 1997; Pearson and Rengel, 1997; Sadana et al., 2002; Sadana et al., 2003; Chatzistathis et al., 2009), as well as in the use of fertilizers having the ability to oxidize the environment of rhizosphere (e.g. ammonium sulphate), while in some other cases soil disinfection in glasshouses should not be excluded, since it increases Mn availability for plants (Reuter et al., 1988). According to Jiang and Ireland (2005) and Jiang (2006), Mn efficient wheat cultivars own this ability to a better internal use of Mn, rather than to a higher plant Mn accumulation. In addition to this, Chatzistathis et al. (2009) found that between olive cultivars 'Koroneiki' and 'Kothreiki' the first one had better internal Mn use efficiency (better transport from root system to shoot), compared to the second one, so it was Mn-efficient, while 'Kothreiki' was Mn-inefficient.

CONCLUSIONS AND FUTURE DIRECTIONS

Manganese is an essential nutrient for plant growth, although it is required in very low concentrations in plants. On the other hand, in most soils Mn is found in abundance, so deficiency is less frequent nutritional disorder than toxicity. However, there are enough cases where Mn deficiency occurs (in alkaline soils, in soils rich in $CaCO_3$, in very acid, highly leached, soils, in soils with great moisture fluctuations e.t.c.). The soil factors influencing Mn solubility are those of pH, redox potential, soil moisture, soil microorganisms e.t.c. Under conditions of low, or excess, Mn availability, it is of great importance to choose the suitable (tolerant to Mn deficiency or toxicity) genotypes for cultivation. Apart from the differences existing between vegetative species, concerning their resistance to Mn deficiency/toxicity, great differences also exist among different genotypes of the same species. In this case, under Mn toxicity, genotypes with enhanced capacity: i) to increase rhizosphere pH, or to convert Mn^{2+} into Mn^{4+}, ii) to accumulate Mn and compartmentalize it in tolerant to Mn-toxicity tissues/organs, iii) to develop AMF symbiosis for the protection of plants from Mn toxicity, and/or, iv) to produce specialized enzymes, proteins, or aminoacids for genetic or biochemical defense to Mn toxicity should be preferred. On the other hand, under Mn deficiency (for example under alkaline conditions) the selected genotypes for cultivation should have high ability: i) to acidify their rhizosphere (convertion of Mn^{4+} to Mn^{2+}), ii) to produce organic root exudates in order to facilitate Mn uptake, through the formation of Mn-organic complexes, iii) to modify root morphology in order to be able to absorb more easily Mn^{2+}, iv) to develop AMF symbiosis in order to increase Mn uptake. Finally, in the near future, particular emphasis should be given in the genetic selection, improvement and production of genotypes: i) having the above described characteristics under Mn toxicity and deficiency, ii) being highly Mn-efficient, especially under the recent advances in genetics, biotechnology, soil science, plant nutrition and ecology. Of great significance is, of course, to integrate the laboratory- produced- results with field settings in order to find in each case the best soil and natural environment for each new genotype.

REFERENCES

Alam S., Kamei S., Kawai S., 2001. Amelioration of manganese toxicity in barley with iron. *Journal of Plant Nutrition*, 24(9), 1421-1433.

Alifragis, D.A., Papamichos, N., 1995. *Soil sampling and chemical analyses of forest soils and vegetative tissues*. Dedousi publications. Thessaloniki, Greece. In Greek.

Anderson J.M., Pyliotis N.A., 1969. Studies with manganese-deficient spinach chloroplasts. *Biochimica et Biophysica Acta (BBA)-Bioenergetics*, 189(2), 280-293.

Asrar Z., Khavari-Nejad R.A., Heidari H., 2005. Excess manganese effects on pigments of *Mentha spicata* at flowering stage. *Archives of Agronomy and Soil Science*, 51(1), 101-107.

Aubert H., και Pinta M., 1977. *Trace elements in soils*. Elsevier Scientific Publishing Company, pp. 43-53.

Boyd H.W., 1971. Manganese toxicity to peanuts in autoclaved soil. *Plant and Soil,* 34, 133-144.

Bromfield S.M., 1958. The properties of a biologically formed manganese oxide, its availability to oats and its solution by root washings. *Plant and Soil*, IX(4), 325-337.

Candan N., Tarhan L., 2003. The correlation between antioxidant enzyme activities and lipid peroxidation levels in *Mentha pulegium* organs grown in Ca^{2+}, Mg^{2+}, Cu^{2+}, Zn^{2+} and Mn^{2+} stress conditions. *Plant Science*, 165, 769-776.

Chatzistathis, T., Alifragis, D., 2004. *The influence of soil liming on the limitation of Mn and Cu toxicity,* in *Populus, Pseudoacacia, Junglans and Eucalyptus* plantations. Proc. 10th Panhellenic Soil Science Symposium. Volos, Greece 22-25 September 2004. p. 465-478. In Greek.

Chatzistathis, T., 2008. *Investigation of the role of Mn on olive trees' mineral nutrition.* Doctoral Thesis. Aristotle University of Thessaloniki, Greece. In Greek.

Chatzistathis, T., Therios, I., Alifragis, D., 2009. Differential uptake, distribution within tissues, and use efficiency of Manganese, Iron and Zinc by olive cultivars 'Kothreiki' and 'Koroneiki'. *Hort Sciene* 44: 1994-1999.

Chatzistathis, T., Therios, I., Dimantidis G., 2010. Polyphenol oxidase activity of two olive cultivars as an early criterion of Mn toxicity. *Journal of the Science of Food and Agriculture* 90: 676-682.

Chatzistathis, T., Papadakis I., Therios, I., Giannakoula, A., Dimassi, K., 2011. Is chlorophyll fluorescence technique a useful tool to assess Manganese deficiency and toxicity stress in olive plants? *Journal of Plant Nutrition* 34: 98-114.

Clarkson D.T., 1988. *The uptake and translocation of manganese by plant roots*. In 'manganese in soils and plants'. Proceedings of the International symposium on 'Manganese in soils and plants', edited by R.D. Graham, R.J. Hannam and N.C. Uren. Kluwer Academic Publishers, pp. 101-111.

Cotter D.J., Mishra U.N., 1968. The role of organic matter in soil manganese equilibrium. *Plant and Soil*, XXIX(3), 439-448.

Davis J.G., 1996. Soil pH and magnesium effects on manganese toxicity in peanuts. *Journal of Plant Nutrition*, 19(3and4), 535-550.

Douka, A.E., 1973. *Oxidation of Mn^{2+} through the action of two Mn-oxidazing bacteria.* Taxonomy and study of their morphology and physiology. Doctoral Thesis. National and Kapodestrian University of Athens.

El-Jaoual T., Cox D.A., 1998. Manganese toxicity in plants. *Journal of Plant Nutrition*, 21(2), 353-386.

Fecht-Christoffers M.M., Maier P., Horst W.J., 2003. Apoplastic peroxidases and ascorbate are involved in manganese toxicity and tolerance of *Vigna unguiculata. Physiologia Plantarum,* 117(2), 237.

Foy C.D., Webb H.W., Jones J.E., 1981. Adaptation of cotton genotypes to an acid, manganese toxic soil. *Agronomy Journal,* 73, 107-111.

Foy C.D., Weil R.R., Coradetti C.A., 1995. Differential manganese tolerances of cotton genotypes in nutrient solution. *Journal of Plant Nutrition*, 18(4), 685-706.

Foy C.D., Farina M.P.W., Oakes A.J., 1998. Iron-Manganese interactions among clones of nilegrass. *Journal of Plant Nutrition,* 21(5), 978-1009.

Ghasemi-Fasaei R., Ronaghi A., Maftoun M., Karimian N., Soltanpour P.N., 2002. Influence of FeEDDHA on Iron-Manganese Interaction in Soybean Genotypes in a Calcareous Soil. *Journal of Plant Nutrition*, 26(9), 1815-1823.

Gherardi M.J., Rengel Z., 2003. Genotypes of lucerne (*Medicago sativa* L.) show differential tolerance to manganese deficiency and toxicity when grown in bauxite residue sand. *Plant and Soil*, 249(2), 287-296.

Ghiorse W.C., 1988. *The biology of manganese transforming microorganisms in soil.* In 'manganese in soils and plants'. Proceedings of the International symposium on 'Manganese in soils and plants', edited by R.D. Graham, R.J. Hannam and N.C. Uren. Kluwer Academic Publishers, pp. 75-85.

Gilkes R.J., McKenzie R.M., 1988. *Geochemistry and mineralogy of manganese in soils.* In 'manganese in soils and plants'. Proceedings of the International symposium on 'Manganese in soils and plants', edited by R.D. Graham, R.J. Hannam and N.C. Uren. Kluwer Academic Publishers, pp. 23-35.

Gonzalez A., Steffen K.L., Lynch J.P., 1998. Light and excess manganese. Implications for oxidative stress in common bean. *Plant Physiology*, 118, 493-504.

Gotoh S., Patrick W.H., 1972. Transformation of manganese in a waterlogged soil as affected by redox potential and pH. *Soil Science Society of America Proceedings*, 36, 738-741.

Graham R.D., 1988. *Genotypic differences in tolerance to manganese deficiency.* In 'manganese in soils and plants'. Proceedings of the International symposium on 'Manganese in soils and plants', edited by R.D. Graham, R.J. Hannam and N.C. Uren. Kluwer Academic Publishers, pp. 261-276.

Grasmanis V.O., Leeper G.W., 1966. Toxic manganese in near-neutral soils. *Plant and Soil,* XXV(1), 41-48.

Grass L.B., MacKenzie A.J., Meek B.D., Spencer W.F., 1973. Manganese and Iron solubility changes as a factor in tile drain clogging: I. Observations during flooding and drying. *Soil Science Society of America Proceedings,* 37, 14-17.

Hannam R.J., Ohki K., 1988. *Detection of manganese deficiency and toxicity in plants.* In 'manganese in soils and plants'. Proceedings of the International symposium on 'Manganese in soils and plants', edited by R.D. Graham, R.J. Hannam and N.C. Uren. Kluwer Academic Publishers, pp. 243-259.

Heenan D.P., Carter O.G., 1976. Tolerance of soybean cultivars to manganese toxicity. *Crop Science,* 16, 389-391.

Horiguchi T., 1988. Mechanism of manganese toxicity and tolerance of plants. VII. Effect of light intensity on manganese-induced chlorosis. *Journal of Plant Nutrition*, 11, 235-246.

Horst W.J., Marschner H., 1978. Effect of silicon on manganese tolerance of bean plants (*Phaseolus vulgaris* L.). *Plant and Soil*, 50, 287-303.

Horst W.J., 1988. *The physiology of manganese toxicity.* In 'manganese in soils and plants'. Proceedings of the International symposium on 'Manganese in soils and plants', edited by R.D. Graham, R.J. Hannam and N.C. Uren. Kluwer Academic Publishers, pp. 175-188.

Houtz R.L., Nable R.O., Cheniae G.M., 1988. Evidence for effects on the *in vivo* activity of ribulose-bisphosphate carboxylase/oxygenase during development of Mn toxicity in tobacco. *Plant Physiology*, 86, 1143-1149.

Huang C., Graham R.D., 1997. Efficient Mn uptake in barley is a constitutive system. *Plant nutrition-for sustainable food production and environment,* 269-270.

Jiang W.Z., Ireland C.R., 2005. Characterization of manganese use efficiency in UK wheat cultivars grown in a solution culture system and in the field. *Journal of Agricultural Science,* 143, 151-160.

Jiang W.Z., 2006. Mn use efficiency in different wheat cultivars. *Environmental and Experimental Botany,* 57, 41-50.

Jucker E.I., Foy C.D., De Paula J.C., Centeno J.A., 1999. Electron paramagnetic resorance studies of manganese toxicity, tolerance, and amelioration with silicon in snapbean. *Journal of Plant Nutrition* 22(4-5), 769-782.

Kabata-Pendias A., 2001. *Trace Elements in Soils and Plants*. Third Edition, CRC Press, NY, pp. 286-295.

Keramidas, V., 1997. Soil Fertility. Aristotle University of Thessaloniki, Greece. In Greek.

Khabaz-Saberi H., Graham R.D., Rathjen A.J., 1997. Genotypic variation for Mn efficiency in durum wheat (*Triticum turgidium* L. var. *durum*). *Plant Nutrition-for sustainable food production and environment,* 289-290.

Khan A.A., McNeilly T., Collins J.C., 2000. Accumulation of aminoacids, proline and carbohydrates in response to aluminum and manganese stress in maize. *Journal of Plant Nutrition* 23(9), 1303-1314.

Kitao M., Lei T.T., Koike T., 1997a. Effects of manganese toxicity on photosynthesis of white birch (*Betulla platyphylla* var. *japonica*) seedlings. *Physiologia Plantarum*, 101, 249-256.

Kitao M., Lei T.T., Koike T., 1997b. Comparison of photosynthetic responses to manganese toxicity of deciduous broad-leaved trees in northern Japan. *Environmental Pollution,* 97(1-2), 113-118.

Kitao M., Lei T.T., Koike T., 1999. Effects of manganese in solution culture on the growth of five deciduous broad-leaved tree species with different successional characters from northern Japan. *Photosynthetica,* 36(1-2), 31-40.

Kothari S.K., Marschner H., Romheld V., 1991. Effect of a vesicular arbuscular mycorrhizal fungus and rhizosphere microorganisms on manganese reduction in the rhizosphere and manganese concentrations in maize (*Zea mays L.*). *New Phytologist,* 117(4), 649-655.

Lidon F.C., 2000. Rice adaptation to excess manganese: Nutrient accumulation and implications of the quality of crops. *Journal of Plant Physiology*, 156, 652-658.

Lidon F.C., Teixeira M.G., 2000. Rice tolerance to excess Mn: Implications in the chloroplast lamellae and synthesis of a novel manganese protein. *Plant Physiol. Biochem.,* 38, 969-978.

Lidon F.C., 2001a. Modulation of macronutrients uptake and translocation in Mn-treated rice in early stages of vegetative growth. *Journal of Plant Nutrition*, 24(1), 61-74.

Lidon F.C., 2001b. Tolerance of rice to excess manganese in the early stages of vegetative growth. Characterisation of manganese accumulation. *Journal of Plant Physiology* 158, 1341-1348.

Lidon F.C., 2002. Micronutrient uptake and translocation in Mn-treated rice. *Journal of Plant Nutrition,* 25(4), 757-768.

Lindsay W.L., 1979. *Chemical equilibria in soils*. pp.151-160.

Loneragan J.F., 1988. *Distribution and movement of manganese in plants*. In 'manganese in soils and plants'. Proceedings of the International symposium on 'Manganese in soils and plants', edited by R.D. Graham, R.J. Hannam and N.C. Uren. Kluwer Academic Publishers, pp. 113-124.

Macfie S.M., Taylor G.J., 1992. The effects of excess manganese on photosynthetic rate and concentration of chlorophyll in *Triticum aestivum* grown in solution culture. *Physiologia Plantarum,* 85(3), 467-475.

Mandal L.N., 1961. Transformation of Iron and Manganese in water-logged rice soils. *Soil Science,* 91, 121-126.

Marschner H., 1988. *Mechanisms of manganese acquisition by roots from soils*. In 'manganese in soils and plants'. Proceedings of the International symposium on 'Manganese in soils and plants', edited by R.D. Graham, R.J. Hannam and N.C. Uren. Kluwer Academic Publishers, pp. 191-204.

Marschner H., 1995. *Mineral Nutrition of Higher Plants*. 2nd edition. Academic Press, London. pp. 324-333.

Maruyama, T., Higuchi, K., Yoshida M., Tadano, T., 2005. Comparison of iron availability in leaves of barley and rice. *Soil Science and Plant Nutrition* 51: 1037-1042.

Marsh K.B., Peterson L.A., McCown B.H., 1989. A microculture method for assessing nutrient uptake. II. The effect of temperature on manganese uptake and toxicity in potato shoots. *Journal of Plant Nutrition*, 12, 219-232.

McKenzie R.M., 1977. *Manganese oxides and hydroxides*. In 'Minerals in soil environments'. Published by Soil Science Society of America, Madison, Wisconsin USA, pp. 181-193.

McQuattie C.J., Schier G.A., 2000. Response of sugar maple (*Acer saccharum*) seedlings to manganese. *Canadian Journal of Forest Research*, 30(3), 456-467.

Monnet F., Vaillant N., Vernay P., Coudret A., Sallanon H., Hitmi A., 2001. Relationship between PSII activity, CO_2 fixation, and Zn, Mn and Mg contents of *Lolium perenne* under zinc stress. *Journal of Plant Physiology*, 158(9), 1137-1144.

Nable R.O., Houtz R.L., Cheniae G.M., 1988. Early inhibition of photosynthesis during development of Mn toxicity in tobacco. *Plant Physiology*, 86, 1136-1142.

Nautiyal N., Chatterjee C., 2002. Copper-Manganese interaction in cauliflower. *Journal of Plant Nutrition,* 25(8), 1701-1707.

Nogueira M.A., Cardoso E.J.B.N., Hampp R., 2002. Manganese toxicity and callose deposition in leaves are attenuated in mycorrhizal soybean. *Plant and Soil*, 246(1), 1-10.

Nogueira M.A., Magalhaes G.C., Cardoso E.J.B.N., 2004. Manganese toxicity in mycorrhizal and phosphorus soybean plants. *Journal of Plant Nutrition*, 27(1), 141-156.

Nogueira M.A., Nehls U., Hampp R., Poralla K., Cardoso E.J.B.N., 2007. Mycorrhiza and soil bacteria influence extractable iron and manganese in soil and uptake by soybean. *Plant and Soil*, 298, 273-284.

Norvell W.A., 1988. *Inorganic reactions of manganese in soils*. In 'manganese in soils and plants'. Proceedings of the International symposium on 'Manganese in soils and plants', edited by R.D. Graham, R.J. Hannam and N.C. Uren. Kluwer Academic Publishers, pp. 37-58.

Papadakis, I., 2004. *Reaction of Citrus plants to Manganese*. Doctoral thesis. Aristotle University of Thessaloniki, Greece. In Greek.

Papadakis I.E., Bosabalidis A.M., Sotiropoulos T.E., Therios I.N., 2007. Leaf anatomy and chloroplast ultrastructure of Mn-deficient orange plants. *Acta Physiologiae Plantarum*, 29(4), 297-301.

Patrick W.F., Turner F.T., 1968. Effect of redox potential on manganese transformation in waterlogged soil. *Nature*, 220(2), 476-478.

Pearson J.N., Rengel Z., 1997. Genotypic differences in the production of carbohydrates between roots and shoots of wheat, grown under zinc or manganese deficiency. *Annals of Botany*, 80, 803-808.

Posta K., Marschner H., Romheld V., 1994. Manganese reduction in the rhizosphere of mycorrhizal and nonmycorrhizal maize. *Mycorrhiza*, 5(2), 119-124.

Qing-Ren W., Xiu-Mei L., Yan-Shan C., Yi-Ting D., Peter C., 2002. Responses of legume and non-legume crop species to heavy metals in soils with multiple metal contamination. Journal of Environmental Science and Health. *Part A, toxic/hazardous substancesand environmental engineering*, 37(4), 611-621.

Quartin V., Ramalho J.C., Nunes M.A., 1998. Responses of biomass and several photosynthetic indicators to manganese excess in Triticales. *Journal of Plant Nutrition*, 21(8), 1615-1629.

Quartin V.M.L., Antunes M.L., Muralha M.C., Sousa M.M., Nunes M.A., 2001. Mineral imbalance due to manganese excess in Triticales. *Journal of Plant Nutrition*, 24(1), 175-189.

Reisenauer H.M., 1988. *Determination of plant-available soil manganese*. In 'manganese in soils and plants'. Proceedings of the International symposium on 'Manganese in soils and plants', edited by R.D. Graham, R.J. Hannam and N.C. Uren. Kluwer Academic Publishers, pp. 87-98.

Rengel Z., 2001. Genotypic differences in micronutrient use efficiency in crops. *Communications in Soil Science and Plant Analysis*, 32(7-8), 1163-1186.

Reuter D.J., Alston A.M., McFarlane J.D., 1988. *Occurrence and correction of manganese deficiency in plants*. In 'manganese in soils and plants'. Proceedings of the International symposium on 'Manganese in soils and plants', edited by R.D. Graham, R.J. Hannam and N.C. Uren. Kluwer Academic Publishers, pp. 205-224.

Rogalla H., Romheld V., 2002. Role of leaf apoplast in silicon-mediated manganese tolerance of *Cucumis sativus* L. *Plant, Cell and Environment*, 25, 549-555.

Rout G.R., Samantaray S., Das P., 2001. Studies on differential manganese tolerance of mung bean and rice genotypes in hydroponic culture. *Agronomie*, 21, 725-733.

Sadana U.S., Lata K., Claassen N., 2002. Manganese efficiency of wheat cultivars as related to root growth and internal manganese requirement. *Journal of Plant Nutrition*, 25(12), 2677-2688.

Sadana U.S., Samal D., Claassen N., 2003. Differences in manganese efficiency of wheat (*Triticum aestivum* L.) and raya (*Brassica juncea* L.) as related to root-shoot relations and manganese influx. *Journal of Plant Nutrition and Soil Science*, 166, 385-389.

Sarkar D., Pandey S.K., Sud K.C., Chanemougasoundharam A., 2004. In vitro characterization of manganese toxicity in relation to phosphorus nutrition in potato (*Solanum tuberosum* L.). *Plant Science*, 167, 977-986.

Schier G.A., McQuattie C.J., 2000. Effect of manganese on endomycorrhizal sugar maple seedlings. *Journal of Plant Nutrition*, 23(10), 1533-1545.

Shi Q., Zhu Z., 2008. Effects of salicylic acid on manganese toxicity, element contents and antioxidative system in cucumber. *Environmental and Experimental Botany*, 63, 317-326.

Sirkar S., Amin J.V., 1974. The manganese toxicity of cotton. Plant Physiology, 54, 539-543.

Smith K.A., Paterson J.E., 1990. Manganese and Cobalt. In '*Heavy metals in soils*', pp. 225-243.

Subrahmanyan D., Rathore V.S., 2000. Influence of manganese toxicity on photosynthesis in ricebean (*Vigna umbellate*) seedlings. *Photosynthetica,* 38(3), 449-453.

Taylor, R.M., McKenzie, R.M., 1966. The association of trace elements with manganese minerals in Australian soils. *Aust. J. Soil Res.*, 4, 29-39.

Therios, I., 2009. Olives. Crop Production Science in Horticulture. CABI, U.K., p. 409.

Tong Y., Rengel Z., Graham R.D., 1997. Interactions between nitrogen and manganese nutrition of barley genotypes differing in manganese efficiency. *Annals of Botany*, 79(1), 53-58.

Tracy J.E., 1991. *Manganese toxicity in Avocado* (*Persea americana* Mill.). California Avocado Society. Yearbook, 75, 147-158.

Vadez V., Sinclair T.R., Serraj R., Purcell L.C., 2000. Manganese application alleviates the water deficit-induced decline of N_2 fixation. *Plant, Cell and Environment*, 23(5), 497-505.

Welkie G.W., Pound G.S., 1958. Manganese nutrition of *Nicotiana tabacum* L. in relation to multiplication of tobacco mosaic virus. *Virology,* 5(1), 92-109.

Wissemeier A.H., Horst W.J., 1991. Simplified methods for screening cowpea cultivars for manganese leaf-tissue tolerance. *Crop Science*, 31, 435-439.

Wissemeier A.H., Horst W.J., 1992. Effect of light intensity on manganese toxicity and callose formation in cowpea (*Vigna unguiculata* L.). *Plant and Soil*, 143, 299-309.

Wu S., 1994. Effect of manganese excess on the soybean plant cultivated under various growth conditions. *Journal of Plant Nutrition*, 17(6), 991-1003.

In: Soil Nutrients
Editor: Mohammad Miransari
ISBN: 978-1-61324-785-3

Chapter 9

SUSTAINABLE USE OF RHIZOSPHERIC MICROFLORA IN RECLAMATION OF SOIL HEALTH

Devendra Kumar Choudhary*
Department of Science, Faculty of Arts, Science and Commerce (FASC),
Mody Institute of Science and technology (MITS), Lakshmangarh-332311,
Sikar (Rajasthan), India

ABSTRACT

Understanding the complex microbial community in the rhizosphere environment has proven to be a challenging task because of the vast diversity and the enormity of the population inhabiting this unique habitat. Extensive studies have investigated perturbation of microbial community equilibrium population by changes in environmental conditions and soil management practices. It has long been recognized that the activity of soil microorganisms play an intrinsic role in residue decomposition, nutrient cycling, and crop production. Any shift in microbial community structure can be reflected in implementation of various land use and management systems that lead to development of best management practices for an agro-ecosystem.

Diversity of the soil community reflects the capacity of a soil to maintain multi-functionality. It is widely accepted that soil biodiversity which is displayed in the complexity of the soil food web, is fundamental for the resistance, resilience, and recovery of soil functions to environmental perturbation in the short-term. The characteristics of the soil health emphasize and include high biological diversity, high community stability that can provide resilience to disturbance (short-term) or stress (long-term), ability to maintain the integrity of nutrient cycling and energy flow, suppression of multiple pests and pathogens, and improved plant health. Therefore, the purpose of this chapter is to provide a current multidisciplinary view of the study of the

* For Correspondence: Dr. Devendra Kumar Choudhary, Asstt. Prof., Department of Science, Faculty of Arts, Science and Commerce (FASC), Mody Institute of Science and technology (MITS), Lakshmangarh-332311, Sikar (Rajasthan), INDIA. E-mail devmicro@rediffmail.com.

soil health along with a brief description of the current chemical and biological techniques that are used to analyze soil composition.

PROLOGUE

Soil is a dynamic living matrix which is an outstanding example of a natural system where living organisms play a major role in the development of its physico-chemical properties. Soil health describes the biological, chemical, and physical features of soil that are essential for long-term sustainable agricultural productivity with minimal environmental impact. Soil health can be inferred by measuring specific soil properties e.g., organic matter content, and soil status i.e., fertility. Soil represents the largest C pool on the Earth's surface (2157-2293 Pg) and the amount of this element is two times higher in soil compared to the atmosphere and three times larger than the amount in all living matter (Prentice et al., 2001). Small modifications in soil C status might have a significant effect on the global C balance and it reflects on the climate change which is indicative of large quantity of C stored in soils (Gonázlez-Pérez et al., 2004). Soil contains an intricate network of plants and microbes in a heterogeneous solid medium wherein physico-chemical conditions vary at the scale of molecule and the cell. Therefore, it is difficult to understand the variations in soils without appropriate knowledge that can be derived from both chemical and biological approaches because microorganisms affect the environment and vice versa. Soil microorganisms despite their small volume play a key role in the global cycling of organic matter (OM), renewal of organic residues or mineralizating them to CO_2, H_2O, N, P, S, and other nutrients. The functions of soil biota are central to decomposition processes and nutrient cycling, therefore, it is considered a store house of microbial activity wherein the space occupied by living organism is estimated to be less than 5% of the total space. Therefore, profound microbial activity is confined to the 'hot-spot', i.e., aggregates with accumulated OM in the rhizosphere (RS) (Pinton et al., 2001).

SOIL NUTRIENTS

There are unintended consequences of modern agriculture that extend well beyond agricultural landscapes. Biogeochemical cycles have been profoundly altered at multiple scales and exceed soil loss over soil formation. Nitrogen (N) and phosphorus (P) are the two most important nutrients that limit biological production and are the most extensively applied nutrients in managed agro-ecosystem in the form of soluble inorganic fertilizers (F_i). Nutrient enrichment reflects detrimental effects in agro-ecosystem wherein global N and P fluxes are projected to increase substantially with concomitant inorganic fertilizers production capacity (Galloway, 2000). Therefore, a new approach to nutrient supply in intensively managed ecosystem is required to control their process of eutrophication wherein an integrated nutrient management (INM) strategy based on ecological concepts has been applied. As a result, nutrient management research continues to emphasize improved delivery of F_i to the root zone during the period of crop uptake through several modification viz., banding, fertigation, and split fertilizer applications (Cassman et al., 2002).

SOIL ORGANIC MATTER (SOM)

SOM can be considered as a mixture of biogenic components and comprises of microorganisms and non-decomposed plant material (1-10%) in variable proportions and evolutionary stages. SOM can be either active e.g., fast recycling, corresponding mainly to carbohydrate, amino acid and lipid functions, remaining in the soil for months or even decades (10-40%), or passive or refractory (humic fraction), remaining in the soil for centuries to millennia (40-60%) that is dependent in the nutshell on the turnover time in soil. When we consider qualitative aspects of SOM, it influences the physico-chemical properties of the soil including the availability of nutrients for microbial and plant growth. It accumulates over long period of time and its current distribution in soil profile reflects upon continuous reprocessing by microbes, recombination by chemical reactions, physical movement by soil animals, disturbances such as tree fall, and movement of the soil solution. Consequently, C cycling and its solubilization in soils are intimately associated with soil structure (Kögel-Knabner, 2000). Changes in the nutrient status of the soil reflect decrease in the OM content that directly affects the microbial diversity. Several resident soils are found in grassland pastures where bacteria are associated with root material that attaches to clay particles. Soil productivity in well drained soil is sustained by a healthy bacterial microbiota. Therefore, knowledge of the biodiversity of microbes in soil is essential to maintain plant productivity.

SOIL PHYSICO-CHEMICAL INDICATORS

One of the key objective in determining soil health is to acquire indicators that can be used to evaluate the soil's current status and hence to develop sustainable agricultural system. Among the variables proposed to assess soil health, physical indicators are of prime importance. However, their site-specific interpretation with respect to soil quality will depend on specific land use and crop tolerance. Several physico-chemical indicators are however used:

Water infiltration rate (WIR): It is subjected to significant changes with soil use, management, and time and affected by the development of plant roots, earthworm burrows, soil aggregation and overall increase in stable OM. WIR improves immediately after tillage that is reflected by soil type, texture, and soil water content and is a consequence of loosening of surface crusts/compacted areas. Nevertheless, tillage disrupts aggregates, and soil structure and creates the potential for renewed compaction and surface crusting which finally leads to loss of continuous surface-connected pores. Other components of relative importance in soil health are, soil pH, bulk density, ion-exchange capacity, electrical conductivity and aggregate stability.

Aggregate is usually formed by interaction of soil biota and the plant community and their products with soil mineral components. Aggregates play a major role in several aspects of soil health that include: (i) The movement and storage of water, (ii) Soil aeration, (iii) Physical protection of SOM, (iv) The prevention of erosion, (v) Root development, and (vi) Microbial community activity (Arias et al., 2005). External destructive forces disrupt soil integrity and therefore aggregate stability is a measure of the vulnerability of soil. Soil aggregation can naturally develop, disintegrate, and reform periodically. When aggregates are suddenly immersed in water they fragment because of their inability to withstand the stresses of rapid water uptake, the so called 'slaking'. Soil slaking is used as a measure of the ability of the soil to maintain its structure and is affected by water content, rate of wetting, texture, clay mineralogy, and OM content.

In addition to the above, topsoil depth, root growth, and penetration resistance are other important indicators of soil health. Several practices viz., ploughing, burning, overgrazing, and other management practices remove the protective vegetative cover that reflects on changes in topsoil thickness that leads to erosion. These changes result in loss of both the most fertile soil layer and its water holding capacity including organic C content and productivity. Physico-chemical restraints in soil occur because of anomalies observed in root growth along with the soil profile that includes compaction and the presence of area with higher penetration resistance, deficiency in soil structure, high salt content, the stone layer, the frozen layer, and the water table. All of these factors reflect on plant stress, and reduced crop growth and ultimately the productivity. The most important property of soil is soil texture, i.e., the size distribution of primary soil particles smaller <2mm (sand, silt, and clay). Texture is significantly modified by cultivation and other practices that cause mixing of the different soil layers. Finally, texture influences most of the soil health indicators that help in determination of water intake rates, water storage in soil, ease of tillage, and soil aeration (Arias et al., 2005).

Functional Activity of Bacteria in Rhizosphere and Soil Health

There are a number of key indicators related to microbial functional activities that are used to estimate biomass and activity viz., soil respiration and microbial quotient. Various methods have been employed to determine microbiological activity in the rhizosphere along with bulk soil, wherein several measure the rate of entire metabolic processes e.g., the evolution of CO_2 reflects the catabolic degradation under aerobic condition, the rate of NH_4^+ oxidation to NO_3^- is represented by nitrification activity, the amount of DNA synthesis can be determined by measuring the incorporation of ^{3}H- or ^{14}C-thymidine into bacterial DNA. Similarly, the amount of incorporation of ^{3}H- or ^{14}C-leucine (an amino acid that is incorporated into proteins only), reflects the level of bacterial protein synthesis, and dehydrogenase activity represents the intracellular flux of electrons to O_2 because of redox reaction (Baath, 1998; Nannipieri et al., 2003). Hydrolysis of fluorescein diacetate (FDA) which is a colourless compound, converts into fluorescein (coloured compound) is a measure of the contribution of several enzymes e.g., non-specific esterases, proteases, and lipases, that are involved in the decomposition of organic matter in soil (Adam and Duncan, 2001).

FDA hydrolysis reflects on the overall rhizospheric bacterial activity in soil because more than 90% of energy flow in a soil system passes through microbial decomposers that is reflected in the populations of heterotrophic microorganisms which are predominant in rhizosphere soil. The measured enzyme activities on FDA analysis depend on the contribution of both extracellular and intracellular enzyme activities. The stable extracellular enzyme activities are associated with soil colloids and persist even in harsh environment that limit intracellular microbiological activity. More importantly, the intracellular enzyme activities truly reflect microbial activity because the contribution of free extracellular enzymes released by active microbial cell is quite negligible and its life span is short because they are degraded by proteases when they are adsorbed by clays or immobilized by humic molecules. Nannipieri et al (2002) described that the present enzyme assays do not distinguish properly the contribution of intracellular from extracellular and stabilized enzyme activities, and they do not reflect valid information on the distribution and comparative importance of reactions conducted by microbes. It has been emphasized that there is need to measure the activities of a large number of enzymes and to reunite these measured activities in a single index, to provide information on microbial activity at the ecosystem level. There are many well documented assays that are available for several soil enzyme activities that include cellulase, urease, phosphatase, and phenol oxidase. The mineralization of soil organic nitrogen through nitrate to gaseous nitrogen by soil microorganism is a major component of global nitrogen cycling. By measuring the activities of enzymes involved in these processes viz., urease is an important component of determining overall microbial activity (Arias et al., 2005).

Rhizosphere Functionality and Soil

The resilience of the soil is associated with biodiversity i.e., microbial diversity that increases its resilience capacity. The functional diversity of microbial population in soils is determined by measuring the expression of different enzymes with respect to carbon utilization pattern. The most important aspect of soil biodiversity is soil suppressiveness that is an indicator of the capacity of soils to suppress specific plant pathogens with the help of inherent biotic and abiotic factors (Domingguez et al., 2001).

The link between microbial diversity and the function of soil ecosystem has been described and its importance studied mainly in the above-ground system. Loreau et al (2001) reported the well-known relation between microbial diversity and function by describing the formation of hump-shaped curve wherein there is an increase in plant production i.e., the function, concurrent with increasing microbial diversity until a certain point is reached when a sharp reduction occurs in plant production because of further increase in microbial biodiversity. It is difficult to measure stability (resistance and resilience) in soil. The capacity of soil to recover from perturbation can be assessed by monitoring microbial activities. It is quite tedious to measure microbial diversity as a whole and it does not reflect the link between microbial diversity and function as such. The multifunctionality of soil measured by determining the rates of microbial processes without knowing the microbial species effectively involved in the measured process does not do justice to stability within the system. Griffith et al (2001) have described two approaches i.e., destructive and constructive to describe the link between microbial diversity and soil function. In the destructive approach, repeated $CHCl_3$ fumigation of soils decreases microbial diversity and specific biocides are responsible for killing specific soil microorganisms whereas constructive approach includes sterile soils inoculated with soil microorganisms. Degens et al (2001) investigated the relation between microbial diversity and soil functioning by measuring catabolic evenness in two silt clay loam soils subjected to three different stresses (decline in pH, increase in EC and increasing Cu concentration), and two cyclic disturbances viz., wetting and drying, and freezing and thawing. These workers reported that pasture soil was more resistant to cell stresses than all other perturbations. Finally, the effect of microbial diversity on microbial function in soils depends on the measured function (Griffith et al., 2001). The central problem with the link between microbial diversity and soil function is to understand the relation between genetic diversity and community structure and between community structure and function (O' Donnell et al., 2001). At present various molecular techniques have been employed to determine the composition of soil microflora wherein recent advances include bacterial artificial chromosome (BAC) cloning libraries that allow the functional and taxonomic analysis of soil DNA i.e., metagenome having insights into several other microbial processes viz., pathogenicity, competitiveness, substrate range and bioactive molecules produced by soil microorganisms.

Rhizosphere and Soil Health

Physico-chemical properties of soil are fundamental to soil health including soil texture which is one of the most influential factors. Soil particles held together cohesively influence the precise pore structure of the soil. Soil texture stability reflects the prevention of soil erosion when the soil is exposed to climatic stresses. A well aggregated soil structure ensures soil tilth, soil-plant water relation, water infiltration rates, soil aeration, root penetrability, and organic matter accumulation which all contribute to soil health (Miller and Jastrow, 2000; Buscot, 2005).

It has been demonstrated that microbial cooperation in the rhizosphere reflects the formation and stabilization of soil aggregates wherein soil particles are held together by bacterial products followed by hyphae of saprophytic and arbuscular mycorrhizae (AMF) which form stable microaggregates of size 2-20μm in diameter. These microaggregates are bound by the microbial products again into quite large microaggregates (20-250 μm in diameter) with bacterial polysachharides acting as the binding agents. Finally, microaggregates are then bound into macroaggregates of size > 250 μm in diameter with bacterial polysachharides and AM mycelia that increase the size of microaggregates. The branching habit and three dimensional structure of the external mycelium of AM that colonizes the soil surrounding the roots allows persistence upto 22 weeks after the plant has died (Miller and Jastrow, 2000).

The formation of water-stable soil aggregates is evident in different ecological situations as a result of the effect of AM fungi in co-operation with other microbes, and the involvement of glomalin, a glycoprotein produced by the external hyphae of AM fungi. Glomalin participates in the initiation and stabilization of soil aggregates because of its glue-like hydrophobic nature (Requena et al., 2001). Distribution of natural plant communities is accompanied by loss of physico-chemical and biological properties of soil e.g., soil texture, plant nutrient availability, OM content and microbial activity which is ultimate result of degradation/desertification processes. It has been investigated frequently that management of AM fungi together with rhizobacteria can restore soil traits (Jeffries and Barea, 2001; Requena et al., 2001). The increase in N content in the rhizosphere of the legumes considerably accounts for improvement in nodulation and N-fixing capacity, resulting from cooperative interaction of the symbionts e.g., *Rhizobium* and AM fungi (Barea et al., 2005). There is considerable experimental evidence to show that several bacteria and fungi can colonize the root-soil environment where they carry out a variety of interactive activities known to benefit plant growth and health, as also soil quality.

The varied genetic and functional activities of the microbial populations impart critical impact on soil functions based on the fact that microbes are driving forces for fundamental metabolic processes which involve specific enzyme activities. Most of the microbial interactions in rhizosphere are responsible for key environmental processes viz., the biogeochemical cycling of nutrients and matter and the maintenance of plant and soil health (Nannipieri et al., 2003; Barea et al., 2004). Several investigators of investigations have reported that soil-borne microbes interact with plant roots and soil constituents at the root-soil interface wherein C fluxes are crucial determinants of rhizosphere function (Toal et al., 2000). The release of root exudates provides sources of C compounds for the heterotrophic soil biota whereby microbial activity in the rhizosphere affects rooting pattern and the supply of available nutrients to plants (Gryndler, 2000).

Soil Health and Plant Growth

Soil microorganisms play an important role in soil processes that determine plant and soil productivity. Exhaustive efforts have been made to explore soil microbial diversity of indigenous community, their distribution and behaviour in soil habitats to understand the successful functioning of introduced microbial bio-inoculants and their influence on soil health. Improvement in agricultural sustainability requires optimal use and management of soil fertility and soil physical property and relies on soil biological processes and soil biodiversity (Tilak et al., 2005; Roesti et al., 2006). Plants play an important role in selecting and enriching the type of bacteria by the constituents of their root exudates. The bacterial community develops in the rhizosphere which is a result of diverse nature and concentration of organic constituents of exudates and the corresponding ability of the bacteria to utilize these as sources of energy. Therefore, rhizosphere bacterial community has an efficient system for uptake and catabolism of organic compounds present in root exudates (Barraquio et al., 2000).

Link between Belowground Diversity and Plant

It has been described frequently that a plant obtains almost everything directly from the soil to support growth. The soil must have a structure that is physically capable of supporting the above-ground half of the plant through its developing root system as it grows. The soil needs to be maintained at an appropriate pH, provide protection from toxic substances and pathogens and contain suitable water table. In addition to this all the essential mineral elements that a plant requires are obtained from the soil. Most of these elements are taken from the soil solution in their ionic form (White, 2003). The interaction between plant roots and organisms within rhizosphere assists in acquiring essential mineral nutrients and prevents the accumulation of toxic elements. The essential mineral element that most frequently limit plant growth is P; it is taken up in the form of inorganic phosphate (P_i , $H_2PO_4^-$) from the soil solution. The concentration of P_i in the soil solution (2-10μm) is very low which limits P_i diffusion to the root system with the resultant P_i depletion in the rhizosphere.

Plants have evolved several strategies against limiting nature of P_i , to release and acquire P_i from the soil wherein plant increases its carbohydrate allocation to the roots which results in an increased root: shoot ratio and alters the morphology of root system by accelerating lateral root growth and produces long root hairs to increase the volume of soil explored. In addition, under P deficient conditions the abundance of P_i transporter proteins increases and the exudation of organic acids, RNases, and phosphatases to mobilize P from organic/insoluble compounds is enhanced (Raghothama, 2005). It is not surprising that a series of generalized and specific plant-microbe associations in rhizosphere exist that allow efficient solubilization of all the minerals that a plant requires.

ROLE OF CROP ROTATION IN RESTORATION OF SOIL HEALTH

The most staple food crops of India are rice and wheat that have been cultivated as wheat-rice cropping rotation worldwide by farmers. The cultivation practices of both crops are different where rice requires water logging that creates microaerophilic conditions followed by anaerobic environment that changes the rhizosphere microbial community. The microbial community structure changes to aerobic when wheat is sown in the same field and reflects on the alteration in soil biological equilibrium (Tilak et al., 2005).

The rhizosphere community structure of wheat crop and influence of genotype on community structure has been studied extensively for the Indo-Gangetic region in India (Germida and Siciliano, 2001; Roesti et al., 2006). In a study involving three Indian rain-fed wheat fields which differed in fertilizer management and yield, PGPR/AMF bioinoculation influenced bacterial community structure in the rhizosphere (RS) and rhizoplane/ endorhizosphere (RE) niches (Roesti et al., 2006). Variation partitioning analysis of the community data derived from DGGE analysis showed a predominant effect of stage of wheat growth (30.4% of the variance, P=0.001) over field type (9.0%, P=0.027) in RE fraction. Effect of plant age in RS was much less dominant. However, the influence of bio-inoculation was evident in both, RS and RE. Plant yield or grain quality was either increased or remained unaffected while root colonization by AMF was significantly higher in various treatments; this influence did not change with PGPR inoculation. The authors' have concluded that wheat rhizobacterial community structure is highly dynamic and is influenced by factors such as age of the plant, fertilizer input and type of bio-inoculant. Combined bio-inoculation of diacetyl-phloroglucinol (DAPG) producing PGPR strains and AMF can synergistically improve the nutritional quality of the grain without negatively affecting mycorrhizal growth. Considering a major role played by DAPG in disease management in wheat, search for *phl* D gene as a marker in the wheat root ecosystem was monitored (Imfeld et al., 2006). Available data showed predominance of one or two *phl* D alleles in wheat fields cultivated as rice-wheat rotation for two decade, suggesting that continuous rice-wheat cropping leads to enrichment of particular *phl* D genotypes. New sequences recovered in this study exhibited no close relative among known *phl* D sequences, indicating that part of the *phl* D allelic diversity could have been missed on standard media culture conditions. The effect of plant type on community composition was investigated in a study of rhizosphere microbial population of wheat (*Triticum aestivum*) and mandua (*Eleusine coracana*) grown at Chaukhutia, Almora region in India wherein *Pseudomonas* and *Bacillus* were the predominant rhizosphere inhabitants of both the crops. Recently, raised based management practices of wheat cultivation has been employed by researchers in field to achieve sustainable agriculture and to maintain soil health wherein the bacterial diversity data has been found to corroborate with soil nutrients and soil health parameters (Mittal et al., 2005).

RECLAMATION OF CONTAMINATED SOIL

Heavy metal pollution of soil is a significant environmental problem and has its negative impact on human health and agriculture. Rhizosphere plays a significant role in phytoremediation of contaminated soil by heavy metals wherein microbial population is known to affect heavy metal mobility and availability to the plant through release of chelating agents, acidification, phosphate solubilization and redox changes and therefore, have the potential to enhance phytoremediation (Jing et al., 2007). Excessive metal concentration in contaminated soils results in decreased soil microbial activity and soil fertility and ultimately yields losses. Contamination of soil with Cd negatively affects biodiversity and the activity of soil microbial community. Heavy metal can not be destroyed biologically (i.e., no 'degradation' but convert in the nuclear structure of element) but transformed from one oxidation state/organic complex to other (Garbisu and Alkorta, 2001).Methods employed for their remediation viz., excavation and land fill, thermal treatment, acid leaching, and electroreclamation are not suitable for practical application because of their high cost and low efficiency, large destruction of soil texture and fertility and high dependence on the contaminants of concern, soil properties, and site condition. Therefore, the development of phytoremediation strategies for heavy metal contaminated soil is necessary (Lasat, 2002).

Phytoremediation has been reported to be an effective *in situ*, non-intrusive, low cost, aesthetically pleasing, ecologically benign and socially accepted technology to remediate polluted soils that prevent landscape destruction and enhances the activity and diversity of soil microorganisms to maintain healthy ecosystem (Garbisu et al., 2002). Remedy of contaminated soil through phytoremediation involves phytoextraction, rhizofiltration, phytostabilization, and phytovolatilization. Glick (2003) reported that rhizobacteria play an important role in phytoremediation technology. They affect heavy metal mobility and availability to the plant through release of chelating agents, acidification, phosphate solubilization, and redox changes (Abou-Shanab et al., 2003).

EPILOGUE

This chapter has highlighted the importance of addressing restoration of soil health in the context of multifunctionality of soil along with soil biodiversity. Soil biodiversity indicators can be employed by farmers and governments to assess and monitor soil health and ecosystem functioning under different land use system and management practices. They can help understand impacts of land use change and land degradation processes brought about by various driving forces. More importantly, the assessment and monitoring of soil life and soil health can be used to encourage the development and adaptation by farmers to develop a more sustainable and productive farming system. It is clear that a range of appropriate indicators of soil health are required to account for the multiple dimensions of soil ecosystem functions. Indicators should be selected that show a close link viz., between the soil characteristics and primary production wherein a positive indicator of soil quality could be organic matter and a negative one, incidence of crop damage by soil pathogens.

There is a need for holistic consideration of soil health with transdisciplinary soil management approaches that integrate biological and physico-chemical strategies to achieve soils supporting a sustainable agriculture. The environmental and economic benefits of sustainable soils are enormous which include increased resource efficiency, decomposition and nutrient cycling, nitrogen fixation, and water holding capacity as well as prevention of pollution and land degradation. The adoption of agricultural practices reduces soil biodiversity mainly because of extensive use of chemicals that lead to compaction and other perturbations and hence irreversible adverse ecological alteration resulting in loss of agricultural productivity. Several reports suggest that organic/sustainable system can increase both SOM accumulation and microbial activity. In addition, the organic C lost during intensive agriculture could be regained through sustainable management practices thus contributing to mitigating climate change.

The development of enormous approaches that do not include establishment of microbial cultures will undoubtedly enhance our knowledge of biodiversity and promote the invention of new microorganisms with unique capacities for bioremediation, soil restoration, and other applications.

References

Abou-Shanab, R.A., Angle, J.S., Delmore, T.A., Chaney, R.L., van Berkum, P., Moawed, H., Ghanem, K and H.A. Ghozlan 2003 Rhizobacterial effects nickel extraction from soil and uptake by *Alyssum Murale*. *N. Phytol.* 158 (1):219-224.

Adam, G. and H. Duncan 2001 Development of a sensitive and rapid method for the measurement of total microbial activity using fluorescein diacetate (FDA) in a range of soils. *Soil Biol. Biochem.* 33:943-951.

Arias, M.E., González-Pérez, J.A., González-Vila, F.J. and A.S. Ball 2005 Soil health-a new challenge for microbiologists and chemists. *Int. Microbiol.* 8(1):3-21.

Baath, E. 1998 Growth rates of bacterial community in soils at varying pH: a comparision of the thymidine and leucine incorporation techniques. *Microbiol. Ecol.* 36: 316-327.

Barea, J.M., Azcón, R. and C. Azcón-Aguilar 2004 *Mycorrhizal fungi and plant growth promoting rhizobacteria*. In: Plant surface microbiology (eds. L. Abbott, D. Werner and R. Hampp) pp. 351-371, Springer-Verlag, Heidelberg, Germany.

Barea, J.M., Azćon, R. and C. Azćon-Aguilar 2005 *Interactions between mycorrhizal fungi and bacteria to improve plant nutrient cycling and soil structure*. In: Microorganism in soils: roles in genesis and function (eds F. Buscot and S. Varma) pp. 195-212, Springer-Verlag, Heidelberg, Germany.

Barraquio, W.L., Segubre, E.M., Gonzalez, M.S., Verma, S.C., James, E.K., Ladha, J.K. and A.K. Tripathi 2000 In: *The quest for nitrogen fixation in rice*. Pp. 93-118, IRRI, Los Banos, Phillipins.

Buscot, F. 2005 *What are soils?* In: Microorganism in soils: roles in genesis and function (eds. F. Buscot and S. Varma) pp. 3-18, Springer-Verlag, Heidelberg, Germany.

Cassman, K.G., Dobermann, A. and Walters, D.T. (2002). Agroecosystems, nitrogen-use efficiency, and nitrogen management. *Ambio*. 31, 132-140.

Degens, B.P., Schipper, L.A., Sparling, G.P. and L.C. Duncan 2001 Is the microbial community in a soil with reduced catabolic diversity less resistant to stress or disturbance? *Soil Biol. Biochem*. 33:1143-1153.

Dorninguez, J., Negrin, M.A., and C.M. Rodriguez 2001 Aggregate water stability, Particle-size and soil solution properties in conductive and suppressive soil to *Fusarium* wilt of banana from Canary Islands (Spain) *Siol Biol. Biochem*. 33: 499-455.

Galloway, J.N. (2000). Nitrogen mobilization in Asia. *Nutrient Cycling in Agroecosystem*. 57, 1-12.

Garbisu, C. and I. Alkorta 2001 Phytoextraction: a cost-effective plant-based technology for the removal of metals from the environment. *Bioresource Technol*. 77(3):229-236.

Germida, J.J. and S.D. Siciliano 2001 Taxonomic diversity of bacteria associated with the roots of modern recent and ancient wheat cultivar. *Bio Fertl. Soils* 33:410-415.

Glick, B.R. 2003 Phytoremediation: synergistic use of plants and bacteria to clean up the environment. *Biotechnol. Adv*. 21(5):383-393.

González-Pérez, J.A., González-Vila, F.J., Almendrous, G. and H. Knicker 2004 The effect of fire on soil organic matter.-*a review Environ. Int.* 30:855-870.

Griffiths, B.S., Bonkowski, M., Riy, J. and K. Ritz 2001 Functional stability, substrate utilization and biological indicators of soils following environmental impacts. *Appl. Soil Ecol.* 16:49-61.

Imfeld, G., Shani, N., Roesti, D., Fromn, N., Johri, B.N., Gaur, R., Rossi, P., Locatelli, L., Poly, F. and M. Aragno 2006 Diversity of *phl* D alleles in the rhizosphere of wheat cropped under annual rice-wheat rotation in fields of the Indo-Gangetic plains: influence of cultivation conditions. *Curr. Sci*. 90(11): 1521-1525.

Jeffries, P. and J.M. Barea 2001 *Arbuscular mycorrhiza: a key component of sustainable plant-soil ecosystem.* In: The mycota: fungal associations (eds. B. Hock) pp. 95-113, vol IX, Springer, Berlin, Heidelberg, Germany.

Jing, Y-de, He, Z-li and X-e, Yang 2007 Role of soil rhizobacteria in phytoremediation of heavy metal contaminated soils. *J. Zhejiang Univ. Sci*. 8(3): 192-207.

Kögel-Knabner, I. 2000 Analytical approaches for characterizing soil organic matter *Org. Geochem.* 31: 609-625.

Lasat, H.A. 2002 Phytoextraction of toxic metals: a review of biological mechanisms. *J. Environ. Quality* 31(1):109-120.

Loreau, M., Naeem, S., Inchausti, P., Bengtsson, J., Grime, J.P. and Hector, A., Hooper, D.U., Huston, M.A., Raffaelli, D., Schmid, B., Tilman, D. and D.A. Wardle 2001 Biodiversity and ecosystem functioning: current knowledge and future challenges. *Science* 294:804-808.

Miller, R.M. and J.D. Jastrow 2000 *Mycorrhizal fungi influence soil structure*. In:Arbuscular mycorrhizas: physiology and functions (eds. Y. Kapulnik and Jr D.D. Douds) pp. 3-18, Dordrechet, Kluwer Academic Publisher, The Netherland.

Miller, R.M. and J.D. Jastrow *2000 Mycorrhizal fungi influence soil structure*. In:Arbuscular mycorrhizas: physiology and functions (eds. Y. Kapulnik and Jr D.D. Douds) pp. 3-18, Dordrechet, Kluwer Academic Publisher, The Netherland.

Nannipieri, P., Ascher, J., Ceccherini, M.T., Landi, T., Pietramellara, G. And G. Renella 2003 Microbial diversity and soil functions. *Eur. J. Soil Sci*. 54:655-670.

Nannipieri, P., Kandeler, E. and P. Ruggiero 2002 *Enzyme activities and microbiological and biochemical processes in soil*. In: Enzymes in the environment (eds. R. G. Burns and R. Dick) pp. 1-33, Marcel Dekker, New York.

Ó Donnel, A.G., Seasman, M., Macrae, A., Waite, I. and J. T. Davis 2001 Plants and fertilizers as drivers of change in microbial community structure and functions in soils. *Plant and Soil* 232:135-145.

Pinton, R., Varanini, Z., and P. Nannipieri *2001 The rhizosphere*: Biodiversity and organic substances at the soil-plant interface. Marcel Dekker, New York.

Prentice, I.C., Farquhar, G.D., Fasham, M.J.R., Goulden, M.L., Heimann, M. and V.J. Jaramillo 2001 *The carbon cycle and atmospheric carbon dioxide*. In: Climate change: the scientific bases (eds. J.T. Houghton, Y. ding, D.J. Griggs, M. Noguer, P.J. van der Linden, X. Dai, K. Marskell and C.A. Johson) pp. 183-237, Cambidge University Press, Cambridge, UK.

Raghothama, K.G. 2005 *Phosphorus*. In: Plant nutritional genomics (eds. M.R. Broadly and P.J. White) pp. 112-126, Blackwell, Oxford.

Requena, N., Prez-soils, E., Azcón-Aguilar, C. Jeffries, P. and J.M. Barea 2001 Management of indigenous plant-microbe symbioses aids restoration of desertified. *Appl. Environ. Microbiol.* 67:495-498.

Roesti, D., Gaur, R., Johri, B.N., Imfeld, G., Sharma, S., Kawaljeet, K. and M. Aragno 2006 Plant growth stage, feriliser management and bio-inoculations of arbuscular mycorrizal fungi and plant growth promoting rhizobacteria affect the rhizobacterial community structure in rain-fed wheat fields. *Soil Biol. Biochem*. 38:1111-1120.

Tilak, K.V.B.R., Ranganayaki, N., Pal, K.K., De, R., Saxena, A.K., Nautiyal, C.S., Mittal, S., Tripathi, A.K. and B.N. Johri 2005 Diversity of plant growth and soil health supporting bacteria. *Curr. Sci.* 89(1):136-150.

White, P.J. 2003 Ion Transport. In: *Encyclopaedia of applied plant sciences* (eds. B. Thomas, D.J. Murphy and B.G. Murray) pp. 625-634, Academic Press, London.

In: Soil Nutrients
Editor: Mohammad Miransari
ISBN: 978-1-61324-785-3

Chapter 10

The Use of Vermicompost in Sustainable Agriculture: Impact on Plant Growth and Soil Fertility

Cristina Lazcano* and Jorge Domínguez
Departamento de Ecoloxía e Bioloxía Animal, Universidade de Vigo, Vigo, Spain

Abstract

Vermicomposting is a low-technology, environmentally-friendly process used to treat organic waste. The resulting vermicompost has been shown to have several positive impacts on plant growth and health. This organic fertilizer is therefore increasingly considered in agriculture and horticulture as a promising alternative to inorganic fertilizers and/or peat in greenhouse potting media. However, the effects of vermicompost on plant-soil systems are not yet fully understood. In this chapter we summarize the research carried out during the last few decades, and the proposed mechanisms explaining the effects of vermicompost on soil quality and plant growth. Although much effort has been dedicated to the investigation of biologically mediated mechanisms of promoting plant growth, the conflicting results indicate the need to open up new lines of research, defining a clear and objective concept of vermicompost, and clarifying the conditions and sources of variability in the biological effects. A case study is presented in which the direct and indirect effects of vermicompost on plant growth, as well as variability in the plant responses, are examined in a field experiment with sweet corn.

Keywords: sustainable agriculture; earthworms; soil fertility; plant-growth promoting substances; humic acids

* Corresponding autor. Present adress: Centro Tecnológico del Mar - Fundación CETMAR. C/ Eduardo Cabello s/n - E-36208 Bouzas - Vigo (Pontevedra). Tel. +34 986 247 047 Ext. 607 Fax. +34 986 296 019. E-mail: clazcano@cetmar.org ; cristina@uvigo.es.

1. INTRODUCTION

1.1. The Use of Composted Materials in Sustainable Agriculture

In recent years, increasing consumer concern about issues such as food quality, environmental safety and soil conservation has lead to a substantial increase in the use of sustainable agricultural practices. Sustainable agriculture can be defined as a set of practices that conserve resources and the environment without compromising human needs, and the use of organic fertilizers such as animal manure has been indicated as one of its main pillars (Tilman et al., 2002). Animal manure is a valuable resource as a soil fertilizer because it provides large amounts of macro- and micronutrients for crop growth and is a low-cost, environmentally-friendly alternative to mineral fertilizers. However, the use of manure in agriculture is being abandoned because of increasing transportation costs and environmental problems associated with the indiscriminate and inappropriately-timed application to agricultural fields (Hutchison et al., 2005). Processing of this waste material through controlled bio-oxidation processes, such as composting, reduces the environmental risk by transforming the material into a safer and more stable product suitable for application to soil (Lazcano et al., 2008), and also reduces the transportation costs because of the significant reduction in the water content of the raw organic matter. Composted materials are therefore gaining acceptance as organic fertilizers in sustainable agriculture, and there has been a considerable increase in research dedicated to the study of the effects of compost-like materials on soil properties and plant growth.

The results of several long-term studies have shown that the addition of compost improves soil physical properties by decreasing bulk density and increasing the soil water holding capacity (Weber et al., 2007). Moreover, in comparison with mineral fertilizers, compost produces significantly greater increases in soil organic carbon and some plant nutrients (García-Gil et al., 2000, Bulluck et al, 2002, Nardi et al., 2004, Weber et al., 2007). Long-term beneficial effects of composted materials are also observed in soil humic substances (due to an increase in the complexity of their molecular structure, which increases the humic/fulvic acid ratio), as well as in soil sorption properties (with increased cation exchange capacity and base saturation) (Weber et al., 2007). In addition to the changes exerted on the chemical and physical properties, composted materials have a clear impact on soil biological properties, such as increases in microbial biomass and activity (Knapp et al. 2010), as well as changes in the activity of soil enzymes (Garcia-Gil et al., 2000, Ros et al., 2006) and in the structure of the soil microbial community (Ros et al., 2006).

1.2. Vermicomposting and Vermicompost Properties

Vermicompost is a nutrient-rich, microbiologically-active organic amendment that results from the interactions between earthworms and microorganisms during the breakdown of organic matter. It is a stabilized, finely divided peat-like material with a low C:N ratio, high porosity and high water-holding capacity, in which most nutrients are present in forms that are readily taken up by plants (Domínguez, 2004).

Unlike compost, vermicompost is produced under mesophilic conditions, and although microorganisms degrade the organic matter biochemically, earthworms are the crucial drivers of the process, as they aerate, condition and fragment the substrate, thus drastically altering the microbial activity. Earthworms act as mechanical blenders, and by fragmenting the organic matter they modify its physical and chemical status by gradually reducing the ratio of C:N and increasing the surface area exposed to microorganisms - thus making it much more favourable for microbial activity and further decomposition (Domínguez et al., 2010).

As a result of the different processes involved in the production of compost and vermicompost, they exhibit different physical and chemical characteristics that affect soil properties and plant growth in diverse ways. Vermicomposting generally converts organic matter to a more uniform size, which gives the final substrate a characteristic earthy appearance, whereas the material resulting from composting usually has a more heterogeneous appearance (Ndegwa and Thompson, 2001; Tognetti et al., 2005).

The use of compost in horticulture has occasionally been shown to be limited by the high electrical conductivity and the excessively high amount of certain ions that cause phytotoxicity (García-Gómez et al., 2002), as a consequence of the chemical properties of the initial waste and /or inadequate composting procedures. These adverse effects, although possible, are less likely to occur when vermicompost is used as a potting amendment (Chaoui et al., 2003). Nevertheless, the most remarkable differences between compost and vermicompost are related to their biological properties. Composting and vermicomposting are two rather different biological processes, which condition the biological properties of the final substrate and result in important differences between compost and vermicompost, in both the bacterial community composition (Vivas et al., 2009) and fungal abundance (Lazcano et al., 2008), even when the same organic waste is used as a feedstock material. These differences may lead to rather different effects on plant growth and morphology that must be investigated. Although the composting process and the effects of compost on soil and plant growth have been fairly well studied, this is not the case for vermicompost and the vermicomposting process.

2. How Vermicompost Influences Plant Growth: Effects and Proposed Mechanisms

2.1. Effects of Vermicompost on Plant Growth

Vermicompost significantly stimulates the growth of a wide range of plant species including several horticultural crops such as tomato (Atiyeh et al., 1999; Atiyeh et al., 2000a, Atiyeh et al., 2000b; Atiyeh et al., 2001; Hashemimajd, et al., 2004; Gutiérrez-Miceli et al., 2007), pepper (Arancon et al., 2004a, Arancon et al., 2005a), garlic (Argüello et al., 2006), aubergine (Gajalakshmi and Abbasi, 2004), strawberry (Arancon et al., 2004b), sweet corn (Lazcano et al, 2011) and green gram (Karmegam et al., 1999). Vermicompost has also been found to have positive effects on some aromatic and medicinal plants (Anwar et al., 2005; Prabha et al., 2007), cereals such as sorghum and rice (Bhattacharjee et al., 2001; Reddy and Ohkura, 2004; Sunil et al., 2005), fruit crops such as banana and papaya (Cabanas-Echevarria, et al., 2005; Acevedo and Pire, 2004), and ornamentals such as geranium (Chand et al. 2007), marigolds (Atiyeh et al., 2002), petunia (Arancon et al., 2008), chrysanthemum (Hidalgo and Harkess, 2002a) and poinsettia (Hidalgo and Harkess, 2002b). Positive effects of vermicompost have also been observed in forestry species such as acacia, eucalyptus and pine tree (Donald and Visser, 1989; Lazcano et al., 2010a, 2010b).

Vermicompost has been found to have beneficial effects when used as a total or partial substitute for mineral fertilizer in peat-based artificial greenhouse potting media and as soil amendments in field studies. Likewise, some studies show that vermicomposting leachates or vermicompost water-extracts, used as substrate amendments or foliar sprays, also promote the growth of tomato plants (Tejada et al., 2008), sorghum (Gutiérrez-Miceli et al., 2008), and strawberries (Singh et al., 2010).

Positive effects of vermicompost include stimulated seed germination in several plant species such as green gram (Karmegam et al., 1999), tomato plants (Atiyeh et al., 2000b; Zaller, 2007), petunia (Arancon et al., 2008) and pine trees (Lazcano et al., 2010a). Vermicompost also has a positive effect on vegetative growth, stimulating shoot and root development (Edwards et al., 2004). The effects include alterations in seedling morphology such as increased leaf area and root branching (Lazcano et al., 2009). Vermicompost has also been shown to stimulate plant flowering, increasing the number and biomass of the flowers produced (Atiyeh et al., 2002; Arancon et al., 2008), as well as increasing fruit yield (Atiyeh, et al., 2000b; Arancon et al., 2004a, 2004b; Singh et al., 2008). In addition to increasing plant growth and productivity, vermicompost may also increase the nutritional quality of some vegetable crops such as tomatoes (Gutiérrez-Miceli et al., 2007), Chinese cabbage (Wang et al., 2010), spinach (Peyvast et al., 2008), strawberries (Singh et al., 2008), lettuce (Coria-Cayupán et al., 2009), and sweet corn (Lazcano et al., 2011).

Nevertheless, despite the large body of scientific evidence showing the positive effects of vermicompost on plant growth and yield, there is also strong evidence that these effects are not general or constant, and that there is great variability in the magnitude of the effects reported in different studies. In fact, some studies report that vermicompost may decrease growth and even cause plant death (Roberts et al., 2007; Lazcano et al., 2010c). The variability in the effects of vermicompost may depend on the cultivation system into which it is incorporated, as well as on the physical, chemical and biological characteristics of vermicompost, which vary widely depending on the original feedstock, the earthworm species used, the production process, and the age of vermicompost (Rodda et al., 2006, Roberts et al., 2007; Warman and AngLopez, 2010).

There is also a large variation in the effects of vermicompost depending on the plant species or even the variety considered. This was observed in tomato plants where the replacement of a fertilized commercial potting media with vermicompost had different effects on germination, seedling elongation, biomass allocation, fruit morphology and chemical properties of three tomato varieties (Zaller, 2007). Similar variation was observed in an experiment studying the effects of vermicompost and vermicompost extracts on the germination and early growth of six different progenies of maritime pine Lazcano et al. (2010a). In this experiment, the speed of maturation increased, relative to the control without vermicompost, in three out of the six pine progenies, decreased in two of the progenies and was unaffected in the other. It may be expected that different hybrids or plant genotypes will respond differently to vermicompost, considering that plant genotype determines important differences in nutrient uptake capacity, nutrient use efficiency and resource allocation within the plant. Different genotypes may therefore enhance root growth or modify root exudation patterns in order to increase nutrient uptake (Kabir et al., 1998; Cavani and Mimmo, 2007), and all of these strategies will determine the establishment of different interactions with the microbial communities at the rhizosphere level. In fact, after the application of vermicompost to sweet corn crops, the different genotypes assayed showed important differences in their rhizosphere microbial community (Aira et al., 2010).

In light of this evidence, it is clear that vermicompost constitutes a promising alternative to inorganic fertilizers in promoting plant growth. However, further research into the exact mechanisms and circumstances that stimulate plant growth by this organic substrate is necessary in order to maintain consumer confidence in this type of fertilizer.

2.2. Plant Growth Regulating Mechanisms

Vermicompost is a microbiologically-active, nutrient-rich, peat-like material, which when added to plant growing media may influence plant growth directly or indirectly through different chemical, physical and biological mechanisms (Figure 1).

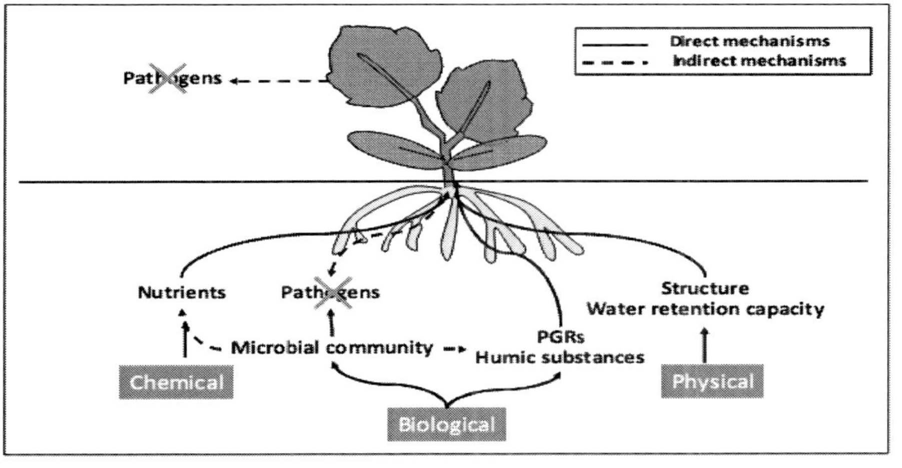

Figure 1. Some of the proposed chemical, biological and physical mechanisms by which vermicompost may directly or indirectly influence plant growth and development.

2.2.1. Direct Mechanisms

As regards the direct effects on plant growth, vermicompost constitutes a source of plant macro- and micronutrients. Although some of these nutrients are present in inorganic forms and are readily available to plants, most are released gradually through mineralization of the organic matter, thus constituting a slow-release fertilizer that supplies the plant with a gradual and constant source of nutrients (Chaoui et al., 2003). However, in contrast to chemical fertilizers, the amount of nutrients provided may vary greatly depending on the original feedstock, processing time and maturity of the vermicompost (Campitelli and Ceppi, 2008).

Vermicompost is a finely-divided peat-like material and because of this fine structure, the addition of vermicompost to plant potting media causes significant changes in the physical properties, altering water and air availability in the substrates and conditioning root growth. Nonetheless, the type and magnitude of these effects depends on the physical characteristics of the original growing medium. For example, the bulk density of vermicompost is usually higher and the particle size lower than in some of the most commonly used peat-based substrates (Atiyeh et al., 2001); mixing of these two substrates produces a significant increase in the bulk density and water holding capacity, while decreasing particle size and total porosity (Atiyeh et al., 2001; Bachmann and Metzger, 2007). Yet Hidalgo and Harkess (2002a) and Hidalgo et al. (2006) reported a significant increase in total porosity and water holding capacity after addition of vermicompost to a greenhouse potting medium comprising a mixture of sand, pine bark and peat. Vermicompost may also have significant effects on soil physical properties. Ferreras et al. (2006) observed that addition of 20 tn ha^{-1} of vermicompost to an agricultural soil in two consecutive years significantly improved soil porosity and aggregate stability. The number of large, elongated soil macropores increased significantly after a single application of a dose of vermicompost equivalent to 200 kg ha^{-1} of N to a corn field (Marinari et al., 2000). Similarly, Gopinath et al. (2008) reported a significant decrease in soil bulk density and a significant increase in soil pH and total organic carbon after application of vermicompost in two consecutive growing seasons, at a rate equivalent to 60 kg ha^{-1} of N. Together these changes in soil properties improve the availability of air and water, thus encouraging seedling emergence and root growth.

Despite the above-mentioned physical and chemical mechanisms, there is much experimental evidence showing that vermicompost enhances plant growth further than expected because of nutrient supply and improvements in the physical condition of substrates. This was first suggested by Scott (1988) and Edwards and Burrows (1988), who observed that small doses of vermicompost added to the potting media of several ornamental species, produced a much larger increase in plant growth than the equivalent dose of nutrients. These effects were maintained even when vermicompost was diluted 1:20 with other potting media, resulting in a dose of vermicompost that would be expected to have negligible physical effects (Edwards and Burrows, 1988). This was also observed more recently in field experiments carried out by Arancon et al. (2004b) and Singh et al. (2008). Both studies indicated a significant increase in the growth and productivity of strawberries cultivated with 5 and 7.5 t ha^{-1} of vermicompost respectively, in comparison with strawberries cultivated with equivalent doses of mineral fertilizers. Furthermore, many studies have shown that increases in growth and yield often involve changes in plant development and/or plant morphology such as increased leaf area, root volume and root branching (Singh et al., 2008; Lazcano et al., 2009). It has been postulated that such enhancement of plant growth may be attributed to the existence of biologically-mediated plant growth promoting mechanisms.

Vermicompost may influence plant growth directly via the supply of plant growth regulating substances (PGRs). This was first proposed by Tomati et al. (1983, 1987, 1988), Tomati et al. (1990); Grapelli et al. (1987) and Tomati and Galli (1995). In these experiments the authors compared the effects of different vermicompost extracts on the growth of Begonia, Petunia and Coleus with the effects produced by auxins, gibberellins and cytokinins, and concluded that there was a strong evidence of hormonal activity caused by the earthworms. Hormonal activity has also been associated with the humic substances present in vermicompost. Canellas et al. (2002) and Zandonadi et al. (2006) reported that the humic substances extracted from earthworm compost were capable of inducing lateral root growth in maize plants by stimulation of the plasma membrane H^+-ATPase activity, thus producing similar effects such as the exogenous application of indole-3-acetic acid (IAA). Furthermore, the induction of lateral root initiation by vermicompost-derived humic substances in arabidopsis has been related to the activation of the transcription of some auxin responsive genes (Trevisan et al., 2010). The hypothesis of the auxin activity of vermicompost humic substances is reinforced by the presence of exchangeable auxin groups in their macrostructure (Canellas et al., 2002).

It has been suggested that earthworms may be important agents capable of influencing the production of PGRs by microorganisms through the stimulation and promotion of microbial activity in organic substrates (Nielson, 1965; Springett and Syers, 1979; Grappelli et al., 1987; Tomati et al., 1983, 1987, 1988; Tomati et al., 1990; Tomati and Galli, 1995, Nardi et al., 1988; Dell'Agnola and Nardi, 1987). Nevertheless, some authors suggest that earthworms, and not microorganisms, are responsible for the production of PGRs. Nielson (1965) reported the first evidence of the presence of indole compounds in the tissues of *Aporrectodea caliginosa*, *Lumbricus rubellus*, and *Eisenia fetida*. More recently, El Harti et al. (2001a, 2001b) showed that a crude extract of the earthworm *Lumbricus terrestris* was able to stimulate rooting in bean seeds due to the presence of indole compounds of endogenous origin.

2.2.2. Indirect Mechanisms

Vermicompost has also been found to have a wide range of indirect effects on plant growth such as the mitigation or suppression of plant diseases (Figure 1). Suppression of plant diseases has been extensively investigated in other organic amendments such as manure and compost (Noble and Coventry, 2005; Termorshuizen et al., 2006; Trillas et al., 2006). Likewise, some studies have shown that vermicompost can suppress a wide range of microbial diseases, insect pests and plant parasitic nematodes.

As regards the suppression of fungal diseases, Orlikowski (1999) observed that the addition of vermicompost extracts to three ornamental plant species significantly reduced sporulation of the pathogen *Phytophthora cryptogea*. Similarly Nakasone et al. (1999) observed that aqueous extracts of vermicompost were capable of reducing the growth of pathogenic fungi such as *Botrytis cinerea, Sclerotinia sclerotiorum, Corticium rolfsii, Rhizoctonia solani* and *Fusarium oxysporum*. The addition of solid vermicompost to tomato seeds significantly reduced infection caused by *Fusarium lycopersici* (Szczech, 1999) and *Phytophthora nicotianae* (Szczech and Smolinska, 2001). Nevertheless, Szczech and Smolinska (2001) did not find any significant suppressive effects of a sewage sludge vermicompost on *Phytophthora nicotianae*, in comparison with peat. Edwards et al (2006) observed that the suppressive effect exerted by several types of vermicompost on several plant pathogens such as *Pythium*, *Rhizoctonia, Verticillium,* and *Plectosporium*, disappeared after sterilization of the vermicompost, and concluded that disease suppression may be related to the presence of biological suppressive agents in vermicompost.

As regards the effects of vermicompost on insect pests and mites, field studies have shown that the addition of vermicompost to soil significantly reduces the incidence of the psyllid *Heteropsylla cubana* (Biradar et al., 1998), the sucking insect *Aproaerema modicella* (Ramesh, 2000), jassids, aphids, beetles, and spider mites (Rao, 2002).

Furthermore, some studies show that vermicompost reduces not only the degree of plant infestation, but also the populations of plant parasitic insects in soils. This has been reported by Arancon et al. (2005b), who showed that vermicompost produced from food waste significantly reduced the populations of two species of beetles in soil (*Acalymma vittatum*, *Diabotrica undecimpunctata*), in comparison with inorganic fertilization. Similarly Yardim et al. (2006) observed a significant decrease in the larvae of the worm *Manduca quinquemaculata* after addition of vermicompost to a cucumber crop. Arancon et al. (2007) observed a significant reduction in the populations of spider mites (*Tetranychus urticae*), mealy bugs (*Pseudococcus* sp.) and aphids (*Myzus persicae*) after the addition of food waste vermicompost to several vegetable crops (tomatoes, cucumber, cabbage, bush beans and eggplants). Vermicompost may also have significant effects on both the incidence and abundance of plant-parasitic nematodes in soil. This was first reported by Swathi et al. (1998), who found that addition of vermicompost to soil at a rate of 1kg m^{-2} significantly inhibited the incidence of the parasite nematode *Meloidogyne incognita* in tobacco plants. Similar reductions in the degree of plant infestation by *Meloidogyne incognita* were observed by Morra et al. (1998), while Ribeiro et al. (1998) reported a significant decrease in the egg mass of *Meloidogyne javanica* after the application vermicompost to the growth medium. Arancon et al. (2003) indicated significant reductions in the abundance of plant-parasitic nematodes in soil plots amended with different doses of two types of vermicompost, in comparison with the effects of inorganic fertilizers.

According to the mechanisms proposed for compost (Hointink and Bohem, 1999; Noble and Coventry, 2005), disease suppression by vermicompost may be attributed to either direct suppression of pathogens or to the induction of systemic resistance in the plant (Figure 1). Direct suppression of the pathogen by the vermicompost-associated microflora and/or microfauna may be general or specific, depending on the existence of a single suppressive agent or the joint action of several agents. In both cases, the proposed mechanisms are competition, antibiosis and parasitism. The mechanisms responsible for disease suppression by vermicompost are not yet well known, and the effects are usually attributed to general suppressive effects. Vermicompost increases microbial biomass in soil and changes the diversity and abundance of soil fauna (Gunadi et al., 2002; Arancon et al., 2006), and thus a broader range of organisms may act as biocontrol agents. Furthermore, there is recent evidence showing that the use of vermicompost extracts as foliar sprays in different crop plants effectively reduces the incidence of fungal diseases such as *Phytophthora infestans* (Zaller, 2006), *Erysiphe pisi* and *Erysiphe cichoracearum* (Singh et al., 2003). Reductions in disease incidence are sometimes accompanied by an increase in the production of defence substances by the plant (Singh et al. 2003), thus suggesting the induction of plant systemic resistance by vermicompost.

Some of the indirect effects of vermicompost have been related to the change in the microbiological properties of the soil or the potting media. Processing by earthworms during vermicomposting has a strong effect on the microbial community of the initial waste (Dominguez et al., 2010). Vermicompost therefore has a rather different microbial community structure than the parent waste, with lower biomass and activity but enhanced metabolic diversity (Lores et al., 2006; Aira et al., 2007). Application of such a microbiologically active organic substrate may have important effects on the microbial properties of soil or greenhouse potting media thereby influencing plant growth. However, information regarding the impacts of vermicompost on soil microbial properties is still limited. A single application of vermicompost to a strawberry crop has been shown to produce a significantly higher increase in soil microbial biomass than application of an inorganic fertilizer, independently of the dose used (Arancon et al., 2006). As well as increasing microbial biomass, vermicompost increases microbial activity (Ferreras et al., 2006) promotes the establishment of a specific microbial community in the rhizosphere different from that of plants supplemented with mineral fertilizers or other type of organic fertilizers such as manure (Aira et al., 2010). Kale et al., (1992) observed that the application of vermicompost may produce significantly greater increases in the abundance of N-fixers, actinomycetes and spore formers than in soil supplemented with inorganic fertilizers. Soil enzyme activity is also significantly increased by vermicompost addition as compared to equivalent rates of mineral fertilizers (Marinari et al., 2000; Arancon et al., 2006, Saha et al., 2008).

3. Effects of Vermicompost on Plant Growth and Soil Fertility in a Sweet Corn Stand: A Case Study

As mentioned earlier in this chapter, vermicompost has been shown to increase the growth and yield of a wide range of species due to several direct and indirect beneficial effects. Furthermore, the use of organic fertilizers such as vermicompost helps maintain soil fertility, and has evident environmental benefits as it enables on-farm recycling of organic waste. However, agricultural producers often claim that crop yields are much lower with this type of fertilizer than with inorganic fertilizers. This is generally attributed to the fact that the amount of nutrients provided by organic fertilizers is very unreliable in comparison with those supplied by inorganic fertilizers (Trewavas, 2001). Furthermore, most of the benefits of organic fertilizers on soil fertility have mainly been reported to be long-term effects, so that repeated applications over several years are necessary to achieve the desired steady state that will guarantee crop productivity.

We carried out a field experiment in which we tested whether the replacement of 25% of the applied inorganic NPK fertilizers with rabbit manure vermicompost could maintain the yield of a sweet corn crop at the same levels. Crop yield was assessed three months after vermicompost application to soil. We hypothesized that for the same amounts of nutrients supplied, vermicompost would produce similar or higher plant growth yields than inorganic fertilizers. The effects of vermicompost were compared with those of the equivalent unprocessed organic matter (rabbit manure). Thus, three types of fertilization regimes were applied: (i) a conventional fertilization regime with inorganic fertilizer, and integrated fertilization regimes in which 75% of the nutrients were supplied by the inorganic fertilizer and 25% of the nutrients were supplied by either (ii) rabbit manure, or (iii) vermicompost. All three fertilizers were supplied at two different doses, a *normal* dose (80:24:20 kg ha^{-1} of N:P:K) for an expected final crop yield of 4 t ha^{-1} of dry weight grain, and a *high* dose of 120:36:30 kg ha^{-1} of N:P:K for an expected final yield of 6 t dry grain ha^{-1}, according to the dry grain yield reported for sweet corn (Ordás et al., 2007). The amounts of each fertilizer required to supply these nutrient requirements were calculated taking into account the nutrient content of each and the nutrient content of the soil at the experimental site. Thus, the normal doses were 5.4 and 4.2 t substrate ha^{-1} for manure and vermicompost, respectively, and the high doses were 8.2 and 6.3 t substrate ha^{-1} for manure and vermicompost, respectively. The field trial was repeated over two consecutive years. In both years, fertilizers were applied manually to the plots one week before sowing, and incorporated by mixing into the upper 20 cm of the soil.

As already explained, much of the observed variability in the effects of vermicompost can be attributed to the different plant species and genotypes used. Therefore in this experiment, four different sweet corn hybrids (H1, H2, H3 and H4) were included, in order to explore the interactions between the type of fertilizer and plant genotype. Experimental units consisted of 10 m^2 plots each with one combination of hybrid, fertilizer and dose. Each plot contained two central and two border rows spaced 0.80 m apart, with 25 two-plant hills spaced 0.21 m apart; plots were overplanted and thinned to a final density of approximately 60000 plants ha^{-1}. Plots were arranged in the field following a randomized complete block design with two replications per year.

At harvest, three months after sowing, five plants were selected at random from the two central rows of each plot, and their heights were measured. Plants were subsequently cut to soil level and transported to the laboratory for determination of total yield (total ear biomass per plant), quality of the marketable ear (grain percentage, NPK content), leaf NPK nutrient content and total plant biomass. In order to evaluate the effects of the fertilizers on soil biochemical and microbiological properties, one composite soil sample, consisting of 3 randomly selected sub-samples, was collected in each plot. The structure of the soil microbial community was assessed by phospholipid fatty acid (PLFA) analysis; certain PLFAs were used as biomarkers to determine the presence and abundance of specific microbial groups. Microbial community function was determined by measuring bacterial and fungal growth rates, by incorporating radioactively labelled leucine into proteins and radioactively labelled acetate into the fungal-specific lipid ergosterol, respectively (Bååth et al., 2001). The activities of some of the main soil enzymes (β-Glucosidase, protease and alkaline phosphomonoesterase) were also assessed, as were the available N (NH_4^+ and NO_3^-) and P (PO_4^-) contents. Data were analyzed on a plot mean basis, by linear mixed models. Post hoc comparisons of means were performed with the LSMEAN statement in the SAS 9.1. software program.

3.1. Effects of Vermicompost on Sweet Corn Growth and Yield

In the present study vermicompost produced significant improvements in the growth and yield of sweet corn, as also reported by Atiyeh et al. (2000b), Arancon et al. (2004b) and Singh et al. (2008). Replacement of 25% of the nutrients provided to the plants by vermicompost produced significant greater increases in plant height and marketable crop yield than those produced by 100% inorganic fertilization (Figure 2). Nevertheless the magnitude of such effects was relatively small and not general, with a large degree of variability between the different hybrids assayed. Furthermore, there was little or no difference from the effects produced by manure on plant height and crop yield. Interestingly, increases in sweet corn growth and marketable yield were produced by both the normal and the high doses of vermicompost and manure, indicating that these effects were independent of the amount of N, P and K provided.

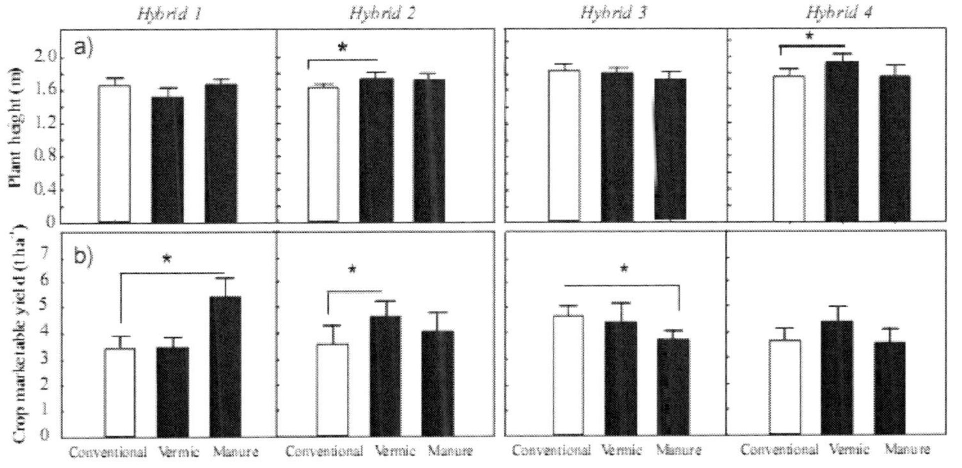

Figure 2. Plant height, crop yield, N content in grain and grain yield of the marketable ear of the four sweet corn hybrids grown under the different fertilization treatments evaluated: conventional (100% inorganic NPK), vermicompost (75% inorganic + 25% organic NPK), and manure (75% inorganic + 25% organic NPK). Asterisks and bars indicate significant differences at P< 0.05. Values are means ± standard error (Modified from Lazcano et al. 2011).

Table 1. Leaf nutrient content of the sweet corn hybrids subjected to normal and high doses of the different fertilization regimes. Values are means ± standard error. Different letters within the same column indicate significant differences at P<0.05

Dose	Treatment	N (mg kg^{-1}dw)	P (mg kg^{-1}dw)	K (mg kg^{-1}dw)
Normal	Conventional	30751 ± 1206	2625 ±167 b	13875 ± 3809
	Vermicompost	30931 ± 1222	3044 ±171 a	14634 ± 3816
	Manure	29950 ± 1206	2908 ± 167 ab	13976 ± 3810
High	Conventional	31394 ± 1206	2546 ± 168 b	12987 ± 3809
	Vermicompost	31574 ± 1206	2766 ± 168 a	13960 ± 3809
	Manure	29545 ± 1212	2737 ± 168 ab	14262 ± 3813

Foliar analysis revealed that the differences in macronutrient content between the plants cultivated with the three types of fertilizers were minor (Table 1). However, plants cultivated with vermicompost showed a slightly higher foliar P content. The same amounts of nutrients (NPK) were supplied with each dose of the three fertilizers and therefore the higher foliar phosphorus content in plants amended with vermicompost presumably corresponds to greater availability of this nutrient in the soil.

3.2. Effects of Vermicompost on the Quality of the Marketable Ears

The fertilizers applied had a significant effect on the quality of the marketable ears in the sweet corn plants. Grain content of the marketable ear was higher in response to vermicompost and manure amendment than in response to 100% inorganic fertilization (Figure 3).

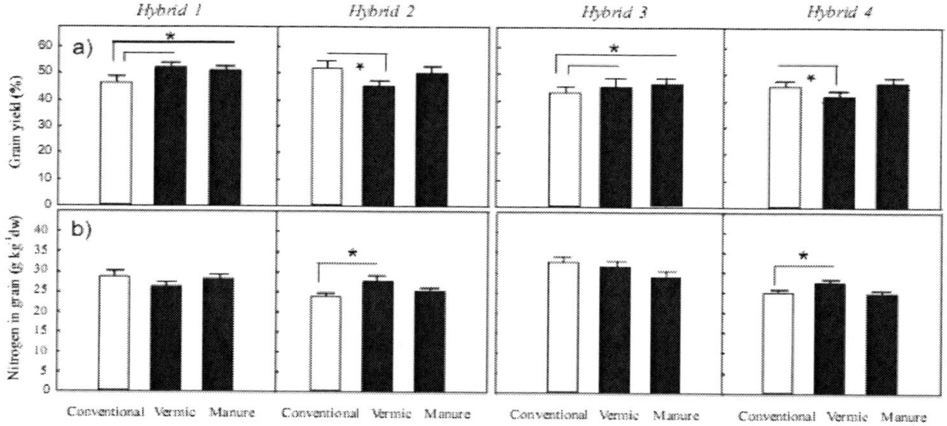

Figure 3. Grain yield (a) and N content (b) of the marketable ear of the sweet corn hybrids grown under the different fertilization treatments evaluated: conventional (100% inorganic NPK), vermicompost (75% inorganic + 25% organic NPK), and manure (75% inorganic + 25% organic NPK). Asterisks and bars indicate significant differences at P< 0.05. Values are means ± standard error (Modified from Lazcano et al. 2011).

However, this depended on the hybrid considered, and significant decreases in grain content of the ears were also observed after the application of vermicompost. As well as the effects on grain content, the addition of vermicompost resulted in different grain chemical composition than in response to inorganic fertilizer. Plants cultivated with vermicompost had higher N contents in the grain although this was observed only in two of the four hybrids assayed (Figure 3). There were no differences in the P and K contents of the grain between the different fertilization treatments.

3.2. Effects of Vermicompost in Soil Biochemical and Microbiological Properties

Regarding the nutrient content of the soil at harvest there were no differences in N-NH_4^+, N-NO_3^- or K_2O_5 between the fertilizing treatments and doses applied (Table 2). However, the PO_4^- content of the soil was significantly higher in the plots to which vermicompost was added than in the plots that received only inorganic fertilizers or manure (Table 2). However, this only took place at the high doses applied, whereas there were no differences between the treatments at the normal dose.

β-Glucosidase, protease and alkaline phosphomonoesterase enzyme activity rates were significantly higher in the soil plots amended with vermicompost and manure than in those treated with inorganic fertilizer (Figure 4). Fertilization with vermicompost and manure also favored growth of microorganisms in the soil. Bacterial growth increased significantly after application of the high doses of vermicompost and manure, although fungal growth was not affected (Figure 5). Organic amendments promoted microbial growth, which resulted in a higher soil microbial biomass, although only in the plots treated with manure (Figure 6a). The increase in soil microbial biomass was due exclusively to an increase in Gram-negative bacteria, as shown by the increased concentration of the PLFAs biomarkers for this microbial group (Figure 6b).

Table 2. Content of available plant macronutrients at harvest in the soils subjected to the different fertilization regimes. Values represent least square means ± standard error. Different letters within the same column indicate significant differences at P<0.05

Dose	Treatment	$N\text{-}NH_4^+$ (µg g dw^{-1})	$N\text{-}NO_3^-$ (µg g dw^{-1})	K_2O (%)	PO_4^- (µg g dw^{-1})
	Conventional	4.66 ± 1.1a	28.82 ± 8.8a	4.66 ± 0.0a	161.39 ± 4.0bc
	Vermicompost	5.19 ± 1.1a	30.22 ± 8.8a	4.69 ± 0.0a	164.32 ± 4.0b
Normal	Manure	4.70 ± 1.1a	30.19 ± 8.8a	4.65 ± 0.0a	162.38 ± 4.0bc
	Conventional	6.23 ± 1.1a	38.32 ± 8.8a	4.65 ± 0.0a	152.4 ± 4.0c
	Vermicompost	5.25 ± 1.1a	30.24 ± 8.8a	4.63 ± 0.0a	177.17 ± 4.0a
High	Manure	4.60 ± 1.1a	25.66 ± 8.8a	4.65 ± 0.0a	166.55 ± 4.0 ab

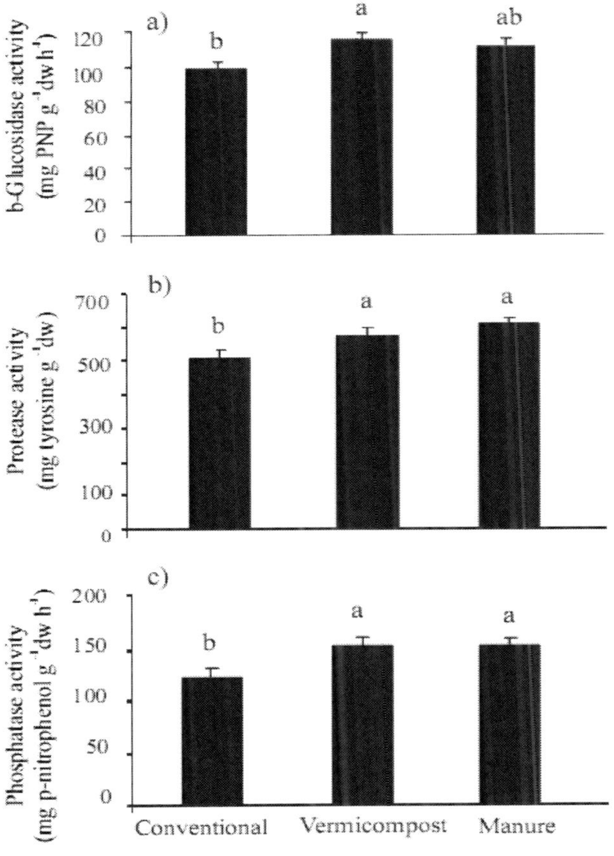

Figure 4. β-Glucosidase (a), protease (b), and phosphatase (c) enzyme activity of the soil amended with the different fertilization treatments evaluated: conventional (100% inorganic NPK), vermicompost (75% inorganic + 25% organic NPK), and manure (75% inorganic + 25% organic NPK). Bars are means ± standard error. Different letters indicate significant differences at P<0.05.

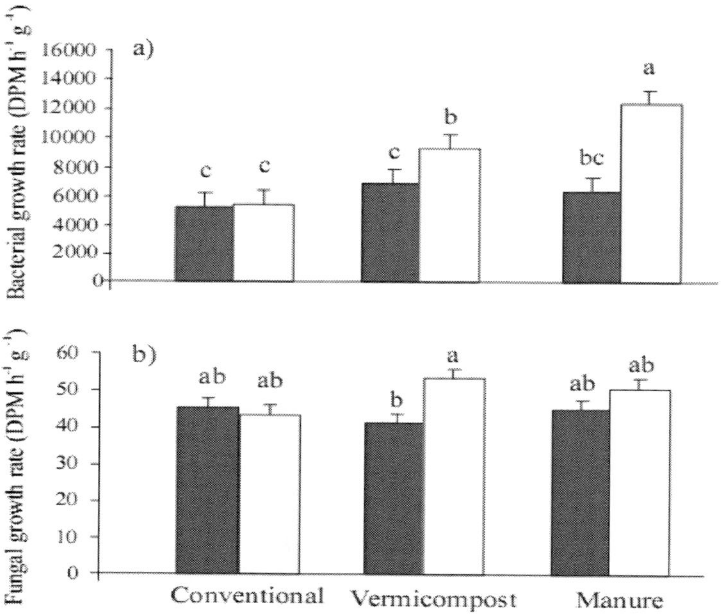

Figure 5. Bacterial (a) and fungal growt (b) h of the soil amended with the different fertilization treatments evaluated : conventional (100% inorganic NPK), vermicompost (75% inorganic + 25% organic NPK), and manure (75% inorganic + 25% organic NPK). Bars are means ± standard error. Different letters indicate significant differences at P<0.05.

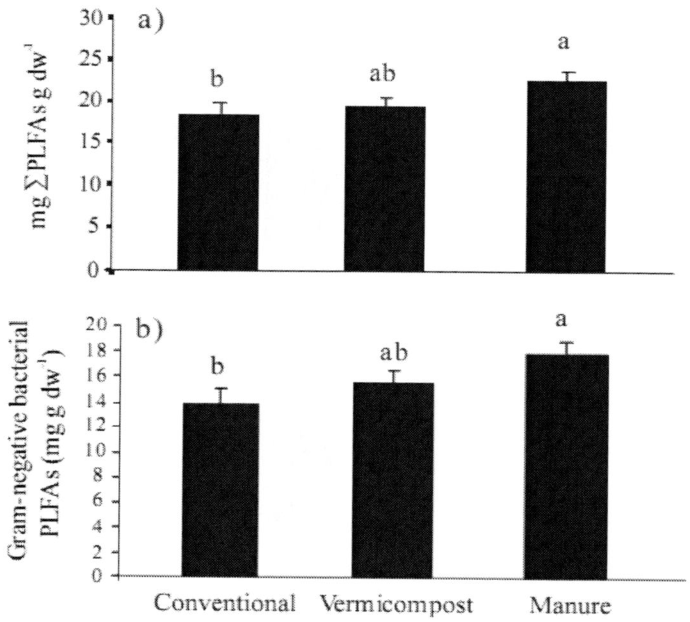

Figure 6. Microbial biomass (a) and Gram negative bacteria (a) of the soil amended with the different fertilization treatments evaluated : conventional (100% inorganic NPK), vermicompost (75% inorganic + 25% organic NPK), and manure (75% inorganic + 25% organic NPK). Bars are means ± standard error. Different letters indicate significant differences at P<0.05.

Replacement of 25% of the nutrients provided to the corn plants by Vermicompost maintains plant productivity at the same levels than 100% inorganic fertilization. In addition, vermicompost produces significant changes in soil biochemical and microbial properties promoting bacterial growth and increasing enzyme activity. In spite of the fact that the same amounts of N, P and K were provided with the three types of fertilizers at each dose, vermicompost amended plots exhibited higher phosphatase and phosphate levels, and the P content of the plants was higher than in soils with inorganic fertilization. The results presented here confirm that there is a large variation in vermicompost effects depending on the plant genotype. Therefore, within a given crop certain genotypes or varieties may be more suitable for organic or combined inorganic-organic cropping systems, than others.

Conclusions

Vermicompost can be described as a complex mixture of earthworm faeces, humified organic matter and microorganisms, which when added to the soil or plant growing media, increases germination, growth, flowering, fruit production and accelerates the development of a wide range of plant species. The enhanced plant growth may be attributed to various direct and indirect mechanisms, including biologically mediated mechanisms such as the supply of plant-growth regulating substances, and improvements in soil biological functions. Use of this type of organic fertilizer therefore has great potential; however some recent studies raise serious doubts about the general applicability of these results and propose a more complex model of action for these types of effects. Stimulation of plant growth may depend mainly on the biological characteristics of vermicompost, the plant species used, and the cultivation conditions. Extensive research on inorganic fertilization and plant breeding, carried out within the framework of conventional agriculture, has allowed agricultural producers to fine-tune nutrient inputs and plant needs in order to maximize yields. However, such detailed knowledge has not yet been attained as regards the interactions between plants and organic fertilizers in sustainable agriculture. Given the complex and variable composition of vermicompost in comparison with inorganic fertilizers and the myriad of effects that it can have on soil functioning, a clear and objective concept of vermicompost is required, and the complex interactions between vermicompost-soil-plant must be unraveled in order to maintain consumer confidence in this type of organic fertilizer.

Acknowledgments

This research was financially supported by the Xunta de Galicia (07MRU023383PR) and by the Spanish Ministerio de Ciencia e Innovación (CTM2009-08477). The authors thank the personnel at the Misión Biolóxica de Galicia for their support and assistance in the field experiments in the case study. Cristina Lazcano is financially supported by an 'Ángeles Alvariño' fellowship from the Xunta de Galicia.

REFERENCES

Acevedo, I.C. and Pire, R. (2004). Effects of vermicompost as substrate amendment on the growth of papaya (*Carica papaya* L.). *Interciencia* 29(5), 274-279.

Aira, M., Gómez-Brandón, M., Lazcano, C., Bååth, E. and Domínguez, J. (2010). Plant genotype strongly modifies the structure and growth of maize rhizosphere microbial communities. *Soil Biology and Biochemistry* 42 (12), 2276-2281.

Aira, M., Monroy, F. and Domínguez, J. (2007). Earthworms strongly modify microbial biomass and activity triggering enzymatic activities during vermicomposting independently of the application rates of pig slurry. *Science of the Total Environment* 385, 252-261.

Anwar, M., Patra, D.D., Chand, S., Kumar, A., Naqvi, A.A. and Khanuja, S.P.S. (2005). Effect of organic manures and inorganic fertilizer on growth, herb and oil yield, nutrient accumulation, and oil quality of French basil. *Communications in Soil Science and Plant Analysis* 36 (1-14), 1737-1746.

Arancon, N.Q., Edwards, C.A., Babenko, A., Cannon, J., Galvis, P. and Metzger, J.D. (2008). Influences of vermicomposts, produced by earthworms and microorganisms from cattle manure, food waste and paper waste, on the germination, growth and flowering of petunias in the greenhouse, *Applied Soil Ecology* 39, 91-99.

Arancon, N.Q., Edwards, C.A. and Bierman, P. (2006). Influences of vermicomposts on field strawberries: Part 2. Effects on soil microbiological and chemical properties. *Bioresource Technology*, 97, 831-840.

Arancon, N.Q., Edwards, C.A., Bierman, P., Metzger, J.D. and Lucht, C. (2005a). Effects of vermicomposts produced from cattle manure, food waste and paper waste on the growth and yield of peppers in the field. *Pedobiologia,* 49, 297-306.

Arancon, N.Q., Edwards, C.A., Bierman, P., Welch, C. and Metzger, J.D. (2004b). Influences of vermicomposts on field strawberries: 1. effects on growth and yields. *Bioresource Technology* 93, 145-153.

Arancon, N.Q., Edwards, C.A., Yardim, E.N., Oliver, T.J., Byrne, R.J. and Keeney, G. (2007). Suppression of two-spotted spider mite (*Tetranychus urticae*), mealy bug (*Pseudococcus* sp.) and aphid (*Myzus persicae*) populations and damage by vermicomposts. *Crop Protection* 26, 29-39.

Arancon, N.Q., Edwards, C.E., Atiyeh, R.M. and Metzger, J.D. (2004a). Effects of vermicompost produced from food waste on the growth and yields of greenhouse peppers. *Bioresource Technology* 93, 139-144.

Arancon, N.Q., Galvis, P., Edwards, C.A. and Yardim, E. (2003). The trophic diversity of nematode communities in soils treated with vermicompost. *Pedobiología* 47, 736-740.

Arancon, N.Q., Galvis, P.A., Edwards, C.A. 2005b. Suppression of insect pest populations and damage to plants by vermicompost. *Bioresource Technology* 96, 1137-1142.

Argüello, J.A., Ledesma, A., Núñez, S.B., Rodríguez, C.H. and Díaz Goldfarb, M.D.C., (2006). Vermicompost effects on bulbing dynamics nonstructural carbohydrate content, yield, and quality of 'Rosado Paraguayo' garlic bulbs. *Hortscience* 41 (3), 589-592.

Atiyeh, R.M., Arancon, N., Edwards, C.A. and Metzger, J.D. (2002). The influence of earthworm-processed pig manure on the growth and productivity of marigolds. *Bioresource Technology* 81, 103-108.

Atiyeh, R.M., Arancon, N.Q., Edwards, C.A. and Metzger, J.D. (2000b). Influence of earthworm- processed pig manure on the growth and yield of green house tomatoes. *Bioresource Technology* 75, 175-180.

Atiyeh, R.M., Edwards, C.A., Subler, S. and Metzger, J.D. (2001). Pig manure vermicompost as a component of a horticultural bedding plant medium: Effects on physicochemical properties and plant growth. *Bioresource Technology* 78, 11-20.

Atiyeh, R.M., Subler, S., Edwards, C.A., Bachman, G., Metzger, J.D. and Shuster, W. (2000a). Effects of vermicomposts and compost on plant growth in horticultural container media and soil. *Pedobiologia* 44, 579-590.

Atiyeh, R.M., Subler, S., Edwards, C.A. and Metzger, J. (1999). Growth of tomato plants in horticultural media amended with vermicompost. *Pedobiologia*, 43, 724-728.

Bååth, E. (2001). Estimation of fungal growth rates in soil using 14C-acetate incorporation into ergosterol. *Soil Biology and Biochemistry* 33, 2011-2018.

Bachman, G.R., Metzger, J.D., 2007. Physical and chemical characteristics of a commercial potting substrate amended with vermicompost produced from two different manure sources. *HortTechnology* 17(3), 336-340.

Bhattacharjee, G., Chaudhuri, P.S. and Datta, M. (2001). Response of paddy (Var. TRC-87-251) crop on amendment of the field with different levels of vermicompost. Asian *Journal of Microbiology, Biotechnology and Environmental Sciences*, 3 (3), 191-196.

Biradar, A.P., Sunita, N.D., Teggelli, R.G. and Devaranavadgi, S.B. (1998). Effect of vermicomposts on the incidence of subabul psyllid. *Insect Environment*, 4, 55-56.

Brussaard, L., de Ruiter, P.C. and Brown, G.G. (2007). Soil biodiversity for agricultural sustainability. *Agriculture Ecosystems and Environment*, 121, 233-244.

Bulluck, L.R., Brosius, M., Evanylo, G.K. and Ristaino, J.B. (2002). Organic and synthetic fertility amendments influence soil microbial, physical and chemical properties on organic and conventional farms. *Applied Soil Ecology*, 19, 147-160.

Cabanas-Echevarría, M., Torres –García, A., Díaz-Rodríguez, B., Ardisana, E.F.H. and Creme-Ramos, Y. (2005). Influence of three bioproducts of organic origin on the production of two banana clones (*Musa* spp AAB.) obtained by tissue cultures. *Alimentaria,* 369, 111-116.

Campitelli, P. and Ceppi, S. (2008). Chemical, physical and biological compost and vermicompost characterization: A chemometric study. *Chemometrics and Intelligent Laboratory Systems,* 90, 64-71.

Canellas, L.P., Olivares, F.L., Okorokova, A.L. and Facanha, A.R. (2002). Humic acids isolated from earthworm compost enhance root elongation, lateral root emergence, and plasma H^+-ATPase activity in maize roots. *Plant Physiology*, 130, 1951-1957.

Cavani, N. and Mimmo, T. (2007). Rhizodeposition of *Zea mays* L. as affected by heterosis. *Archives of Agronomy and Soil Science*, 53, 593 – 604.

Chand, S., Pande, P., Prasad, A., Anwar, M. and Patra, D.D. (2007). Influence of integrated supply of vermicompost and zinc-enriched compost with two graded levels of iron and zinc on the productivity of geranium. *Communications in Soil Science and Plant Analysis,* 38, 2581–2599.

Chaoui, H.I., Zibilske, L.M. and Ohno, T. (2003). Effects of earthworm casts and compost on soil microbial activity and plant nutrient availability. *Soil Biology and Biochemistry*, 35, 295-302.

Coria-Cayupán, Y.S., De Pinto, M.I.S. and Nazareno, M. A. (2009). Variations in bioactive substance contents and crop yields of lettuce (*Lactuca sativa* L.) cultivated in soils with different fertilization treatments. *Journal of Agricultural and Food Chemistry*, 57(21), 10122-10129.

Dell'Agnola, G. and Nardi, S. (1987). Hormone-like effect and enhanced nitrate uptake induced by depolycondensed humic fractions obtained from *Allolobophora rosea* and *A. caliginosa* feces. *Biology and Fertility of Soils*, 4,115-118.

Domínguez, J. (2004). *State of the art and new perspectives on vermicomposting research*. In: C.A. Edwards (Ed.). Earthworm Ecology (2nd edition). CRC Press LLC. Pp. 401-424.

Domínguez, J., Aira, M. and Gómez Brandón, M. (2010). *Vermicomposting: earthworms enhance the work of microbes*. In: H. Insam, I. Franke-Whittle and M. Goberna, (Eds.), Microbes at Work: From Wastes to Resources (pp. 93-114). Springer, Berlin Heidelberg,

Donald, D.G.M. and Visser, L.B. (1989). Vermicompost as a possible growth medium for the production of commercial forest nursery stock. *Appl. Plant Sci.* 3, 110-113.

Edwards, C.A., Arancon, N.Q. and Greytak, S. (2006). Effects of vermicompost teas on plant growth and disease. *BioCycle* 47, 28-31.

Edwards, C.A. and Burrows, I. (1988). *The potential of earthworm composts as plant growth media.* In: C.A. Edwards and E.F. Neuhauser (Eds). Earthworms in Waste and Environmental Management. (Pp. 211-219). SPB Academic Publ. Co. The Hague, Netherlands.

Edwards, C.A., Domínguez, J. and Arancon, N.Q. (2004). *The influence of vermicomposts on plant growth and pest incidence*. In: S.H Shakir and W.Z.A. Mikhaïl, (Eds). Soil Zoology for Sustainable Development in the 21st century. (pp 397-420), Cairo.

El Harti, A., Saghi, M., Molina, J-A.E. and Téller, G. (2001a). Production d'une substance rhizogène à effet similaire à celui de l'acide indole acètique par le ver de terre *Lumbricus terrestris*. *Canadian Journal of. Zoology* 79:1911-1920.

El Harti, A., Saghi, M., Molina, J-A.E. and Téller, G. (2001b). Production des composés indoliques rhizogénes par le ver de terre *Lumbricus terrestris*. *Canadian Journal of. Zoology* 79:1921-1932.

Ferreras, L., Gomez, E., Toresani, S., Firpo, I. and Rotondo, R. (2006). Effect of organic amendments on some physical, chemical and biological properties in a horticultural soil. *Bioresource Technology* 97, 635-640.

Gajalakshmi,S. and Abbasi, S.A. (2004). Neem leaves as a source of fertilizer-cum-pesticide vermicompost. *Bioresource Technology* 92, 291-296.

Garcia-Gil, J.C., Plaza, C., Soler-Rovira, P. and Polo, A. (2000). Long-term effects of municipal solid waste compost application on soil enzyme activities and microbial biomass. *Soil Biology and Biochemistry* 32 (13), 1907-1913.

García-Gómez, A., Bernal, M.P.and Roig, A. (2002). Ornamental plants growth in substrates using composts from agroindustrial wastes. *Bioresource Technology* 83, 81–87.

Gopinath, K.A., Supradip, S., Mina, B.L., Pande, H., Kundu, S. and Gupta, H.S. (2008). Influence of organic amendments on growth, yield and quality of wheat and on soil properties during transition to organic production. *Nutrient Cycling in Agroecosystems,* 82 (1), 51-60.

Grappelli, A., Galli, E. and Tomati, U. (1987). Earthworm casting effect on *Agaricus bisporus* fructification. *Agrochimica*, 31, (4-5), 457-461.

Gunadi, B., Edwards, C.A. and Arancon, N.A. (2002). Changes in trophic structure of soil arthropods after the application of vermicomposts. *European Journal of Soil Biology*, 38, 161-165.

Gutiérrez-Miceli, F.A., García-Romero, R.C., Rincón-Rosales, R., Abud-Archila, M., Oliva-Llaven,M.A., Guillen-Cruz,M.J. and Dendooven, L. (2008). Formulation of a liquid fertilizer for sorghum (*Sorghum bicolor* (L.) Moench) using vermicompost leachate. *Bioresource Technology*.

Gutiérrez-Miceli, F.A., Santiago-Borraz, J., Montes Molina, J.A., Nafate, C.C., Abdud-Archila, M., Oliva Llaven, M.A., Rincón-Rosales, R. and Deendoven L. (2007). Vermicompost as a soil supplement to improve growth, yield and fruit quality of tomato (*Lycopersicum esculentum*). *Bioresource Technology* 98, 2781-2786.

Hashemimajd, K., Kalbasi, M., Golchin, A. and Shariatmadari, H. (2004). Comparison of vermicompost and composts as potting media for growth of tomatoes. *Journal of Plant Nutrition*. 6:1107-1123.

Hidalgo, P.R. and Harkess, R.L. (2002a). Earthworm casting as a substrate amendment for Chrysanthemum Production. *Hortscience* 37 (7), 1035- 1039.

Hidalgo, P.R. and Harkess, R.L. (2002b). Earthworm casting as a substrate for Poinsettia production. *Hortscience* 37 (2), 304-308.

Hidalgo, P.R., Matta, F.B. and Harkess, R.L. (2006). Physical and chemical properties of substrates containing earthworm castings and effects on marigold growth. *HortScience*, 41, 1474-1476.

Hointik, H.A.J. and Bohem, M.J. (1999). Biocontrol within the context of soil microbial communities: a substrate-dependent phenomenon. *Annual Review of Phytopathology* 37, 427-46.

Hutchison, M.L., Walters, L.D., Avery, S.M., Munro, F. and Moore, A. (2005). Analyses of livestock production, waste storage, and pathogen levels and prevalence in farm manures. *Applied and Environmental Microbiology* 71, 1231–1236.

Kabir, Z., O'Halloran, I. P., Fyles, J. W. and Hamel, C. (1998). Dynamics of the mycorrhizal symbiosis of corn (*Zea mays* L.): effects of host physiology, tillage practice and fertilization on spatial distribution of extra-radical mycorrhizal hyphae in the field. *Agriculture, Ecosystems and Environment*, 68, 151-163.

Kale, R.D., Mallesh, B.C., Kubra, B. and Bagyaraj, D.J. (1992). Influence of vermicompost application on the available macronutrients and selected microbial populations in a paddy field. *Soil Biology and Biochemistry*, 24, 1317-1320.

Karmegam, N., Alagumalai, K. and Daniel, T. (1999). Effect of vermicompost on the growth and yield of green gram (*Phaseolus aureus* Roxb.). *Tropical Agriculture* 76, 143-146.

Knapp, Brigitte A., Ros, M. and Insam, H. (2010). *Do Composts Affect the Soil Microbial Community*? In: H. Insam, I. Franke-Whittle and M. Goberna, (Eds.), Microbes at Work: From Wastes to Resources. (pp 93-114). Springer, Berlin Heidelberg.

Lazcano, C., Arnold, J., Tato, A., Zaller, J.G. and Domínguez, J. (2009). Compost and vermicompost as nursery pot components: Effects on tomato plant growth and morphology. *Spanish Journal of Agricultural Research* 7, 944-951.

Lazcano, C. and Domínguez, J. 2010c. Effects of vermicompost as a potting amendment of two comercially-grown ornamental plant species. *Spanish Journal of Agricultural Research* 8 (4), 1260-1270.

Lazcano, C., Gómez-Brandón, M. and Domínguez, J. (2008). Comparison of the effectiveness of composting and vermicomposting for the biological stabilization of cattle manure. *Chemosphere*, 72, 1013-1019.

Lazcano, C., Revilla P., Malvar, R.A. and Domínguez, J. (2011). Yield and fruit quality of four sweet corn hybrids (*Zea mays*) under conventional and integrated fertilization with vermicompost. *Journal of the Science of Food and Agriculture*. In press.

Lazcano, C., Sampedro, L., Zas, R. and Domínguez, J. (2010a). Vermicompost enhances germination of the maritime pine (*Pinus pinaster* Ait.). *New Forest*. 39, 387-400.

Lazcano, C., Sampedro, L., Zas, R. and Domínguez, J., (2010b). Assessment of plant growth promotion by vermicompost in different progenies of maritime pine (*Pinus pinaster* Ait.). *Compost Science and Utilization* 18, 111-118.

Lores, M., M. Gómez-Brandón, D. Pérez-Díaz and J. Domínguez. (2006). Using FAME profiles for the characterization of animal wastes and vermicomposts. *Soil Biology and Biochemistry* 38, 2993-2996.

Marinari, S., G. Masciandaro, B. Ceccanti and Grego, S. (2000). Influence of organic and mineral fertilizers on soil biological and physical properties. *Bioresource Technology*, 72, 9–17

Morra, L., Palumbo, A.D., Bilotto, M., Ovieno, P. and Picascia, S. (1998). Soil solarization: organic fertilization and grafting contribute to build an integrated production system in a tomato-zucchini sequence. *Colture-Protette*, 27, 63-70.

Nakasone, A.K., Bettiol, W. and de Souza, R.M. (1999). The effect of water extracts of organic matter on plant pathogens. *Summa Phytopathologica*, 25, 330-335.

Nardi, S., G. Arnoldi and Dell'Agnola, G. (1988). Release of hormone-like activities from *Alloborophora rosea* and *Alloborophora caliginosa* feces. *Journal of Soil Science*, 68:563-657.

Nardi, S., Morari, F., Berti, A., Tosoni, M. and Giardini, L. (2004). Soil organic matter properties after 40 years of different use of organic and mineral fertilisers. *European Journal of Agronomy* 21, 357-367.

Ndegwa, P.M. and Thompson, S.A. (2001). Integrating composting and vermicomposting in the treatment and bioconversion of biosolids. *Bioresource Technology* 76, 107-112.

Nielson, R.L. (1965). Presence of plant growth substances in earthworms demonstrated by paper chromatography and the Went pea test. *Nature*, 208, 1113-1114.

Noble, R. and Coventry, E. 2005. Suppression of soil-borne plant diseases with composts: A review. *Biocontrol Science and Technology* 15, 3-20.

Ordás, B., Romay, M. C. and Revilla, P. 2007. Maíz dulce ¿porqué no? *Horticultura*, 202, 14-20.

Orlikowski, L.B. 1999. Vermicompost extract in the control of some soil borne pathogens. *International Symposium on Crop Protection*, 64, 405-410.

Peyvast, G., Olfati, J.A., Madeni, S. and Forghani, A. (2008). Effect of vermicompost on the growth and yield of spinach (*Spinacia oleracea* L.). *Journal of Food Agriculture and Environment*, 6, 110-113.

Prabha, M.L., Jayraay, I.A., Jayraay, R. and Rao, D.S. (2007). Effect of vermicompost on growth parameters of selected vegetable and medicinal plants. *Asian Journal of microbiology, Biotechnology and Environmental Sciences*, 9(2), 321-326.

Ramesh, P. (2000). Effects of vermicomposts and vermicomposting on damage by sucking pests to ground nut (*Arachis hypogea*). *Indian Journal of Agricultural Sciences*, 70 (5), 334.

Rao, K.R. (2002). Induced host plant resistance in the management of sucking insect pests of groundnut. *Annals of Plant Protection Science* 10, 45-50.

Reddy, M.V. and Ohkura, K. (2004). Vermicomposting of rice-straw and its effects on sorghum growth. *Tropical Ecology* 45, (2), 327-331.

Ribeiro, C.F., Mizobusi, E.H., Silva, D.G., Pereira, J.C.R. and Zambolim, L. (1998). Control of Meloidogyne javanica on lettuce with organic amendments. *Fitopatologia Brasileira* 23, 42-44.

Roberts, P., Jones, D.L. and Edwards-Jones, G. (2007). Yield and vitamin C content of tomatoes grown in vermicomposted wastes. *Journal of the Science of Food and Agriculture* 87:1957-1963.

Rodda, M.R.C., Canellas, L.P., Façanha, A.R, Zandonadi, D.B., Guerra, J.G.M., De Almeida, D.L. and De Santos, G.A. (2006). Improving lettuce seedling root growth and ATP hydrolysis with humates from Vermicompost. II- Effect of Vermicompost source. *Revista Brasileira de Ciencia do Solo*, 30, 657-664.

Ros, M., Pascual, J.A., García, C., Hernández, M.T. and Insam, H. (2006). Hydrolase activities, microbial biomass and bacterial community in a soil after long-term amendment with different composts. *Soil Biology and Biochemistry*, 38, 3443-3452.

Saha, S., Mina, B.L., Gopinath, K.A., Kundu, S. and Gupta, H.S. (2008). Relative changes in phosphatase activities as influenced by source and application rate of organic composts in field crops. *Bioresource Technology*, 99, 1750-1757.

Scott, M.A. (1988). *The use of worm-digested animal waste as a supplement to peat in loamless composts for hardy nursery stock*. In: C.A. Edwards and E.F. Neuhauser, (Eds.). Earthworms in Environmental and Waste Management. (Pp. 231-229). SPB Acad. Publ. The Netherlands.

Singh, R., Gupta, R.K., Patil, R.T., Sharma, R.R., Asrey, R., Kumar, A. and Jangra, K.K., (2010). Sequential foliar application of vermicompost leachates improves marketable fruit yield and quality of strawberry (*Fragaria x ananassa* Duch.). *Scientia Horticulturae*, 124, 34-39.

Singh, R., Sharma, R.R., Kumar, S., Gupta, R.K. and Patil, R.T. (2008). Vermicompost substitution influences growth, physiological disorders, fruit yield and quality of strawberry (*Fragaria x ananassa* Duch.). *Bioresource Technology*, 99, 8507-8511.

Singh, U.P., Maurya, S. and Singh, D.P. (2003). Antifungal activity and induced resistance in pea by aqueous extract of vermicompost and for control of powdery mildew of pea and balsam. *Zeitschrift fur Pflanzenkrankheiten und Pflanzenschutz*, 110, 544-553.

Springett, J.A. and Syers, J.K. (1979). *The effect of earthworm casts on ryegrass seedlings*. In: Crosby T.K. and R.P. Pottinger, editors. Proceedings of the 2nd Australasian Conference on Grassland Invertebrate Ecology (Pp. 44-47). Government Printer, Wellington.

Sunil, K., Rawat, C.R., Shiva, D. and Suchit, K.R. (2005). Dry matter accumulation, nutrient uptake and changes in soil fertility status as influenced by different organic and inorganic sources of nutrients to forage sorghum (*Sorghum bicolor*). *Indian Journal of Agricultural Science,* 75 (6), 340-342.

Swathi, P., Rao, K.T. and Rao, P.A. (1998). Studies on control of root-knot nematode *Meloidogyne incognita* in tobacco miniseries. Tobaco Research 1, 26-30.

Szczech, M. (1999). Supressiveness of vermicompost against *Fusarium wilt* of tomato. *Journal of Phytopathology* 147, 155-161.

Szczech, M., Smolinska, U. (2001). Comparison of suppressiveness of vermicompost produced from animal manures and sewage sludge against *Phytophthora nicotianae* Breda de Haar var. *nicotianae*. *Journal of Phytopathology*, 149, 77-82.

Tejada, M., Gonzalez, J.L., Hernandez, M.T. and Garcia, C., (2008). Agricultural use of leachates obtained from two different vermicomposting processes, *Bioresource Technology*, 99, 6228-6232.

Termorshuizen, A.J., van Rijn, E., van der Gaag, D.J., Alabouvette, C., Chen, Y., Lagerlöf, J., Malandrakis, A.A.,. Paplomatas, E.J, Rämert, B., Ryckeboer, J., Steinberg, C. and Zmora-Nahum, S. (2006). Supressiveness of 18 composts against 7 pathosystems: Variability in pathogen response. *Soil Biology and Biochemistry* 38, 2461-2477.

Tilman, D., Cassman, K.G., Matson, P.A., Naylor, R.and Polasky, S. (2002). Agricultural sustainability and intensive production practices. *Nature*, 418, 671-677.

Tognetti, C., Laos F., Mazzarino, M.J. and Hernández, M.T. (2005). Composting vs. vermicomposting: A comparison of end product quality. *Compost Science and Utilization*, 13, 6-13.

Tomati, U. and Galli, E. (1995). Earthworms, Soil Fertility and Plant Productivity. Proceedings of the International Colloquium on Soil Zoology. *Acta Zoologica Fennica*, 196:11-14.

Tomati, U., Galli, E., Grapppelli, A. and Dihena, G. (1990). Effect of earthworm casts on protein synthesis in radish (*Raphanus sativum*) and lettuce (*Lactuca sativa*) seedlings. *Biology and Fertility of Soils*, 9, 288-289.

Tomati, U., Grappelli, A. and Galli, E. (1983). *Fertility factors in earthworm humus*. Pp. 49-56. In: Proc. Int. Symp. Agric. Environ. Prospects in Earthworm Farming. Publication Ministero della Ricerca Scientifica e Technologia, Rome.

Tomati, U., Grappelli, A. and Galli, E. (1987). *The presence of growth regulators in earthworm worked wastes.* In: A.M. Bonvicini Paglioi, P. Omodeo (Eds). On Earthworms. (pp. 423-436) Selected Symposia and Monographs 2. Mucchi Editore, Modena, Italy.

Tomati, U., Grappelli, A. and Galli, E. (1988). The hormone-like effect of earthworm casts on plant growth. *Biology and Fertility of Soils*, 5, 288-294.

Trevisan, S., Pizzeghello, D., Ruperti, B., Francioso, O., Sassi, A., Palme, K., Quaggiotti, S. and Nardi, S. (2010). Humic substances induce lateral root formation and expression of the early auxin-responsive IAA19 gene and DR5 synthetic element in Arabidopsis. *Plant Biology,* 12, 604-614.

Trewavas, A. (2001). Urban myths of organic farming. *Nature* 410, 409-410.

Trillas M.I., Casanova, E., Cotxarrera, L., Ordovás, J., Borrero, C. and Avilés, M. (2006). Composts from agricultural waste and the *Trichoderma asperellum* strain T-34 suppress *Rhizoctonia solani* in cucumber seedlings. *Biological Control* 39, 32-38.

Vivas, A., Moreno, B., Garcia-Rodriguez, S. and Benitez, E. (2009). Assessing the impact of composting and vermicomposting on bacterial community size and structure, and microbial functional diversity of an olive-mill waste, *Bioresource Technology*, 100, 1319-1326.

Wang, D., Shi, Q., Wang, X., Wei, M., Hu, J., Liu, J. and Yang, F. (2010). Influence of cow manure vermicompost on the growth, metabolite contents, and antioxidant activities of Chinese cabbage (*Brassica campestris* ssp. chinensis). *Biology and Fertility of Soils*, 46, 689-696.

Warman, P.R. and AngLopez, M.J. (2010). Vermicompost derived from different feedstocks as a plant growth medium. *Bioresource Technology*, 101, 4479-4483.

Weber, J., Karczewska, A., Drozd, J., Licznar, M., Licznar, S., Jamroz, E. and Kocowicz, A. (2007). Agricultural and ecological aspects of a sandy soil as affected by the application of municipal solid waste composts. *Soil Biology and Biochemistry*, 39, 1294-1302.

Yardim, E.N., Arancon, N.A., Edwards, C.A., Oliver, T.J. and Byrne, R.J. 2006. Suppression of tomato hornworm (*Manduca quinquemaculata*) and cucumber beetles (*Acalymma vittatum* and *Diabotrica undecimpunctata*) populations and damage by vermicomposts. *Pedobiologia,* 50, 23-29.

Zaller, J.G. (2006). Foliar spraying of vermicompost extracts: effects on fruit quality and indications of late-blight suppression of field-grown tomatoes. *Biological Agriculture and Horticulture,* 24, 165-180.

Zaller, J.G. (2007). Vermicompost as a substitute for peat in potting media: Effects on germination, biomass allocation, yields and fruit quality of three tomato varieties. *Scientia Horticulturae*, 112, 191-199.

Zandonadi, D.B., Canellas,L.P. and Rocha Façanha, A. (2006). Indolacetic and humic acids induce lateral root development through a concerted plasmalemma and tonoplast H^+ pumps activation. *Planta*, 225, 1583-1595.

In: Soil Nutrients
Editor: Mohammad Miransari
ISBN: 978-1-61324-785-3

Chapter 11

OPTIMUM SOIL NUTRIENTS: A *SINE QUA NON* FOR SUSTAINABLE AGRICULTURE FOR FOOD AND NUTRITION SECURITY

***Emmanuel Uzoma Onweremadu*[1*] *and Angela Chika Udebuani*[2]**
[1]Department of Soil Science and Technology, Federal University of Technology, Owerri, Nigeria
[2]Department of Biotechnology, Federal University of Technology, PMB 1526 Owerri, Nigeria

ABSTRACT

Agriculture is largely soil-based, suggesting the need for adequate soil nutrients supply for its sustenance. Soil nutrients (mineral elements) vary spatially, vertically and temporally, requiring a good understanding of these variabilities for good management. Soil nutrients are classified based on relative need into macronutrients and micronutrients, and each group has a specific role it plays in crop performance. Plants take these nutrients by diffusion, mass flow and interception by roots; and uptake is governed by soil, crop and climatic characteristics of a given agroecology. Soil nutrient depletion is obvious and common in Africa where soil mining is a major cause. Continuous nutrient withdrawal without replacement amidst excruciating climate and increasing population accelerates rate of soil depletion in soils. Factors such as soil climate, soil organisms, tillage techniques, aeration, moisture, soil pH and pedogenic processes interact to influence soil fertility. Fertilizer use is practiced outside traditional soil fertility enhancement strategies but is governed by crop, soil, climatic, economic and management factors in the sub-Saharan Africa. An integration of traditional and modern soil fertility regeneration practices could be an antidote to soil nutrient depletion especially in tropical agriculture.

* E-mail: uzomaonweremadu@yahoo.com.

Keywords: Degradation, nutrients, soil quality, soil fertility enhancement

INTRODUCTION

Agriculture is largely soil-based except say in hydroponics. Soils are fragile and vulnerable. Its quality changes over time hence Seybold et al. (1998) defined *Dynamic Soil Quality* as the effects of human use and management on soil functions. The capacity of a soil to function within its ecosystem boundaries for the purpose of sustaining biological productivity and diversity is vital to environmental quality and food security.

Many farmers in sub-Saharan Africa and elsewhere are worried about declining soil productivity which is largely traced to nutrients deficiency and imbalance in the soil ecosystem. The challenge for these farmers is to adapt to farming systems in order to conserve their soils and nutrients contained therein. It is therefore a major goal of researchers, policy-makers, development practitioners, farmers and other natural resource users to restore soil productivity, sustain it while building soil resilience with regard to their interrelated biophysical functions and nutrient balances.

Soil nutrients remain one of the most important essential growth factors apart from heat (temperature), light, air, water and mechanical support. The growth and development of crops (plants) is among other factors a measure of soil nutrients potential. These nutrients are supplied to crops based on the prevailing controlling factors particularly soil and climatic attributes of the agroecology.

Soil nutrients vary geospatially, vertically and temporally, that is, soil fertility changes in space, depth of profile pit and over seasons, respectively. This variability affects the soil's capacity to supply nutritional needs of plants, and indirectly man. While few agroecosystems are undisturbed and need no intervention from soil managers, many agroecosystems are highly disturbed, thus need supplemental provision of nutrients in order to sustain life forms. Some nutrients are translocated and leached beyond reach of plant roots especially in highly degraded sandy soils. Yet, in some areas with marked temporal variations, soil nutrients respond to such seasonality. Furthermore, crops have varying responses to nutrients in the soil. Based on the above using scientific knowledge and ecological wisdom will be helpful in understanding nutrient dynamics in both soil and aquacultural (hydroponics) environments.

MEANING AND CLASSIFICATION OF SOIL NUTRIENTS

Nutrients are mineral substances absorbed by plants to enable them grow and develop. Nutrients are essential elements necessary for a plant to complete its life cycle. Such essential elements could be of direct involvement in the nutrition of the said plant, apart from the possible effects in correcting some unfavourable condition in the soil or culture medium (Foth, 1984).

Sixteen essential elements are known to be important in crop performance and survival. Two main groupings can be achieved viz: mineral and non-mineral plant nutrients. The mineral nutrients include thirteen essential elements, which originate from the soil for crop uptake. They are dissolved by soil water and absorbed via crop root hairs. These mineral nutrients are further subdivided into macronutrients and micronutrients.

The macronutrients are those essential elements required in relatively large proportions for good performance of crops. Macronutrients are broken into two groups: primary and secondary macronutrients. The primary macronutrients include nitrogen (N), phosphorus (P) and potassium (K). These trinity of macronutrients are readily deficient in soils since plants need them in large quantities. The secondary macronutrients are calcium (Ca), magnesium (Mg), and sulphur (S). These secondary macronutrients are not as readily inadequate as in primary macronutrients. Impliedly, secondary macronutrients do not require regular fertilization as in primary macronutrients. But, in acidic soils, large amounts of calcium and magnesium are added in order to neutralize soil acidity common in most humid tropical soils. Sulphur is usually found in sufficient quantity in soils due to slow release by the decomposing soil organic matter. Consequently, forested areas rarely suffer from sulphur deficiency unlike sparsely vegetated areas of the world.

Another group of mineral elements are the micronutrients, which are those essential elements required in relatively small quantities by plants. The remaining 7 elements which include boron (B), copper (Cu), manganese (Mn), molybdenium (Mo), iron (Fe), chlorine (Cl) and zinc (Zn) are micronutrients. Brady and Weil (1999) included nickel and cobalt as micronutrients, and reported that other elements such as silicon, vanadium and sodium appear to improve the growth of certain plants. Brady and Weil (1999) noted increasing attention being directed toward micronutrient deficiencies due to:

- Intensive plant production practices have increased crop yields resulting in greater removal of micronutrients from soils,
- The trend towards high analysis fertilizers has reduced the use of impure salts and organic manures, which formerly supplied significant amounts of micronutrients,
- Increased knowledge of plant nutrition and improved methods of analyses in the laboratory are helping in the diagnosis of micronutrient deficiencies that might formerly have gone unnoticed,
- Increasing evidence indicates that food grown on soils with low levels of trace elements may provide insufficient human dietary levels of certain elements, even though the crop plants show no signs of deficiency themselves.

FUNCTIONS OF SOIL NUTRIENTS

Nitrogen is part of all living cells and a necessary component of all proteins, enzymes and metabolic processes involved in the synthesis and transfer of energy. Nitrogen promotes rapid and luxuriant growth as well as increases seed and fruit production.

Nitrogen confers succulence in leaves and forage crops. Nitrogen is part of chlorophyll underlining its importance in photosynthesis. It is sourced from legumes, other N-fixers and N-fertilizers.

Phosphorus is commonly found in bone meal, superphosphate and P-fertilizers. Phosphorus is involved in photosynthesis, and helps in the transformation of solar energy into chemical energy. It enhances maturation crops and helps plants to withstand stress. Phosphorus encourages root growth.

Potassium is absorbed in large amounts in soils and helps in the building of proteins. It is relevant in photosynthesis , potassium improves fruit quality and reduces disease infections in plants.

Calcium is an essential secondary macronutrient that is absorbed by plants in large quantities. Calcium is an essential part of cell wall structure. Calcium provides for normal transport and retention of other elements. It gives strength to plants. Calcium balances the effects of alkali salts and organic acids within the plant. Sources of calcium include dolomitic lime, calcium superphosphate and gypsum.

Magnesium is part of the chlorophyll in all green plants hence essential in photosynthesis. Magnesium is an enzyme activator; and this activation function is necessary for plant growth. Magnesium is sourced from fertilizers dolomitic limestone and organic materials.

Sulphur is essential in protein synthesis, promoting activity and development of enzymes and vitamins. Sulphur helps in chlorophyll formation. It improves root growth and seed production. Sulphur promotes vigorous plant growth and helps plants to withstand coldness. Main sources of sulphur include lower grade fertilizers, rain water and gypsum.

Iron is essential in chlorophyll formation. Iron is obtained from the soil, iron sulphate and iron chelate.

Boron regulates other nutrients. Boron helps in the production of carbohydrates. Boron is essential for seed and fruit development. Organic matter and borax are good sources of boron.

Chlorine aids in plant metabolism. It is found in soils.

Molybdenum helps in the use of nitrogen. The soil is main source of molybdenum.

Manganese combines with enzyme systems to breakdown carbohydrates. It functions with enzymes to metabolize nitrogen.

Copper is important for reproductive growth and helps in root metabolism. Copper helps in the utilization of proteins.

Zinc assists in the transformation of carbohydrates. It regulates the consumption of sugar. Zinc is part of enzyme systems which regulate plant growth, zinc oxide, soil zinc sulphate and zinc chelates are major sources of zinc.

Non-mineral nutrients are carbon (C) hydrogen (H) and oxygen (O). They are found in water and air. Their sources make might be difficultly controlled by man. In a photosynthetic process; plants utilize energy from the sun (solar energy) to alter carbon dioxide and water leading to the formation of starches, sugars and emission of oxygen thus,

$$6H_2O + 6CO_2 \underset{Chlorophyll}{\overset{Solar\,\text{Energy}}{\rightleftarrows}} C_6H_{12}O_6 + 6O_2 \uparrow$$

NUTRIENT UPTAKE

Processes involved in the movement of nutrients to root surfaces are important in plant nutrition. Nutrient absorption processes determine to a great extent plant growth and yield.

Three key processes identified as interception, mass flow and diffusion, govern availability of soil nutrients for plant uptake and utilization.

By interception, plant roots extend through soil environment, thereby exploring new supplies nutrients and water. There is contact between plant roots and nutrients in the soil system. The assumption in nutrient uptake by interception is that ions attached to the root surfaces exchange with those held on the surfaces of soil organic matter and clays. These ions are held electrostatic and van der waal forces at specific sites of colloids and roots. The attainment of ions to certain oscillation volumes cause them to overlap, leading to ion exchange. The intimate contact at certain points along root surfaces with soil particles result in direct exchange of ions between soil medium and root surfaces hence uptake by interception is often referred to as contact exchange.

Diffusion refers to the movement of ions via soil water to the outer space of root cells before their migration into the vacuole for further synthesis. It is an important process since mass flow and interception may not contribute much to total nutrients absorbed by plants. Movement of phosphorus ions to the root surfaces is greatly achieved by diffusion. Diffusion occurs when there is a concentration or diffusion gradient between root surfaces and soil solution. Diffusion of nutrient ions occurs when plants deplete nutrients in their vicinity, thereby creating a concentration gradient between bulk soil solution and the root surfaces, resulting to the continuous migration of additional ions to the root surfaces for absorption.

Mass flow occurs commonly when roots have invaded a soil region, moving nutrients over short distances within the pedosphere. Mass flow can be defined as the movement of water together with dissolved nutrient ions in the soil system. Mass flow is facilitated by rainfall and irrigation. It occurs when diffusion gradient is created resulting from moisture or water intake by plant roots. Nutrient uptake via mass flow depends on the amount of water moving to the roots as well as concentration of electrolytes (nutrient ions) in the soil water.

Mass flow and diffusion play important role in the movement of potassium, magnesium ad nitrogen. Mass flow dominates in the movement of calcium and sulphur to root surfaces as diffusion governs to a great extent the movement of phosphorus to root surfaces.

There is an intimate contact between plant roots and soil particles at many points. The roots are surrounded by soil solution but given a certain level of metabolic energy, plants roots selectively absorb ions over time and over a range of soil and plant conditions. Diffusion of ions occurs at the outer space of root cells and this is reversible as plants are passive about the activity. But, this action precedes an irreversible transport of nutrient ions across an impermeable membrane after a metabolic energy has been expended to the inner space of the cell.

Nutrient uptake is governed by root characteristics especially root density and distribution. Generally, roots are less able to penetrate deeply enough to access the water and nutrients stored in the soil (Williams and Weil, 2004). Shallow rooting pattern leaves greater residual soil water (Menill et al., 2003) and nutrients below root zone. This is unlike deep-rooted plants that explore nutrients in the subsoil. Such exploration of nutrients can reach 1.2 m deep in soils under no-tillage systems (Moroke et al., 2005). It implies that crops with deep-rooting patterns need less fertilization when compared with shallow-rooted crops. Density and root ramification is important in soils when nutrients deplete and when their mobility decreases. This underscores the relevance of endotrophic mycorrhizae and ectomycorrhizae in increasing effective root surface with increased efficacy in the absorption of nutrients (particularly phosphorus) and water. Mycorrhizal symbiotic associations render plant nutrients more available. The microbiology of the rhizosphere (root surface and its adhering soil) become very important factor in nutrient availability. It is important to note that the production of CO_2 in the rhizosphere and the formation of organic and inorganic acids aid in the solubilization of inorganic plant nutrients (Alexander, 1977)).

Other factors which affect nutrient uptake include the supply of respiratory substrate, temperature, increase in the concentration of nutrient ions in an environment external to the root and soil moisture (Foth, 1984). In addition to the above, management, climatic, environmental and crop factors (Soon et al., 2001), differences among hybrids (Ma et al., 1999), growth stage (Rozas et al., 1999), accumulates in root (Jiang and Sullivan, 2004), fertilization (Jiang et al., 2000) and ionic balance (lngrid, et al., 2006) affect nutrient uptake in soils.

The chemical processes and activities of microorganism which convert plant nutrients into available forms are influenced by temperature. High temperatures status affects rock weathering, pedogenesis, and consequent release of plant nutrients for uptake. High temperatures promote N-loss by volatilization as cold temperatures slow down activity in the rhizosphere. Cold temperatures slow the rate of decomposition process in organic matter, thereby retarding release of nutrients contained. Duirnal and nocturnal temperature changes affect nutrient uptake (Tyson et al., 2007).

Soil water and flow dynamics are primary drivers of crop growth, nutrient cycling and contaminant transport (Haws et al., 2004). Soil water influences rate of nutrient ion movement and diffusion into root cells.

Most arable soils are intensively cultivated leading to soil compaction (Siegel –Issem et al., 2005). Soil compaction alters oxygen supply by reducing pore spaces in the soil. It implies that soil compaction can reduce nutrient uptake through its effect on the availability of oxygen for root respiration (Foth, 1984). Poorly aerated soils can cause physiological imbalances that lead to nutrient deficiencies (Siegel–Issem et al., 2005). Nitrogen and other minerals were found deficient in shoots of *Pinus contorta* growing in compacted, remoulded soil cores (Conlin and Ven den Driessche, 1996). Again, compacted soils reduce aeration porosity, and soil oxygen is necessary for aerobes, majority of which are involved in nutrient element transformations in the soilsphere. Root development and plant growth are related to aeration porosity (Baver et al., 1972). It is important to note that oxygen-diffusion rate (ODR) in soils influences greatly the ecology of the rhizosphere especially as it concerns root respiration and activity of aerobes. A limiting factor in the oxygen supply to the rhizosphere is the rate of diffusion of oxygen through soil moisture film around roots; implying that ODR defines soil oxygen conditions. Soil temperature becomes a vital factor which affects ODR by increasing rate of respiration and diffusion coefficient of oxygen in water, and the solubility of oxygen in the liquid phase decreases (Baver et al., 1972).

New mechanisms for nutrient uptake have been discovered (Science Daily, 2007). According to reports biologists at the Camegie Institution's Department of Plant Biology reported a new way that plant cells govern nutrient regulation, involving neighbouring pore-like structures at the cell's surface which physically interact to control uptake of nitrogen. It was concluded that neighbouring molecules are essential in regulating nitrogen uptake.

Table 1. Percent of the nutrients normally taken up by a corn crop which could be supplied through root interception, mass flow and diffusion

Nutrient	Interception	Mass Flow	Diffusion
Nitrogen	< 1	80	19
Phosphorus	2	5	93
Potassium	2	18	80
Calcium	150	375	0
Magnesium	33	600	0
Sulphur	5	300	0

Adapted from Mengel, 1995.

Table 2. Differential effects of changes in soil or root temperature on NH_4+ and NO_3 uptake rate in various species

Species	Soil or Root Temperature (^{0}C)	$NH^+_4 : NO^-_3$ uptake ratio	Source
Lotium multiflorum	5	3.20	Carkson and Warner (1979)
	10	2.50	
	20	1.60	
Brassica inapus	9	1.87	Macduff et al. (1987)
	5	1.20	
	17	0.87	
Picea abies	10	5.00	Marchner et al. (1991)
	20	4.200	
Lycopersicon esculentum (chilling sensitive)	5	1.10	Smart and Bloom (1991)
	20	0.69	
Lycopersion esculentum (chilling resistant)	5	0.90	Smart and Bloom (1991)
	20	0.79	
Fagus sylvatica	10	9.00	Gesster et al. (1998)
	20	3.10	
	25	2.80	

Table 3. Effect of applied N on the growth, N^- uptake and yield of two corn hybrids

Hybrid and N treatment	Days to maximum root growth days	Root length per plant feet/plant	N-uptake 16/acre	Grain yield bu/acre
B73 x $M_0$17				
0N	91	2880	103	79
200N	91	4670	177	114
P3732				
0N	75	24060	103	79
2900N	75	2550	159	100

Adapted from Mengel (1995).

SOIL NUTRIENT DEPLETION IN AFRICA

It is not a myth but a reality that African soils are steadily depleted of mineral nutrients. A soil's potential for producing crops is largely determined by the environment that the soil provides for root growth as roots need nutrients, air, moisture and adequate space in which to develop. Changes in soil quality affect health and productivity of the plant. In developing countries and continents, harsh climatic conditions, increasing population, conflictive land use and decline of traditional soil management practices do not favour soil fertility regeneration. Soils of the area are not only depleted of nutrients but are characterized by nutrient imbalances (Stoorvogel et al., 1993).

Soil productivity constraints have been identified as physical, chemical and socioeconomic (CTA, 1993). Soil physical constraints in Africa include reduced aggregate size and stability, crusting, reduced porosity, infiltration and compaction, low water holding capacity, increased rates of water and wind erosion, poor water retention, presence of hardpans, poor drainage, salinity, high water table and varying levels of toxicities. Socioeconomic problems are poor awareness on modern farming techniques, disincentives, land tenure, poor credit facilities, unavailability of inputs, increasing population, gender disparity, poverty, social taboos, pests and diseases. A major soil chemical constraint is low soil fertility due to low nutrient levels especially nitrogen and phosphorus. Low nutrient level and increasing nutrient depletion is attributed to soil mining (Bationo et al., 1998; Onweremadu, 2006).

Continual withdrawal from soils by cultivation without replacement has led to severe soil erosion forms ranging from inter-rill, rill and gully erosion in the area, which has further depleted soil nutrients by wearing away soils and nutrients contained in them. In soils of Southeastern Nigeria, soil erosion by water has devastated the landscape (Igwe, 2003) resulting in nutrient imbalances in the Ca : Mg and Ca+Mg / Al+H ratios (Oti, 2002). The Ca : Mg ratio imbalance could be a limiting factor in magnesium nutrition of maize crop (Osemwota et al. (2003). Base saturation is known for influencing the availability of plant nutrients apart from affecting soil moisture (Logsdon and Laird, 2004). Onweremadu (2006) reported preponderance of aluminium in eroded soils of Southeastern Nigeria characterized by low organic matter content. With substantial proportion of aluminium occupying the exchange site at low pH, plant nutrients become unavailable to meet crop needs. The pH range of 3.7 to 4.9 is limiting to plant performance and may cause aluminium toxicity although this depends on species of the acidic cation (Gillman et al., 1989). With low organic matter content of degraded soils, retention of plant nutrients becomes minimal, and some mineral elements like phosphorus are lost to runoff to cause non-point-source pollution such as eutrophication (Zhang et al., 2005).

Soil calcium is a secondary macronutrient which is adversely affected in degraded soils of forest ranges in Southeastern Nigeria (Onweremadu and Uhuegbu, 2007). High rainfall amount and duration encourage leaching of basic cations from epipedons of soils making them unavailable to arable crops and grasses. Decline in calcium has been reported in European forest soils (Thimonier et al., 2001) as well as in North American forest soils (Raben et al., 2000). It is possible that a lot of organic acids in these forest soils resulted in acidification and acid leaching in aforementioned forested ecosystems.

Table 4. Changes in chemical properties in an ultisol (Acrisol) in Southeastern Nigeria induced by water erosion (Ap horizon)

Property	Unit	Mbaise		Umuahia		Akwette		Oguta	
		Uneroded	Eroded	Uneroded	Eroded	Uneroded	Eroded	Uneroded	Eroded
pH (KCl)	-	4.3	3.7	5.0	4.1	4.9	3.9	5.2	4.6
ECEC	cmol/kg	10.8	5.6	12.1	6.3	11.2	6.6	13.2	7.0
Base saturation	%	56.6	38.9	58.0	4.0	57.8	42.0	60.0	32.0
Aluminium saturation	%	43.4	61.1	42.0	54.0	42.2	57.0	40.0	68.0
Organic matter	%	3.2	1.1	3.3	1.5	3.6	1.4	3.5	1.3
Total nitrogen	%	0.28	0.09	0.29	0.11	0.3	0.1	0.3	1.1
Available phosphorus	mg/kg	18.6	8.3	18.3	9.8	26.1	9.6	20.1	10.1

ECEC = Effective Cation Exchange Capacity.
Source: Onweremadu, 2006.

Soil ions vary in time and space (Onweremadu and Akamigbo, 2007) due to landscape and land use (Wing et al., 2001), parent material (Kosmas et al., (1993), local carbon distribution (Hobbie, 1996), pedogenesis (Brubaker et al., 1993), deforestation activities (Maineville et al., 2006) and the extent of these changes is scale-dependent (Shitter et al., 2004). In a forested hilly landscape of Southeastern Nigeria, exchangeable base cations occurred in order of abundance as follows:

Ca > Mg > K >Na … (Onweremadu and Akamigbo, 2007).

In that study, basic cations had higher values in the subsurface horizons irrespective of geomorphic setting while exchangeable acidic cations had lower values, suggesting pronounced pedogenic processes of leaching and eluviation. Calcium, Mg, K and Na were readily transported out of the eluvial to illuvial horizons of soils. The agronomic implication is that such soils become less suitable for shallow-rooted crops, thus tree and deep-rooted crops are recommended for such degraded soils.

Nugget values expressed in percentages showed marked differences in the degree of spatial dependence of Ca, Mg, K, Na, CEC and exchangeable acidity in all studied physiographic land units and at both surface and subsurface horizons (Onweremadu and Akamigbo, 2007). In epipedons (surface horizons), variables in soils of the interfluve indicated strong spatial dependence while soils of colluvial and channel bed exhibited strong to moderate spatial dependence.

Soil magnesium varied significantly ($p=0.05$) among geomorphic surfaces of soils of varying lithological origin (Onweremadu, 2009), and this result agreed with the findings of Mueller et al., (2008) who reported variability of magnesium activity on a soilscape. The spatial distribution of magnesium affects availability of calcium and phosphorus. A decrease of Ca/Mg ratio to a level below 3.0 results in unavailability of Ca and Ps (Landon, 1984); csonfirming the relevance of magnesium in improving phosphorus availability (Nearman and Singer, 2004).

Phosphorus is less mobile than nitrogen hence leaching is considered as not so important in phosphorus transportation in most arable soils (Elliot et al., 2005). However, Ham et al.(2000) reported phosphorus leaching in sandy soils, and its movement is facilitated by particles transport (Djo djic et al., 2000). Constituents of phosphorus are easier carried away through surface waters. Phosphorus loss and movement have been reported in inter aggregate or larger pores developed from earthworm burrows or old roots (de Jonge et al., 2004). Yet, micropores in clay soils, cracks, capillary action, and animal burrows can form subsoil as well as epipedal connectivities for movement and loss of soil phosphorus. Soils with less phosphorus–sorbing capacity promote leaching loss of phosphorus (Kleinman et al., 2003).

Table 5. Variability within surface and subsurface horizons using geostatistical analysis

Variables	Surface Model	T/U	Range (m)	Nugget (%)	R^2	Subsurface Model	T/U	Range (m)	Nugget (%)	R^2
Interfluve										
Ca	S	T	3.3	21.6	0.92	E	T	5.5	65.0	0.18
concentration	E	U	4.4	11.9	0.90	S	T	0.9	0.0	0.82
Mg	S	T	1.4	30.1	0.95	S	T	0.8	13.0	0.61
concentration	S	U	1.5	36.2	0.80	S	T	0.5	35.0	0.38
K concentration	S	U	1.7	0.1	0.91	S	U	0.9	41.0	0.71
Na	S	T	1.9	0.2	0.89	E	U	0.5	35.0	0.79
concentration										
Exchangeable										
acidity										
CEC										
Colluvial Footslope										
Ca	L	T	0.4	22.0	0.90	L	U	-	-	0.12
concentration	L	U	0.6	10.6	0.89	S	U	-	-	0.15
Mg	L	U	0.3	25.0	0.93	S	T	-	-	0.17
concentration	E	T	1.1	45.0	0.76	E	T	-	-	0.09
K concentration	E	U	0.4	-	0.88	S	U	38.0	43.0	0.66
Na	L	T	0.1	-	0.86	S	T	31.0	41.0	0.57
concentration										
Exchangeable										
acidity										
CEC										
Channel Bed										
Ca	S	T	0.8	12.1	0.25	L	T	-	-	0.11
concentration	E	U	0.3	9.6	0.11	L	T	-	-	0.12
Mg	S	T	0.4	21.0	0.61	L	T	0.6	48.0	0.13
concentration	S	U	0.2	26.9	0.13	L	U	-	-	0.16
K concentration	L	U	0.8	48.0	0.28	E	U	3.3	-	0.58
Na	L	T	0.2	-	0.08	E	U	2.6	45.0	0.47
concentration										
Exchangeable										
acidity										
CEC										

S = Spherical, E = Exponential, L = Linear, T = Transformed, U = Untransformed.
Source: Onweremadu and Akamigbo, 2007.

Microbial growth requires presence of available form of phosphorus, which are assimilated into microbial nucleic acids, phospholipids and other protoplasmic substances. If carbonaceous residue is deficient in phosphorus, the microbial assimilation of phosphorus will create a temporary phosphorus unavailability and this depresses crop yield.

Soil phosphorus depletion in soils is critical since its total content is low compared to nitrogen and potassium. Again, most soil phosphorus forms and compounds are unavailable for plant uptake because of high insolubility, and soluble forms say from fertilizers and manures when added to soils are easily fixed. Phosphorus deficiency becomes of great agronomic importance especially in developing countries of sub-Saharan Africa, where the decline in per capita food production will not likely be reversed until the critical phosphorus deficiency problems are solved (Brady and Weil, 1999). But, if phosphorus is over applied in soils, there will be accumulation (Sharpley et al., 1999) and its runoff is constituting a primary source of non-point source of phosphorus pollution (Daniel et al., 1994) leading to eutrophication of water bodies (Sharpley et al 1999). Soil properties such as clay content, organic carbon, oxalate forms of iron and aluminium are highly correlated with phosphorus in soils (Zhang et al., 2005).

Application of swine manure to meet nitrogen needs of crops can result in the accumulation of inorganic phosphorus because nitrogen–phosphorus ratio in swine is lower than that needed by most crops (Wienhold, 2005). The researcher explained that low nitrogen–phosphorus ratio in swine results from the inability of swine to efficiently utilize phytate phosphorus, which is a primary form of phosphorus stored in traditional corn.

Nitrogen is a major macronutrient, and characterized by high degree of spatial heterogeneity, thus of ecological significance (McClain et al., 2003). Several factors interact to influence soil nitrogen variability but Monis and Boerner (1998) attributed it to landscape scale patterns of nitrification, which they related to soil moisture.

Because of the critical position of nitrogen supply in crop production and soil fertility, its deficiency markedly reduces yield as well as quality of crops and pastures. Nitrogen is among the few soil nutrients that is lost by the process of volatilization as well as by leaching. Nitrogen can be transformed to organic inorganic and volatile forms, and these transformations are mediated by the microflora. Nitrate–nitrogen is lost in the soil in several ways. High solubility of NO_3–N_2 increases its mobility out of the rhizosphere, beyond root penetration. Plant removal of nitrate and ammonium is common. Yet, nitrogen biologically leaks via denitrification. Denitrification leads to the formation of gaseous nitrogen which is unavailable to most microorganisms and macroorganisms. Nitrate leaching is probably more severe in high rainfall or irrigated areas. Nitrogen depletion in tropical soils is mainly caused by leaching and lost nitrates lead to pollution and eutrophication of surface and underground water systems (Dinnes et al., 2002). Leaching loss is a serious and common event in intensively managed cropping systems, and may be due to high rate of nitrogen fertilizer application (Othman and Pope, 2000). When nitrogen fertilizers are over-applied or applied at wrong times, it results in much accumulation of plant available nitrogen (Myungon at al., 2006) and this promotes leaching losses of nitrates. However, ammonium–nitrogen forms are not readily leached as nitrate–nitrogen as it is easily adsorbed to soil colloids thus soils dominated by ammonium – nitrogen form are less leached (Jones and Shannon, 2000). Nitrates are lost through subsurface soil pores, cracks and drainage tiles (Dinnes et al., 2002) and this depletes soil nitrogen.

Factors Affecting Soil Fertility

Sustainable soil systems depend on agroecological processes that promote to soil fertility. Soil properties including soil climate provide preliminary information related to soil quality. A good quality soil is one with the capacity of sustaining crop and livestock production with minimal damage. Soil temperature influences biological and chemical activities in the pedosphere. These biochemical processes convert plant nutrients into available forms. It is a well-established fact that the rate of organic matter decomposition and the mineralization of organic forms of nitrogen increase with temperature. The effect of temperature on nitrogen mineralization is critical since each biochemical step is catalyzed by temperature–sensitive enzymes generated by soil microorganisms. Ammonium accumulation in composts and manure piles is promoted at thermophilic temperature range while there is no increase in both ammonium and nitrate when soils are frozen (Alexander, 1977).

Soil moisture, which expresses the order of soil water changes over time, governs the availability of soil water for weathering and leaching. Soil moisture affects nitrogen degradation as considerable proportion of organic nitrogen is degraded to inorganic nitrogen with an occurrence of wetting and drying cycles. Soil moisture influences nitrate leaching and uptake in mature and plant seedlings (Jiang and Sullivan, 2004).

Soil aeration governs the distribution of obligate and facultative soil organisms involved in transformation, lead to release of mineral nutrients. The ammonifying populations include aerobes and anaerobes (Alexander, 1977). Ammonification is not eliminated by soil submergence but becomes rapid in fields where oxygen level is quite low. However, soils active in aerobic mineralization also form ammonium cations readily in the absence of oxygen. Soil oxygen is necessary in the nitrification step involving *Nitrosomonas europaea,* which convert ammonium–nitrogen to nitrite–nitrogen. Oxygen availability is a critical environmental factor in denitrification. Denitrification proceeds only when oxygen supply is insufficient to satisfy the microbiological demand. However, oxygen is necessary for the formation of nitrite and nitrate. Anaerobic microenvironments exist at microscopic sites in well-drained soils whenever the biological oxygen demand exceeds supply.

Soil pH affects nutrient mineralization. For instance, the production of inorganic nitrogen – ammonium plus nitrate is greater in neutral than acidic environment. Acidification depresses mineralization. Most soil organisms involved in nutrient transformations have a rather small tolerance to variations in soil pH. Perhaps, the greatest general influence of pH on plant growth is its effect on the availability of nutrients. At low pH, molybdenum forms insoluble compounds with iron and is rendered unavailable. Potassium is available at neutral and alkaline soils. Solubility of some plant nutrients decreases with increase in soil pH. Phosphorus and boron tend to be unavailable in very acidic soils while copper and zinc have reduced availability in both acidic and alkaline soils.

Land preparation technique influences soil nutrient status. Cultivation of certain virgin soils causes a massive decline in their content of organic nitrogen. The promotion of mineralization is associated with and dependent on the decline in organic carbon content of virgin lands brought into cultivation for food production. For the conservation of mineral nutrients, conservation tillage techniques are preferred to conventional tillage technologies. There was a decline in soil organic carbon following cultivation (Follet, 2001) while residue application and reduced tillage enhance soil quality and sequester atmospheric carbon dioxide

(Uri, 2001). Downslope movement of soils by tillage has been reported (Lindstrom et al., 1992; Li et al., 2004). Conservation agriculture aims to keep soils alive. A healthy soil is a living soil rich in plant available nutrients and soil organic matter which make a variety of soil organisms such as earthworms, beetles and others to thrive.

Soil organisms especially the micro-flora and fauna are indicators of soil fertility. As earlier stated soil organisms mediate most transformations taking place in the soil environment. Soil contains five major groups of microorganisms, namely bacteria, algae, fungi, actinomycetes and protozoa. Soil ecosystem includes these organisms, as well as the organic and inorganic constituents of a given site. Some of these organisms are autochthonous or indigenous soil species while others are allochthonous organisms or invaders. While these soil organisms depend on soil organic materials as their substrate, they help in the decomposition of soil organic materials leading to the release of essential plant nutrients. The ability of these biota to perform soil ecosystem functions depends on the suitability of prevailing environmental factors, especially soil moisture, soil aeration, soil pH, soil temperature, soil depth, soil toxicity and cultivation or tillage techniques.

Fertilizer Consumption

Nigeria, and in indeed sub-Saharan Africa use traditional methods of soil fertility regeneration such as bush fallowing, and farm yard manure. In the bush fallowing technique, arable soils are allowed to revert to fallow after 3 to 4 years of cultivation. Efficacy of this system depends on the length of fallow as well as the plant species. With increasing demographic and other socio-economic pressures, fallow lengths have drastically shortened. Following this, soil damage increases hence the spate of widespread forms of land degradation such as crusting soil erosion, sedimentation of farmlands, siltation and landslides.

Replacement of fallow system becomes necessary. Alternatives such as the use of animal manures depends on the animal population that yields these droppings. Compound farmlands benefit from household sweepings and then movement to distant farms become burdensome. Mixed farming was a strategy though still in use . Yet, these soil fertility regeneration measures are not coping with intensive agriculture and the introduction of high yielding and high nutrient-demanding crop varieties.

Consequent upon the above, inorganic fertilizers came into Nigeria agriculture in 1937 (FPDD, 1989). Inorganic fertilizer consumption became popular in Nigeria on the assumption that it is the most efficient approach to sustaining soil nutrient needs of crops and food demands by the increasing population. This era of inorganic fertilizer use led to increase in cultivable land area as well as intensive agriculture by continuous cropping.

The inorganic fertilizer was formulated to supply nitrogen, phosphorus and potassium, and thus is referred to as complete fertilizer. In Nigeria, need for inorganic fertilizer exceeds availability. For instance, in the savanna agroecosystem characterized by low organic matter, fertilizer recommendation of 100 – 120 kgN/ha, 40 – 50 KgP_2O_5/ha and 30 – 45 KgK_2O/ha is required for good yield of maize (*Zea mays*). But, the average application per hectare is less than 30 Kg/ha compared to world average of 68 Kg/ha and 80 Kg/ha in Kenya (FPDD, 1989).

Application of these inorganic fertilizers is most often done without soil testing to ascertain the nutrient needs of soils. These indicate possibility of over-application or under-application of nutrient elements with attendant environmental problems.

Again, inorganic fertilizer consumption in Nigeria is unacceptable to many farmers who lean towards tradition. It is the belief of most farmers in the countryside that application of inorganic fertilizers damages the natural taste and flavour of farm products. This is in addition to some superstitions and misinformation associated with inorganic fertilizer use in Nigeria and other sub-Saharan African countries.

In synopsis, factors influencing consumption of inorganic fertilizers – include soil factors, crop factors, climatic factors, management factors and economic factors. There is spatial variability in the supply of plant nutrients by soils. Temporal and depth differences in nutrient supply by soils exist. Every soil has its inherent productive capacity which is different from its neigbhouring soil on the landscape. Such spatial, vertical and temporal differences in plant nutrient supply are associated with soil texture, soil depth, soil structure, soil reaction, soil moisture, soil climate, soil aeration, abundance and distribution of soil organisms. Sandy textures retain less nutrients while clayey soils retain much nutrients which may be difficultly available. Some soils are repletes with abundant nitrifiers' and nitrogen–fixing bacteria, and these ensure high nitrogen content in addition to external inorganic fertilizer input. Heavy applications of inorganic fertilizers is profitable on soils with high productive capacity. It shows that soils respond differently to fertilizer types. Southeastern Nigeria is dominated by ultisols. According to Mbagwu (1992), over 60% of soils of the Southeastern Nigeria agroecology is occupied by ultisols, which is characterized with high acidity, fragility, coarse textures and multiple nutrient deficiencies (Enwezor et al., 1990). Ultisols respond differently to inorganic fertilizers compared with responses of other soil groups such as Alfisols, Inceptisols, Aridisols, Vertisols, Entisols, Histosols, Andisols, Mollisols, Spodosols, Oxisols and Gelisols. Most tropical soils lack nitrogen and phosphorus (Kogbe and Adediran, 2003). While nitrogen deficiency in tropical soils is attributed to leaching losses, ammonium fixation by clay minerals, soil erosion, volatilization, microbial immobilization and denitrification (Malpassi et al., 2008), phosphorus deficiency is caused by scarcity of phosphorus forming rocks, intense weathering, phosphorus fixation, immobilization as well as leaching and runoff losses (Sharpley et al., 2008).

Crop factors govern fertilizer use. Certain crops use larger amounts of particular nutrients than others. Legumes require large amount of phosphorus whereas cereals require proportionately more nitrogen. Tuber crops require ample supply of potassium in the soils. Acha (*Digitaria exilis*) gives appreciable yield on poor nutrient soils that do not support growth of maize, sorghum, rice and cassava but responded positively to inorganic fertilizer application (Dachi and Kwon Ndung, 2007). Soils under date palm (*Phoenix dactylifera*) require more phosphorus fertilizers applied in splits for adequate yield (Ugbah, 2007).

Okra (*Abelmoschus esculentus*) varieties gave higher yield when fertilizer was combined with benlate (fungicide) in soils of Bida, Nigeria (Ajari et al., 2002). Tuber crops also respond to fertilizer levels and differently. Ano and Orkwor (2006) reported that fertilizer application at 200 kg/ha NPK 20-10-10 compared to other level levels (0, 400, 60 kg/ha) significantly ($p=0.05$) increased yam (*Dioscorea rotunda* CV obiaturugo) tuber yield and productivity in soils of Umudike, Southeastern Nigeria. Indirectly, Nitrogen-fixing plants require calcium containing fertilizers to promote nodulation. Onweremadu et al. (2005) reported significant ($p \leq 0.05$) improvement in the nodulation of coropea (*Vigna unguicalata*) in acidic soils of

Owerri, Nigeria. However, responses inorganic fertilizers by crops vary among crop varieties as improved varieties are more responsive to higher doses of fertilizers.

Agrometeorological attributes influence fertilizer consumption. In areas with low rainfall amount, leaching is less, implying her concentration of nutrient elements especially at the epidedons. But, soils of humid tropics characterized by high intensity, long duration and high amount of rainfall witness much of its soil nutrients lost by leaching and weathering. There is ample supply of soil moisture during the cropping seasons but nutrient supply is inadequate. Fertilizer application before rainfall is susceptible to runoff losses and leaching. Runoff losses increase if topography is sloping. Application of fertilizers after rains will be more beneficial to crops.

Table 6. Total nutrient content in soils under date palm

Location	Detpth (cm)	Total Nitrogen (Kg/ha)	N-Mineralized (Kg/ha/yr)	Total Available P. (Kgha0	Total Exchangeable K (Kg/ha)
Gudurba	0-15	1260	12.6+	2.10	139.2
	15-30	1200	12.0	5.76	103.0
	30-60	2880	28.8	13.92	243.4
	60-90	2400	24.0	13.92	205.9
	90-120	2400	24.0	2.40	168.5
	Total	10140	101.4	38.10	860.0
Potiskum	0-15	1260	12.6	11.97	98.3
	15-20	1240	24.0	7.92	112.3
	30-60	3360	33.6	9.12	149.8
	60-90	2400	24.0	6.72	131.0
	90-120	7200	72.0	4.80	168.5
	Total	16620	166.2	40.53	659.9
Zuga	0-15	2730	27.3	19.95	286.7
	15-30	1920	19.2	29.76	271.4
	30-60	3840	38.4	32.16	673.9
	60-90	3840	38.4	50.40	468.0
	90-120	3840	38.4	23.04	879.8
	Total	16170	161.7	155.31	2578.8
Kafia-Madaki	0-15	2310	23.1	64.89	991.0
	15-.30	2400	24.0	4.56	402.5
	30-60	4800	48.0	27.36	2340.0
	60-90	4320	43.2	173.28	2003.0
	90-120	480	4.8	287.52	2901.6
	Total	14310	142.1	557.61	8638.1

Adapted from Ugbah, 2007.

Soil moisture and soil temperature affect nitrification with optimum soil moisture for nitrification being about 60% of the soil pore space (Brady and Weil, 2002) and at an optimum temperature of 30 ^{0}C in the soil (Gobat et al., 2004). However, inundations resulting from excess rainfall will promote denitrification in soils. Hot dry weathers will favour volatilization losses from the soil. It follows that application of fertilizers especially nitrogenous formulations will be vulnerable to volatilization losses in dry seasons of the tropical Africa. Application of nitrogen containing fertilizers of lower temperatures reduces volatilization (Shi and Norton, 2000). Seasonal variations in climate affect phosphorus immobilization (Salas et al., 2003) and uptake. Chassot and Richner (2002) reported effect of temperature on maize phosphorus uptake, saying that at high temperatures phosphorus uptake increases due to rise in the activity of the meristem.

Management influences fertilizer consumption. Increasing crop yields requires higher fertilizer inputs. Management coordinates soil type, crop type, climate, fertilizer application technique, tillage practices, timing operations, pest and disease control as well as water management in order to optimize yield. Literacy level influences managerial efficiency. Tiamiyu et al. (2002) pointed at high cost and non-availability of farm inputs such as inorganic fertilizer as constraining management factors in Nigeria. Management factor in fertilizer consumption involves a consideration of best management practices which may vary among nutrient types. There are pros and cons of different application strategies. Given a specific field situation, foliar and soil analysis are recommended in order to identify the respective nutritional disorder as the application of nutrients which are non-deficient may aggravate crop's performance. Cultural methods such as proper tillage, drainage, water supply, liming, rootstock or variety selection, planting density, etc., should be employed to stabilize or build up inherent soil fertility. Method of application differs among soil nutrients. In citrus orchards, zinc and manganese chelates tend to be uneconomical and ineffective when applied through the soil. Phytotoxicity is high with soil applications of certain micronutrients such as boron. This is common in citrus orchards on a light-textured soil after prolonged dry seasons. However, fertigation may overcome such obstacles as application of acid fertilizers on calcareous soils improved soil environment in the rhizosphere. This leads to the development of a *sweet zone* which temporarily increases soil nutrient availability (Ristimaki and Papadoipoulus, 2000), thereby increasing uptake of micronutrients applied. Although nutrient uptake via leaves declines with plant age (Marschner, 1995), foliar application is more effective as the rate of nutrient uptake by leaves is by far less than by roots as leaves have very small pores in the leaf cuticle (Pulschen, 2004).

Increased use of fertilizer is encouraged by low fertilizer prices and vice versa. Good responses encourage farmers hence the urge to use more. High cost of inorganic fertilizers as well as other agrochemicals reduce their consumption which may result in lower yields.

SOIL FERTILITY REGENERATION

Declining soil productivity resulting from soil acidity, salinity and alkalinity; and consequent adverse effects on soil macro and micro-mineral nutrients calls for proactive attention towards soil quality improvement. Soil fertility regeneration most consider soil biodiversity since that constitutes agents for soil organic matter build-up. Practices should recognize soil organisms (termites, earthworms, beetles, and so on), physical environment (soil texture, temperature, moisture and soil air) and quality of organic matter in regenerating soil fertility. Organic fertilizers vary and include crop residues, cover crops, night soil, mulch, animal manure and composts. Inorganic fertilizers exist, occurring in single or compound forms. They contain primary macronutrients such as nitrogen (N), phosphorus (P) and potassium (K). Sometimes, they contain secondary macronutrients particularly calcium (Ca) and Magnesium (Mg).

Inorganic fertilizers are expensive while organic fertilizers are of poor nutrient quality. It becomes necessary to combine inorganic and organic fertilizers in soil fertility regeneration. Some farmers in Western Kenya fortify their compost manure with about 2-5 Kg rock phosphate, and sometimes with leaves and stems of wild sunflower (*Thithonia diversifolia*), which decomposes very fast, releasing high levels of nitrogen and potassium, and substantial quantity of phosphorus to soils (IIRR and Act, 2005).

Inorganic fertilizers include nitrogen–containing fertilizers (ammonium sulphate, ammonium nitrate, ammonium phosphates, ammonium chloride, sodium nitrate, calcium ammonium nitrate, urea, anhydrous ammonia, urea-sulphur, potassium nitrate, calcium cynamide, etc), phosphate–containing fertilizers (single superphosphate, triple superphosphate, dicalcuim phosphate, enriched superphosphates ,rock phosphate, basic slag, etc) and potassium–containing fertilizers (nuriate of potash, potassium suphate, potassium magnesium phosphate, potassium nitrate, potassium metaphosphate) (Ahn, 1979). Inorganic calcium sources include calcium oxide, calcium hydroxide, dolomite, calcium carbonate and calcium silicate slag.

Sequestering agents or chelates are popularly used in curing micronutrient deficiencies especially in micronutrients such as iron (Fe), Zinc (Zn), Copper (Cu) and Manganese (Mn). Fertilizer trials and demonstrations are associated with tropical agricultural research. In Nigeria, maize, yam and rice are common test crops while maize, groundnut, rice, yams and guinea are popular test crops in Ghana. Moderate and good crop responses were reported in Gambia where rice is a dominant test crop with trials on millet common in Senegal.

In the organic sources, significant ($P = 0.05$) increases in Ca and Na were recorded in sewage sludge (Mbagwu et al., 2001). Improved organic carbon, total nitrogen and available phosphorus was reported in degraded ultisols of southeastern Nigeria when they were amended with residues from water hyacinth (Oguike and Mbagwu, 2001). Swine slurry contains high amount of nitrogen and phosphorus (Wienhold, 2005). Treated municipal solid wastes contain high amount of nitrogen (Claasen and Carey, 2004). Other organic fertilizer sources include palm bunch refuse ash, bone meal, oyster shells, periwinkle shells, wood ash, night soil, etc .

Because agriculture is a soil-based industry that extracts nutrients from the soil, effective and efficient techniques for slowing down removal will be necessary while nutrients return to the soil should be encouraged in order to sustain agriculture on soils on a long-term basis. Plant growth is the result of a complex process whereby the plant synthesizes solar energy, carbon dioxide and water as well as nutrients from the soil to that extent that 21 to 24 elements are involved (Gruhn et al., 2000).

Soil fertility management or regeneration practices include green manuring, green fallow periods/improved fallows, biomass transfer technology, mulching, agroforestry, farm yard manure, use of municipal solid wastes, composting, night-soil technology, crop rotation, legume intercropping, use of liquid soil conditioners/amendments, legume relay cropping, use of inorganic fertilizers, among others.

Mulches in form of organic materials release nutrients contained therein on decomposition. Stubble mulch tillage attempts to plow without turning under the surface residues. It is a system of management of crop residues that uses subsurface tillage to loosen the soil and at the same time leave the residues on the surface for soil erosion control.

Mulches are primarily meant to control soil temperature, especially insulating the soil against excessive heat and conserving soil moisture. This creates an auspicious soil environment for soil biodiversity, a majority of which play vital roles in nutrients transformations. According to an estimate, 113.6 million Mg of rice and wheat residues containing about 1.90 million Mg of nutrients are available in Indo-Gangetic plains of India (Sarkar et al., 1999) while about 12 million Mg of rice straw burnt annually in India Punjab caused 0.7 million Mg of Nitrogen loss (Singh et al., 2004). This is in addition to emission of 70 % CO_2, 7 % CO, 0.66 % CH_4 and 2.09 % N_2O from burning of rice straw (Samra et al., 2003), gases of which have obvious adverse warming effects on the atmospheric temperature. Shallow incorporation of residues accelerate aerobic decomposition, leading to increased nitrogen availability (Witt et al., 2000) and reduced methane emissions (Wassmann et al., 2000). Apart from retaining high organic matter in the soil, mulches improve utilization of nutrients from chemical fertilizers (Von Scholl 1998). Layers of mulch materials prevent and /or minimize contact between phosphatic fertilizers and soil particles that promote phosphate fixation. However, mulch materials encourage proliferation of termites and other soil fauna which can damage crops.

Green manures supply nutrients to the main crop, improves soil structure, increases organic matter content, protects the soil against degradative forces of wind, water and ice, minimizes dehydration and extreme temperature fluctuations, increases available water capacity and fixes nitrogen if leguminous crops are used in the greening for soil fertility project. Green manuring has been known to improve soil fertility, and reduce nitrogen loss through leaching and soil erosion (Sultani et al., 2007). Green manures increases soil organic matter thereby increasing the distribution and stability of soil aggregates and decreasing soil bulk density (McRae and Mehuya, 1988). Green manuring increases the productivity of the succeeding crop (Carlssom and Huss-Danell, 2003) since leguminous green manures fix nitrogen to the soil (Mayer et al., 2003). Vegetative cover built up by green manures reduce splash erosion and consequent crushing of soils.

Intercropping involving multiple crops on the same soil promotes efficient utilization of plant nutrients. Multiple intercropping includes crops with varying root patterns, which absorb nutrients at various soil horizons, thereby promoting nutrient pump (a situation where plants with deep roots bring nutrients back to the topsoil where they can be used again by shallow rooting crops). *Gliricidia* or *leucaena* with maize are good inter crops.

Crop rotation is practiced, in which legume components such as groundnuts and beans are meant to fix nitrogen in the soil. When the green parts and roots die and decompose, the nutrients released are used by other crops.

Cover crops in addition to protecting soils from degradative forces enrich the soils with nutrients particularly when leguminous crops are used. Members of the cucurbitaceae family provide good cover crops as well as green manures.

Relay cropping technology enhances soil fertility. Interplanting of maize and *sesbania* seedlings concurrently at desired maize population is efficacious in nutrient supply. The results from Makoka Research Station Malawi demonstrated higher maize yields from relay-cropped plots that had been continuously monocropped with maize without inorganic fertilizers (CTA, 2002). Relay cropping is best suited to densely populated areas where farm sizes are too small to accommodate normal rotational fallows.

Bush fallows are natural systems of regenerating poor soils. The total nutrients in circulation include those in the soil and those in the associated vegetation, and both constitute the nutrient capital available for possible use by crops. The improvement in soil fertility and productivity in natural fallows are connected with the increase in soil organic matter content during the fallow period. In the countrysides with low population, bush fallowing may have longer fallow lengths but this may not be possible in urban and peri-urban centres with high demographic pressure as well as other competing land use types. Anthropogenic activities such as deforestation for fuelwood, hunting, mining and other socioeconomic activities have most often disrupted nutrient cycling and efficacy of natural fallows in sustaining crop and rangelands for optimum productivity. These problems coupled with the adverse climate changes (global warming) tend to suggest natural fallows as unsustainable methods cf replenishing soil nutrients.

Table 7. Nutritional value of *Moringa* parts per 100 g edible portion

Nutrient	Unit	Pods	Leaves	Leaf Powder
Calcium	Mg	30.0	440.0	2003.00
Magnesium	Mg	24.0	24.0	368.00
Phosphorus	Mg	110.0	70.0	204.00
Potassium	Mg	259.0	259.0	1324.00
Copper	Mg	3.1	11.0	0.57
Iron	Mg	5.3	7.0	28.20
Sulphur	mg	137.0	137.0	870.00

Adapted from Flugie, (2000).

To overcome the challenges of natural fallows, high quality fallow plants can be planted to regenerate soil fertility at a faster pace. These are often referred to as improved or plant fallows. In Zambia, two to three-year *Sesbania*-based fallow have proved effective in restoring soil fertility (CTA, 2002). Maize grain yields following a 3-year *Sesbania sesban* fallow without N-fertilizer in Chipata, Zambia were 2.27, 5.59 and 6.02 t/ha after 1, 2 and 3 years fallow, respectively compared with the control plots with 1.6, 1.2 and 1.8 t/ha after 1, 2 and 3 years of continuous cropping, respectively (Kwesiga and Chisumpa, 1992). *Moringa oleifera* could serve as fallow tree crop in highly degraded soils following its tolerance to sandiness and drought (Auwalu, 2009) and can grow in alleys with cereals and legumes (Muhamman, 2010).By these attributes ,the multipurpose plant qualifies as a phyto-remediating resource in highly degraded soils of the sub-Saharan Africa. Litter from *Morniga oleifera* can enrich soils tremendously after mineralization given the mineral nutrients composition.

Mixed farming is traditionally practiced, where droppings from farm animals are used to enrich soils for crop production. Leys are grazed for one or two seasons by livestock, after which they are ploughed for crop production. The soil will have been enriched with urine and faeces from grazing animals. Manure storage in deep litter bedding pack under cover that is refreshed daily to maximize nutrient retention.

Crop residue, weeds, farm and kitchen sweepings and animal droppings can be rotted down in stacks or pits to make compost. Composting of bedding packs to a proper carbon-nitrogen ratio to maximize organic matter production, nutrient stabilization and retention is a good practice. Composting promotes nutrient cycling in soils.

Table 8. Some macronutrient elements in fauna-modified and non-modified soils

Soil Type	N (%)	P K (mg//kg)		Ca	Mg cmol/kg	Na
Earthworm cast	0.31	24.39	0.17	14.20	7.20	0.15
Nasute termite	0.30	12.17	0.38	5.00	3.40	0.33
Macrotermes termite	0.05	7.50	0.32	3.00	1.00	0.30
Ordinary soil	0.16	12.34	0.20	2.40	0.90	0.16
$LSD_{0.05}$	0.05	3.10	NS	10.70	4.39	NS

Adapted from Asawalam and Johnson (2003).

Table 9. Some micronutrient elements (ug/g) in fauna-modified and no-modified soil

Soil type	Zn	Cu	Mo	Mn	Fe
Earthworm cast	96.54	68.33	223.32	675.44	1280.6
Nasute termite	99.17	47.92	20.90	62.78	1292.7
Macrotermes termite	92.64	49.17	8.48	25.56	1271.9
Ordinary soil	104.86	47.92	12.99	37.78	1327.5
$LSD_{0.05}$	NS	15.39	24.47	83.64	NS

Adapted from Asawalam and Johnson, (2003).

Table 10. Nutrient content of sewage sludges from 10 different waste water treatment plants

Sewage sludge	pH water	C %	N %	CN	P %	Ca %	K Mg mgkg	
1	4.45	48.21	1.67	28.87	0.238	0.627	1205	445
2	3.92	6.13	0.68	9.03	0.321	0.422	1021	521
3	6.43	32.10	1.54	20.84	0.410	1.17	941	1690
4	3.92	11.15	2.70	4.13	0.652	0.323	462	278
5	3.57	56.67	1.52	37.94	0.821	0.160	401	389
6	5.61	41.36	2.27	18.22	0.778	1.12	1209	112
7	5.24	37.41	2.82	13.26	0.634	0.832	728	870
8	5.89	18.28	2.65	6.90	1.20	2.16	621	2062
9	6.02	33.35	2.90	11.50	1.62	0.926	972	2902
10	5.72	41.23	2.64	15.62	0.470	0.640	678	1575
mean	5.08	32.59	2.14	16.63	0.714	0.838	824	1084

Adapted from Rosenani et al. (2008).

Compositing can be mediated by earthworms, a process often referred to as vermin-composting. Earthworms rework soils and injest them leading to the formation of casts. Soils modified by earthworms and termites gave better crop performance (Asawalam and Hauser, 2001). Earthworm casts contain higher concentrations of copper, molybdenum and manganese than ordinary soils (Asawalam and Johnson, 2003). Earthworm(*Pontoscolex corethrurus)* bioremediated physical chemical and biological soil properties and improved soil fertility in tropical soils (Mba, 1994).

Night-soil technique where human wastes are composted brings benefits to soil resource. People decline the use of human excreta on account of hygiene and lack of workforce for the task. In addition to this, there is concern for social status in the use of night-soil in soil fertility restoration and enhancement. Dry toilets are converted to night-soil compost. A rectangular hole is dug in the toilet floor and the night-soil drops down to the ground floor, where composting takes place. During composting, water is disallowed and after defecating, the faeces are covered with other materials (*fot*) which enriches the compost. Fot includes dry cattle dung, ash, dry grass and leaves. The night soil compost is stored in vaults for about six months. The composting can be done in different vaults for continuous composting and supply. It is usually used in the night due to social stigma attached to it. Night-soils increase available plant nutrients and organic matter content which promotes macroaggregation. Keeping and ageing the night-soil decreases carbon-nitrogen ratio during ageing, eliminates harmful substances accumulated at first stage of decomposition and reduces nutrient loss by volatilization and runoff.

Municipal wastes in form of refuse and sewage are becoming very popular in urban and peri-urban centres especially in horticultural and floricultural plants. Refuse (municipal solid wastes) are composted to influence soil chemical properties such as soil pH (Aggelides and Londa, 2000; Martinez et al., 2002; Claudio et al., 2007). Municipal solid wastes supplies soil nutrients (Myungou et al., 2006) and increases soil organic matter (Hangreaves et al., 2008) and cation exchange capacity (Ito, 2006). Application of municipal solid wastes to soils increases soil nitrogen, phosphorus, potassium, sulphur, calcium, magnesium, sodium, iron and copper (Claassen and Carey, 2004). Municipal waste water sewage (biosolids) are applied to soils to provide mainly nitrogen phosphorus and potassium (Kulling, et al., 2001).

Tropical soils are improved and nutrient content enhanced using sewage sludge. There were significant ($p<0.05$) difference between organic matter content, total nitrogen, exchangeable calcium and cation exchange capacity among sewage and no-sewage treated Arenic Kandiustult of Southeastern Nigeria (Mbagwu et al., 2001).

Agroforestry are known to conserve soil nutrients. Most agroforestry practices are multi-purpose in function, and include increase organic matter, nitrogen fixation, nutrient pump in its extensive root system and shade for livestock. Agroforestry systems involves processes which augment additions to the soil, processes which reduce losses from the soil, processes which improve soil physical conditions, processes which improve chemical properties and processes which relate to the quality of plant residues and the timing of their transfer to the soil.

Agroforestry practices which promote soil fertility regeneration and soil conservation include improved tree fallow, trees on cropland, plantation crop combinations, multistorey tree gardens, hedgerow intercropping (alley cropping), trees on erosion control structures, windbreaks and shelterbelts, trees on pastures, etc. The nitrogen-fixing trees such as Acacia, Casuarina, Erythrina, Gliricidia, Inga and Prosopsis are capable of fixing 50 – 200 kg N/ha/yr while Leucaena leucocephala fixes 500 kg N/ha/yr (Dommergues, 1987). Soil fertility maintenance and improvement can be achieved substantially for sustainable crop production through agroforestry (Young, 1989).

Farmers in Southern Africa apply plant biomass as fertilizer to soils. Biomass transfer technology involves applying leguminous trees and shrubs biomass to soils as a source of plant nutrients.

It is established in situ in hedges or woodlot, and Leucaena biomass at 1 to 5 t/ha improved yields of maize in sandy soils of Tabora, Tanzania (CTA, 2002). Among the Bemba of Northern Zambia, the chitemene system is practiced involving collection of tree biomass and burning into heaps, followed by cultivation of maize or millet on resulting calcium – and potassium-rich soils (CTA, 202). In Masvingo province in Zimbabwe, farmers collect plant litter from woodlots which is applied on very sandy soils to sustain soil fertility (Nyathi and Campbell, 1994), Sesbania sesban Indigofera errata, Senna siamea, Glincidia sepium and Leucaena leucocephala are suitable for biomass transfer technology.

Table 11. Some properties of soils affected by sewage application

Horizon	Depth (cm)	Ca	Mg Cmol/kg	K	Na	Exchangeable acidity	Cation exchange capacity
S/NSK/1 (Sewage Soil)							
Ap	0-15	1.8	1.0	0.11	0.12	2.6	7.5
AB	15-35	2.8	2.0	0.23	0.18	5.6	8.0
Bt_1	35-55	1.0	0.8	0.09	0.09	3.8	4.0
Bt_2	55-105	0.8	0.6	0.06	0.06	4.0	4.0
Bt_3	105-160	0.9	0.6	0.06	0.06	4.0	5.6
Mean		1.5	0.8	0.11	0.10	4.0	5.8
S/NSK/2 (Sewage Soil)							
Ap	0-15	.2	1.2	0.12	0.16	3.2	9.0
AB	15-35	2.6	3.0	0.16	0.20	2.4	4.0
Bt_1	35-65	1.2	0.6	0.09	0.13	1.6	6.5
Bt_2	65-80	2.2	1.0	0.14	0.09	9.4	6.0
Bt_3	80-160	1.0	0.8	0.05	0.06	4.4	6.7
Mean		1.9	1.3	0.11	0.13	4.2	6.0
NS/NSK (Non-Sewage Soil)							
Ap	0-14	1.8	0.8	0.09	0.10	3.6	6.0
AB	14-35	1.6	0.8	0.12	0.12	3.0	6.0
Bt_1	35-76	1.2	0.6	0.08	0.09	3.6	4.5
Bt_2	76-90	1.0	0.6	0.08	0.08	3.2	4.0
Bt_3	90-160	0.8	0.8	0.05	0.06	2.8	4.0
Mean		0.6	0.6	0.08	0.09	3.2	4.9

Adapted from Mbagwu et al. (2001).

Table 12. Yields of maize at Msekera, Zambia in response to fertilizer and different fallows (Kwesiga and Beniest, 1998)

Treatment	Maize yield (ton/ha)
Maize with fertilizer	3.96
2-year *Sesbania sesban* fallow	5.36
1-year *Sesbania sesban* fallow	3.43
2-year *Tephrosia vogelli* fallow	3.18
2-year *Sesbania macrantha* fallow	2.97
1-year *Tephrosia vogelli* fallow	2.80
2-year *Cajanus cajan* fallow	2.78
1-year *Cajanus cajan* fallow	2.40
1-year *Sesbania macrantha* fallow	2.07
Groundnut-maize rotation	1.87
Grass fallow	1.84
Maize without fertilizer	1.09

Table 13. Effect of fallow species and fallow duration on soil fertility

Fallow	Duration of fallow (YAP)								
	1	2	Mean	1	2	Mean	1	2	Mean
	P(mg/kg)			K(cmol/kg)			N(%)		
Aeschynomene histrix	27.5	21.4	24.5	0.11	0.11	0.11	0.12	0.06	0.09
Centrosema	31.4	37.0	34.2	0.11	0.08	0.10	0.07	0.09	0.08
brasilianum	31.4	26.9	29.2	0.10	0.14	0.12	0.09	0.09	0.09
Chamacrista	33.7	31.7	32.7	0.10	0.09	0.10	0.11	010	0.11
rotundifolia	35.2	24.0	29.6	0.09	0.11	0.10	0.11	0.08	0.09
Centrosema	61.5	50.6	56.1	0.19	0.16	0.18	0.07	0.07	0.07
pascuorum	37.9	19.9	28.9	0.12	0.10	0.11	01.	0.09	0.10
Mucuna georgia	34.8	24.9	29.9	0.09	0.09	0.09	0.11	0.08	0.10
Mucuna ghana	36.0	25.1	30.6	0.11	0.11	0.11	0.08	0.07	0.08
Pueraria phaseoloides	36.6	29.1	32.9	0.11	0.11	0.11	0.10	0.08	0.09
Stylosanthes guanensis									
Natural fallow									
Mean									

$LSD_{0.05}$ for 2-year fallow species (T) Means = $\frac{P}{6.27}, \frac{K}{0.044}, \frac{N}{NS}$

$LSD_{0.05}$ for 2-year fallow duration (D) Means = $\frac{P}{13.31}, \frac{K}{NS}, \frac{N}{0.014x}$

Adapted from Okpara, et al. (2009).

The use of chemical fertilizers is still popular especially in sub-Saharan Africa since the nutrients are readily available to crops. High cost of inorganic fertilizers is a major constraint facing resource-power farmers especially in Africa. There are different types, each with its peculiar characteristics. Fertilizer use is often governed by not only socioeconomic factors but edaphic, climatic, crop and institutional factors. Low use of mineral fertilizer is related to soil degradation which is a threat to food security in sub-Saharan Africa (De Jager et al., 2003). Some scholars (Smaling et al., 1996; Mbagwu et al., 2001) suggest a combination of organic sources of nutrients with chemical fertilizers for enhancing crop performance. While organic sources may not provide readily available plant nutrients compared with inorganic fertilizers, they improve soil physical properties especially macroaggregation.

Such combinations tend to be very desirable given heightened soil nutrient mining in Africa. Defoer (2003) reported negative nutrient balances suggesting the use of resource flow mapping (RFM) and ResourceKit as decision support tools.

Fertilization of soils using inorganic fertilizer is done by placement, bedding, foliar application, row application, broadcast, top dressing, etc. However, liquid fertilizers have assumed considerable importance in advanced countries. Some liquid fertilizers are applied to young vegetables at the time of planting, and these solutions are referred to as *starter solutions*. The liquid fertilizers reach the plant roots immediately unlike solid fertilizer .Some are applied via leaves (foliar spray), and nutrients are absorbed through the leaves of such plants. Fertilizers are applied through irrigation water, a process known as FERTIGATION. It is a simple and efficient way of supplying mineral nutrients especially nitrogen and potassium to the soils. Such fertilizers should be injected during the last half of the irrigation to minimize leaching of nutrients. However, some nutrients like phosphorus tend to precipitate out and block the orifices of irrigation emitters hence unsuitable in fertigation. Performance of fertilizers whether organic or inorganic can be improved using bioactivators (which include inducers of self- defence in crops), biostimulants as well as amino acids (from synthesis and hydrolysis). Other nutrient enhancement techniques include the application of lime in acidic soils with the consequent release of nutrients which were fixed and difficultly available to plants. Liming is also necessary when acidic fertilizers are applied in less acidic soils to reduce acidification.

There is a relationship between bioavailability of nutrients and uptake by crops. In contaminated soils, pH is a crucial factor especially in the extractability of cadmium. Under low pH, cadmium become available but if such acidic soils are limed with calcium carbonate raising the pH to say 7.0 to 7.3, less cadmium is extracted by plants. Application of zeolite significantly reduces the solubility of lead. The use of lime materials, organic manures increases soil pH and reduces solubility of toxicants. At slightly acidic to neutral pH, most soil nutrients become available for uptake by crops.

Conclusions

Soil nutrients remain vital in agriculture. Optimum supply of soil nutrients is necessary for sustainable agriculture for food and nutrition security. Soil nutrients vary spatially and temporally but are taken by diffusion, mass flow and root interception. Soil nutrients in the tropics are highly depleted due to soil nutrient mining. Fertilizer consumption is low but influenced by economic, management,institutional, crop, edaphic and climatic factors Soil restoration technologies should consider both traditional and modern methods, and indeed a blend of both which could be necessitated by local circumstance.

There is need for assessing nutrient content of soils, nutrient balance and flows for the purpose of developing a soil fertility capability map. This map will help in developing appropriate decision support tools for each locality since soils are of diverse properties requiring a consideration of such uniqueness in precision agriculture for food and nutrition security in a period of changing atmosphere.

References

Aggelides, S.M. and Londa, P.A. (2000). Effects of compost produced town wastes and sewage sludge on the properties of a loamy and clay soil. *Biores. Technol.*, 71(3):252-259.

Ahn, P.M. (1979). *West African soils*. Oxford University Press, Oxford, Great Britain.

Ajari, O.J., Tsado, E.K., Oladiran, J.A. and Salako, E.A. (2002). *Effects of fertilizer and pre-harvest fungicided application on plant height of four okra vonietics at Bida, Nigeria.* Proceedings of the 35th Annual Conference of Agricultural Society of Nigeria at UNAAB, Abeokuta Nigeria, 16 – 20th September, 2002. pp. 64-66.

Alexander, M. (1977). *Introduction to soil microbiology second edition*. John Wiley and Sons, New York.

Ano, A.O. and Orkwor, G.C. (2006). Effect of fertilizer and intercropping with pigeon pea (Cajanus cajan) on the productivity of yam minisett (Dioscorea rotundata) based system. *Niger. Agric. J.*, 37:65-73.

Asawalam, D.O. and Hauser, S. (2001). Effects of earthworm casts on crop production. *Trop. Agric.*, 78(2):76-81.

Asawalam, D.O. and Johnson, S. (2003). *Macro – and Micronutrient characteristics of soils modified by earthworms and termites.* Proceedings of the 28th Annual Conference of Soil Science Society of Nigeria at NRCRI, Umudike, Nigeria, 4 – 7 November, 2003 pp. 147 – 152.

Auwalu, B.M. (2009). *Agronomic management practices for commercial production of moringa* (moringa oleifera lam). Paper presented at a 2-day workshop on sustainable production and commercialization of moringa oleifera lam at Imam Wali Conference Hall, Kano, 16-17 November 2009.

Bassirad, H. (2000). Kinetics of nutrient uptake by roots: responses to global change. *New Phytoll,* 147:155-169.

Bationo, A., Lompo, F. and Koala, S. (1998). Research on nutrient flows and balances in West Africa: State of the art. *Agric. Ecosyst. Environ.*, 71:19-35.

Baver, L.D., Gardner, W.H. and Gardner, W.R. (1972). *Soil physics 4th edition.* John Wiley and Sons Inc, New York.

Brady, N.C. and Weil, R.R. (1999). *The nature and properties of soils.* Twelveth edition. Prentice Hall, Upper Saddle River, USA.

Brady, N.C. and Weil, R.R. (2002). *The nature and properties of soils.* 13th edition. Prentice Hall Inc. New Jersey.

Brubaker, S.C., Jones, A.J., Lewis, D.T. and Frank, K. (1993). Soil properties associated with landscape position. *Soil Sci. Soc. Am. J.,* 57:235-239.

Carlsson, G., and Huss-Danell, K. (2003). Nitrogen fixation in perennial forage legumes in the field. *Plant Sci.,* 253:353-372.

Chassot, A. and Richner, W. (2002). Root characteristics and phosphorus uptake of maize seedlings in a bilayered soil. *Agron. J.,* 94:118-127.

Claasen, V.P. and Carey, J.L. (2004). Regeneration of nitrogen fertility in disturbed soils using composts. *Compost Sci. Util.,* 22(2):180-186.

Clarkson, D.T. and Warner, A.J. (1999). Relationship between roof temperature and transport of amounuim and nitrate ions by Indian and perennial ryegrass (Lolium multifoluim and lolium perenne). *Plant Physiol.,* 64:557-561.

Claudio, P.I., Nascentes, C.C., Fontes, R.L.F., Cecon, P.R. and Pereira, J.L. (2007). Effect of composted urban solid waste addition on yield and metal contents of lettuce. *J. Brazi. Chem. Soc.,* 18(1):347-361.

Conlm, T.S. and VenDen Driessche, R. (1996). Short-term effects of soil compaction on growth of Pinus contorta seedlings. *Can. J. For. Res.,* 26:727-739.

CTA (*Technical Centre for Agricultural and Rural Cooperation*) (1993). Sustaining soil productivity in intensive African agriculture. Seminar Proceedings on Sustaining Soil Productivity in Intensive African Agriculture edited by Ahenkorah, Y., Owusu – Bennoah, E. and Dowuona, G.N.N., Accra, Ghana, 15-19 Nov. 1993.

CTA (*Technical Centre for Agricultural and Rural Cooperation*) (2002). Agroforestry in Malawi and Zambia. Summary Report of a CTA/MAFE Study Visit, 20-30 Jan., 2002. Wagennigen, The Netherlands. 32pp.

Dachi, S.N. and Kwon-Ndung, E.H. (2007). Response of acha (Digitaria exilis stapf) to NPK fertilizer at Badeggi, Southern Guinea Savannah in Nigeria. *Prod. Agric. Techn.,* 3(1):131-146.

De Jager, A. Onduru, D. and Walaga, C. (2003). *Using NUTMON to evaluate conventional and low external input farming practice in Kenya and Uganda.* In: Struif-Bontkes, T.E. and Wopereus, M.C.S. (eds.). decision support tools for smallholder agriculture in Sub-saharan Africa: A practical guide. IFDC and CTA, Wagenrrigen, The Netherlands. Pp. 40-53.

Defoer, T. (2003). *Assessing changes in soil fertility management in Southern Mali using resource flow mapping and resource kit.* In: Struif-Bontkes T.E. and Wopereis, M.C.S. (eds.). Decision support tools for smallholder agriculture in Sub-saharan Africa: A practical guide. IFDC and CTA, Wagenmgen, The Netherlands. Pp. 24-39.

Dinnes, D.L., Douglas, L.K., Jaynes, D.B., Kasper, T.C., Hartfield, J.L., Colvin, T.S. and Cambardella, C.A. (2002). Nitrogen management strategies to reduce nitrate leaching in tile-drained Midwestern Soils. *Agron. J.,* 94:153-171.

Dommergues, Y.R. (1987). The role of biological nitrogen fixation in agroforestry. In: Steppler, H.A. and Nair, P.K.R. (eds.): Agroforestry: A decade of development. Nairobi, Kenya. *ICRAF*. Pp. 245-271.

Enwezor, W.O., Ohiri, A.C., Opowaribo, E.E. and Udo, E.J. (1990). *A review of soil fertilizer use in crops.* 1: Southeastern zone of Nigeria. Federal Ministry of Agriculture and Natural Resources, Lagos. Pp. 144-158.

FFDD (Fertilizer Procurement and Distribution Division) (1989). *Fertilizer use and management practices for crops in Nigeria.* The FPDD, Federal Ministry of Agriculture Water Resource and Rural Development Lagos. 163pp.

Flugie, L.J. (2000). *New uses of Moringa*, studies in Nicaragua ECHO Development Notes, http://www.echotech.org/network/mdules.php.

Follete, R.F. (2001). Soil management concepts and carbon sequestration in cropland soils. *Soil Till. Res.,* 61:77-92.

Foth, H.D. (1984). *Fundamentals of soil science seventh edition.* John Wiley and Sons Ltd. New York.

Gessler, A., Schneider, S., VonSengbusch, D., Weher, P., Hanemann, W., Huber, C., Roche, A. Krentzer, K., and Rennenberg, G. (1998). Field and laboratory experiments on net uptake of nitrate and ammonium by the roots of spruce (Pices abies) and beach (Fagus sylvaticus) trees. *New Phytologist*, 147:13-31.

Gillman, G.F., Shamshuddin, J. and Bell, L.C. (1989). *Soil chemical parameters and organic matter in soil management.* Proceedings of a workshop organized by the International Board for Soil Research and Management (IBSRAM), Bangkok, Thailand. Pp. 141-153.

Gobat, J.M., Aragno, M.and Matthey, W. (2004). *The living soil: Fundamentals of soil science and microbiology*. Science Publications Inc., Enfield (NH) USA. Pp. 337-364.

Gruhn, P. Gdetti, F. and Yudelman, M. (2000). *Integrated management, soil fertility and sustainable agriculture*: Current issues and future challenges. IFPRI, Washington DC. 31pp.

Hangreaves, J.C., Adi, M.S. and Warman, P.R. (2008). A review of the use of composted municipal solid waste in agriculture. *Agric. Ecosyst. Environ*., 123:1-14.

Haws, N.W., Lin, B., Boast, C.W., Rao, P.S.C., Kladivko, E.J. and Franzmeir, D.P. (2004). Spatial variability and measurement scale of infiltration rate on an agricultural landscape. *Soil Sci., Soc. Am. J.,* 68:1818-1826.

Hobbie, S.E. (1996). Temperature and plant species control over litter decomposition in Alaskan tundra. Ecol. Monogr., 66:503-522.

Igwe, C.A. (2003). *Soil degradation response to soil factors in Central Eastern Nigeria*. In: Proceedings of the 28th Annual Conference of Soil Science Society of Nigeria held at Umudike, Nigeria, 4-7 Nov. 2003. pp. 228-234.

IIRR (International Institution of Rural Reconstruction) and ACT (African Conservation Tillage Network). (2005). Conservation agriculture. *A manual for farmers and extension workers I Africa.* Majestic Printing Works Ltd., Nairobi Kenya.

Ingrid, W., Martinez, F. and Cueva, G. (2006). Plant and Soil responses to the application of composted MSW in a degraded, semi-arid shrub land in Central Spain. *Compost Sci., and Util.*, 14(2):147-154.

Jiang, Z. and Sullivan, W.M. (2004). Nitrate uptake of seedling and mature Kentucky bluegrass plants. *Crop Sci.,* 44:567-574.

Jiang, Z., Sullivan, W.M. and Hull, R.J. (2000). Nitrate uptake and nitrogen use efficiency by Kentucky bluegrass cultivars. *Hort. Sci.,* 35:1350-1354.

Jonnes, D.L. and Shannon, D. (200). Mineralization of amino acids applied to soils. *Soil Sci. Soc. Am. J.*, 63:1199-1206.

Kogbe, J.O.S. and Adediran, J.A. (2003). Influence of nitrogen phosphorus and potassium application on the production of maize in the zone of Nigeria. *Afr. J. Biotechnol.,* 2(10):345-349.

Kosmas, C., Danalatos, N., Moustakas, N., Tsatiris, B., Kallianou, C., and Yassoglou, N. (1993). The impact of parent material and landscape position and drought and biomass production of wheat under semi-arid conditions. *Soil Technol.*, 6:337-349.

Kullnig, D., Stadelmann, F. and Herter, U. (2001). Sewage sludge – fertilizer or waste? *UKWIR Conference*, Brussels.

Kwesiga, F. and Beniest, J. (1998). *Sesbania improved fallows for eastern Zambia*. An extension guide. ICRAF, Nairobi, Kenya. 57pp.

Landon, J.R. (1984). Booker tropical manual: A handbook for soil survey and agricultural land evaluation I the tropics and subtropics. Longman Inc. New York.

Li, Y., Tian, G., Lindstrom, M.J. and Bork, H.R. (2004). Variation of surface soil quality parameters by intensive donkey – drawn tillage on steep slope. *Soil Sci. Soc. Am. J.,* 68:907-913.

Lindstrom, M.J. Nelson, W.W. and Schumacher, T.E. (1992). Quantifying tillage erosion rates due to moldboard plowing. *Soil Till. Res.*, 24:243-255.

Ma, B.L., Dwyer, L.M. and Gregorich, E.G. (1999). Soil nitrogen amendment on nitrogen uptake and grain yield of maize. *Agron. J.,* 91:650-656.

Maduff, J.H., Hopper, M.J. and Wild, A. (1987). The effect of temperature on growth and uptake of ammonium and nitrate by Brassica napus L.W. Bien venu in flouring solution culture. *J. Expt. Bol.,* 39:53-66.

Mainville, N., Webb, J., Lucotte, M., Davidson, R., Betanconrt, O., Cueva, E. and Mergler, D. (2006). Decrease of soil fertility and release of mercury following deforestation in the Andean Amazon, Napo River Valley, Ecuador. Sci. Total Environ., 368:88-98.

Malpassi, R.N., Kaspar, T.C., Parkin, T.B. and Cambardella, C.A. (2008). Nitrogen management in leaching tile. *Soil Bio. Biochem.*, 39(6):1277-1283.

Marschner, H. (1995). *Mineral nutrition of higher plats*. 2nd ed., Academic Press, Amsterdam.

Marshner, H., Haussling, M. and George, E. (1991). Ammonium and nitrate uptake rates and rhizosphere pH in non-mycorrhizal roots of Nonway Spruce (pices abies L.) Karst. *Trees,* 5:14-21.

Martinez, F., Gabriela, C., Teresa, I. and Ingrid, W. (2002). *Urban organic wastes effects on soil chemical properties in a degraded semand ecosystem*. The 17th WCSS, Thailand, 14 – 21 August 2002. Paper No. 2103-1 to 2103-9.

Mayer, J., Buegger, F., Jensen E.S.,and Scholter, M. (2003). Residual nitrogen contribution from grain legumes to succeeding wheat and rice and related microbial process. *Plant Soil,* 255:541-554.

Mba, C.C. (1994). Rock phosphate solubilizing and cellulolytic actinomycete isolates of earthworm casts. *Environ. Manage*., 18(2):257-261.

Mbagwu, J.S.C. (1992). Improving the productivity of degraded ultisols in Nigeria using organic and inorganic amendment. Part 1. Chronical properties and maize yield. *Biores. Echnol.,* 42:149-154.

Mbagwu, J.S.C., Oti, N.N., Agbim, N.N. and Udom, B.E. (2001). Effects of sewage sludge application on selected soil properties and yield of maize and bambara groundnut. *Agro-Sci.,* 2(2):37-43.

McClain, M.E., Boyer, E.W., Dent, C.L., Gorgel, S.E., Grimin, N.B., Groffrnan, P.M., Hart, S.C., Harvey, J.W., Johnston, C.A., Mayorga, E., McDowell, W.H. and Pinay, G. (2003). Biogeochemical hotspots and hotspots and hot moments of the interface of terrestrial and aquatic ecosystems. *Ecosystems*, 6:301-312.

McRac, R.J. and Mehuya, G.R. (1988). The effect of green manuring on the physical properties of temperate – area soils. *Adv. Soil Sci*., 3:71-94.

Mengel, D. (1995). *Roots, growth and nutrient uptake*. Department of Agronomy Publication AGRY – 95 – 08, Revised May 1995, Purdue University, West Lafayette, IN.

Merrill, S.D., Tanaka, D.L., Krupinsky, J.M. and Ries, R.E. (2003). Soil water depletion and coverage by residue under sunflower compared to other diverse crops. In: Proceeding of the 25th Sunflower Research Workshop, 16-17 Jan. 2003. *National Sunflower Association,* Fargo ND.

Moroke, T.S., Schwartz, R.C., Brown, K.W.and Juo, A.S.R. (2005). Soil water depletion and root distribution of three dry land crops. *Soil Sci. Soc. Am. J.,* 69:197-205.

Morris, S.J. and Boerner, R.E.J. (1998). Landscape patterns of nitrogen mineralization and nitrification in Southern Ohio hardwood forests. *Landscape Ecol.,* 13:215-224.

Mueller, E.N., Wainright, J. and Parsons, A.J. (2008). Spatial variability of soil and nutrient characteristics of semi-arid grasslands and shrublands, Jornada Basin, New Mexico. *Ecohydrol.*, 1:3-12.

Muhamman, M.A. (2010). Problems and prospects of cultivating Moringa (Moringa oleifera lam) for consumption and industrial uses in Nigeria *Proceedings of the 24th Annual National Conference of Farm Management Mubi, Nigeria,* 11 – 14th October, 2010. pp. 242-247.

Myungou, P.S., Olayvanbi, S., Wansik, K., Eunhee, C., Tuongbae, C. and Togmini, S. (2006). Effects of long term composted and fertilizer application on soil phosphorus status under paddy cropping systems. *Commun. Soil Plant Anal.,* 33(11 and 12):1635-1640.

Neama, A.and Singer, A. (2004). The effects of palygorskite on chemical and physicochemical properties of soils: *A review. Geodema* 123(3):297-303.

Nyathi, P. and Campbell, B. (1994). Lead quality of Sesbania sesban, Leucaenia leucocephala and Brachystegia spiciformis: potential agroforestry species. *For. Ecol. Manage.*, 64:259-264.

Oguike, P.C. and Mbagwu, J.S.C. (2001). Effects of hyacinth residues on chemical properties and productivity of degraded tropical soils. *Agro-Sci.*, 2(2):44-51.

Okpara, D.A., Udo, A.A., Iroegbu, M. and Ikeorgu, J.G. (2009). *Effect of fallow species and duration o soil fertility, biomass and white yam (Dioscorea rotundata poir) production.* Proc. of the 5th National Conference of Org. Aquc. Project in Tertiary Institutions in Nigeria, at FUTO, 15 – 19 Nov. 2001.

Onweremadu, E.U. (2006). Assessment of mined soils in erosion-degraded farmlands in Southeastern Nigeria. *Estud. Biol.*, 28(65):59-67.

Onweremadu, E.U. (2009). Magnesium content of two soil groups in Southeastern Nigeria in relation to selected pedological properties. *American – Euroasian J. Sust. Agric.*, 3(3):481-486.

Onweremadu, E.U.and Akamigbo, F.O.R. (2007). Spatial changes in the distribution of exchangeable cations in soils of a forested hilly lands. *Res. J. For.* 1(2):55-65.

Onweremadu, E.U. and Uhuegbu, A.N. (2007). Pedogenesis of calcium in degraded tropical rangeland soil. *J. Am. Sci.*, 3(2):23-29.

Onweremadu, E.U., Duruigbo, C.I., Onyia, V.N., Dialoke, S., Ogoke, J., Tom, C.T., Ibe, A.E., Ihejirika, G.O. and Obilo, O.P. (2005). Relative contribution of calcium waste to the nodulation of cowpea in an ultisol of Owerri agricultural zone, Nigeria. *Int. J. Enviro. Health Hum. Ev.*, 6(1):8-11.

Osemwofa, I.O., Omueti, J.A.I. and Ogbogbodo, A.I. (2003). Effects of Ca/Mg ratio in the soil on magnesium availability, yield and yield components of maize (Zea mays L.). Proceedings of the 28th Annual Conference of Soil Science Society of Nigeria Umudike, 4-7 November, 2003. pp. 91-98.

Othman, M.J. and Pope, N.V. (2000). Nitrogen fertilizer movement I the soil as influenced by nitrogen rate and timing in irrigated wheat. *Soil Sci. Soc. Am. J.*, 64:1883-1892.

Oti, N.N. (2002). Discriminant functions for classifying erosion degraded lands at Otamiri, Southeastern Nigeria. *Agro. Sci.*, 3:34-40.

Pulschen, L. (2004). *Application of micronutrients: Pros and cons of the different application strategies.* IFA International Symposium on micronutrients, 23 – 25 February 2004, New Delhi, India. Kemira GrowHow, Denmark. 13pp.

Rabens, G., Andreae, G.H. and Meyer-Heisig, M. (2000). Longterm acid load and consequence in forest ecosystems in Saxony Germany. *Water, Air, Soil Pollut.*, 122:93-103.

Ristimaki, L. and Papadopoalcs, I. (2000). *A comparison of the efficiency of conventional soil phosphorus application and phosphorus fertigation.* In: Ryan, J. (ed.) plant nutrition management under pressurized irrigation systems in the mediterianean region. ICARDA, Aleppo, Syria. Pp. 260 – 268.

Rosenani, A.B., Kala, D.R. and Fauziah, C.I. (2008). Characterization of Malaysian sewage sludges and nitrogen mineralization in three soils treated with sewage sludge. *Malay. J. Soil Sci.*, 12:103-112.

Rozas, H.S., Echerriva, H.E., Studdert, G.A. and Andrade, F.H. (1999). No-till maize nitrogen uptake and yield: Effect of urease inhibitor and application time. *Agron. J.*, 91:950-955.

Sarkar, A., Yadav, R.L., Gangwar, B., and Bhatia, P.C. (1999). Crop residues in India. Technical Bulletin. Project Directorate for Cropping Systems Research Modipuram, India.

Science Daily (2007). New mechanism for nutrient uptake discovered. Carnegie Institution, Eurek Alert http//www.sciencedaily.com/releases/2007/02/07Q200654htm.

Seybold, C.A., Grossman, R.B., Hoper, H., Muckel, G. and Karlen, D. (2004). Soil quality morphological index measured in the 1996 NRI pilot study. *Soil Surv. Horiz.,* 45:86-95.

Sharpley, A.N., Kleinman, P.J.A., Heathwaite, A.L., Ghurek, W.J., Weld, J.L. and Folniar, G.J. (2008). Integrating contributing areas and indexing phosphorus loss from agricultural watersheds. *J. Environ. Qual.,* 37:1488-1496.

Shi, W. and Norton, J.M. (2000). Effect of long term biennial fall-applied anhydrous ammonia and nitrapyrin on soil nitrification. *Soil Sci. Soc. Am. J.,* 64:228-234.

Siegel-Issem, C.M., Burger, J.A., Powers, R.F., Ponder, F. and Patterson, S.C. (2005). Seedling root growth as a function of soil density and water content. *Soil Sci., Soc., Am., J.,* 69:215-226.

Singh, Y., Singh, B., Ladha, J.K., Khind, C.S., Khera, T.S. and Bucno, C.S. (2004). Effects of residue decomposition on productivity and soil fertility in rice – wheat rotation. *Soil Sci. Soc. Am. J.,* 68:854-864.

Smaling, E.M.A. Fresco, L.O. and De Jager, A. (1996). Classifying and monitoring soil nutrient stocks and flows in African agriculture. *Ambio,* 25:492-496.

Smint, D.R. and Bloom, A.S. (1991). Influence of root NH4- and NO3- content on the temperature response of net NH4+ and NO3- uptake in chilling sensitive and chull. resistant..

Soon, Y.K., Clayton, G.W. and Rice, W.A. (2001). Tillage and previous crop effect on dynamics of nitrogen in wheat-soil system. *Agron. J.,* 93:842-849.

Stoorvogel, J.J., Smalmg, E.M.A. and Janssen, B.H., (1993). Calculating soil nutrient balances in Africa at different scales. *Fert. Res.,* 35:227-335.

Stutter, M.I., Decks, L.K. and Billet, M.F. (2004). Spatial variability in soil ion exchange chemistry in granitic upland catchment. *Soil Sci. Soc. Am. J.,* 68:1304-1314.

Sultani, M.I., Gill, M.A., Anwar, M.M. and Athar, M. (2007). Evaluation of soil physical properties as influenced by various green manuring legumes and phosphorus fertilization under ramified conditions. *Int. J. Environ. Sci. Techn*., 4(1):109-118.

Sumra, J.S., Singh, S. and Kumar, K. (2003). *Managing crop residues in the rice-wheat system of the Indo-Gangeti Plain.* In: Ladha, J.K. (ed.) Improving the productivity and sustainability of rice – wheat systems: Issues and impact ASA Spec. Publ. 65., ASA, Madison, Wis. Pp. 173-195.

Thimonier, A., Dupouey, J.L. and Le Tacon, F. (2001). Recent losses of base cations from soils of Fagus sylvatica L. Stands in Northeastern France. *Ambio* 29:314-321.

Tiamiyu, S.A., Umar, A., Umar, A.B. and Olaniyan, G.O. (2002). Economic assessment of crop management technologies: A case of lowland rice farmers in Niger State *Proc. Of the 35th Ann. Conf. of ASN, UNAAB, Abeokuta*. Sept. 16-20, 2001. pp. 344-347.

Tyson, R.V., Simome, E.H., Davis, M., Lamb, E.M., White, J.M. and Treadwell, D.D. (2007). Effect of nutrient solution; nitrate concentration and pH on nitrification rate in perlite solution. *J. Plant Nutr.,* 30(6):901-913.

Ugbah, M.M. (2007). Fertility evaluation of soils under date palm in Northern Nigeria. *Prod. Agric. Techn.* 3(1):52-59.

Uri, N.D. (201). Conservation practices in US agriculture and their impact on carbcn sequestration. *Environ. Monit. Assess.,* 70:323-344.

Von School, L. (1998). *Soil fertility management.* Third Revised Edition, CTA Agromisa, Wagennigen, The Netherlands. 80pp.

Wassmann, R., Lantin, R.S., Nene, H.U., Buendia, L.V., Corton, T.M. and Lu, Y. (2000). *Characteristic of methane emission from rice fields in Asia.* In: Wassmann, R. (ed.). Methan emissions from major rice ecosystems in Asia, Klurver Academic Publishers Dordrecht. Pp. 23-36.

Waug, J., Fu, B., Qiu, Y. and Chen, L. (2001). Soil nutrients in relation to land use and landscape position in the semi-arid small catchment on the loess plateau in China. *J. Arid Environ.,* 48:537-550.

Wienhold, B.J. (2005). Changes in soil attributes following low phosphorus swine slurry application to no-tillage sorghum. *Soil Sci. Soc. Am. J.,* 69:206-214.

Williams, S.M. and Weil, R.R. (2004). Crop cover root channels may alleviate soil compaction effects on soybean crop. *Soil Sci. Soc. Am. J.,* 68:1403-1409.

Witt, C. Cassman, K.G., O/K, D.C., Biker, V., Libon, S.P., Samson, M.I. and Ottow, J.C.G. (2000). Crop rotation and residue management effects on carbon sequestration, nitrate cycling and productivity of irrigated rice systems. *Plant Soil* 225:263-278.

Young, A. (1989). Agroforestry for sustainability on steeplands. *IBSRAN Proceedings* No. 8, Thailand. Pp. 37-49.

Zhang, H., Schroder, J.L., Fuhrnan, J.K., Basta, N.T., Storm, D.E. and Payton, M.E. (2005). Path and multiple regression analysis of phosphorus sorption capacity. *Soil Sci. Soc. Am.* J., 69:96-106.

In: Soil Nutrients
Editor: Mohammad Miransari
ISBN: 978-1-61324-785-3

Chapter 12

CHEMICAL SOIL QUALITY AND MYCORRHIZA

***Marcela Claudia Pagano**[*,1] **and Aba C. Persiano**[2]*
[1]Biology Department, Federal University of Ceará, Brazil
[2]Physics Department, Federal University of Minas Gerais, Brazil

ABSTRACT

Land degradation due to salinity, waterlogging, erosion, as well as agricultural intensification and soil pollution can adversely affect soil quality. Key nutrient transformations such as nitrification and nitrogen fixation are conducted by only specialized microbes and this is a fact nowadays more accepted among researchers as well as farmers. However, knowledge on e.g. mycorrhizas is still very scarce. One type of mycorrhizas, arbuscularmycorrhizal fungi (AMF) associating with terrestrial and also with some aquatic plants, has drawn the interest of productive studies in agriculture, agroforestry, plant and soil ecology and restoration of degraded lands. Among the roles of AMF, to increase plant uptake of immobile macro- and micronutrients by extending the volume of soil accessible is the most familiar. In addition, AMF soil hyphae are probably important in nutrient cycling by helping to prevent nutrient losses, nevertheless, AMF are affected by disturbances in the ecosystems, like global change, especially fertilization.Most important propagules of AMF in soils are generally unknown, but they are transported by wind, water erosion and human activities. This review builds upon previous papers to outline when soil fungi are useful for nutrient ecosystem process focusing on crops, most of which are mycorrhizal dependent. Evidence for researcher and decision-makers is provided; however, we do not know enough about mycorrhizal associations and nutrient uptake in natural, disturbed, or managed ecosystems to assess their potential for applied use.

Keywords: Chemical proprieties, Nutrient transformations, Mycorrhizal symbioses, Soil fungi

[*]Marcela Claudia Pagano. Present Address: Researcher from Federal University of Ceará, Fortaleza, Ceará, Brazil. Email: pagano@netuno.lcc.ufmg.br.

Introduction

In native soils plants often deal with nutrient deficiency, especially for macronutrients such as nitrogen (N) and phosphorus (P) which are necessary for plant growth and production. Land degradation due to intensive agriculture, which has alleviated nutrient deficiencies, present imbalances of nutrients, and the majority of world agriculture occurs with low fertility inputs on soils (reviewed by Lynch and St.Clair 2004).

Among edaphic stresses to plant growth chemical factors (sub-optimal availability of mineral nutrients, acidity, alkalinity, salinity, and metal toxicity) are more important than physical factors. Root adaptations to edaphic stress are crucial to explore the soil, acquire nutrients and water, and interact with symbionts and biota (Lynch and St.Clair 2004). But it will depend, on many proprieties of the system, including the relative importance of nutrients that are enriched in surface (P), available at greater depth (N), and root densities in soil horizons (see Lynch and St.Clair 2004). Topsoil is the region of greatest P availability; but mechanisms to counter nutrient stress are poorly understood, being important in future environmental scenarios.

Among parameters for the evaluation of soil quality (physical, chemical and biological), the last have gained particular importance because organisms respond more rapidly than most chemical and physical parameters to changes in land use or contamination (Burns et al. 2006 and references therein). Therefore the abundance of fungi is used as an indicator of soil quality (included in many processes that sustain soil quality).

One type of soil fungi, arbuscularmycorrhizal fungi (AMF) associating with terrestrial and also with some aquatic plants (see Pagano 2011a), has drawn the interest of productive studies in agriculture, agroforestry, plant and soil ecology and restoration of degraded lands (Miransari2010, Pagano and Covacevich 2011, Pagano et al. 2011a,). Among soil microbiota, AMF increase the uptake of P and hence plant growth, in glasshouse experiments using P-deficient soils, therefore this is generally extrapolated for agricultural fields, revegetation experiments and natural ecosystems (Smith and Read 2008).In this chapter we present the most recent advances on the uptake of mineral nutrients by plants in particular referred to symbiotic associations.

Mycorrhizal Symbioses

Mycorrhizal symbioses participate in mobilization of nutrients from organic matter (ectomycorrhizas and ericoid mycorrhizas) and acquisition of limited nutrients (especially P and Zn) (AMF) (Lynch and St.Clair 2004).

Mycorrhizasincrease plant uptake of immobile macro- and micronutrients by extending thevolume of soil approachable, provide plants with more nutrients (that are poorlyavailable to plants) and increase yield, nutrient accumulation and/or reproductive success (Brundrettand Abbott 2002).

Recent research suggested that ectomycorrhizas(EcM) mobilize essential nutrients from insoluble mineral sources and it affectsnutrient cycling in forest systems (Landeweert et al. 2010). For a review ontheroles of EcM associations in plantnutrition, biogeochemical cycles and pedogenesis see Van Schöll et al. (2008).

Table 1. Database survey conducted on January 2011 (SCOPUS) for journal articles dealing with mycorrhizas and chemical soil quality

Key words	Number of journal articles
AMF	2,871
AMF + soil	895
AMF + soil quality	63
AMF + chemical soil quality	0
EcM	2,147
EcM + soil quality	33
EcM + chemical soil quality	0

AMF = arbuscularmycorrhizal fungi,EcM = ectomycorrhizas.

Most of the economically important agronomic plant crops as well as commercial fruit and economically important forest trees form mycorrhizas (See Pagano and Covacevich 2011, Pagano and Cabello 2011); however, few research was published in relation to soil properties (Table 1).

In addition, AMF soil hyphae are probably important in nutrient cycling by helping to prevent nutrient losses, nevertheless, AMF are affected by disturbances in the ecosystems, like global change, especially fertilization.Most important propagules of AMF in soils are generally unknown, but they are transported by wind, water erosion and human activities (Harner et al. 2010).

Some potential interactions of global change variables with mineral stress denote AMF as important for P and Zn uptake (see Lynch and St.Clair 2004). However, the P uptake and translocation by AMF associated with perennial plants was scarce explored. For general conceptual models of the response of individual plants to resource availability (models that must be improved to predict plant response to global change) see Lynch and St.Clair (2004).

Moreover, mycorrhizas influence plant uptake of pollutants such as metals, affecting theirconcentrations in food plants and animals we consume (Joner and Leyval 1997). For an outline of present knowledge of AMF and soil health (current methods and their advantages) see Pagano et al. (2011b).

Mycorrhizal Symbioses and Biological Nitrogen Fixation

Key nutrient transformations such as nitrification and N-fixation are conducted by only specialized microorganisms and this is a fact nowadays more accepted among researchers as well as farmers.Plant capable of symbiotic N-fixation, such as legume–grass pastures, soybean and most legume trees, which are generally more mycotrophic than other plants (Plenchette et al. 2005) are considered profitables for many uses.

In tropical cropping systems nodulation and N-fixation evidence enough variation, therefore, those legumes deplete or addition N to the ecosystem, depending on the occurrence of compatible and effective rhizobia and AMF progagules (Cardoso and Kuyper2006, Marques et al. 2001, Matias et al. 2009). In general, tripartite symbiosis (AMF, rhizobium and legumes) enhance P uptake and biological N-fixation (reviewed by Miransari

2010).Additionally, P plays a key role in N fixation and legume response to CO_2 (since legume nodules have a high P requirement) (Lynch and St.Clair 2004).

Mycorrhizal Symbioses and Soil Factors

In field studies and glasshouse experiments soil factors can influence both the diversity of AMF and also levels of mycorrhizal root colonization and sporulation (See Brundrettand Abbott 2002).

Low soil pH, high aluminum levels or high soil P levels inhibit some AMF influencing their distribution. This is also for NaCl (Juniper and Abbott 1993). Generally, isolates of AMF are often most infective when used in the soil from which they were collected (See Brundrettand Abbott 2002). For an overview of AMF and mineral nutrition see Jakobsen et al. (2003) and Cardoso and Kuyper(2006), who focused on the tropics showing that P can be mobilized through plants and microorganisms.

Rhizospheric interactions, which can benefit soil properties such as enhanced availability of nutrients, and in increased plant growth are essential for agriculture (reviewed by Miransari2010).

Previous research by Egerton-Warbunton and Allen (2000) and Treseder and Turner (2007) showed that *Glomus*could be an indicator of eutrophization (N enrichment) and different ecological traits of AMF genus suggested that *Glomus* dominates the AM community following N additions. For further discussion of functional diversity of AMFrefer to Pagano et al. (2011b) and articles cited within.

With regard to mycorrhizal symbioses and pH, previous report byClark (1997) explained that someAMFoccurs in acid soil(*Acaulospora*), and by the other hand, *Gigaspora* appears to be common in acid soils. Some AMF are more tolerant to acid conditions and high Al than others. Moreover, root colonization is generally less in low than in high pH soils.Some AMF isolates are effective in overcoming soil acidity factors, especially Al toxicity, that decreases plant growth at low pH (Clark 1997).

Nowadays,the need to acknowledge the role of the AM (arbuscularmycorrhizal) mycelium, and to consider feedbacks between soil structure and AMF, is urgently required. Studies focusing on AMF and soil compaction by Miransari and focusing on AMF propagulesbank in soils by Schalamuk and Cabello have showed that AMF inoculation can enhance corn growth in greenhouse and variations in AMF species composition in wheat crops under different tillage systems, respectively (reviewed by Pagano et al. 2011b).

MycorrhizalSymbioses and Soil Factors In Brazil

Recent attention has been done to the change in composition of the AMF propagule banks in agricultural soils, soil compaction and glomalin content over the world.

Most Brazilian native plants were highly dependent on AM, and many were incapable of absorbing P without mycorrhizas even in highly fertile soils (Siqueira and Saggin-Júnior 2001), these associations are often the case in natural ecosystems (Brundrett 2004). Also in natural and revegetated dry forest in Brazil (sandy soils with lower levels of clay, silt and P content) most native plants were colonized by AMF (Pagano et al. 2008a, 2009) and P content in leaves of plants inoculated with AMF was higher than in non-inoculated ones (Figure 1). Those soils (QuartzarenicNeosoil) presented high infiltration rates, were moderately acid and have small amounts of soil organic matter.

In riparian soils from Minas Gerais State, Brazil (pH neutral, organic matter content 0.77%) presenting high base saturation andmoderated P content, high plant colonization and abundance of spores of AMF were obtained (Pagano et al. 2008b).

Also in Brazil, highland field's soils contained natural AMF species that can be affected by mining, what supplies important information for restoration programs. Vegetation growing on rock outcrops at the Serra do Cipó and at an ironstone outcrop in Minas Gerais State (Figure 1), showed abundant AMF spores in rhizospheric soils, but low AMF diversity (Pagano and Scotti 2009b, Matias et al. 2009, Pagano et al. 2010a). Soils presented low base saturation and P content and acidity was moderated. The texture of the fine soil showed high content of sand and low content of clay, belonging to the sand textural class.

The elemental composition of spore wall of AMF was analyzed by energy dispersive X-ray spectrometry (EDS) in native Brazilian dry forest (Table 2). Spore populations analyzed (*Acaulospora*, *Gigaspora*, *Glomus*and *Scutellospora*)suggested that Glomeraceae spores sequester more metallic ions (Al, Fe) as well as more nonmetallic ions, metalloids and halogens such as P, Si and Cl, respectively, and less C than Acaulosporaceae andGigasporaceae; also the presence of titanium was observed in two species. Elements sequestered by AMF spores in soils presenting different conditions of disturbance suggest that Glomeraceae spores have more nutrients than Acaulosporaceae and Gigasporaceae, followed this order: *Glomusbrohultii*>*Acaulosporascrobiculata*>*Scutellosporagregaria*>*Gigaspora margarita*. *Scutellosporagregaria*occured only in disturbed sites, whereas the other species were detected also in the natural ecosystem.

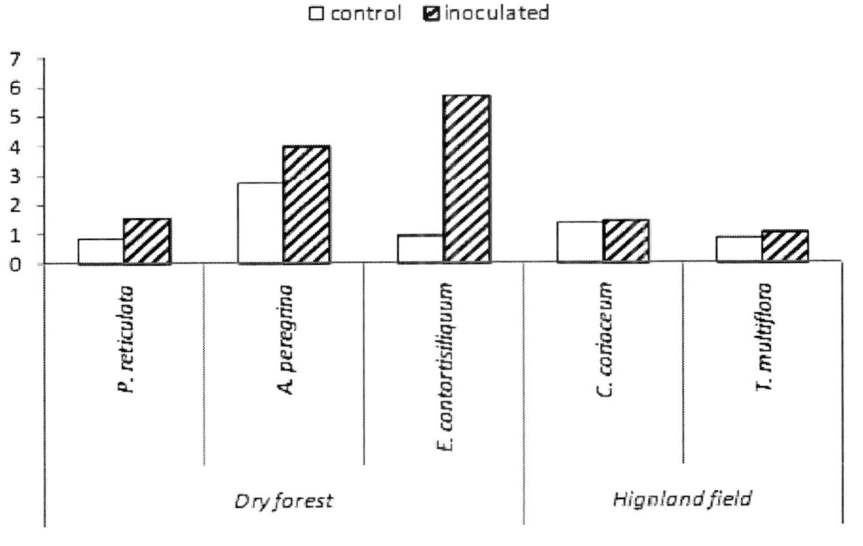

Figure 1.P content in leaves (mg g^{-1}) in Brazilian native plants (*A. peregrina, P. reticulata, E. contortisiliquum, C. coriaceum*and *T. mutiflora*) inoculated with AMF. (Adapted from Pagano et al. 2007, Matias et al. 2009).

Table 2. Semi quantitative spore analyses by EDS of some AMF species (spores) at Minas Gerais-Brazil

AMF species	Elements												
	C	O	Na	Al	Si	P	S	Cl	Ca	Fe	K	Ti	Mg
Acaulosporaceae													
*Acaulosporascrobiculata*Trappe	90.5*	2.9	<0.01	3.04	3.27	ND	0.02	0.05	0.15	0.04	0.09	0.17	ND
Gigasporaceae													
Gigaspora margarita Becker and Hall	97.8	1.7	ND	0.16	0.17	ND	ND	ND	0.04	ND	ND	ND	ND
*Scutellosporagregaria*Walker and Sanders	90.68	7.1	ND	0.69	0.81	0.43	0.06	ND	0.1	ND	0.1	ND	ND
Glomeraceae													
*Glomusbrohultii*Sieverd. and Herrera	64.6	7.1	0.05	13.08	13.6	0.25	0.02	0.32	0.04	0.12	0.3	0.2	0.05

*Spore wall in atomic %. Data are average values of three replicate entire spores; accelerating voltage = 10kV, electron beam current = 20nA, EDS time = 200 s. ND= not detected.

Moreover, in riparian sites presenting different conditions of disturbance the presence of nickel was observed in one *Scutellospora*species(Pagano et al. 2010b).

The extent of mycorrhizal diversity in soils is not known, so it is presently not possibleto model all scenarios of mycorrhizal community structure shifts. Moreover, soil microbial communities are so vastly diverse that complex interactions may occur among microbial species and functional groups.

Application of AMF in Agriculture

AMF can be used in agriculture to increase crop yields while minimizing the requirements on chemical fertilizers. Inoculating plants with AMF for enhancing the efficiency of crop plants (See Pagano and Covacevich 2011) and also forest trees (See Pagano et al. 2009, Pagano and Cabello 2011) is crucial also when few propagules are present in native soils. In Argentina for example, although soils presented high native AMF that colonize crop plants under different management systems they have to be appropriately administrate to increase plant productivity (reviewed by Pagano 2011b).However, AMF technology is difficult to routinely apply in agriculture due to problems in identifying and tracking fungal species in the field, the poor understanding of the basic biology of AMF and the inability to grow AMF in pure cultures.On the other hand, fertilizer may select for less beneficial AM species, and mildly nutrient-stressed vegetal species can hold better symbionts (reviewed by Pagano and Covacevich 2011).

Other vantage is that AMF can survive soil freezing, or wetting and drying cycles, their spores being more resistant to topsoil disturbance and storage than root fragments and much more resistant than hyphae. However, it was highlighted that mycorrhizalpropagules can be severely influenced by damage to vegetation and soils resulting from human intervention, as well as intense fires, topsoil removal and flooding. Moreover, AMF affect plant water relations in both water-limiting and well-watered conditions (reviewed by Pagano and Covacevich 2011).

Moreover, as different microorganisms are capable of increasing the absorption of P by the plant, the inoculation of plants with different functional groups of fungi (Mycobization) was proposed (Velázquez and Cabello 2011).

Agricultural AM inoculum production demand certain parameters for his accomplishment (appropriate AMF species, the quality of AMF inoculum, and their ecological properties); however, agricultural practices such as chemical fertilization, soil tillage, biocides, monocultures and nonmycorrhizal plants inhibit AM populations in soil (Miransari 2010). Tillage affects soil physical and chemical properties, influencing the population, diversity and activities of soil microbes, including both free and symbiotic fungal populations (Reviewed by Pagano 2011b).

Model plants such as rice, which is mycotrophic can help increase the knowledge regarding AM symbiosis, and the selection of adequate agricultural practices which affect AM colonization in the field must be promoted, especially for organic farming where biological fertilization is used (Miransari 2010, Pagano 2011b).

Outlook and Conclusion

In South America the culture collections of AMF are maintained by researchers; in spite of the fact that more research is needed in relation to germplasm banks of AMF for increase the production of agricultural plants (an urgent issue to fulfill requirements of ecosystem services created by a modern agriculture) (Pagano and Covacevich 2011).

In the present chapter it was also tried to highlight the challenges and the perspectives that come in terms of nutrients in the agriculture. To maintain ecosystem function, we must better investigate the key role of AMF ecosystem services and soil health, which was explained in recent reviews (Gianinazzi et al. 2010, Pagano et al. 2011b), as they are a functional component in natural and agricultural systems.

Excessive chemical fertilization is not agriculturally and environmentally sustainable, while low fertilizer inputs and plant root symbioses may greatly benefit agriculture. Moreover, AMF are resistant to stress, and can help their host plants (multiple benefits), being biotechnologically as important as rhizobium (Miransari 2010).

A number of reviews have assessed the relationship between soil quality and AMF, exemplifying the increasing interest in this subject. Several of these reviewshighlighted the relevance of this topic for researchers and decision-makers such as understanding the relationship between soil health in agriculture(Jeffries and Barea 2001, Jeffries et al. 2003,Gianinazzi et al. 2010, Pagano et al. 2011b) and soil structure (Rillig and Mummey 2006). Evidence for researcher and decision-makers is provided; however, we do not know enough about mycorrhizal associations and nutrient uptake in natural, disturbed, or managed ecosystems to assess their potential for applied use.

It is also important to mention other endophytic plant symbionts (binucleate *Rhizoctonia*spp., *Piriformasporaindica*, various plant growth-promoting rhizobacteria and *Trichoderma* spp.) colonizing plant roots (Harman 2011).

In addition to the synergistic interactions of AM fungiand soil microbes' interactions in agriculture, fungi which have adapted to local soil conditions are especially needed to increase nutrient efficiency use (simulating natural ecosystems).

ACKNOWLEDGMENTS

Dr Marcela Pagano is grateful to Council for the Development of Higher Education at Graduate Level, Brazil (CAPES), and to FAPEMIG (Minas Gerais Research Funding Agency - Process 311/07) for Post-doctoral scholarships granted.

REFERENCES

Burns, R.G.; Nannipieri, P.; Benedetti A.; Hopkins D.W. Defining Soil Quality. In: Bloem J, Hopkins DW, Benedetti A, editors. *Microbiological methods for assessingsoil quality*.CABI Publishing; 2006; 15–22.

Brundrett, M.C.;AbbottL.K. ArbuscularMycorrhizas in PlantCommunities. In: Sivasithamparam K., Dixon K.W., Barrett R.L., editors. *Microorganisms in Plant Conservation and Biodiversity*; Kluwer Academic Publishers; 2002; 151–193.

Brundrett, M. Diversity and classification ofmycorrhizal associations.*Biol. Rev.* 2004, 79, 473–495.

Cardoso, I.M.; Kuyper, T.W. Mycorrhizas and tropical soil fertility.*Agric. Ecosyst. Environ.*, 2006, 116, 72–84.

Clark, R.B. Arbuscularmycorrhizal adaptation, spore germination, root colonization,and host plant growth and mineral acquisition at low Ph. *Plant and Soil,* 1997, 192, 15–22.

Gianinazzi, S.; Gollotte, A.; Binet, M.; van Tuinen, D.; Redecker, D.; Wipf D.Agroecology: the key role of arbuscularmycorrhizasin ecosystem services. *Mycorrhiza*, 2010, 20, 8, 519–30.

Harman, G.E. Multifunctional fungal plant symbionts: new tools to enhance plant growth andproductivity. *New Phytologist*, 2011, 189, 647–649.

Harner, M.J.; Opitz,N.; Geluso,K.; Tockner,K;RilligM.C. Arbuscularmycorrhizal fungi on developing islands within a dynamic river floodplain: an investigation across successional gradients and soil depth.*Aquatic Sciences,*2011, 73, 1,35–42.

Jakobsen, I.; Smith, S.E.; Smith, F.A. Function and diversity of Arbuscularmycorrhizae in carbon and mineral nutrition.In: van der Heijden M.G.A., Sanders I.R., editors. *Mycorrhizal Ecology*. Berlin: Springer; 2003;75–92.

Jeffries, P.; Barea, J.M. Arbuscularmycorrhiza– a key component of sustainable plant-soil ecosystems.In: *The Mycota*, Vol. IX: Fungal Associations (Ed. B. Hock), Berlin, Springer-Verlag; 2001; 95–113.

Jeffries, P.; Gianinazzi, S.; Perotto, S.; Turnau, K.; Barea, J.M.The contribution of arbuscularmycorrhizal fungi in sustainable maintenance of plant health and soil fertility.*Biology and Fertility of Soils*, 2003, 37, 1–16.

Landeweert, R.; Hoffl, E.; Finlay, R.D.; Kuyper, T.W.; Breemen, N. Linking plants to rocks: ectomycorrhizal fungi mobilize nutrients from minerals.*Trends Ecol. Evol.*, 16, 248–254, 2001.

Joner, E.J.; Leyval C. Uptake of ^{109}Cd by roots and hyphae of a *Glomusmosseae/Trifoliumsubterraneum*mycorrhiza from soil amended with high and low concentration of cadmium. *New Phytol.*, 135, 353–360.

Juniper, S.; Abbott, L. Vesicular-arbuscularmycorrhizas and soil salinity.*Mycorrhiza,* 1993, 4, 45–57.

Lynch, P.; St.Clair, S.B. Mineral stress: the missing link in understanding how global climate change will affect plants in real world soils. *Field Crops Research,* 2004, 90, 101–115.

Marques, M.S.; Pagano, M.C.; Scotti, M.R.M. Dual inoculation of woody legume (*Centrolobiumtomentosum*) with rhizobia and mycorrhizal fungi in southeastern Brazil.*Agrofor. Syst.*, 2001, 52, 107–117.

Matias, S.R.; Pagano, M.C.; Muzzi, F.C.; Oliveira, C.A.; Carneiro, A.A.; Horta, S.H.; Scotti, M.R.Effect of rhizobia, mycorrhizal fungi and phosphate-solubilizing microorganisms in the rhizosphere of native plants used to recover an iron ore area in Brazil. *Eur. J. Soil Biol.*, 2009, 45, 259–266.

Miransari,M. Interactions between arbuscularmycorrhizal fungi and soil bacteria.*Appl. Microbiol. Biotechnol.*, 2010, DOI 10.1007/s00253-010-3004-6.

Pagano, M.C.;Cabello,**M.N.**; Scotti, M.R. Phosphorus response of three native Brazilian trees to inoculation with four arbuscularmycorrhizal fungi.*J. Agric. Technol.*, 2007, 3, 231–240.

Pagano, M.C.; Cabello, M.N.; Bellote, A.F.; Sá, N.M.H.; Scotti, M.R. Intercropping system of tropical leguminous species and *Eucalyptus camaldulensis*, inoculated with rhizobia and/or mycorrhizal fungi in semiarid Brazil. *Agroforestry Systems*. 2008a, 74, 3, 231–242.

Pagano, M.C.; Marques, M.S.; Sobral, M.; Scotti, M.R. Screening for arbuscularmycorrhizal fungi for the revegetation of eroded riparian soils in Brazil. In: VI Latinoamerican Congress of Mycology, Mar del Plata, Argentine, 2008b.

Pagano, M.C.; Scotti, M.R.A survey of the arbuscularmycorrhizaoccurrence in *Paepalanthusbromelioides*and *Bulbostylis*sp. in rupestrian fields, Brazil. *MicologiaAplicada International*, 2009a, 21, 1–10.

Pagano M.C., Scotti M.R., Cabello M.N. Effect of the inoculation and distribution of mycorrhizae in *Plathymeniareticulata*Benth, under monoculture and mixed plantation in Brazil.*New Forests*. 2009a, 38, 197–214.

Pagano, M.C.; Cabello, M.N.; Scotti, M.R. Agroforestry in Dry Forest, Brazil: Mycorrhizal Fungi Potential. In: LR Kellymore, ed. *Handbook on Agroforestry: Management Practices and Environmental Impact*. Nova Science Publishers, New York; 2010a; 367–388.

Pagano, M.C.; Cabello, M.N.; Scotti, M.R.Arbuscularmycorrhizal colonization and growth of *Eremanthusincanus* Less.in a highland field. *Plant Soil Environ.,*2010a,56, 9, 412–418.

Pagano, M.C.; Persiano, A.I.C.; Cabello, M.N.; Scotti, M.R. Elements sequestered by arbuscularmycorrhizal spores in riverine soils: a preliminary assessment. *Journal of Biophysics and Structural Biology*. 2010b, 2, 2, 16–21.

Pagano, M.C. Soil Tillage in Agroforestry and Agroecosystems: Mycorrhizal Benefits. In: Miransari M., editor. *Soil Tillage and Microbial Activities Research Signpost Publications*, India. 2011b, ISBN: 978-81-308-0444-6.

Pagano, M.C.; Covacevich, F.ArbuscularMycorrhizas in Agroecosystems. In: *Mycorrhizal Fungi: Soil, Agriculture and Environmental Implications*. Nova Science Publishers, New York, 2011, ISBN: 978-1-61122-659-1.

Pagano, M.C. Mycorrhiza: Occurrence and Role in Natural and Restored Environments. In: Mycorrhiza: occurrence in natural and restored environments. Pagano, M. C. editor, Nova Science Publishers, Hauppauge, NY.ISBN: 9781612092263, 2011a.

Pagano, M.C.;Utida, M.K., Gomes, E.A.; Marriel, I.E.; Cabello,M.N., Scotti, M.R. Plant-type dependent changes in arbuscularmycorrhizal communities as soil quality indicator in semi-arid Brazil. *Ecological Indicators*. 11,2, 643–650, 2011a.

Pagano, M.C.;Schalamuk,S.;Cabello, M.N. Arbuscularmycorrhizal parameters and indicators of soil health and functioning: Applications for agricultural and agroforestal systems. Miransari M., editor. *Soil Microbes and Environmental Health*, Nova Science Publishers, New York, USA, Chapter 10, ISBN: 9781612096476, 2011b.

Pagano, M.C.; Cabello, M.N. Importance of mycorrhizal interactions for reforestation.ISRN ECOLOGY.Hindawi Publishing Corporation. ISSN: 2090-4614, e-ISSN: 2090-4622, doi:10.5402, 2011.

Pagano, M. C.;Scotti, M. R.Restoration of degraded riparian forest at Sabará River - Minas Gerais. Proceeding of VII National symposium of Recuperation of degraded lands.Annals. Curitiba: FUPEF, 2008, 1, 337–346 (in Portuguese).

Plenchette, C.; Clermont-Dauphin, C.; Meynard, J.M.; Fortin, J. A.Managing arbuscularmycorrhizal fungi in cropping systems.*Can. J. Plant. Sci.*, 2005, 85, 31–40.

Rillig, M.C.; Mummey, D. L. Mycorrhizas and soil structure.*New Phytologist*, 2006, 171, 41–53.

Siqueira, J.O.; Saggin-Júnior, O. J.Dependency on arbuscularmycorrhizal fungi and responsiveness of some Brazilian native woody species.*Mycorrhiza*, 2001, 11, 245–255.

Smith, S.E.; Read, D.J. *Mycorrhizal Symbiosis*.New York, Elsevier; 2008.

Van Schöll, L.; Kuyper, T. W.; Smits, M. M.; Landeweert, R.; Hoffland, E.; van Breemen, N. Rock-eating mycorrhizas: their role in plant nutritionand biogeochemical cycles. *Plant Soil*, 2008, 303, 35–47.

Velázquez, M.S.; Cabello, M.N. Mycobization as a Biotechnological Tool: A Challenge, In: Thangadurai D., Busso C.A., Hijri M., editors. *Mycorrhizal Biotechnology*, 2010, 140–151.

In: Soil Nutrients
Editor: Mohammad Miransari
ISBN: 978-1-61324-785-3

Chapter 13

MICRONUTRIENT AVAILABILITY IN CROP SOILS OF THE PAMPAS REGION, ARGENTINA

***Silvana Irene Torri*[1*], *Susana Urricariet*[1] and *Raúl Silvio Lavado*[1,2]**
[1]College of Agronomy, University of Buenos Aires, Argentina
[2]INBA CONICET/FAUBA, Av San Martín 4453, Ciudad Autónoma de Buenos Aires, Argentina

ABSTRACT

Micronutrient deficiencies occur all over the world but variations in soil conditions, climate, crop genotypes and management result in marked variations in the incidence of these problems. Due to the naturally high soil fertility of the Pampas region, Argentina, macro and micronutrients have not been applied in a massive way to soils through fertilization. As a consequence, some micronutrient deficiencies were noticed in the last years as a result of nutrient depletion of soils, adoption of high-yield genotype crops, improved agricultural mechanization and soil degradation. Different soil test surveys indicate that Zn and B would be the most frequently deficient micronutrients with probabilities of economic responses as a result of fertilizer application. Micronutrient deficiencies are an emerging limiting factor for annual crop production. In annual crops such as maize and wheat, deficiencies of Zn have been reported, whereas sunflower is one of the most sensitive field crops to low B supply. Responses to Cl are not economically significant and deficiencies of Cu, Fe, Mn and Mo are exceptional. Crop micronutrient fertilization is a new issue in the Pampas region, and consequently, diagnostic techniques and application technology issues require a great of deal of research for improvement.

Keywords: Micronutrient availability, climate, crop genotypes, soil properties, geographical region

---

* E-mail: torri@agro.uba.ar.

INTRODUCTION

Essential nutrients are defined as those elements without which plants cannot complete the vegetative or reproductive stage of their life cycle, are irreplaceable by other elements, or are directly involved in plant metabolism (Arnon and Stout 1939). Based on the quantity required, essential nutrients are divided into macro and micronutrients. Micronutrient requirements by crops for normal growth and high yield are small compared to those of macronutrients (Epstein and Bloom, 2005). The term "trace element" is widely used for those elements that occur at relatively low concentrations in the dry matter of living organisms, usually under 100 mg kg^{-1}. Several trace elements have been reported to be essential for certain plant species. Some others are known to have stimulating effects on plant growth (Asher, 1991), although their functions have not been recognized yet (Barker and Pilbeam, 2007). Micronutrients are those trace elements which are essential for the normal healthy growth and reproduction of all plant species, and are: boron (B), chlorine (Cl), copper (Cu), iron (Fe), manganese (Mn), molybdenum (Mo), and zinc (Zn) (Welch and Graham, 2005). Chlorine is referred to as micronutrient, even though its concentration in plant tissue is often equivalent to those of macronutrients. Although the list of essential micronutrients for plants remains unchanged since 1987, this list may grow as convincing evidence of the essentiality of other trace elements will reported in the future, as a result of advances in plant physiology and the availability of highly sensitive analytical instruments. Nickel (Ni), for example, is cited as an essential micronutrient in some publications; however, the essentiality of this element has not been proved for all plants (Brown et al., 1987; Eskew equivalent to those of macronutrients. Although the list of essential micronutrients for plants remains et al., 1983). Cobalt (Co) is also known to be essential for the bacterial fixation of atmospheric nitrogen in leguminous plants, although it is not considered to be essential for all higher plants (Marschner, 1995).

FUNCTIONS AND GENERAL SITUATION OF MICRONUTRIENTS

Micronutrients play many complex roles in plant nutrition, and are involved in virtually all metabolic and cellular functions, like energy metabolism, primary and secondary metabolism, cell protection, gene regulation, hormone perception, signal transduction and reproduction, among others (Marschner (1995), Kabatas-Pendias and Pendias, 2001).

Worldwide, high crop yields associated with greater nutrient removal and the use of high doses of macronutrients have resulted in the depletion of available micronutrients in soils. Moreover, the increase in grain yield over the last four decades has been accompanied by a decrease in mineral concentrations in grain (Garvin et al., 2006; Fan et al, 2008). In addition to yields, the contents of micronutrients in crop products are of great importance. By 1990, it was clear that micronutrient deficiencies in humans not only was a major and still increasing problem, especially in underdeveloped countries, but it also caused an enormous health cost in developed countries as well. In 2002, the World Health Organization (WHO) reported that more than half of the world population was affected by Fe and Zn deficiencies. Therefore, there is increasing interest in fortifying crop products with micronutrients, as discussed by Zhao and McGrath (2009), and Waters and Sankaran (2011, *in press*).

Micronutrient deficiencies occur all over the world but variations in soil conditions; climate, crop genotypes and management result in marked variations in the incidence of these problems. In Argentina, some micronutrient deficiencies were noticed in these last years as a result of the degradation and nutrient depletion of soils, adoption of high-yield genotype crops and improved agricultural mechanization which have resulted in higher crop production per unit area, the use of fertilizers with a higher degree of purity or simply due to nutritional imbalances (Andrade and Sadras 2000; Urricarriet and Lavado 1999). The objectives of this chapter are to discuss micronutrient problems raised in the last years in the Pampas region, Argentina, and analyse measures to correct deficiencies for annual crop production.

CHARACTERISTICS OF THE STUDY AREA

The Pampas Region, Argentina, is one of the largest temperate field cropland areas of the Southern Hemisphere. It is located between 32° to 39°S and 56 to 67°W. This zone covers more 52 Mha of agriculturally prime quality land. The region also includes some areas which are marginal or unsuitable for cropping and are devoted to husbandry. The climate of the region is humid, characterized by long warm summers and mild winters. Mean annual rainfall ranges from 600 mm in the west to 1200 mm in the east, whereas mean annual temperature ranges from 14 °C in the south to 21 °C in the north. Predominant soils are Mollisols formed on loess like materials of eolian origin, and to a lesser degree from fluvial sediments (Imbellone et al., 2010). The humid and semiarid subregions are characterized by Udolls and Ustolls, respectively, with minor occurrence of Aquolls in wet flat areas (Moscatelli and Pazos, 2000). These soils have naturally adequate concentration of trace elements and show no signs of contamination with heavy metals and other trace metals (Lavado et al., 1998; 2004). Agriculture is performed on well drained soils, both in the semiarid and humid portions of the region, with nearly 50% of the whole pampas area devoted at present to soybean (*Glycine max*). Other important crops are maize (*Zea mays*), wheat (*Triticum aestivum*) and sunflower (*Helianthus annuus*). Minor crops are barley, rye, canola, sorghum, flax, and others. Because of its extension and yield potential, the region is considered as one of the most suitable areas for grain crop production in the world (Satorre and Slafer, 1999). From the end of XIX century, the region is a net exporter of grains to the world.

The Pampas region maintained its native grasslands until the last quarter of 19th century and the beginning of the 20th century, when the establishment of perennial pastures and annual crops by European immigrants started (Viglizzo et al., 2001). More than 70% of the cropping area in the region is nowadays under no-tillage systems (AAPRESID, 2011). Grain production in the Pampas region, has sharply increased in the last decade. For soybeans, this increase was mainly due to area expansion rather than yield increases. Production increases in wheat and maize showed a greater equilibrium between area expansion and yield increases. Sunflower moderated its decrease in sown area since 1998 with higher national average yields.

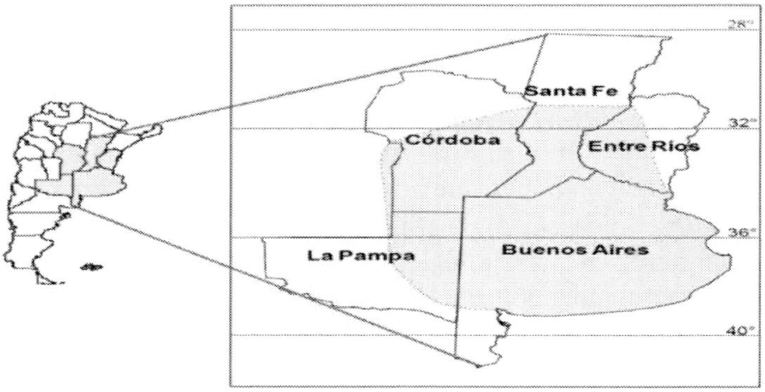

Figure 1. Map of the Pampas Region, Argentina.

Fertilizer Consumption in the Area

Natural high soil fertility and wrong political issues affecting the economy of farmers prevented widespread chemical fertilization in the area (Lavado and Taboada, 2009). Until the 1960s the use of fertilizers on the extensively grown crops in the Pampas region was ignore or denied. It was during the 1970s, especially with the introduction of genetically improved crops, that the response to nitrogen became evident. Unfavourable fertilizer: grain price relationships prevented widespread chemical fertilization until the 1990s, when a new economical scheme and the swift adoption of continuous agriculture with higher fertilization rates and minimum or no tillage use emerged. This was accompanied by a high level of investment in infrastructure. The highest level of fertilizer consumption was reached during the 1996/97 season, when large areas sown to wheat and maize coincided with a favourable ratio of fertilizer: cereals price. Although the use of fertilizers greatly increased from 1990, doses of N, P, potassium (K) and sulfur (S) were applied at lower levels than crop nutrient extraction. Ciampiti and Garcia (2008) reported that during 2006/2007, the doses of fertilizers applied replenished only 31%, 53% 1% and 27% of the N, P, K and S removed by wheat, maize, soybean, sunflower at harvest. Less attention has been paid to micronutrients fertilization in the region. Therefore, deficiencies in zinc and boron were identified, especially on soils with low organic matter content (Melgar, 2001).

Micronutrients in Soils and Occurrence of Micronutrient Deficiencies in the Pampas Region

Total soil micronutrient levels usually exceed crop demands by several orders of magnitude. Like so, crop micronutrient deficiency can be viewed as a combination of soil inefficiency in supplying micronutrients in plant-available forms and/or plant inefficiency in micronutrient assimilation. A soil is considered deficient in a given nutrient when its addition as a fertilizer produces increased growth, even though the quantity of nutrient added may be small compared with its total amount in soil.

Table 1. Total concentration of micronutrients in the E of the Pampas region (Lavado et al., 2004)

Boron	Copper	Manganese	Molybdenum	Zinc
mg kg^{-1}				
25.67	11.75	451.56	2.00	38.39

Severe deficiency symptoms for most micronutrients appear on the young leaves at the top of the plant because most micronutrients are not readily translocated. Molybdenum is an exception in that it is readily translocated, and its deficiency symptoms generally appear on the whole plant (Gupta 1997).

The total micronutrient content of the Pampas soil are within the values find in normal non contaminated soils of the world (Lavado et al., 2004). Table 1 shows the total concentration (hydrochloric, sulfuric and nitric acids digestion) of five micronutrients in soils of the eastern portion of the Pampas region.

Total micronutrient content of soil usually includes fractions that are not immediately available to plants, micro-organisms and soil fauna (Zhang et al., 2001). From the agricultural point of view, the availability of micronutrients is more important than their total concentration. A wide variety of extractants, including acids, salts, and chelates, have been used for estimating the available level of micronutrients in soils, with variable degrees of success. Among chelates, diethylenetriaminepentaacetic acid (DTPA) and ethylenediaminetetraacetic acid (EDTA) have been considered the most suitable extracting agents (Moral et al., 2002; Quevauviller et al., 1998). In addition to single extractions, sequential extraction is a useful technique for determining the nature of the association of micronutrients with soil components. Such information is potentially valuable for predicting bioavailability, metal leaching rates and transformations between physicochemical forms in agricultural and polluted soils (Torri and Lavado 2008 a,b). The use of sequential extraction procedures in soil analysis as a complement to single extractions has increased, although this technique is usually used for scientific purposes. Thus, soil testing for plant micronutrients using single extraction techniques has become a standard practice to ascertain if a soil is deficient, borderline or sufficient with regard to a given micronutrient for a particular crop (or group of crops)

Due to the naturally high soil fertility of the Pampas region, macro and micronutrients have not been applied in a massive way to soils through fertilization. The lack of nutrients reposition resulted, with time, in a depletion of soil phytoavailable micronutrients (Urricariet and Lavado 1999). In the centre of Santa Fe province, Silva et al. (2001) found decreases in soil bioavailable concentrations of B, Cu, Fe and Mn, caused by continuous agriculture. Figure 1 shows the effect of years of agriculture on DTPA available B y Zn on pristine soils in the same area (C. Miretti, *personal communication*). Soil losses by intense erosion under conventional tillage systems also produced a net loss of nutrients (Bernardos et al, 2001). Around 1990, the soils which had suffered erosion and nutrient depletion were unable to sustain high crop yields (Lavado and Taboada, 2009). These soils exhibited markedly lower contents of organic matter, total nitrogen, phosphorous and sulfur when compared with pristine soils. In the case of micronutrients, the picture was less defined due to the lower concentrations found in soils and the lower crop requirements.

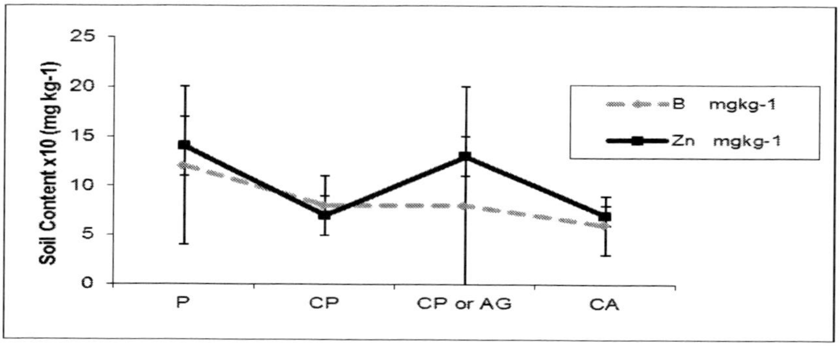

Figure 2. The effect of agriculture on DTPA extractable B and Zn. P: Pristine soils; CP: cattle production grazing pastures all year round in the field, as usual in the region; CP or AG: alternate animal grazing and agriculture and CA: continuous agriculture (C. Miretti, *personal communication*).

Different soil test surveys in the Pampas region indicate that B and Zn would be the most frequently deficient micronutrients. Response to the application of B and Zn were found on alfalfa, maize, soybean and sunflower in different parts of the Pampas region (Melgar et al., 2001; Sainz Rozas et al., 2003). Reports of some micronutrient deficiencies in the main crops cultivated in the Pampas Region are provided below.

Zinc (Zn)

Total Zn concentration in agricultural soils ranged from 50 to 300 mg kg−1 (Adriano, 2001). Almost 50% of soils used for cereal production in the world have low availability of Zn, which not only reduces yield but also affects crop nutritional quality (Graham and Welch, 1996). Weathering of Zn minerals produces mobile Zn2+, especially in acid, oxidizing environments. This species is easily adsorbed by mineral and organic components and thus, in most soil types, accumulation of Zn in the surface horizons is observed. Mizuno et al. (1988) studied the vertical distribution of Zn in two typical agricultural soils of the Pampas, with contrasting soil particle distribution, an Hapludoll and an Argiudoll (U.S. Soil Taxonomy). Compared with the upper 20 cm, both underlying layers exhibited a significant decrease in Zn content, with a subsequent increase in depth in the soil profile. Zinc vertical distribution suggested a depletion of available Zn in the rhizospheric horizons, with a enrichment in the upper soil layer due to stubble accumulation and decomposition. Comparing with deeper soil layers, Lavado et al. (2004) lately reported that top soils were not contaminated from anthropogenic sources. Marked variations in the range of Zn availability in soils of the Pampas region have been observed (Table 2).

Torri and Lavado (2008 b) studied the distribution of Zn among different fractions in the superficial horizon of three typical pristine soils of the Pampas region. The concentration of Zn in soil fractions was determined by the sequential chemical extraction procedure described by McGrath and Cegarra (1992). Native Zn was mainly found as RES-Zn, exceeding 84% of the total content. The order of Zn concentration in the fractions of the three pristine soils was: RES-Zn > INOR-Zn > OM- Zn > EXCH-Zn (Table 3). This result is typical for Zn found in non-contaminated soils (Shuman, 1999) where it is mostly found in unreactive forms in the crystal lattices of the minerals.

Table 2. Zinc availability in representative soils of the Pampas region

Extractant	Soils	Zone	Range (mg kg^{-1})	References
EDTA	Argiudolls	N Buenos Aires	2.1-13.2	Urricariet and Lavado (1999)
	Haplustolls	La Pampa	1.4-15.0	Gonzalez and, Buschiazzo (1998)
DTPA	Argiudolls	N Buenos Aires	1.47	Lavado and Porcelli (2000)
	Argiudolls	NW Buenos Aires and S Santa Fé	1.1-5.4	Ratto de Miguez and Fatta (1990)
	Typic Argiudoll	N Buenos Aires	0.30-2.60	Ratto et al. (1997)
	Argiudolls and Haplustolls	S Córdoba	0.40-2.50	Buffa and Ratto (2005)

Table 3. Soil distribution of endogenous Zn in the three pristine soils of the Pampas region, among water-soluble and exchangeable fraction (EXCH), organic matter bound fraction (OM), inorganic precipitate fraction (INOR) and residual fraction (RES) (adapted from Torri and Lavado (2008 b)

Soils	EXCH-Zn	OM- Zn	INOR-Zn	RES-Zn
	mg kg^{-1}			
Typic Hapludoll	1.61 ±0.07	2.62 ±0.28	3.90 ±0.40	46.87 ± 0.75
Typic Natraquoll	1.07±0.07	1.75 ±0.04	4.14 ±0.07	40.04 ± 0.11
Typic Argiudoll	1.00 ±0.03	1.97 ±0.06	6.22 ±0.26	49.80 ± 0.29

Although Zn fertilization is not a common practice in cropping systems of the Pampas region, different soil test surveys indicate that Zn would be, together with B, the most frequently deficient micronutrients with probabilities of economic responses as a result of fertilizer applications (Melgar et al., 2001). Maize and wheat are the most frequent crops in the Pampas region which were reported to be deficient in Zn. Zinc requirement and crop removal of maize are 53 gr Zn/t and 239 g Zn/ha respectively for yields of 9 t ha^{-1}. In the case of wheat, yields of 5 t ha^{-1} require 52 gr Zn/t, removing 114 g Zn/ha by crop. Zinc deficiencies in maize occur especially in the early stages, where the low temperature and/or excessive soil moisture reduces the availability of this nutrient. Deficiency symptoms were related to low Zn mineralization and reduced root growth (Moraghan and Mascagni, 1991). High doses of phosphorus fertilization also induce Zn deficiency at low soil temperature (Gregory and Frink, 1995). These situations were frequently reported in the W of Buenos Aires, SW of Santa Fe and S of Cordoba, where a vertical striping of newer, upper leaves between the midrib and leaf margin were observed in the first two weeks after emergence (Melgar et al., 2001). Zinc deficiency symptoms do not appear in newer growth later in the season, as the temperature increases and the root system explores a larger volume of soil

(Melgar et al., 1997). Temporal Zn deficiency symptoms have spread since the introduction of direct seeding technology package, where high amounts of P-band are placed near the seed.

The Pampas region produces approximately 90% of the wheat area of the country. Wheat productivity greatly increased in the last two decades as a result of more intensive management practices. Although macronutrient fertilization is nowadays a usual practice for wheat, addition of micronutrients like zinc (Zn) or copper (Cu) were never considered a limiting factor. Sainz Rozas et al. (2003) evaluated wheat yield response to Zn and Cu fertilization in 19 farmer fields located in the southeast of Buenos Aires province. These soils are classified as Typic Argiudolls (US Soil Taxonomy) and have high contents of organic matter (4.9-7.4 % OM), moderately acidic pH values to sub-acid (5.5 to 6.4) and low available P (Echeverría and Ferrari 1993). Wheat grain yield responses to Zn addition were positive in only 4 sites, which had soil pH above 6, whereas no yield response to Cu addition was observed. Soil pH is more important than any other single property for controlling Zn availability in soils (Torri and Lavado, 2008 b). As soil pH increases above pH 5.5, Zn is adsorbed on hydrous oxides of Al, Fe, and Mn (Moraghan and Mascagni, 1991).

Boron (B)

Total concentration of B in soils ranges from 2 to 100 mg kg^{-1} (Goldberg et al., 1993). The B proportion in the soil solution is generally low, and varies from 3.2 to 5.3% of its total content (Kabatas-Pendias and Pendias, 2001). Boron is unique among micronutrients because it exists as a non-dissociated molecule under most soil pH (Power and Woods, 1997). Therefore, it is the most mobile element among micronutrients. Boron movement in soils follows the water flux, especially in well drained, sandy soils with neutral to acidic pH and low organic matter content located in humid regions are prone to B deficiency (Díaz-Zorita 2000). Different soil test surveys in the Pampas region indicate that boron (B) would be deficient, with probabilities of economic responses as result of fertilizer applications (Table 4). Furthermore, Rivero et al. (2006) reported that 9. 10^6 ha of the Pampas are considered of low availability of B.

Table 4. Boron availability in representative soils of the Pampas region

Extractant	Soils	Zone	Range (mg kg^{-1})	References
Azomethine-H method	Ustolls, Udolls	SW Buenos Aires	0.20-0.58	Ron et al. (1999)
$CaCl_2$, colorimetric	Argiudolls	N Buenos Aires	0.31-0.82	Urricariet and Lavado (1999)
	Argiudolls	N Buenos Aires	0.41-1.28	Ratto de Miguez and
	Hapludolls	W Buenos Aires	0.30-1.03	Fatta ((1990)
	Argiudolls	SE Buenos Aires	0.64-2.48	
Hot water, ICP-AES	Typic Argiudoll	N Buenos Aires	0.46	Lavado and Porcelli (2000)
Melich 3, ICP	Typic Argiudoll	NW Buenos Aires, S Santa Fé	0.10-0.80	Melgar et al. (2001)

Boron is generally required in greater amounts by dicot plant species than by monocotyledons (Srivistava and Gupta, 1996). Like other micronutrients, B is immobile in the plant tissues and its deficiency symptoms first appear on the younger plant tissues. A common result of boron deficiency in all crops is the interruption in flowering. Yields are reduced, and the fruit or grain is deformed or discoloured..

Sunflower is one of the most sensitive field crops in the Pampas to low B supply. Boron requirement and crop removal of sunflower are 165 gr B/t and 130 g B/ha respectively for yields of 3.5 t ha^{-1}. Ratto (2006) reported that 80% of the area planted with sunflower in the Pampas region is deficient in B. Deficiency symptoms first become evident on the younger leaves which have a bronze colour and become hardened, malformed and necrotic. The capitulum's is often malformed with poor seed set (Blamey et al., 1997). Boron deficiencies in sunflower are also induced by water stress, most likely because of the reduced mobility of B by mass flow to roots (Barber, 1995). In this way, soil drying reduces B diffusion by reducing the mobility of soil solution and increasing the diffusion path length (Scott et al., 1975). Therefore, under drought conditions, the availability of B decreases, increasing responses to foliar applications. In sandy soils of the W of Buenos Aires province, yield responses to B fertilization in sunflower were related to soil available B before planting, estimated by Melich-III (Díaz Zorita and Duarte, 1998). Quiñónez et al., (2003) also observed B deficiencies in the centre of Santa Fe Province.

Forage legumes such as alfalfa also have relatively high boron requirements. In the past 20 years, forage production of alfalfa in the Pampas region increased in at least 40%, and consequently the rate of removal of boron from soils increased. Boron requirement and crop removal of alfalfa are 450 gr B/t and 350 g B/ha respectively for yields of 15 t ha^{-1}. Boron shortage on clay loam and clay soils was not reported. Deficiencies show up mainly on sandy loam soils, resulting in a yellowing or reddening of the upper leaves of the plant. Stem growth between leaves becomes shortened, giving plants a stunted appearance. Yield response to boron application varies, depending on soil type, temperature, moisture, and nutrient balance in the soil. Several trials on alfalfa pastures in the west of the province of Buenos Aires and in the center of Santa Fe, among others, showed yield responses to boron application, increasing root **nodulation** (C. Amador, *personal communication*).

Molybdenum (Mo)

Molybdenum is the least abundant of the micronutrients in the lithosphere (Mortvedt, 2000), and soil concentrations range from about 0.2 to 5 mg kg^{-1} (mean of 2 mg kg^{-1}). Unlike the other plant micronutrients, Mo is most readily taken up by plants in soils with a pH above 7 and is relatively unavailable in acid soils. This maybe explained because Mo solubility in well-aerated soils is low due to the formation of insoluble ferric molybdates. In poorly drained soils, formation of soluble ferrous molybdates or molybdites may lead to high Mo availability to plants (Welch et al., 1991). Plants take up Mo mainly as molybdate ions, and its absorption is proportional to its concentration in the soil solution. Molybdenum is highly important for legumes for biological fixation of atmospheric N (Srivastava, 1997). In N_2-fixing systems, Mo acts as a co-factor of the proteins responsible for electron transfer in the synthesis of the nitrogenase complex, responsible for the conversion of N_2 into ammonium (Martens and Westermann, 1991).

Molybdenum deficiency frequently limits biological N-fixation under field conditions (Campo et al., 2000), resulting in poor growth and activity of root nodules. In the Pampas region, the availability of Mo is very low, ranging from 0.01 to 0.10 mg kg^{-1} (Lavado and Porcelli, 2000). Responses in soybean to molybdenum (Mo) and cobalt (Co) were observed when these elements were applied to seeds or as foliar applications at early stages (Melgar et al., 2001). Molybdenum and Co applied to seeds together with inoculants which provide the nitrogen fixing bacteria of the genus *Bradyrhizobium japonicum* improved yields by 8% over the single inoculation of fixing bacteria, and by 15% over the no inoculated control in the N of Buenos Aires province and W of Santa Fe (Ferraris and Couretot 2008). However, the results are variable and there have been situations of non-response in the same region. It seems as if the response to Co-Mo would increase when the soil pH drops below 5.4/5.5, a frequent situation in soils with long agricultural history.

Chlorine (Cl)

Chlorine is a common halogen element under terrestrial conditions. Like other halogens, the distribution of Cl in soils exhibits a clear trend of decreasing concentration with increasing distance from the sea. Rainfall is usually an adequate source of Cl but in inland areas, remote from the sea, inputs from this source are generally low. Amounts of Cl deposited from rain range from 18 to 36 kg ha^{-1} in continental areas to greater than 100 kg ha^{-1} in coastal areas (Heckman, 2007). In soils of humid climate zones, Cl is leached down the profile and transported with drainage water, whereas in soils of arid and semiarid climates, this nutrient is known to be concentrated in surface horizons. In general, soils do not adsorb chloride ions in significant amounts. Therefore, regular inputs of this element from rainfall, weathering minerals, groundwater or fertilisers are necessary to maintain an adequate supply for crops. Several research studies demonstrate the importance of fertilizing crops with Cl to maintain grain yields, especially in soils where Cl availability is low, such as sandy and sandy loam soils (Sillanpaa 1982).

Díaz-Zorita et al. (2004) studied the effects of two chloride fertilizer sources (KCl and NH_4Cl) on dryland spring wheat (*Triticum aestivum* L.) on the sandy Pampas region. These soils contained high to very high extractable K levels. Their research indicated that chloride fertilization increased grain yields in 50% independently of the fertilizer source. Averaged over 10 locations, the grain yield benefit due to chloride fertilization was 253 kg ha^{-1} and it was mostly explained by a greater number of grains per square meter. They concluded that soil Cl^- levels of 13.2 mg kg^{-1} (0.0 to 0.2 m) were adequate for maximum spring wheat grain yields in the sandy Pampas region.

Responses to Cl fertilization have been also observed in wheat in the Rolling Pampas. In 10 of 26 field trials, soil applications of 46 kg Cl ha^{-1} increased crop yields by 319 kg ha^{-1}. Concentrations of soil Cl (soil depth 0.20 m) greater than 35 mg kg^{-1} or amounts of soil Cl (soil depth 0.60 m) greater than 65-70 kg Cl ha^{-1} showed yield responses lower than 10% (García 2008). Several research documented the benefits of Cl fertilization in lowering the negative effects of disease during the development of the crops and the subsequent improvement in yield (Xu et al., 2000; Mann et al, 2004; Chen et al., 2010). Conversely, foliar disease reduction was not observed in the Pampas region (García, 2008).

Copper (Cu)

The total concentration of Cu in soils varies from 2 to 100 mg kg-1 (Mortvedt, 2000). Since cropping areas of the Pampas region are located far from contamination sources such as mining, factories or large cities, the two main factors that govern the initial Cu status in soils are parent material and soil formation processes. In the Pampas region, average soil availability of copper is 2.13 mg kg-1, although lower values have been reported in some areas in the centre of Buenos Aires province and centre-south of Córdoba (Cruzate et al. 2006). Copper is tightly bound to compounds in organic matter even more so than the other micronutrients (Mathur and Levesque, 1983). Chelating and the formation of organic complexes are the key reactions governing Cu behavior in most soils. Accordingly, common distribution of Cu in soil profiles indicates a light accumulation in the top horizons, as a result of bioaccumulation. Table 5 summarizes concentrations of Cu concentration in top soils of the Pampas region.

Table 5. Cu availability in representative soils of the Pampas region

Extractant	*Soils*	Zone	Range (mg kg^{-1})	References
DTPA	Argiudolls	NW Buenos Aires and S Santa Fé	0.81-1.85	Ratto de Miguez and Fatta (1990)
	Typic Argiudoll	N Buenos Aires	0.60-1.33	Ratto et al. (1997)
	Argiudolls	N Buenos Aires	1.38	Lavado and Porcelli (2000)
	Argiudolls and	S Córdoba	0.60- 2.10	Buffa and Ratto (2005)
	Haplustolls	N Buenos Aires	2.25-3.40	
EDTA	Haplustolls	La Pampa	0.60-2.40	Gonzalez and Buschiazzo (1998)
	Argiudolls	N Buenos Aires	2.20-3.40	Urricariet and Lavado (1999)
$CaCl_2$ + NaOH	Typic Hapludoll	W Buenos Aires	3.60-4.10	Torri and Lavado (2000)
	Typic Argiudoll	N Buenos Aires	6.30-7.70	Torri and Lavado (2000)
	Typic Natraquoll	S Buenos Aires	3.00-3.90	Torri and Lavado (2000)

Table 6. Soil distribution of Cu (water-soluble and exchangeable fraction (EXCH), organic matter bound fraction (OM), inorganic precipitate fraction (INOR) and residual fraction (RES)) in three pristine representative soils of the Pampas region (adapted from Torri and Lavado (2008 a)

Soils	EXCH-Cu	OM- Cu	INOR-Cu	RES-Cu
	mg kg^{-1}			
Typic Hapludoll	< 0.5 mg Cu kg	3.85 ±0.29	2.73 ±0.10	15.35 ± 0.30
Typic Natraquoll	< 0.5 mg Cu kg	3.46 ±0.15	3.86 ±0.31	3.60 ± 0.46
Typic Argiudoll	< 0.5 mg Cu kg	7.01 ±0.03	6.56 ±0.04	2.35 ± 0.07

Although the most common mobile species in the surface environment is believed to be the Cu^{2+}, this species was below the analytical detection limit in three representative pristine soils (< 0.5 mg Cu kg^{-1}) of the Pampas region. The largest proportion of Cu was specifically adsorbed to organic matter and, to a lesser extent, precipitated with primary or secondary minerals or in residual fractions (Table 6). Contrary to what happens with other cationic micronutrients, the increase in pH increases the precipitated forms of Cu, while decreasing the concentration of Cu bound to organic fraction (Alva et al., 2000, Torri, 2001).

According the literature, the soils most often associated with Cu deficiency include sandy soils (Fagería et al., 2002). Soils of the Semiarid Argentinean Pampas include Typic Ustipsamments, Entic Haplustolls and Entic Hapludolls. However, no Cu deficiencies were reported in these soils or in the rest of the Pampas region in recent years (Sainz Rozas et al., 2003; Melgar, 2006).

Iron (Fe)

Iron is one of the major constituents of lithosphere. Iron is believed to occur in soils mainly in the forms of oxides and hydroxides as small particles or associated with mineral surfaces. Iron stage of oxidation in weathering processes is largely dependent on the Eh-pH system of the environment: oxidizing and alkaline conditions promote the precipitation of Fe^{3+} whereas acid and reducing conditions promote the reduction of Fe^{3+} to Fe^{2+}, increasing the solubility of Fe compounds. The content of soluble Fe in soils is extremely low in comparison with total Fe content. In the majority of soils, the concentration of soluble Fe in the rhizosphere is often far below the level required for adequate plant growth (Marschner, 1995). The solubility of Fe in well-aerated soils is controlled by dissolution and precipitation of Fe^{3+} (Moraghan and Mascagni, 1991). The concentration of Fe in soil solutions within common soil pH levels ranges from 30 to 550 μg L^{-1}, whereas in very acid soil it can exceed 2000 μg L^{1} .

Iron deficiency may be induced by heavy application of P fertilizers. Although P may precipitate Fe, root internal immobilization of Fe - probably due to the formation of Fe phosphate - appears to be the cause of Fe chlorosis (Fageria et al., 1990). Neither deficiencies nor responses to Fe addition were registered in the Pampas soils (Sainz Rozas et al. 2003; Melgar, 2006). Symptoms of Fe deficiencies were found in some lime-rich soils, i.e. Calciustoll in the semiarid area (R. Lavado, *personal communication*).

Manganese (Mn)

Manganese is one of the most abundant trace elements in the lithosphere, and its common concentration range in soils is 20 to 3000 mg kg^{-1} (mean of 600 mg kg^{-1}) (Lindsay, 1979). In soil, the complex mineralogical and chemical behaviour of Mn results in the formation of the large number of oxides and hydroxides. The solubility of Mn in soils is highly dependent on the pH and redox potential (Lindsay, 1979; Mortvedt, 2000). Microbiological soil activity is also known to be largely responsible for the oxidation and reduction of Mn compounds, as well as for the formation of Mn concretions (Le Mer and Roger 2001).

Weighted average availability of Mn in the Pampas region is 85.38 mg kg^{-1} (Cruzate et al., 2006). Although there are three areas in the centre of Santa Fe, S and E of Buenos Aires with higher availability values, the likelihood of response to Mn fertilization covers much of the Pampas region area. Manganese deficiency in the Pampas region has been detected in soybeans (G. Ferraris, *personal communication*)

PRODUCTS TO USE AND APPLICATION FORMS

Correction of micronutrient deficiencies shows some differences from that of macronutrients. The easiest and most simple practice to correct micronutrient deficiency is to apply fertilizers to soils or foliar sprays during vegetative growth (Torri et al., 2006; Melgar et al., 2001, Sainz Rozas et al., 2003). In turn, fertilizers with micronutrients also show a complex picture, as they can be applied to soils or plants in many chemical and physical forms. Inorganic compounds, oxides or salts (carbonates, chlorides, nitrates, sulfates, borates, molybdates, etc) are applied to soils. Although this kind of application is economically more affordable, most soil-applied micronutrients are quickly fixed into plant-unavailable forms. Thus, soil-applied micronutrient requires adding more fertilizer along growth stage to maintain a sufficient level of phytoavailable element. Furthermore, many studies reported that soil application of micronutrient fertilizers produces minimal or no increases in grain micronutrient content even when yield was increased (Khoshgoftarmanesh et al, 2010).

Foliar applications are used to supply micronutrients more rapidly in order to correct severe deficiencies commonly induced during the early stages of growth but they are temporary solutions to the problem. Greater absorption by leaves is favored under low light, optimum temperature, and high humidity conditions (Mortvedt, 2000). To increase the effectiveness of foliar uptake, wetting agents are usually added to sprays. These chemicals are neutral nonionic compounds which reduce surface tension and increase wetting of leaf surfaces to enable larger amounts of solution to be absorbed (Kannan, 1990). Multiple sprays may be required for season-long correction (Mortvedt, 2000). However, most foliarly-applied micronutrients are not efficiently transported toward roots, which may remain deficient. For foliar fertilization, soluble salts and organic compounds, like chelates are used. Chelates are expensive and are mainly used in orchards. Application of micronutrients to either soil or foliage may be successful in increasing yield only in well-defined circumstances. Micronutrient fertilization is crop specific and varies with the growth stage and place. The application technology is also very important. That is why extrapolating results to a new situation (i.e. to generalize a dose and moment of application) is unlikely to be successful without a specific adjustment in the fertilizer practice. Consequently, only few examples will be considered.

Sunflower suffers B deficiencies in and B foliar application is recommended at two times of the crop cycle (Diggs et al., 1992): in vegetative stages, applying 1.2 kg B ha^{-1} and at the blooming start (0.6 kg B ha^{-1}). In western Buenos Aires province B responses were associated to Melich III-B extractable concentration at crop seeding (Díaz Zorita and Duarte, 1998). Soybean is also usually deficient, and responds to B fertilization. The addition of coadjutants to the borate solutions increased the crop responses to foliar fertilization (Ferraris et al. 2005).

In NW Buenos Aires and S Santa Fe provinces, Melgar et al. (2001) evaluated the response of maize to Zn applications, relating these responses to its leaf concentration and soil availability. These authors found a lineal response to Zn application: 109 kg maize/ kg of applied Zn (0 to 6 kg Zn ha^{-1} as oxisulfate in the sowing line). Ferraris and Couretot. (2009 a) found similar results in maize in the west of Buenos Aires province. They also experimented foliar application of Zn with 4-7 leaf of maize. Ferraris and Couretot (2009 b) also studied the incorporation of Zn together with nitrogen-sulfur liquid fertilizers, and its application in early vegetative stages with 3-6 leaf of maize. The application of Zn in the seed treatment of maize (seed priming) was recently studied (Quintero et al. 2006). These authors reported enhanced germination speed and synchrony of primed seeds compared to control.

Final Considerations

Micronutrient availability has historically been considered suitable for crops in the Pampas region. However, some micronutrient deficiencies had been observed in last years as well as responses to fertilizer application. The lack of nutrients reposition resulted, with time, in a depletion of soil phytoavailable nutrients. At present, there are responses of Zn in maize and B in sunflower, although their application is not yet generalized. Responses to Cl are not economically significant and deficiencies of Cu, Fe, Mn and Mo are exceptional. In vegetables, fruits and other intensive crops micronutrient deficiency are more common, in special those irrigated with alkaline waters. As a synthesis, micronutrient fertilization is a new issue in commercial farms of the Pampas region, with lower development on diagnostic techniques and application technology.

References

AAPRESID. 2011. *No-tillage evolution in Argentina*. AAPRESID. Rosario, Santa Fe, Argentina. Available at www.aapresid.org.ar. Accessed January 30, 2011.

Adriano, D C. 2001. *Trace elements in terrestrial environments*: biogeochemistry, bioavailability and risks of metals, 2nd edn., p. 860. Springer, New York.

Alva A, B Huang and S Paramasivam. 2000. Soil pH affects copper fractionation and phytotoxicity. *Soil Sci. Soc.Am. J.* 64: 955-962.

Andrade, F y V Sadras. 2000. Bases para el manejo del maíz, el girasol y la soja. *Ed Médica Panamericana*. pp 207-232. Balcarce.

Arnon, DI and PR Scout. 1939. The essentiality of certain elements in minute quantity for plants with special reference to copper. *Plant Physiology* 14:371-375.

Asher, CJ. 1991. Beneficial elements, functional nutrients, and possible new essential elements. In Mortvedt, J.J., Cox, F.R., Shuman, L.M., and Welch, P.M., eds., Micronutrients in Agriculture, 2nd edn. Madison, WI: *Soil Science Society of America*, pp. 703–723.

Barber, S A 1995. *Soil nutrient bioavailability*: A mechanistic approach, 2nd edition. Wiley,New York.

Barker AV and DJ, Pilbeam. 2007. *Handbook of Plant Nutrition*, Taylor and Francis Group Press, Boca Raton, FL.

Bernardos, JN, EF Viglizzo, V Jouvet, FA Lértora,AJ Pordomingo, FD Cid,. 2001. The use of EPIC model to study the agroecological change during 93 years of farming transformation in the Argentine Pampas. *Agric. Syst.* 69: 215–234.

Blamey FPC, RK Zollinger and AA Schneiter.1997. *Sunflower production and culture*. In Sunflower Science and Tehnology. Ed Schneiter AA. American Society of Agronomy, Madison.

Brown, PH, RM Welch, and EE Cary. 1987. Nickel: A micronutrient essential for higher plants. *Plant Physiol.* 85: 801–803.

Buffa, EV y SE Ratto. 2005. Disponibilidad de cinc, cobre, hierro y manganeso extraíble con DTPA en suelos de Córdoba (Argentina) y variables edáficas que la condicionan. *Ciencia del Suelo* 23:107-114.

Campo, RJ, UB Albino and M Hungria. 2000. *Importance of molybdenum and cobalt to the biological nitrogen fixation*. In: F.O. Pedrosa, M. Hungria, G. Yates and W.E. Newton, Editors, Nitrogen Fixation: From Molecules to Crop Productivity, Springer, pp. 597–598.The Netherlands.

Ciampitti, IA y FO García. 2008. Requerimientos nutricionales. Absorción y extracción de macronutrientes. II. Hortalizas, frutales y forrajeras. *Archivo Agronómico* 12, Informaciones Agronómicas del Cono Sur 37.

Cruzate, GA,, E Rivero and R Turati. 2006. *Cobre, hierro y manganeso*: mapas de disponibilidad y reposición en suelos de la Región Pampeana, XX Congreso Argentino de la Ciencia del Suelo, I Reunión de Suelos de la Región Andina, Salta..

Díaz-Zorita, M and G Duarte. 1998. Diagnóstico y manejo de la fertilización de trigo en el Oeste bonaerense. *Publicación Técnica* 26. INTA. -15 p.

Díaz-Zorita, M, GA Duarte, and GA Barraco, M. 2004 Effects of chloride fertilization on wheat (Triticum aestivum L.) productivity in the Sandy Pampas Region, Argentina. *Agronomy Journal* 96: 839-844..

Diaz-Zorita,M. 2000. *Nutrición del girasol*. Suplemento Girasol. Revista Agromercado

Diggs C, S Ratto de Miguez y V Schorrocks. 1992. La evaluación de síntomas de deficiencias de B. El método más confiable para decidir fertilizaciones de B en girasol. 13th *International Sundflower Conference*, Palazzo dei Congressi Paccini Editore, Pisa, Italy.

Echeverría, HE and J Ferrari. 1993. *Relevamiento de algunas características de los suelos agrícolas del sudeste bonaerense*. INTA, Ce.R.B.A.S. , E.E.A. Balcarce, Boletín Técnico 112.

Epstein, E and A Bloom. 2005. *Mineral nutrition of plants: Principles and perspectives*. 2nd ed. Sinauer Associates, Sunderland, Maryland.

Eskew, DL, RM Welch, and EE Cary. 1983. Nickel: An essential micronutrient for legumes and possibly all higher plants. *Science* 222: 621–623.

Fageria, NK, VC Baligar and RB Clark. 2002. Micronutrients in crop production. *Advances in Agronomy* 77: 185-268.

Fan, MS, FJ Zhao, SJ Fairweather-Tait, PR Poulton, SJ Dunham and SP McGrath. 2008. Evidence of decreasing mineral density in wheat grain over the last 160 years. *J. Trace Elements Med. Biol.* 22 : 315–324.

Ferraris, G y L Couretot. 2008. *Estrategias de inoculación en soja*. En: Soja en Siembra Directa. ISSN 1850-0633. AAPRESID.

Ferraris, G y L Couretot. 2009 a. Respuesta del maíz a la fertilización complementaria con zinc, boro y otros nutrientes. En: Revista Agromercado. ISSN 1514-2213. *Cuadernillo Fertilidad y Fertilizantes.* 28:22-24.

Ferraris, G y L Couretot. 2009 b. Tecnologías para la aplicación de microelementos en maíz. Dosis y sistemas de aplicación de zinc en combinación con fuentes nitrógeno-azufradas. *Revista Fertilizar* 12:21-26.

Ferraris, G, L Couretot, y J Ponsa. 2005. Evaluación de la utilización de molibdeno, cobalto, boro y otros nutrientes en soja de primera. En: Soja. Resultados de Unidades demostrativas del Proyecto Regional Agrícola, año 2005. *CERBAN. Áreas de Desarrollo Rural EEA INTA Pergamino y General Villegas*. pp 62-65

Garcia, F. 2008. Cloro en trigo: Resultados de las experiencias en la región pampeana argentina. Años 2001 a 2006. *Informaciones Agronómicas* 38: 17-21. IPNI Cono Sur. Acassuso, Buenos Aires.

Garvin, DF, RM Welch and JW Finley. 2006. Historical shifts in the seed mineral micronutrient concentration of US hard red winter wheat germplasm, *J. Sci. Food Agric.* 86:2213–2220.

Goldberg, S, HS Forster and EL Heick. 1993. Boron adsorption mechanisms on oxides, clay minerals, and soils inferred from ionic strength effects. *Soil Sci. Soc. Am. J.* 57:704-708.

Gonzalez, G y D Buschiazzo. 1998. Contenido de hierro, cobre, manganeso y cinc em suelos de la provincia de La Pampa, rgentina. *XV Congreso Argentino de la Ciencia del Suelo.*

Graham, RD and RM. Welch. 1996. Breeding for staple-food crops with high micronutrient density. Agricultural Strategies for Micronutrients. p 1-72. *Working Paper No. 3.* Washington, DC: International Food Policy Research Institute.

Gregory, J and C Frink. 1995. Phosphorus and zinc fertilization of corn grown in Connecticut soil. *Commun. Soil Sci. Plant. Anal.* 26: 269.276.

Gupta, U C. 1997. *Symptoms of molybdenum deficiency and toxicity in crops*. In: Gupta U C (ed.), Molybdenum in Agriculture. U.K: Cambridge. Pp 160-170. University Pres.

Heckman, JR. 2007 Chlorine. In Barker, A.V., Pilbeam, D.J. (Eds.), *Handbook of Plant Nutrition*, CRC Press, Taylor and Francis. pp. 279–291. Boca Raton, FL.

Imbellone, P A, J E Giménez, JL Panigatti. 2010. Suelos de la Región Pampeana. Procesos de formación. *INTA.* 320 p.

Kabatas-Pendias, A and H Pendias. 2001. *Trace elements in soils and plants.* Third edition. Boca Raton, FL: CRC Press.

Khoshgoftarmanesh, AH, R Schulin, RL Chaney, B Daneshbakhsh, and M Afyuni. 2010. Micronutrient-efficient genotypes for crop yield and nutritional quality in sustainable agriculture. *A review. Agronomy for Sustainable Development* 30: 83-107.

Lavado, RS and CA Porcelli. 2000. Contents and main fractions of trace elements in Typic Argiudolls of the Argentinean Pampas. *Chemical Speciation and Bioavailability*. 12: 67-70.

Lavado, RS and M A Taboada. 2009. The Argentinean Pampas: A key region with a negative nutrient balance and soil degradation needs better nutrient management and conservation programs to sustain its future viability as a world agroresource. *Journal of Soil and Water Conservation* 64:150-153.

Lavado, RS, M Rodríguez,J Scheiner,M Taboada, G Rubio, R Alvarez, and M Alconada. 1998. Heavy metals in soils of Argentina: comparison between urban and agricultural soils. *Commun. Soil Sci. Plant Anal.* 29: 1913–1917.

Lavado, RS, M Zubillaga, R Alvarez and M.A Taboada. 2004. Baseline levels of potentially toxic elements in Pampas Soils. *Soil and Sediment Contamination* 13:427-437.

Le Mer, J and P Roge. 2001. Production, oxidation, emission and consumption of methane by soils: A review. *European Journal of Soil Biology*, 37:25-50.

Lindsay, WL and WA Norvell. 1979. Development of a DTPA soil test for zinc, iron, manganese and copper. *Soil Sci. Soc. Am. J.*, 42: 421–428.

Marschner, H. 1995. *Mineral nutrition of higher plants* (2nd ed), Academic Press, London.

Martens, DC and DT Westermann. 1991.Fertilizers applications for correcting micronutrient deficiencies. In: J.J. Mortvedt, F.R. Cox, L.M. Shuman and R.M. Welch, Editors, Micronutrients in Agriculture (2nd edition), *Soil Science Society of America*, pp. 90–112. Madison, WI.

McGrath, SP and J Cegarra. 1992. Chemical extractability of heavy metals during and after long-term applications of sewadge sludge to soil. *Journal of Soil Science* 43: 313–321.

Melgar R. 1997. Potasio, azufre y otros nutrientes necesarios para considerar en una fertilización. Fertilizar Sup. Trigo. p. 17-24.

Melgar, R, J Lavandera, M Torres Duggan y L Ventimigli.2001. Respuesta a la fertilización con boro y zinc en sistemas intensivos de producción de maíz. *Ciencia del Suelo* 19:109-124.

Melgar, R. 2006. *Uso de micronutrientes en cultivos de gruesa*.www.Fertilizar.com.ar

Mizuno, I, AM. Villa, M Jiménez, M Moretti,V Sanguesa, D Efron, y L Berasategui. 1988. Elementos mayores y menores en algunos perfiles de suelos de la provincia de Buenos Aires. *Rev. Facultad de Agronomía.*

Moraghan, JT and H Mascagni. 1991. Environmental and soil factors affecting micronutrient deficiencies and toxicities. In Micronutrients in Agriculture. Second Edition (Eds. JJ Mortvedt, FR Fox, LM Shuman, RM Welch) pp. 371-425. *Soil Sci. Soc. Am.*, Madison.

Moral, R, RJ Gilke, and J Moreno-Casalles. 2002. A comparison of extractants for heavy metals in contaminated soils from Spain. *Commun. Soil Sci. Plant Anal.* 33:2781-2791.

Mortvedt, J J, FR Cox, LM Shuman, and RM Welch. 1991. Micronutrients in Agriculture. Number 4, *Soil Sci. Soc. of Am. Book Series*, Madison , WI.

Mortvedt, J J. 2000. Bioavailability of micronutrients. In: *Hand book of Soil Science*. M E Sumner, (Ed.), pp. D71–D88. CRC Press, Boca Raton, FL.

Moscatelli, G and MS. Pazos. 2000. Soils of Argentina: Nature and Use. In: I. Kheoruenromne and S. Theerawong (Eds.). Bangkok, Thailand. Proceedings of the International Symposium on Soil Science: *Accomplishments and Changing Paradigm towards the 21st Century, and IUSS Extraordinary Council Meeting* 81-92.

Power, PF and WG Woods. 1997. The chemistry of boron and its speciation in plants. *Plant and Soil* 193: 1-13.

Quevauviller, P, M Lachica, E Barahona, A Gomez, G Rauret, A Ure, and H Muntau. 1998. Certified reference material for the quality control of EDTA and DTPA-extractable trace metal contents in calcareous soil (CRM 600). *Fresen. J. Anal. Chem.* 360: 505-511.

Quiñonez, AG, L Dalla Fontana and AL Mollo.2003. Respuesta de la alfalfa al agregado de fósforo, boro y calcio. *Rev. FAVE* 2:47-53.

Quintero, C.E.; E.S. Arévalo; N.G. Boschietti y N.M. Spinelli. 2006. Clorosis en suelos con calcáreo. En: M Vázquez (Ed.) Micronutrientes en la agricultura. *AACS, Argentina.* pp 113-125.

Ratto de Miguez S, Fatta N, Lamas MC. 1991. Análisis foliar en plantas de cultivo de maíz II. Micronutrientes. *Rev. Facultad de Agronomía* 12: 31-38.

Ratto de Míguez, S y N Fatta. 1990. Disponibilidad de micronutrientes en suelos del área maicera núcleo. *Ciencia del Suelo* 8: 9-15.

Ratto, SE. 2006. Los microelementos en el sistema productivo del área pampeana. En: M Vázquez (Ed.) *Micronutrientes en la agricultura*. pp 79-111. AACS. Buenos Aires.

Ratto, SL, L Giuffre y C Sainato. 1997. Variación espacial de micronutrientes en suelo y planta en un Molisol. *Ciencia del Suelo* 15:39-41.

Rivero, E, GA Cruzate y R Turati. 2006. Azufre, boro y zinc: mapas de disponibiilidad y reposicion en suelos de la Región Pampeana. *Proceedings of XX Congreso Argentino de la Ciencia del Suelo.* I Reunión de Suelos de la Región Andina.

Ron, M, S Bussetti y T Loewy. 1999. Boro extraíble en suelos del sudoeste bonaerense. *Ciencia del Suelo* 17: 54-57.

Sainz Rozas, H, HE Echeverría, PA Calviño, PA Barbieri y M Redolatti. 2003. Respuesta del trigo al agregado de cinc y cobre en suelos del sudeste bonaerense. *Ciencia del Suelo* 21: 52-58.

Satorre EH and GA Slafer. 1999. *Wheat production systems of the Pampas*. In: E.M. Satorre and G.A. Slafer, Editors, Wheat. Ecolocy and Physiology of Yield Determination, The Haworth Press, Inc., pp. 333–348. New York. .

Scott, H. D., Beasley, S. D., and Thompson, L. F. (1975). Effect of lime on B transport to and uptake by cotton. *Soil Sci. Soc. Am. Proc*. 39, 1116–1121.

Shuman, L.M., 1999. Organic waste amendments effect on zinc fractions of two soils. *Journal of Environmental Quality* 28, 1442–1447

Sillanpaa, M. 1982. Micronutrients and the nutrient status of the soils: A global study. *Food and Agriculture Organization* (FAO), Rome.

Silva, AP, S Imhoff, NFB.N.F.B. Giarola, CA Tormena. 2001. Análisis multivariado y univariado en la discriminación de sistemas de uso de suelos del Centro de Santa Fe. *Edafología* (Madrid) 8:21-34.

Srivistava, PC and UC Gupta. 1996. *Trace elements in crop nutrition.* Science Publishers, Lebanon, NH, 356 pp.

Srivistava, PC. 1997. *Biochemical significance of molybdenum in crop plants*. In: Gupta U C (ed.), Molybdenum in Agriculture. pp: 47–70.U.K. Cambridge University Press.

Torri S.I, Urricariet A.S, Ferraris, G, Lavado R. 2010. *Micronutrientes en agrosistemas*. En: Fertilidad de suelos. Caracterización y manejo en la región pampeana. R. Alvarez, G.Rubio, C.Alvarez y R Lavado (eds). EFA, pp 395-423, 496 p. ISBN 978-950-29-1234-9.

Torri S I. 2001. Distribución y biodisponibilidad de Cd, Cu, Pb y Zn en suelos fertilizados con biosolidos. Tesis de Magister Scientiae en Ciencias del Suelo, FAUBA.

Torri, S.I and R Lavado. 2008 a. Dynamics of Cd, Cu and Pb added to soil through different kinds of sewage sludge. *Waste Management*, 28: 821-832.

Torri, SI and R Lavado. 2008 b. Zn distribution in soils amended with different kinds of sewage sludge. *Journal of Environmental Management*, 88: 1571-1579.

Torri, S.I, S Urricariet S and R. Lavado. 2006. Micronutrientes y elementos traza. En: Fertilidad de suelos y Fertilización de cultivos. García F y Echeverría H. Ediciones *INTA*, pp 189-203. Balcarce.

Urricariet, S and RS Lavado. 1999. Indicadores de deterioro en suelos de la Pampa Ondulada. *Ciencia del Suelo* 17: 37-44.

Viglizzo, EF, FA Lertora, AJ Pordomingo, JN Bernardos, ZE Roberto and H Del Valle. 2001. Ecological lessons and applications from one century of low external input farming in the pampas of Argentina. *Agric. Ecosyst. Environ.* 83:65–81

Waters, B M and RP Sankaran. In Press. Moving micronutrients from the soil to the seeds: Genes and physiological processes from a biofortification perspective Plant. *Science*, doi:10.1016/j.plantsci.2010.12.003

Welch, R.M and R.D. Graham, 2005. Agriculture: the real nexus for enhancing bioavailable micronutrients in food crops. *Journal of Trace Elements in Medicine and Biology* 18: 299-307.

Welch, RM, WH Allaway, WA House and J Kubota. 1991. Geographic distribution of trace element problems, In: Mortvedt, J.J., Cox, F.R., Shuman, L.M., Welch, R.M. (Ed.), Micronutrients in Agriculture, 2nd ed, pp. 31–57. *Soil Sci. Soc. Am. Madison,* WI.

WHO: The World Health Report. 2002. Reducing risks, promoting healthy life. Geneva, Switzerland: *World Health Organization*; 2002:1–230.2

Xu, G, H Magen, J Tarchitzky and U Kafkaf. 2000. Advances in chloride nutrition of plants. *Advances in Agronomy* 28: 97–150.

Zhang, F, V Römheld, and H Marschner., 1989. Effect of zinc deficiency in wheat on release of zinc and iron mobilizing root exudates. *Z. Pflanzenernähr. Bodenkd.* 152: 205–210.

Zhang, H, FJ Zhao, B Sun, W Davison, and SP McGrath. 2001. A new method to measure effective soil solution concentration predicts copper availability to plants. *Environ. Sci. Technol.* 35:2602—2607.

Zhao, F and S McGrath. 2009. Biofortification and phytoremediation. *Current Opinion in Plant Biology* 12:373–380.

In: Soil Nutrients
Editor: Mohammad Miransari

ISBN: 978-1-61324-785-3

Chapter 14

SOIL MICROBES AND SOIL NUTRIENTS

M. Miransari*[1*]*, H. Omidi*[2]*, A. Korde Naij*[2]*,
***M. Amini*[2]*, and S. Maleki*[2]**

[1]corresponding author: Department of Soil Science,
College of Agricultural Sciences, Shahed University, Tehran, Iran
[2]Department of Agronomy and Plant Breeding, College of Agricultural Sciences, Shahed University, Tehran, Iran

ABSTRACT

Soil nutrients including micro- and macro-nutrients are necessary components for plant growth and yield production as they regulate different plant activities. There are different ways of supplying plant necessary nutrients including the use of chemical and biological fertilization. Chemical fertilization is a quick method of supplying nutrients to the plant; however it is not economically and environmentally recommendable especially if used at excess amounts. Biological fertilization, which is the use of soil microbes including arbuscular mycorrhizal fungi and plant growth promoting rhizobacteria (PGPR) has been proved to be an efficient method of supplying soil nutrients to plant, while economically and environmentally recommendable. In this chapter some of the most important details regarding the effects of soil microbes on the availability and hence uptake of nutrients by plant is reviewed. Such details can be important for the selection and hence production of appropriate microbial inoculums used for biological fertilization.

Keywords: Biological and chemical fertilization, AM fungi, PGPR, plant growth, soil nutrients

---

* E-mail: Miransari1@gmail.com, Miransari @shahed.ac.ir.

INTRODUCTION

Soil microbes are among the most important ecosystem components and their functioning in the soil result in the enhanced soil efficiency and hence plant growth. There is a wide range of microbes in soil, which are beneficial, neutral or pathogenic. The most applicable soil microbes are the ones, which are beneficial to plant growth and ecosystem health including soil fungi such as arbuscular mycorrhizal (AM) fungi and soil bacteria such as plant growth promoting rhizobacteria (PGPR), which also include the N-fixing bacteria, *rhizobium* (Miransari 2007; 2009; Miransari 2010a,b; 2011a,b).

Although increasing the rate of food production for the world increasing population has made the use of different chemicals including fertilizer and biocides necessary, their excessive use has resulted in some unfavorable effects, economically and environmentally. Hence, with respect to the favorable effects of soil microbes on the soil productivity and ecosystem health, they have been extensively used to improve plant growth and yield production in different parts of the world. The activities of soil microbes include: 1) mineralization of soil organic matter and hence nutrient cycling, 2) improving soil structure, 3) interacting with other soil microbes, 4) production of different plant hormones and 5) controlling pathogens in soil (Artursson et al. 2006; Compant et al. 2010; Egamberdiyeva 2007; Fang et al. 2005; Miransari 2011b).

Plants require at least 16 macro- and micro-nutrients for their growth and yield production. The absence or deficiency of each nutrient in soil can adversely affect plant growth and yield production. Nutrient availability in soil is determined by different parameters including soil and climate properties, plant species and microbial activities. The optimum supplying of soil nutrients can enhance plant growth and yield production, while economically and environmentally, friendly (Marschner 1995; Richardson et al. 2009)

Some of the soil nutrients are present in soil but not available to the plant, due to their chemical properties. Hence, different methods have been tested and proved to be effective on the enhancement of soil nutrients availability in the soil. Such effects can be of significance for the environment and for crop production. Biological methods are among the most important ones. The presence of different microbes in soil and the inoculation of soil with different microbes, which is called biofertilization can efficiently affect the availability of soil nutrients and hence plant growth (Adesemoye and Kloepper 2009; Miransari 2010a; Wang et al. 2010).

The important point about the use of soil microbes is the selection of the right combination of microbes, which have the highest efficiency under different conditions including stress. In other words, for the optimum supplement of nutrients to plant by biofertilization, it is necessary that the most efficient soil microbes be selected and used. Different microbial species and strains may perform differently under different conditions, determining their efficiency and hence their related use (Adesemoye and Kloepper 2009; Miransari 2011b).

PGPR are soil bacteria, which can favorably affect plant growth and yield production by their interesting activities in soil. The soil bacteria, *rhizobia*, which can fix atmospheric N in their symbiosis with their specific host plant, are also among soil PPGR. PGPR can fix nutrients such as N and influence different soil properties including the availability of nutrients by producing a wide range of products. Such products include different enzymes and plant hormones such as phosphatases and auxins, respectively (Lugtenberg and Kamilova 2009; Podile and Kishore 2006; van Loon 2007).

AM fungi can increase the uptake of water and nutrients by their host plant through the process of symbiotic association with most terrestrial plants. In such a symbiosis the host plant supplies the fungi with its necessary carbon in the exchange for water and nutrients. In the presence of the host plant roots, the fungal spore would germinate and the produced fungal hyphae grow into the mycorrhizosphere (the space around the host plant roots, with mycorrhizal presence) and into plant cortical root cells resulting in the formation of vesicles and arbuscules (Smith and Read 2008). In the following some of the latest details regarding the effects of soil microbes on the availability of soil nutrients and hence their uptake by plant is reviewed.

Soil Microbes and Plant N Uptake

Soil microbes such as bacteria can affect plant N uptake by different processes including the symbiotic and non-symbiotic N fixation. *Rhizobium* are soil bacteria, which are able to fix atmospheric N in symbiosis with their specific host plant as the enzyme nitrogenase can reduce N to ammonium. The process of N fixation initiates with some communications between the two symbionts by producing some biochemicals, which are usually flavonoids. Such products determine the specificity of the process of N fixation (Miransari and Smith, 2007; 2009; Ferguson et al. 2010; Van de Velde et al. 2010).

The *rhizobium* bacteria realize the presence of the host plant by the produced flavonoids, resulting in the chemotactic movement of the bacteria toward the plant roots. As a result, the activated bacterial genes produce products such as lipochitooligosaccharides, which are able to alter the morphology of plant roots, especially the root hairs. The bacteria, which are in the form of bacteroids, can enter the root cells by the production of the infection thread and hence reside in the root cortical cells, which their subsequent division produces the nodules, the place of N-fixation by the rhizobium (Miransari et al. 2006; Ferguson et al. 2010; Van de Velde et al. 2010).

The legume host plant provides the bacteria with its necessary carbohydrates in the exchange for the fixed N by the bacteria. Under optimum conditions the bacteria is able to fix significant amounts of atmospheric N, which is of economical and ecological importance. It is because the process of N fixation can significantly decrease the amounts of chemical fertilization, which are, especially under humid conditions, subjected to processes such as leaching and denitrification (Miransari and Mackenzie 2011a; b; Miransari and Smith 2008; Miransari and Smith 2007; 2009).

There are also soil bacteria such as *Azospirillum* spp., which can fix atmospheric N in non-symbiotic association with their host plant. The symbiotic AM fungi can also increase the uptake of different nutrients by their host plant through their extensive hyphal network. Although there are some signs indicating that AM fungi can affect the process of N uptake by the host plant, the greater details regarding the fugal effects on the process of N uptake have yet to be further investigated (Arzanesh et al. 2011; Podile and Kishore 2006; van Loon 2007; Miransari 2011b).

SOIL MICROBES AND PLANT P UPTAKE

Similar to N, P is also among the macronutrients, which are necessary for plant growth and crop production. However in comparison with N, P is absorbed at much less amount by plant and due to its chemical properties, it is subjected to processes such as desolubilization and hence its availability to plant would significantly decrease. For example, during the first year of P application, only 20% of P is available to plant in the field and the remaining part usually precipitates. Accordingly, some methods must be used to increase the efficiency of P availability to plant (Omidi et al. 2008; Zabihi et al. 2011)

For example, at the time of using P fertilization in the field, P has to be used around the plant roots, and the P compounds, which have the highest solubility such as triple super phosphate must be used. In addition, the use of biological methods such as PGPR has also been indicated to be affective on the enhancement of P availability in soil. For example, such bacteria are able to produce enzymes such as phosphatases and products such as carboxylic acids, which can increase P availability in soil by affecting the mineral P sources such as rock phosphate (Houser and Richardson 2010; Niu et al. 2010; Salimpour et al. 2010).

There is a wide range of soil bacteria, which act as PGPR. In addition to the enhanced availability of different nutrients in soil, PGPR can also affect plant growth by the followings: 1) production of enzymes such as 1-aminocyclopropane-1-carboxylate (ACC) deaminase, which are able to alleviate the effects of stress on plant growth, 2) production of different plant hormones, and 3) controlling the adverse effects of pathogens on plant growth by production of products such as HCN (Abbas-Zadeh et al. 2010; Podile and Kishore 2006; Zhao et al. 2010). As enzymes such as ACC deaminase can affect plant root growth by decreasing the production of the stress hormone, ethylene, it can also indirectly influence P affect on root growth and its uptake by plant.

Plants can also affect the availability of soil nutrients by their activities, especially their root activities. Plant roots can affect soil nutrients, in the rhizosphere, through producing different compounds, which influence the solubility and hence availability of nutrients. Such products can affect rhizosphere properties and hence soil nutrients by altering rhizosphere pH, and influencing the activities of rhizosphere microbes including PGPR (Heijden 2010; Hodge 2010; Johnson et al. 2010).

SOIL MICROBES AND PLANT K UPTAKE

K is soluble in soil and hence is subjected to processes such as leaching. It is necessary for plant growth and crop production at high amounts, similar to N. When K is deficient for plant growth, it can be applied by the use of chemical and organic fertilization. However, there are PGPR, which can enhance K availability in soil by their activities. Usually, the amounts of K in soil is higher under arid and semi arid conditions, in comparison with humid areas (Adesemoye and Kloepper 2009; Blaise et al. 2005; Hu et al. 2006; Wu et al. 2005).

As a macronutrient, K has some important functions in the plant such as affecting the activity of different enzymes and adjusting water behavior in plant. Under K deficient conditions, usually plant is more susceptible to drought stress as the efficiency of plant water decreases significantly. It is because K can affect the activity of plant stomata and when K is deficient more water is evaporated from the plant leaf surface decreasing the efficiency of plant water (Ashley et al. 2006; Zhang et al. 2010). When there is low amount of K in plant, K transporters and the related signal molecules including reactive oxygen species and some plant hormones are activated, K channels are up regulated and root developmental stages are affected. Such responses make the plant to survive under variable K amounts in soil (Ashley et al. 2006).

Biological activities in soil can also affect K availability. For example, Hu et al. (2006) isolated the K solubilizing strain, *Paenibacillus mucilaginosus*, with K solubilizing activities even higher than the inoculum strain, *Bacillus mucilaginosus* (Li et al. 2007), which is usually used with K fertilization. By using a range of different mechanisms including acidolysis, chelation and oxidoreduction, and production of different products such as oxalate, lactate, gluconate, citrate, catechol and pseudobactin soil microbes can solubilize nutrients including potassium from minerals (Marschner and Rengel 2007; Uroz et al. 2009).

Soil Microbes and Plant S Uptake

S is also a macronutrient, which is necessary for plant growth and yield production as it can regulate different plant functions such as the production of proteins and amino acids. Similar to some other nutrients, S is also subjected to biological alterations in soil by, for example, the soil bacteria, *Thiobacillus* sp., which are able to oxidize soil S to Sulfate and produce hydrogen, increase soil acidity and hence decrease soil pH. This is particularly of significance in calcareous areas where it is necessary to decrease soil pH by adding elemental S to the soil. Usually, under such conditions, adding S with *Thiobacillus* sp. is a very safe method to decrease soil acidity, which can affect different soil properties including the availability of different soil nutrients.

The most optimum pH at which the uptake of nutrients by plant is at the highest is the neutral pH. Accordingly, one of the suitable methods to increase the availability of soil nutrients to plant is the adjustment of soil pH by adding S or lime to the soil, under calcareous or humid conditiond (Beller et al. 2006; Miransari and Smith 2007; Salimpour et al. 2010). In addition, the biochemical processing in the rhizosphere of plants such as the production of enzymes such as arylsulfatases, which are partly produced by plant roots and mostly produced by soil microbes, can significantly influence the biogeochemical cycling of nutrients such as sulfur in the soil (Hinsinger et al. 2005).

The genome sequence of the obligatory chemolithoautothrophic, *Thiobacillus dentirificans*, as a β-pretobacteria, is among the first sequences, which has been recently recognized for sulfur oxidizing bacteria. Such a discovery can help recognize the molecular pathways by which the bacteria can reduce sulfur compounds under aerobic and anaerobic conditions. Accordingly, it becomes likely to recognize bacterial importance for biogeochemical cycles, related to sulfur, carbon and nitrogen affecting the quality of ground water as well (Beller et al. 2006).

Salimpour et al. (2010) found that treating canola (*Brassica napus* L.) with elemental sulfur inoculated with *Thiobacillus* spp. significantly affected the growth and oil components of plant. They accordingly indicated that use of biological fertilization with chemical fertilization is a favorite way of increasing crop and oil production with economical and environmental significance.

Soil Microbes and Plant Micronutrient Uptake

Iron

Soil microbes can affect the solubility and hence the availability of different micronutrients in soil. There are different mechanisms, utilized by soil microbes to increase the solubility of soil micro-nutrients indicated in the following (Jin et al. 2010; Marschner et al. 2011; Podile and Kishore 2006). 1) Affecting plant metabolism and hence alteration of root exudates by symbiotically or non-symbiotically association with their host plant, 2) affecting the solubility and hence the availability of soil nutrients, and 3) interacting with the other soil microbes (Adesemoye and Kloepper 2009; Fitter et al. 2011; Hodge 2010; Miransari 2011a; b).

Iron (Fe) is an important micronutrient, which becomes especially deficient to plant under dry and calcareous conditions. It is because under such conditions the presence of carbonate and bicarbonate onions in soil can result in the precipitation of Fe in soil and hence significantly decrease its availability to plant. Under Fe deficient conditions the signs of chlorosis appear on plant leaf. The highest Fe solubility is related to humid conditions where the high uptake of Fe by plant may result in some unfavorable effects on plant growth and hence yield production (Marschner 1995).

Plants perform two types of strategies under Fe deficient conditions including strategy I and II. Strategy I plants acidify the rhizosphere and produce organic products such as phenolic compounds resulting in the activation of ferric chelate reductase and Fe (II) transporter in plant roots. However, strategy II plants handle the stress by producing siderophores and activating the Fe^{3+} siderophore transporter in the plasmalemma of root cells (Romheld and Marschner 1986).

The produced siderophores by plant roots can significantly increase Fe availability due to its high affinity to chelate Fe^{3+} and hence its efficient absorption by plant roots (Chen et al. 1998). Interestingly, plant may not sufficiently handle Fe stress using such strategies. For example, it has been indicated that under sterilized conditions, plants are not able to alleviate Fe stress indicating the role of soil microbes under such conditions. Fe concentration in plant may also affect Fe uptake by affecting the quality and quantity of root exudates and hence microbial activities in the rhizosphere (Jin et al. 2010; Yang and Crowley 2000). Similar to plants, PGPR are also able to produce siderophores, which increase the solubility and hence the availability of Fe to the plant (Abbas-Zadeh et al. 2010; Jalili et al. 2009; Zabihi et al. 2011).

The production of root exudates, including organic and amino acids and the secondary metabolites can affect the combination of plant microbes for different plant species as well as the solubility and availability of different nutrients in the plant rhizosphere. In addition, plant morphology and physiological properties may also be altered under nutrient deficiency including iron. For example, under Fe deficient conditions plant may produce more root hairs and products such as carboxylates, water and phenolic compounds (Lamont 2003; Badri and Vivanco 2009).

Plant Fe status can determine soil microbial combination producing siderophores. Jin et al. (2010) found that in the rhizosphere of plants with Fe deficiency the number of soil microbes, which produce siderophores increase. This may be due to the fact that under such conditions phenolic compounds are produced by plant roots enhancing the number of soil microbes producing siderophores in the plant rhizosphere.

Zinc

Zinc (Zn) is also among important micronutrients for plant growth and yield production and is also affected by soil microbial activities. Zn is necessary for plant growth for different functioning such as the activities of different enzymes including the ones, which can alleviate the effects of different stresses by controlling the production of products such as hydrogen peroxide. Enzymes such as dehydrogenase, and RNA and DNA polymerase have also Zn in their structure. Under different stresses enzymes such as super oxide dismutase (SOD) are produced to turn the products of stress into products, which are not unfavorable to plant. Hence, one of the most important effects of AM fungi on plant growth under drought stress is through their alleviating effects on the oxidative stress (Sajedi et al. 2010; 2011). The adverse effects of oxidative stress on plant growth is by peroxidation of lipid, protein denaturation and mutating DNA (Bowler et al. 1992). AM fungi such as *Glomus mosseae* produce the CuZn-SOD enzyme, which is important affecting the host plant tolerance to stresses such as drought. It has been indicated that mycorrhizal plants produce higher amounts of SOD enzyme in comparison with non-mycorrhizal. Zn presence in plants can also influence the activity of plant hormones such as auxin (Sajedi et al. 2010; 2011).

AM fungi can enhance the uptake of different nutrients such as micronutrients including Zn by the host plant (Johnson 2010). Under nutrient deficiency plants use different strategies to handle the stress (Marshner and Romheld 1994; Rengel 1999; Hartmann et al. 2009). He et al. (2009) indicated that the Zn-tolerant bacterial strains including *Bacillus cereus*, *B. subtilis*, *Flavobacterium* spp. and *Pseudomonas aeroginosa* can increase the availability of Zn in soil and hence its uptake by plant, which is also of bioremediation significance. Omidi et al. (2008) found that the availability of Zn in soil is under the influence of different tillage practices, plant varieties and planting date affecting soil biological properties including the production of different enzymes.

Copper

Copper (Cu) is also another necessary micronutrient for plant growth and yield production. However, compared with other micronutrients it is necessary at less amounts by plant. It can regulate different plant activities including the activities of different enzymes as well as the process of symbiotic N fixation in legumes (Makoi and Ndakidemi 2007). The effects of soil microbes on copper availability in soil is directly by the production of different compounds such as carboxylates and phenolic compounds or indirectly through affecting plant growth and hence the production of root exudates (Marschner 1995; Badri and Vivanco 2009).

Manganese

Similar to the other necessary micronutrients for plant growth, optimum amounts of manganese (Mn) is also required for plant growth and yield production and at excess amounts it can adversely influence plant growth and yield production. Unlike the other soil micronutrients, Mn deficiency is less likely under dry climates and as it may be found at high amounts under such conditions. Soil biological activities can also affect Mn availability in soil mostly by affecting plant growth and hence plant root exudates.

Soil microbes such as *Pseudomonas* spp. are able to affect the availability of soil micronutrients by production of products such as carboxylates, which can chelate micronutrients and the production of products such as humic substances, riboflavin, phenazines and quinines, which can dissolve mineral oxides (Hernandez et al. 2004; Johal and Huber 2009; Uroz et al. 2009). While the process of Mn oxidation is biological, its reduction in the soil is both biological and chemical (Ghiorse 1988) indicating that the ratio of soil oxidizing and reducing Mn bacteria can determine the availability of Mn in soil (Rengel 1997). Interestingly, as the reduced form of Mn is available to plant, in the rhizosphere of Mn-efficient plants the number of Mn reducing microbes is higher than the oxidizing ones, which are higher in the rhizosphere of inefficient genotypes (Timonin 1946; Rengel et al. 1996).

There is another important aspect regarding soil micronutrients, especially at high amounts in the presence of soil microbes. It has been indicated that under high concentration of micronutrients, soil microbes are able to alleviate the unfavorable effects of soil micronutrients on plant growth and on the environment using different mechanisms (Zhuang et al. 2007). For example, soil microbes can affect the solubility of micronutrients in soil, their uptake by plant through affecting plant morphological and physiological properties and their translocation to different plant tissues.

Conclusion

The effects of soil microbes on soil nutrients is presented in this review article. Soil microbes can significantly contribute to the availability of soil nutrients to plants, which is of economical and environmental importance. Different soil microbes can differently affect the properties of nutrients in soil. This is particularly important for the production of biofertilizer as the most efficient strains must be selected for inoculum preparation.

REFERENCES

Abbas-Zadeh P, Saleh-Rastin N, Asadi-Rahmani H, Khavazi K, Soltani A, Shoary-Nejati A, Miransari M (2010) Plant growth-promoting activities of fluorescent pseudomonads, isolated from the Iranian soils. *Acta Physiologiae Plantarum* 32: 281-288.

Adesemoye A, Kloepper J (2009) Plant–microbes interactions in enhanced fertilizer-use efficiency. *Applied Microbiology and Biotechnology* 85: 1-12.

Artursson V, Finlay RD, Jansson JK (2006) Interactions between arbuscular mycorrhizal fungi and bacteria and their potential for stimulating plant growth. *Environmental Microbiology* 8: 1-10.

Arzanesh M, Alikhani H, Khavazi K, Rahimian H, Miransari M (2011) Wheat (Triticum aestivum L.) growth enhancement by Azospirillum sp. under drought stress. *World Journal of Microbiology and Biotechnology* 27: 197-205.

Ashley MK, Grant M, Grabov A (2006) Plant responses to potassium deficiencies: a role for potassium transport proteins. *Journal of Experimental Botany* 57: 425-436.

Badri DV, Vivanco JM (2009) Regulation and function of root exudates. *Plant, Cell and Environment* 32: 666-681.

Beller HR, Chain PSG, Letain TE, Chakicherla A, Larimer FW, Richardson PM, Coleman MA, Wood AP, Kelly DP (2006) The Genome Sequence of the Obligately Chemolithoautotrophic, Facultatively Anaerobic Bacterium Thiobacillus denitrificans. *J Bacteriol.* 188: 1473-1488.

Blaise D, Singh JV, Bonde AN, Tekale KU, Mayee CD (2005) Effects of farmyard manure and fertilizers on yield, fibre quality and nutrient balance of rainfed cotton (Gossypium hirsutum). *Bioresource Technology* 96: 345-349.

Bowler C, Van Montagu M, Inzé D (1992) Superoxide dismutase and stress tolerance. *Annual Reviews of Plant Physiology and Plant Molecular Biology* 43: 83-116.

Chen L, Dick W, Streeter J, Hoitink H (1998) Fe chelates from compost microorganisms improve Fe nutrition of soybean and oat. *Plant and Soil* 200: 139-147.

Compant S, Clément C, Sessitsch A (2010) Plant growth-promoting bacteria in the rhizo- and endosphere of plants: Their role, colonization, mechanisms involved and prospects for utilization. *Soil Biology and Biochemistry* 42: 669-678.

Egamberdiyeva D (2007) The effect of plant growth promoting bacteria on growth and nutrient uptake of maize in two different soils. *Applied Soil Ecology* 36: 184-189.

Fang C, Smith P, Smith JU, Moncrieff JB (2005) Incorporating microorganisms as decomposers into models to simulate soil organic matter decomposition. *Geoderma* 129: 139-146.

Ferguson BJ, Indrasumunar A, Hayashi S, Lin M-H, Lin Y-H, Reid DE, Gresshoff PM (2010) Molecular Analysis of Legume Nodule Development and Autoregulation. *Journal of Integrative Plant Biology* 52: 61-76.

Fitter AH, Helgason T, Hodge A (2011) Nutritional exchanges in the arbuscular mycorrhizal symbiosis: Implications for sustainable agriculture. *Fungal Biology Reviews* 25: 68-72.

Ghiorse W C 1988 *The biology of manganese transforming microorganisms in soils*. In Manganese in Soils and Plants. R D Graham, R J Hannam and N C Uren (eds.) pp. 75-85. Kluwer Academic Publishers. Dordrecht, The Netherlands.

Hartmann A, Schmid M, van Tuinen D, Berg G (2009) Plant-driven selection of microbes. *Plant and Soil* 321:235–257.

He C, Tan GE, Liang X, Du W, Chen TL, Zhi GY, Zhu Y (2010) Effect of Zn-tolerant bacterial strains on growth and Zn accumulation in Orychophragmus violaceus. *Applied Soil Ecology* 44:1–5.

Heijden MGAvd (2010) Mycorrhizal fungi reduce nutrient loss from model grassland ecosystems. *Ecology* 91: 1163-1171.

Hernandez ME et al. (2004) Phenazines and other redox-active antibiotics promote microbial mineral reduction. *Applied and Environmental Microbiology* 70:921–928.

Hinsinger P, Gobran G, Gregory P, Wenzel W (2005) Rhizosphere geometry and heterogeneity arising from root mediated physical and chemical processes. *New Phytologist* 168:293–303.

Hodge A (2010) Roots: *The Acquisition of Water and Nutrients from the Heterogeneous Soil Environment*. In: Lüttge U, Beyschlag W, Büdel B, Francis D (eds) Progress in Botany 71. Springer Berlin Heidelberg, pp 307-337.

Houser J, Richardson W (2010) Nitrogen and phosphorus in the Upper Mississippi River: transport, processing, and effects on the river ecosystem. Hydrobiologia 640: 71-88

Hu X, Chen J, Guo J (2006) Two Phosphate- and Potassium-solubilizing Bacteria Isolated from Tianmu Mountain, Zhejiang, China. *World Journal of Microbiology and Biotechnology* 22: 983-990.

Jalili F, Khavazi K, Pazira E, Nejati A, Rahmani HA, Sadaghiani HR, Miransari M (2009) Isolation and characterization of ACC deaminase-producing fluorescent pseudomonads, to alleviate salinity stress on canola (Brassica napus L.) growth. *Journal of Plant Physiology* 166: 667-674.

Jin CW, Li GX, Yu XH, Zheng SJ (2010) Plant Fe status affects the composition of siderophore-secreting microbes in the rhizosphere. *Annals of Botany* 105: 835-841.

Jin CW, Li GX, Yu XH, Zheng SJ (2010) Plant Fe status affects the composition of siderophore-secreting microbes in the rhizosphere. *Annals of botany* 105: 835-41.

Johal GS, Huber DM (2009) Glyphosate effects on diseases of plants. *European Journal of Agronomy* 31: 144-152.

Johnson NC (2010) Resource stoichiometry elucidates the structure and function of arbuscular mycorrhizas across scales. *New Phytologist* 185: 631–647.

Johnson NC, Wilson GWT, Bowker MA, Wilson JA, Miller RM (2010) Resource limitation is a driver of local adaptation in mycorrhizal symbioses. *Proceedings of the National Academy of Sciences* 107: 2093-2098.

Lamont BB (2003) Structure, ecology and physiology of root clusters – a review. *Plant and Soil* 248: 1–19.

Li X,Wu Z, Li W, Yan R, Li L, Li J, et al. (2007) Growth promoting effect of a transgenic Bacillus mucilaginosus on tobacco planting. *Applied Microbiology and Biotechnology.* 74: 1120-1125.

Lugtenberg B, Kamilova F (2009) Plant-Growth-Promoting Rhizobacteria. *Annual Review of Microbiology.* 63: 541-556.

Makoi JH, Ndakidemi PA (2007) Biological, ecological and agronomic significance of plant phenolic compounds in rhizosphere of the symbiotic legumes. *African Journal of Biotechnology*, 6:1358-1368.

Makoi JHJR, Ndakidemi PA (2007). Biological , ecological and agronomic significance of plant phenolic compounds in rhizosphere of the symbiotic legumes. *African Journal of Biotechnology*, 6:1358-1368.

Marschner H (1995) Mineral Nutrition of Higher Plants Academic Press, London.

Marschner H, Römheld V (1994) Strategies of plants for acquisition of iron. *Plant Soil* 165:261–274.

Marschner P, Crowley D, Rengel Z (2011) Rhizosphere interactions between microorganisms and plants govern iron and phosphorus acquisition along the root axis - model and research methods. *Soil Biology and Biochemistry* 43: 883-894.

Marschner P, Rengel Z (2007) Contributions of Rhizosphere Interactions to Soil. In Soil Biological Fertility. Abbott LK, Murphy DV(eds) - *A Key to Sustainable Land Use in Agriculture.* 81-98.

Miransari M (2010a) Biological Fertilization. In: A. Méndez-Vilas, A. (Ed.) *Current Research, Technology and Education Topics in Applied Microbiology and Microbial Biotechnology*. Microbiology Book Series – 2010 Edition, Spain.

Miransari M (2010b) Contribution of arbuscular mycorrhizal symbiosis to plant growth under different types of soil stress. *Plant Biology* 12: 563-569.

Miransari M (2011a) Arbuscular mycorrhizal fungi and nitrogen uptake. *Archives of Microbiology* 193: 77-81.

Miransari M (2011b) Interactions between arbuscular mycorrhizal fungi and soil bacteria. *Applied Microbiology and Biotechnology*: 1-14.

Miransari M, Mackenzie AF (2011a) Development of a Soil N Test for Fertilizer Requirements for Corn Production in Quebec. *Communications in Soil Science and Plant Analysis* 42: 50 – 65.

Miransari M, Mackenzie AF (2011b) Development of a soil n test for fertilizer requirements for wheat. *Journal of Plant Nutrition* 34: 762 – 777.

Miransari M, Smith D (2008) Using signal molecule genistein to alleviate the stress of suboptimal root zone temperature on soybean-*Bradyrhizobium* symbiosis under different soil textures. *Journal of Plant Interactions* 3: 287 – 295.

Miransari M, Smith DL (2007) Overcoming the Stressful Effects of Salinity and Acidity on Soybean Nodulation and Yields Using Signal Molecule Genistein Under Field Conditions. *Journal of Plant Nutrition* 30: 1967 – 1992.

Miransari M, Smith DL (2009) Alleviating salt stress on soybean (Glycine max (L.) Merr.) - Bradyrhizobium japonicum symbiosis, using signal molecule genistein. *European Journal of Soil Biology* 45: 146-152.

Miransari, M, Balakrishnan P, Smith DL, Mackenzie AF, Bahrami HA, Malakouti MJ, Rejali F (2006) Overcoming the stressful effect of low pH on soybean root hair curling using lipochitooligosaccahrides. *Communications in Soil Science and Plant Analysis* 37:1103-1110.

Niu S, Wu M, Han YI, Xia J, Zhang ZHE, Yang H, Wan S (2010) Nitrogen effects on net ecosystem carbon exchange in a temperate steppe. *Global Change Biology* 16: 144-155.

Omidi H, Tahmasebi Z, Torabi H, Miransari M (2008) Soil enzymatic activities and available P and Zn as affected by tillage practices, canola (Brassica napus L.) cultivars and planting dates. *European Journal of Soil Biology* 44: 443-450.

Podile A, Kishore G (2006) *Plant growth-promoting rhizobacteria.* In: Gnanamanickam S (ed) Plant-Associated Bacteria. Springer Netherlands, pp 195-230.

Rengel Z (1999) *Physiological mechanisms underlying differential nutrient efficiency of crop genotypes*. In: Rengel Z (ed) Mineral nutrition of crops: Mechanisms and implications, The Haworth Press, New York, U S A, pp 227–265.

Rengel Z 1997 Root exudation and microflora populations in rhizosphere of crop genotypes differing in tolerance to micronutrient deficiency. *Plant and Soil* 196: 255-260.

Rengel Z, Gutteridge R, Hirsch P and Hornby D 1996 Plant genotype, micronutrient fertilisation and take-all infection influence bacterial populations in the rhizosphere of wheat. *Plant and Soil* 183: 269-277.

Richardson A, Barea J-M, McNeill A, Prigent-Combaret C (2009) Acquisition of phosphorus and nitrogen in the rhizosphere and plant growth promotion by microorganisms. *Plant and Soil* 321: 305-339.

Romheld V, Marschner H (1986) Mobilization of iron in the rhizosphere of different plant species. *Advance in Plant Nutrition* 2: 155-204.

Sajedi N, Ardakani M, Rejali F, Mohabbati F, Miransari M (2010) Yield and yield components of hybrid corn (Zea mays L.) as affected by mycorrhizal symbiosis and zinc sulfate under drought stress. *Physiology and Molecular Biology of Plants* 16: 343-351.

Sajedi NA, Ardakani MR, Madani H, Naderi A, Miransari M (2011) The effects of selenium and other micronutrients on the antioxidant activities and yield of corn (*Zea mays* L.) under drought stress. *Physiology and Molecular Biology of Plants*, in press.

Salimpour S, Khavazi K, Nadian H, Besharati H, Miransari M (2010) Enhancing phosphorous availability to canola (Brassica napus L.) using P solubilizing and sulfur oxidizing bacteria. *Australian Journal of Crop Science* 4: 330-334.

Smith SE, Read DJ (2008) *Mycorrhizal Symbiosis*. Academic Press, London.

Timonin M I (1946) Microflora of the rhizosphere in relation to the manganese -deficiency disease of oats. *Soil Science Society of America Proceedings* 11: 284-292.

Van de Velde W, Zehirov G, Szatmari A, Debreczeny M, Ishihara H, Kevei Z, Farkas A, Mikulass K, Nagy A, Tiricz H, Satiat-Jeunemaître B, Alunni B, Bourge M, Kucho K-i, Abe M, Uroz S., Calvaruso, C., Turpault, M.-pierre, and Frey-klett, P. (2009). Mineral weathering by bacteria : ecology , actors and mechanisms. *Trends in Microbiology* 17:378-387.

van Loon L (2007) Plant responses to plant growth-promoting rhizobacteria. *European Journal of Plant Pathology* 119: 243-254.

Wang X, Pan Q, Chen F, Yan X, Liao H (2010) Effects of co-inoculation with arbuscular mycorrhizal fungi and rhizobia on soybean growth as related to root architecture and availability of N and P. *Mycorrhiza* 21: 173-181.

Wu SC, Cao ZH, Li ZG, Cheung KC, Wong MH (2005) Effects of biofertilizer containing N-fixer, P and K solubilizers and AM fungi on maize growth: a greenhouse trial. *Geoderma* 125: 155-166.

Yang C, Crowley D (2000) Rhizosphere microbial community structure in relation to root location and plant iron nutritional status. *Applied and Environmental Microbiology* 66: 345-351.

Zabihi H, Savaghebi G, Khavazi K, Ganjali A, Miransari M (2011) Pseudomonas bacteria and phosphorous fertilization, affecting wheat (Triticum aestivum L.) yield and P uptake under greenhouse and field conditions. *Acta Physiologiae Plantarum* 33: 145-152.

Zhang J-L, Flowers T, Wang S-M (2010) Mechanisms of sodium uptake by roots of higher plants. *Plant and Soil* 326: 45-60.

Zhao J-L, Zhou L-G, Wu J-Y (2010) Promotion of Salvia miltiorrhiza hairy root growth and tanshinone production by polysaccharide-protein fractions of plant growth-promoting rhizobacterium Bacillus cereus. *Process Biochemistry* 45: 1517-1522.

Zhuang X, Chen J, Shim H, Bai Z (2007) *New advances in plant growth-promoting rhizobacteria for bioremediation Environment International* 33:406–413.

In: Soil Nutrients
Editor: Mohammad Miransari
ISBN: 978-1-61324-785-3

Chapter 15

UPTAKE OF NITROGEN BY ARBUSCULAR MYCORRHIZAL FUNGI

Mohammad Miransari[1]*, M. Habibi[2], A.M. Naji[2], M.H. Fotokian[2], and H. Amirshekari[2]

[1]Department of Soil Science, College of Agricultural Sciences, Shahed University, Tehran, Iran

[2]Department of Agronomy and Plant Breeding , College of Agricultural Sciences, Shahed University, Tehran, Iran

ABSTRACT

Nitrogen (N) is among the most important macro-nutrients significantly affecting plant growth and yield production. Accordingly, N must be supplied adequately so that optimum amounts of yield are resulted. There are different ways of supplying N to the plant including the use of chemical and biological fertilization. The chemical properties of N make it very mobile, especially under humid conditions. Hence, N must not be overfertilized with respect to the economical and environmental points of view. N Biological fertilization includes the use of plant growth promoting rhizobacteria (PGPR) including the N-fixing bacteria, rhizobium. There are also arbuscular mycorrhizal (AM) fungi in the soil, which are symbiotic to most terrestrial plants enhancing plant growth and yield production through increasing the uptake of water and nutrients by the host plant. Numerous experiments have indicated the important role of AM fungi in enhancing P uptake by plant. However, it is yet a matter of debate that how AM fungi may affect soil N dynamic and hence plant N uptake. Some of the most important and recent aspects regarding such effects by AM fungi are highlighted, which can be of significance to health and productivity of the ecosystem.

Keywords: AM fungi, soil N dynamic, plant N uptake, tripartite symbiosis, chemical and biological fertilization

---

* E-mail: Miransari1@gmail.com, Miransari @shahed.ac.ir.

INTRODUCTION

Plant require a range of at least 16 macro- and micro-nutrients for their growth and yield production (Marschner 1995). Among such nutrients nitrogen (N) is of particular significance as it is necessary for many different plant functioning as it is incorporated in the structure of some important macro and micro organic compounds in the plant such as chlorophyll and proteins and amino acids (Marschner 1995). N is a mobile element in the soil and hence under humid conditions is subjected to leaching indicating the importance of appropriate N fertilization under such conditions. On the other hand, under arid and semi arid conditions water deficiency may limit the use of inorganic N by plant.

There are different ways of supplying N to the plant including the use of chemical and biological fertilization. In the case of chemical fertilization proper amounts of N must be supplied with respect to the economical and environmental aspects regarding N compounds (Miransari and Mackenzie 2010, 2011a, b). Biological fertilization can be performed by using soil microbes such as plant growth promoting rhizobacteria (PGPR) including the N-fixing bacteria, rhizobium (Arzanesh et al. 2010; Zabihi et al. 2010; Miransari 2010a, b, c, d).

In addition, Arbuscular mycorrhizal (AM) fungi can also be used a source of biological fertilization. AM fungi are soil fungi, developing symbiotic association with most terrestrial plants. In this symbiosis the fungi provide the host plant with water and nutrients in the exchange for carbon (Smith and Read 2008). The symbiosis between the fungi and host plant is not specific; however some fungi-host plant associations may be more efficient under specific circumstances (Daei et al. 2009). AM fungi are able to alleviate the effects of different stresses on plant growth and yield production by significantly increasing the uptake of water and nutrients including N by the host plant (Miransari et al., 2007; 2008; 2009 a, b; Daei et al 2009).

Different research work has indicated that AM fungi can substantially enhance the uptake of different nutrients, especially phosphorous (P) by the host plant through an extensive network of hypha and some other mechanisms including the production of enzymes such as phosphatases. However, it has also been indicated that AM fungi can also influence the uptake of other nutrients necessary for plant growth and yield production including N. McFarland et al. (2010) indicated that more than 50% of plant N requirement was supplied by mycorrhizal association. As N is among the most important macro nutrients significantly affecting plant growth and yield production, it can be pertinent to determine the contribution of AM fungi to N uptake by the host plant.

Using labeled N it has been indicated that the hypha of AM fungi is able to utilize inorganic N efficiently (Ames et al. 1983; Frey and Schüepp 1992; Johansen et al. 1993, 1994; Subramanian and Charest 1999) and transfer it into the soil in a 10-30 cm distance. This indicates that AM fungi enable the host plant to have access to the N inorganic sources, which are not accessible by non-mycorrhizal plants (Tobar et al. 1994 a,b). According to Johansen et al. (1993) the external hypha of *Glomus intraradices* was able to absorb NO_3 and NH_4 and provide them to the host plant.

The mobility of inorganic P is much lower than inorganic N including NO_3 indicating that the role of AM fungi in absorbing N may be of less significance than P. In other words, plant roots are able to absorb their necessary inorganic N by diffusion and mass flow. In addition plants forming mycorrhizal symbiosis grow in environments with high rates of nitrification. These may all indicate that mycorrhizal symbiosis may not be considered important for the uptake of mineral N by the host plant.

However, there may be situations where plant growth is limited due to reasons such as N limitation. Hence, the growth of fungal hypha in organic patches may be an effective way of supplying N for both the fungi and the host plant (Hodge et al. 2001; Hodge and Fitter 2010). For example, in arid and semi arid climates the mobility of NO_3 in soil is significantly reduced and hence AM fungi may behave more effectively under such conditions to absorb inorganic N (Tillman 1987; Hodge et al. 1999; Subramanian and Charest 1999; Hodge and Fitter 2010). Tobar et al. (1994a, b) indicated that under water stressed conditions AM fungi significantly increased N uptake by the host plant in comparison with well water conditions.

Interestingly, recently a plant ammonium transporter, which is activated in the presence of AM fungi has been identified indicating that the way by which N is transferred in plant may be similar to P transfer (Guether et al 2009). This may indicate that similar to P transporters, AM fungi may also be able to activate N transporters in plant suggesting the significance of the uptake of both nutrients by mycorrhizal plants.

The enhanced activities of nitrate reductase, glutamine synthetase and glutamine synthase in the roots and shoots of mycorrhizal corn (*Zea mays* L.) indicate that the absorbed NO_3 by AM hypha can be directly transferred to the root cells for further utilization and incorporation into the organic structures. Such enzymatic alterations can also enhance plant resistance to drought stress (Cliquet and Stewart 1993; Subramanian and Charest 1999). This indicates that in addition to the direct effects of AM fungi on the alleviation of stresses such as drought (Auge 2001), their indirect effect such as absorbing inorganic N can also contribute to the alleviation of stress. AM fungi are able to alter plant physiological and morphological properties in a way by which plant can handle the stress (Miransari et al. 2008).

Although there are indications regarding the contributions of AM fungi to N uptake by plant, much more details must be elucidated related to such functionality by AM fungi (Atul-Nayyar et al. 2009). Different research work has indicated that AM fungi can absorb N and transfer it to the host plant roots (Tanaka and Yano 2005; Jackson et al. 2008), however the higher rate of N uptake by mass flow may indicate that AM fungi may not be that important in the uptake of N by the host plant (Liu et al., 2007; Atul-Nayyar et al. 2009).

AM fungi can utilize the inorganic N released from organic sources (St. John et al. 1983; Hamel 2004) as well as amino acids (Hawkins et al. 2000). The presence of AM fungi on some plant tissues during the mineralization process indicated that AM fungi are able to enter the tissues through the vascular bundle and utilize the inorganic N released by soil microbes (Aristizábal et al. 2004). The ability of AM fungi to use soil organic matter as a source of inorganic N has yet to be elucidated (Jin et al. 2005; Talbot et al. 2008), however even if AM fungi is not effective in such a process, their influence on the activity of soil mineralizing microbes may indirectly indicate their role in soil N dynamic and uptake by plant (Andrade et al. 1997; Marschner et al. 2001; Hodge 2003a,b; Aneja et al. 2006). Hodge et al. (2001) indicated that the AM fungi, *G. hoi*, can enhance the mineralization process of soil organic matter while earning mineral N.

Atul-Nayyar et al. (2009) found that the C/N ratio of soil organic matter decreased from 17.8 to 13 resulting in higher amounts of mineralized N in the presence of AM fungi. This demonstrates that AM fungi may enhance the mineralization process (by 228%) of organic matter and hence absorbed N uptake by plant. They accordingly indicated that AM fungi may contribute to the enhanced plant growth by increasing the mineralization process of soil organic N and subsequent N uptake by plant.

The rate of N mineralization by AM fungi may be a function of plant demand for mineral N as the high rates of N in soil can decrease the growth and development of AM hypha into organic patches (Liu et al. 2000), which has been attributed to plant N uptake (Hodge et al. 2001; Hodge 2003b). Factors affecting the process of N mineralization in the presence of AM include: 1) the extension of AM hypha, 2) the production of hydrolytic enzymes such as pectinase, xyloglucanase and cellulose, which are able to decompose soil organic matter and 3) the AM fungal effects on the activities of other soil microbes. The effects of AM fungi on soil microbes can be through different mechanisms including enhanced plant growth, induced plant resistance, and altered root exudates and hence signaling pathways (Lioussanne et al. 2008; Toljander et al. 2007). In addition, AM hypha is an important source of organic C for soil microbes (Schimel and Weintraub 2003).

N uptake by AM fungi is of significance from the following aspects: 1) the effects of AM fungi on the process of N cycling, 2) whether AM fungi can reduce the rate of NO_3 leaching, which is an important concern under humid conditions (Miransari and Mackenzie 2010, 2011a, b), 3) AM fungi can be considered as a source of N for plant under conditions with limitations, for example under arid and semi arid conditions where nutrient uptake by plant is limited, and 4) the proper use of chemical fertilization with respect to the use of biological fertilization.

N Uptake in Mycorrhizal Plants

Ames et al. (1983) found that the mycorrhizal fungi *Glomus mosseae* increased N uptake by mycorrhizal celery. In their experiments Tobar et al. (1994) and Azcon et al., (2001; 2008) tested the effects of mycorrhization and P fertilization on the growth of lettuce plants as well as on N assimilation and proline content under drought stress (-0.17 MPa). They also tested the effects of AM fungi on the percentage of N uptake from N fertilization under different levels of soil N. Compared with P fertilization, AM fungi enhanced plant N concentration, the activity of nitrate reductase and proline content in plant. At the medium level of N fertilization (6 mmol N), AM fungi resulted in the higher N uptake from fertilization, relative to the lowest (3 mmol N) and highest (9 mmol N) N levels, which reduced N uptake from N fertilization. These results indicate that mycorrhizal plants can regulate plant N uptake with respect to the amounts of N in the soil. Accordingly, AM fungi can affect plant N uptake from soil and N fertilization, however, high amounts of N fertilization can significantly decrease N uptake by mycorrhizal plants affecting the fertilization strategy.

The tripartite symbiosis between the host plant, AM fungi and N-fixing bacteria rhizobium can affect the uptake of N by the host plant. In such a symbiosis N and P are supplied by the micro-symbionts to the host plant. Accordingly, the association between each micro- symbiont is affected by the interaction effects between the host plant and the micro-symbionts as well as by the interactions effects between the micro-symbionts. Wang et al. (2010) examined the effects of different parameters such as plant genotype and N and P levels on the tripartite symbiosis between soybean (*Glycine max* L.), AM fungi and *Bradyrhizobium japonicum*. They tested the idea that how the nutrient efficiency of soybean plans can be increased. The soybean genotypes differed in their root architecture including shallow and deep rooting genotypes. Co-inoculation of soybean plants with *Bradyrhizobium japonicum* and AM fungi may enhance the symbiotic ability of the two microbes (Xie et al. 1995). In addition to the effects of plant genotype, there may also be some interactions effects between N and P in the field affecting root morphological properties as well as the process of mycorrhization and nodulation (Kaung et al 2005).

Root architecture affected mycorrhizal symbiosis and deep rooting plants had a higher rate of symbiosis with AM fungi at low P levels, compared with shallow rooting plants, however, higher nodulation was resulted at high P levels. This may indicate that N and P levels may affect the tripartite symbiosis between the host plant and the micro-symbiont. Co-inoculation with the two micro-symbionts resulted in higher plant growth and nodulation at low N and P levels and elevating fertilization levels adversely affected the contribution of the two micro-symbionts to the growth of soybean plants (Wang et al., 2010). Such results can be used for proper N and P fertilization in the field with respect to the use of biological fertilization including AM fungi and rhizobium bacteria (Miransari 2010a, b, c, d).

Hodge et al. (2001) indicated that the AM fungi *G. hoi* enhanced the degradation of organic residues and N uptake by the host plant. In the presence of organic matter the growth of fungal hyphal increased whether the host plant was present or not. Hawkins et al. (2000) also tested the effects of *G. mosseae* on N absorption from organic and inorganic sources. Mycorrhizal wheat absorbed organic N in the form of N-glycine by 0.2 and 6% at low and high level of N-fertilization, respectively. The higher organic N uptake at the sufficient N level fertilization was attributed to the more developed hyphal network. The fungi were also able to absorb inorganic N in the form of NO_3 and NH_4. The Ri- T-DNA transformed of mycorrhizal carrot roots were also able to absorb organic N from N-glycine and N-glutamine sources. However, the rate of N absorption in the form of organic and inorganic was not high enough to influence Plant N uptake significantly.

Tian et al. (2010) also indicated that AM fungi is able to absorb both organic and inorganic N. Arginine is synthesized in the extra radical hypha and transferred to the intraradical hypha where it releases N to be absorbed by the host plant. Accordingly, 11 fungal genes related to the N absorption pathway were identified and six of them were sequenced. Such a finding indicated the role of fungal genes in N uptake and the subsequent extra- and intra-radical hypha gene expression as well as plant gene expression.

Hodge and Fitter (2010) tested and proved the hypothesis that organic matter is an important source of N for the fungi themselves. Hence, they grew the host plant both under light and shade so that less amounts of C were transferred to the fungi. Under both conditions the fungi absorbed substantial amounts of its necessary N from the organic patch in addition to the amounts of N, which was transferred to the host plant. Both *G. hoi* and *G. mosseae* indicated such abilities. Hence, AM fungi are an important component of ecosystem in N cycling.

CONCLUSION

Some of the most recent advancements regarding the effects of AM fungi on the process of N uptake by the fungi themselves or by the host plant were reviewed. There are some interesting findings accordingly, which can indicate the significance of N uptake by AM fungi. N cycling, plant growth and ecosystem functioning can be affected by the influence of AM fungi on N dynamic in soil and plant N uptake. This indicates that there is a need for further research regarding the effects of AM fungi on the process of N uptake.

REFERENCES

Ames RN, Reid CPP, Porter KL Cambardella C (1983) Hyphal uptake and transport of nitrogen from two N-labelled sources by *Glomus mosseae*, a vesicular-arbuscular mycorrhizal fungus. *New Phytol.* 95: 381–396.

Andrade G, Mihara KL, Linderman RG, Bethlenfalvay GJ (1997) Bacteria from rhizosphere and hyphosphere soils of different arbuscular-mycorrhizal fungi. *Plant Soil* 192:71–79.

Aneja M, Sharma S, Fleischmann F, Stich S, Heller W, Bahnweg G, Munch J, Schloter M (2006) Microbial colonization of beech and spruce litter—influence of decomposition site and plant litter species on the diversity of microbial community. *Microb. Ecol.* 52:127–135.

Aristizábal C, Rivera EL, Janos DP (2004) Arbuscular mycorrhizal fungi colonize decomposing leaves of *Myrica parvifolia*, *M. pubescens* and *Paepalanthus* sp. *Mycorrhiza* 14:221–228.

Arzanesh MH, Alikhani HA, Khavazi K, Rahimian HA, Miransari M (2010) Wheat (*Triticum aestivum* L.) growth enhancement by *Azospirillum* spp. under drought stress. *World J. Microbiol. Biotechnol.* 27:197-205.

Atul-Nayyar A, Hamel C, Hanson K, Germida J (2009) The arbuscular mycorrhizal symbiosis links N mineralization to plant demand *Mycorrhiza* 19:239–246.

Auge RM. 2001. Water relations, drought and vesicular-arbuscular mycorrhizal symbiosis. *Mycorrhiza* 11: 3–42.

Azcón R, Ruiz-Lozano JM, Rodríguez R (2001) Differential contribution of arbuscular mycorrhizal fungi to plant nitrate uptake (^{15}N) under increasing N supply to the soil. *Can. J. Bot.* 79:1175-1180.

Azcón R, Rodríguez R, Amora-Lazcano E, Ambrosano E (2008) Uptake and metabolism of nitrate in mycorrhizal plants as affected by water availability and N concentration in soil. *Europ. J. Soil Sci.* 59:131–138.

Cliquet JB, Stewart GR (1993) Ammonia assimilation in maize infected with the VAM fungus *Glomus fasciculatum*. *Plant Physiol.* 101 : 65–871.

Daei G, Ardakani M, Rejali F, Teimuri S, Miransari M (2009) Alleviation of salinity stress on wheat yield, yield components, and nutrient uptake using arbuscular mycorrhizal fungi under field conditions. *J. Plant Physiol.* 166:617-625.

Frey B, Schüepp H (1992) Transfer of symbiotically fixed nitrogen from berseem (*Trifolium alexandrium* L.) to maize via vesicular-arbuscular mycorrhizal hyphae. *New Phytol.* 122:447-454.

Guether M, et al. (2009) A mycorrhizal-specific ammonium transporter from *Lotus japonicus* acquires nitrogen released by arbuscular mycorrhizal fungi. *Plant Physiol.* 150:73–83.

Hodge A, Campbell CD, Fitter AH (2001) An arbuscular mycorrhizal fungus accelerates decomposition and acquires nitrogen directly from organicmaterial. *Nature* 413:

Hamel C (2004) Impact of arbuscular mycorrhizal fungi on N and P cycling in the root zone. *Can. J. Soil Sci.* 84:383–395.

Hawkins HJ, Johansen A, George E (2000) Uptake and transport of organic and inorganic nitrogen by arbuscular mycorrhizal fungi. *Plant Soil* 226:275–285.

Hodge A, Robinson D, Griffiths BS, Fitter AH (1999) Why plants bother: Root proliferation results in increased nitrogen capture from an organic patch when two grasses compete. *Plant Cell Environ.* 22:811–820.

Hodge A, Campbell CD, Fitter AH (2001) An arbuscular mycorrhizal fungus accelerates decomposition and acquires nitrogen directly from organic material. *Nature* 413:297–299.

Hodge A (2003a) N capture by Plantago lanceolata and Brassica napus from organis material: The influence of spatial dispersion, plant competition and an arbuscular mycorrhizal fungus. *J. Exp. Bot.* 57:401–411.

Hodge A (2003b) Plant nitrogen capture from organic matter as affected by spatial dispersion, interspecific competition and mycorrhizal colonization. *New Phytol.* 157:303 -314.

Hodge A, Fitter AH (2010) Substantial nitrogen acquisition by arbuscular mycorrhizal fungi from organic material has implications for N cycling. 107:13754–13759.

Jackson LE, Burger M, Cavagnaro TR (2008) Nitrogen transformations and ecosystem services. *Annu. Rev. Plant Biol.* 59:341–363.

Jin H, Pfeffer PE, Douds DD, Piotrowski E, Lammers PJ, Shachar-Hill Y (2005) The uptake, metabolism, transport and transfer of nitrogen in an arbuscular mycorrhizal symbiosis. *New Phytol.* 168:687-696.

Johansen A, Jakobsen I, Jensen ES (1993) Hyphal transport by a vesicular-arbuscular mycorrhizal fungus of N applied to the soil as ammonium or nitrate. *Biol. Fertil. Soils* 16:66-70.

Johansen A, Jakobsen I, Jensen ES (1994) Hyphal N transport by a vesicular-arbuscular mycorrhizal fungus associated with cucumber grown at three nitrogen levels. *Plant Soil* 160:1-9.

Kuang R, Liao H, Yan X, Dong Y (2005) Phosphorus and nitrogen interactions in field-grown soybean as related to genetic attributes of root morphological and nodular traits. *J. Integr. Plant Biol.* 47:549–559.

Liu A, Hamel C, Hamilton R, Smith D (2000) Mycorrhizae formation and nutrient uptake of new corn (Zea mays L.) hybrids with extreme canopy and leaf architecture as influenced by soil N and P levels. *Plant Soil* 221:157–166.

Liu A, Plenchette C, Hamel C (2007) *Soil nutrient and water providers*: How arbuscular mycorrhizal mycelia support plant performance in a resource-limited world. In: Hamel C, Plenchette C (eds) Mycorrhizae in crop production. Haworth, Binghamton, pp 38–66.

Lioussanne L, Beauregard MS, Hamel C, Jolicoeur M, St-Arnaud M (2008) *Interactions between arbuscular mycorrhiza and soil microorganisms*. In: Khasa D, Piché Y, Coughlan A (eds) Advances in mycorrhizal science and technology. NRC, Ottawa.

Marschner H (1995) Mineral nutrition of higher plants, 2nd edn. London: Academic Press.

Marschner P, Crowley D, Lieberei R (2001) Arbuscular mycorrhizal infection changes the bacterial 16 S rDNA community composition in the rhizosphere of maize. *Mycorrhiza* 11:297–302.

McFarland JW, Ruess RW, Kielland K, Pregitzer K, Hendrick R, Allen M (2010) Cross-ecosystem comparisons of in situ plant uptake of amino acid-N and NH_4. *Ecosystems* 13: 177–193.

Miransari M, Bahrami HA, Rejali F, Malakouti MJ, Torabi H (2007) Using arbuscular mycorrhiza to reduce the stressful effects of soil compaction on corn (*Zea mays* L.) growth. *Soil Biol. Biochem*. 39:2014-2026.

Miransari M, Bahrami HA, Rejali F, Malakouti MJ (2008) Using arbuscular mycorrhiza to reduce the stressful effects of soil compaction on wheat (*Triticum aestivum* L.) growth. *Soil Biol. Biochem*. 40:1197-1206.

Miransari M, Rejali F, Bahrami HA, Malakouti MJ (2009a) Effects of soil compaction and arbuscular mycorrhiza on corn (*Zea mays* L.) nutrient uptake. *Soil Till. Res*. 103:282-290.

Miransari M, Rejali F, Bahrami HA, Malakouti MJ (2009b). Effects of arbuscular mycorrhiza, soil sterilization, and soil compaction on wheat (*Triticum aestivum* L.) nutrients uptake. *Soil Till. Res*. 104:48-55.

Miransari M (2010a) Contribution of arbuscular mycorrhizal symbiosis to plant growth under different types of soil stresses. Review article. *Plant Biol*. 12:563-569.

Miransari M (2010b) *Arbuscular Mycorrhiza and Soil Microbes*. In: Mycorrhizal Biotechnology, Editors: Thangadurai D, Busso CA, Hijri M Science, CRC and Taylor and Francis Publishers, 226p, USA.

Miransari M (2010c) *Biological Fertilization*. In: Méndez-Vilas A (Ed.) Current Research, Technology and Education Topics in Applied Microbiology and Microbial Biotechnology. Microbiology Book Series – 2010 Edition, Spain.

Miransari M (2010d) *Mycorrhizal Fungi and Ecosystem Efficiency*, In: Mycorrhizal Fungi: Soil, Agriculture and Environmental Implications, Fulton SM (Ed.) Published by Nova Publishers, USA. ISBN: 978-1-61122-659-1.

Miransari M, Mackenzie AF (2010) Wheat (**Triticum aestivum** L.) grain N uptake as affected by soil total and mineral N, for the determination of optimum N fertilizer rates for wheat production. *Comm. Soil Sci. Plant Anal.* 41:1644-1653.

Miransari M, Mackenzie AF (2011a) Development of a soil N-test for fertilizer requirements for corn (*Zea mays* L.) production in Quebec. *Comm. Soil Sci. Plant Anal*. 42:50-65.

Miransari M, Mackenzie AF (2011b) Development of a soil N test for fertilizer requirements for wheat. *J. Plant Nutr*. 34:762-777.

Schimel JP, Weintraub MN (2003) The implications of exoenzyme activity on microbial carbon and nitrogen limitation in soil: a theoretical model. *Soil Biol. Biochem*. 35:549–563.

Smith SE, Read DJ (2008) *Mycorrhizal Symbiosis*. London: Academic Press.

St. John TV, Coleman DC, Reid CPP (1983) Association of vesicular– arbuscular mycorrhizal hyphae with soil organic particles. *Ecology* 64:957- 959.

Subramanian K, Charest C (1999) Acquisition of N by external hyphae of an arbuscular mycorrhizal fungus and its impact on physiological responses in maize under drought-stressed and well-watered conditions. *Mycorrhiza* 9:69–75.

Talbot JM, Allison SD, Treseder KK (2008) Decomposers in disguise: mycorrhizal fungi as regulators of soil C dynamics in ecosystems under global change. *Funct. Ecol*. 22:955–963.

Tanaka Y, Yano K (2005) Nitrogen delivery to maize via mycorrhizal hyphae depends on the form of N supplied. *Plant Cell Environ.* 28:1247–1254.

Tian C, Kasiborski B, Koul R, Lammers PJ, Bucking H, Shachar-Hill Y (2010) Regulation of the Nitrogen Transfer Pathway in the Arbuscular Mycorrhizal Symbiosis: Gene Characterization and the Coordination of Expression with Nitrogen Flux. *Plant Physiol.* 153: 1175–1187.

Tilman D (1987) Secondary succession and the pattern of plant dominance along experimental nitrogen gradients. *Ecol. Monogr*. 57:189–214.

Tobar RM, Azcon R, Barea JM (1994a) The improvement of plant N acquisition from an ammonium-treated, drought-stressed soil by the fungal symbiont in arbuscular mycorrhizae. *Mycorrhiza* 4:105-108.

Tobar RM, Azcon R, Barea JM (1994b) improved nitrogen uptake and transport from ^{15}N-labelled nitrate by external hyphae of arbuscular mycorrhiza under water-stressed conditions. *New Phytol*. 126:119-122.

Toljander JF, Lindahl BD, Paul LR, Elfstrand M, Finlay RD (2007) Influence of arbuscular mycorrhizal mycelial exudates on soil bacterial growth and community structure. *FEMS Microbiol. Ecol.* 61:295–304.

Wang X, Pan Q, Chen F, Yan X, Liao H (2010) Effects of co-inoculation with arbuscular mycorrhizal fungi and rhizobia on soybean growth as related to root architecture and availability of N and P. *Mycorrhiza* 21: 173-181.

Xie Z, Staehelin C, Vierheili H, Wiemken A, Jabbouri S, Broughton WJ, Vogeli-Lange R, Boller T, Xie ZP (1995) Rhizobial nodulation factors stimulate mycorrhizal colonization of undulating and non-nodulating soybeans. *Plant Physiol.* 108:1519-1525.

Zabihi HR, Savaghebi GR, Khavazi K, Ganjali A, Miransari M (2010) *Pseudomonas* bacteria and phosphorous fertilization, affecting wheat (*Triticum aestivum* L.) yield and P uptake under greenhouse and field conditions. *Acta Physiol. Plant* 33:145-152.

INDEX

A

acetic acid, 136, 217
acid, 2, 3, 6, 14, 16, 17, 35, 37, 42, 61, 78, 79, 83, 93, 94, 96, 99, 100, 101, 103, 104, 106, 111, 112, 117, 118, 119, 120, 122, 123, 124, 126, 130, 134, 135, 136, 137, 142, 146, 154, 155, 170, 171, 178, 179, 186, 187, 189, 190, 191, 192, 197, 207, 212, 220, 242, 250, 263, 270, 271, 281, 282, 284, 285, 288
acidic, 28, 36, 83, 143, 237, 242, 243, 246, 248, 258, 284
acidity, 14, 82, 94, 96, 131, 143, 178, 180, 181, 237, 243, 244, 248, 250, 256, 268, 270, 271, 301
active compound, 11, 15
active site, 78
activity rate, 222
adaptation, 5, 6, 93, 95, 165, 194, 208, 274, 306
adaptations, 268
adenine, 2
adenosine, 2
ADP, 2
adsorption, ix, 42, 44, 94, 99, 105, 112, 115, 122, 134, 160, 292
adsorption isotherms, 99
adverse conditions, 60
adverse effects, vii, 21, 36, 213, 250, 300, 303
aerobic bacteria, 181, 189
Africa, 68, 69, 70, 97, 114, 115, 122, 235, 236, 242, 245, 247, 250, 253, 257, 260, 261, 264
age, 12, 74, 98, 206, 214, 250
aggregation, 112, 201
agricultural producers, 219, 225
agriculture, 1, 17, 23, 57, 63, 73, 82, 87, 88, 127, 128, 129, 143, 153, 154, 176, 179, 200, 207, 208, 211, 212, 225, 235, 247, 251, 258, 259, 260, 261, 264, 265, 267, 268, 270, 272, 273, 280, 281, 282, 292, 305
Agrobacterium, 139, 144, 153, 161
alcohols, 23
alfalfa, 162, 164, 165, 167, 282, 285, 293
algae, 26, 34, 36, 39, 40, 55, 56, 58, 62, 78, 133, 247
alkalinity, 42, 43, 44, 250, 268
alters, 133, 206, 240
aluminium, 172, 242, 245
amines, 79
amino acids, 2, 7, 18, 34, 35, 36, 79, 159, 174, 189, 201, 202, 258, 261, 301, 302, 312, 313, 318
ammonia, 2, 5, 35, 160, 164, 166, 167, 168, 173, 251, 264
ammonium, x, 20, 26, 33, 34, 35, 36, 38, 42, 47, 51, 83, 137, 159, 160, 167, 170, 174, 176, 186, 190, 245, 246, 248, 251, 260, 261, 286, 299, 313, 316, 317, 319
amylase, 3
anaerobic sludge, 88
anatomy, 51, 196
ANC, 146
anoxia, 50
antagonism, 186
anther, 19
antioxidant, 192, 233, 308
APA, 93, 98, 99, 102, 103, 105, 107, 108, 109, 111, 113, 116
apex, 8, 17
aquatic systems, 35, 39, 40, 49, 52
Arabidopsis thaliana, 18
arbuscular mycorrhiza fungus (AMF), x, xii, 120, 121, 123, 124, 178, 189, 190, 191, 204, 206, 267, 268, 269, 270, 271, 272, 273, 297, 305, 307, 308, 316, 317, 319
Argentina, 272, 277, 279, 280, 290, 291, 292, 293, 295
Aristotle, 57, 177, 192, 194, 195
aromatic hydrocarbons, 79
aromatics, 79
arsenic, 67

arthropods, 55, 56, 61, 65, 66, 229
Asia, 209, 265
aspartic acid, 134
assessment, 50, 66, 74, 208, 264, 275
assimilation, 2, 4, 7, 19, 35, 36, 134, 245, 280, 314, 316
atmosphere, 21, 22, 24, 26, 33, 34, 37, 38, 42, 47, 78, 160, 166, 168, 200, 258
atoms, 5
ATP, 2, 4, 39, 81, 231
auxins, 216, 299

B

Bacillus subtilis, 79, 139, 145
bacteria, 27, 31, 35, 36, 42, 55, 56, 58, 59, 60, 61, 63, 65, 68, 78, 85, 86, 102, 107, 121, 129, 133, 134, 135, 138, 139, 140, 141, 142, 144, 145, 148, 150, 151, 154, 156, 157, 160, 161, 162, 163, 164, 165, 167, 168, 170, 171, 173, 174, 175, 176, 181,183, 192, 195, 201, 204, 205, 208, 209, 210, 222, 224, 247, 248, 275, 286, 298, 299, 300, 301, 304, 305, 307, 308, 311, 312, 314, 315, 319
bacterial artificial chromosome (BAC), 204
bacterial strains, 139, 162, 303, 306
bacterium, 83, 137, 138, 141, 144, 152
banks, 270, 273
base, 96, 212, 243, 264, 271
bauxite, 193
bedding, 227, 253, 257
beetles, 61, 62, 217, 218, 233, 247, 250
beneficial effect, 142, 147, 156, 212, 214, 219
benefits, 47, 77, 141, 208, 219, 254, 273, 286
benign, 207
bicarbonate, 42, 51, 113, 302
bioaccumulation, 287
bioassay, 149
bioavailability, 85, 125, 137, 138, 149, 258, 281, 290
biochemical processes, 93, 98, 133, 210, 246
biochemistry, 16, 176
bioconversion, 230
biodegradation, 63, 78
biodiversity, 40, 46, 50, 55, 56, 57, 63, 64, 65, 66, 70, 71, 75, 76, 131, 132, 199, 201, 203, 205, 207, 208, 227, 250, 252
biogenic structures, 55, 56, 58, 62, 112, 117
bioindicators, 69
biological activities, 166, 304
biological activity, 40, 47, 101, 117, 118, 132, 133, 134, 164
biological control, 146, 155
biological processes, 78, 132, 147, 181, 205, 213
biological systems, 77
biomarkers, 220, 222
biomass, 38, 49, 59, 61, 65, 78, 79, 80, 90, 95, 97, 101, 102, 105, 108, 113, 116, 119, 120, 121, 122, 124, 125, 126, 145, 165, 166, 170, 171, 173, 174, 175, 196, 202, 212, 214, 218, 220, 222, 224, 226, 228, 231, 233, 251, 261, 263
biomonitoring, 78
biopores, 55, 56
bioremediation, 73, 74, 78, 208, 303, 309
biosphere, 64, 166
biosynthesis, 15, 18, 187
biotechnology, 150, 191
biotic, x, 56, 73, 89, 105, 156, 159, 160, 203
bonding, 79
bonds, 5
bone, 136, 140, 238, 251
branching, 204, 214, 216
Brazil, 61, 68, 94, 127, 137, 149, 153, 154, 155, 169, 267, 270, 271, 272, 274, 275, 276
breakdown, 23, 63, 133, 138, 149, 160, 212, 238
breeding, 74, 225
Burkina Faso, 119, 122, 124
Butcher, 70
by-products, 112

C

Ca^{2+}, 79, 130, 192
cabbage, 214, 218, 233
cacao, 131
cadmium, 80, 258, 274
Cairo, 228
calcium, 6, 7, 16, 20, 82, 85, 135, 137, 138, 141, 143, 146, 155, 237, 238, 239, 242, 243, 248, 251, 255, 258, 263
calcium carbonate, 138, 141, 251, 258
Cameroon, 60, 65, 150
capillary, 244
carbohydrate, 3, 7, 201, 206, 226
carbohydrate metabolism, 3
carbohydrates, 3, 50, 187, 194, 196, 238, 299
carbon, 4, 7, 15, 17, 19, 21, 45, 48, 64, 66, 69, 82, 86, 95, 101, 112, 120, 121, 134, 137, 155, 165, 171, 172, 173, 203, 210, 212, 216, 238, 243, 245, 246, 251, 253, 255, 260, 265, 274, 299, 301, 307, 312, 318
carbon dioxide (CO2), 4, 7, 19, 21, 22, 23, 28, 32, 33, 38, 45, 63, 67, 125, 130, 133, 135, 136, 174, 176, 195, 200, 202, 210, 238, 240, 246, 251, 252, 270
carboxyl, 3, 79, 112
carboxylic acids, 300
carotene, 7

case study, 49, 149, 171, 211, 225
casting, 70, 228, 229
catabolism, 50, 205
catalysis, 149
category a, 186
cation, 3, 5, 16, 18, 19, 79, 130, 135, 150, 212, 242, 255
cattle, 95, 226, 230, 255, 282
CEC, 96, 243, 244
cell differentiation, 10, 18
cell membranes, 39
cellulose, 63, 68, 164, 314
Central Europe, 49
Chad, 179
chelates, 137, 147, 185, 238, 250, 251, 281, 289, 305
chemical, 1, 3, 22, 23, 27, 37, 42, 62, 67, 70, 73, 74, 77, 78, 79, 82, 85, 86, 87, 97, 114, 118, 120, 122, 123, 124, 128, 129, 130, 131, 132, 133, 136, 137, 147, 150, 151, 164, 176, 181, 190, 191, 200, 201, 202, 204, 208, 212, 213, 214, 215, 216, 222, 226, 227, 228, 229, 238, 240, 242, 243, 246, 252, 254, 255, 257, 260, 262, 263, 268, 269, 272, 273, 280, 282, 288, 289, 297, 298, 299, 300, 302, 304, 306, 311, 312, 314
chemical characteristics, 67, 118, 213, 227
chemical properties, 62, 86, 131, 147, 151, 200, 201, 204, 213, 214, 226, 227, 229, 243, 255, 262, 263, 273, 298, 300, 311
chemical reactions, 27, 128, 201
chemicals, 74, 77, 78, 79, 85, 169, 208, 289, 298
China, 54, 151, 173, 265, 306
chitin, 68
chlorine, 82, 237, 278
chlorophyll, 82, 159, 187, 190, 192, 195, 237, 238, 312
chloroplast, 4, 194, 196
chromatography, 230
chromium, 67, 79, 80
circulation, 22, 252
civilization, 74
classes, 56, 58, 64, 129
classification, 58, 96, 126, 150, 274
clay minerals, 130, 157, 183, 248, 292
cleavage, 143
climate, 23, 47, 49, 51, 98, 128, 132, 177, 200, 208, 235, 246, 248, 250, 253, 277, 279, 286, 298
climate change, 23, 47, 49, 200, 208, 253
climates, 36, 286, 304, 313
climatic factors, 248, 258
cloning, 204
closure, 3, 17
clusters, 306
coal, 42, 129
cobalt, 82, 237, 286, 291
Colombia, 65, 94, 125
colonisation, 120
colonization, 94, 98, 102, 103, 109, 111, 117, 142, 171, 172, 206, 270, 271, 273, 274, 275, 305, 316, 317, 319
combined effect, 44
combustion, 95
commercial, 142, 214, 227, 228, 259, 269, 290
communities, 50, 53, 56, 60, 64, 65, 67, 68, 69, 131, 132, 138, 149, 173, 204, 226, 276
community, 27, 55, 59, 63, 65, 67, 68, 97, 117, 129, 143, 157, 162, 165, 199, 201, 204, 205, 206, 208, 210, 213, 218, 220, 231, 232, 270, 272, 317, 319
compaction, 201, 202, 208, 240, 242, 259, 265, 270, 318
comparative analysis, 67, 139, 151
competition, 42, 44, 59, 112, 115, 144, 167, 168, 174, 183, 218, 317
complex interactions, 132, 143, 225, 272
complexity, 199, 212
composition, 2, 4, 12, 18, 21, 40, 60, 65, 68, 69, 74, 78, 124, 128, 130, 131, 136, 137, 164, 200, 204, 207, 213, 222, 225, 253, 270, 271, 306, 317
compost, 86, 87, 144, 154, 212, 213, 217, 218, 227, 228, 251, 253, 255, 258, 305
composting, 87, 88, 212, 213, 230, 232, 251, 254, 255
compounds, 1, 7, 10, 18, 22, 23, 26, 27, 28, 32, 33, 34, 36, 40, 42, 44, 45, 47, 52, 53, 59, 74, 78, 83, 89, 108, 114, 129, 131, 133, 134, 135, 146, 150, 160, 164, 168, 205, 206, 217, 245, 246, 287, 288, 289, 300, 301, 303, 312
compression, 181
conceptual model, 68, 269
condensation, 2
conditioning, 216
conduction, 18
conductivity, 4, 6
conservation, viii, 64, 73, 74, 86, 89, 212, 246, 292
constituents, 30, 82, 144, 205, 247, 288
construction, 45, 114, 115, 116, 118
consumers, 58
consumption, 113, 128, 238, 247, 248, 249, 250, 258, 262, 280, 293
consumption rates, 113
contaminant, 78, 240
contaminated soils, 80, 88, 207, 209, 258, 281, 282, 293
contamination, 36, 67, 77, 196, 268, 279, 287
copper, 67, 79, 81, 82, 84, 89, 237, 246, 254, 255, 278, 284, 287, 290, 293, 295, 303
correlation, 13, 108, 143, 168, 192

cortex, 8, 17
cost, 77, 85, 130, 207, 209, 212, 250, 257, 278
Costa Rica, 70
cotton, 7, 19, 123, 140, 144, 155, 163, 165, 167, 171, 184, 185, 192, 193, 197, 294, 305
covalent bonding, 79
cracks, 61, 244, 245
critical value, 22, 33, 38, 182
CRM, 293
crop, 75, 77, 80, 81, 82, 83, 85, 86, 87, 88, 111, 134, 137, 145, 147, 148, 160, 162, 164, 165, 166, 167, 168, 169, 173, 175, 196, 199, 200, 201, 202, 206, 208, 212, 218, 219, 220, 221, 225, 227, 228, 235, 236, 237, 240, 241, 242, 245, 246, 247, 248, 249, 250, 251, 252, 253, 254, 257, 258, 259, 264, 265, 272, 277, 278, 279, 280, 281, 282, 283, 285, 286, 289, 291, 292, 294, 298, 300, 302, 308, 317
crop production, 82, 85, 137, 147, 164, 199, 245, 253, 259, 277, 279, 291, 298, 300, 317
crop residue, 160, 166, 250, 251, 264
crop rotations, 87
crops, 1, 7, 74, 77, 81, 82, 83, 85, 86, 87, 127, 128, 129, 130, 145, 147, 159, 162, 164, 165, 167, 171, 178, 190, 194, 196, 206, 207, 213, 214, 215, 218, 236, 237, 238, 239, 242, 243, 245, 247, 248, 249, 250, 251, 252, 257, 258, 260, 262, 267, 269, 270, 277, 278, 279, 280, 281, 282, 283, 285, 286, 290, 292, 295, 308
crowns, 97
crust, 39, 129, 144, 154, 179
crystallization, 133
CTA, 242, 252, 253, 259, 260, 265
cultivars, 13, 14, 152, 184, 185, 186, 187, 189, 190, 192, 193, 194, 196, 197, 261, 307
cultivation, 40, 41, 77, 81, 84, 144, 145, 165, 167, 168, 170, 171, 176, 191, 202, 206, 207, 209, 214, 225, 242, 246, 247
cultivation conditions, 209, 225
cultural conditions, 137
cultural practices, 86
culture, 58, 82, 128, 134, 136, 138, 141, 142, 148, 152, 157, 194, 195, 196, 206, 236, 261, 273, 291
culture conditions, 206
culture media, 136
culture medium, 136, 142, 152, 236
cycles, 22, 31, 39, 44, 47, 64, 95, 132, 169, 200, 246, 268, 272, 276, 301
cycling, 21, 22, 23, 34, 45, 48, 50, 56, 57, 59, 62, 63, 64, 69, 94, 95, 102, 104, 107, 108, 112, 113, 114, 117, 119, 121, 122, 123, 125, 126, 133, 159, 160, 169, 170, 199, 200, 201, 203, 205, 208, 240, 253, 265, 267, 268, 269, 298, 301, 314,315, 316, 317
cysteine, 79
cytokinins, 216
cytoplasm, 2, 3, 4, 5, 7, 8, 9, 11, 14
cytoskeleton, 6

D

damping, 142, 151
database, 151
decay, 23, 42, 63
decomposition, 22, 23, 33, 42, 44, 47, 49, 55, 56, 57, 58, 59, 61, 62, 63, 67, 68, 69, 70, 71, 83, 86, 104, 111, 112, 132, 133, 134, 135, 165, 166, 172, 175, 199, 200, 202, 208, 213, 240, 246, 247, 251, 255, 260, 264, 282, 305, 316, 317
decoupling, 64
defence, 218, 258
deficiencies, 16, 73, 74, 81, 82, 84, 88, 137, 237, 240, 248, 251, 268, 277, 278, 279, 280, 282, 283, 285, 288, 289, 290, 293, 305
deficiency, 16, 17, 18, 19, 20, 27, 47, 74, 81, 83, 84, 89, 98, 103, 104, 108, 111, 118, 123, 151, 177, 178, 179, 181, 183, 184, 185, 186, 188, 189, 190, 191, 192, 193, 196, 202, 236, 237, 245, 248, 268, 280, 281, 283, 284, 285, 286, 288, 289, 290, 292, 295, 298,302, 303, 304, 308, 312
deficit, 1, 2, 3, 4, 5, 6, 7, 11, 14, 197
deforestation, 74, 127, 243, 253, 261
degradation, 23, 34, 39, 46, 63, 64, 65, 77, 79, 88, 89, 95, 133, 169, 202, 204, 207, 208, 246, 247, 257, 261, 267, 268, 277, 279, 292, 315
degradation process, 63, 207
dehiscence, 19
dehydration, 252
demonstrations, 251
denaturation, 303
denitrification, 21, 22, 26, 27, 30, 36, 37, 38, 48, 50, 133, 159, 160, 168, 169, 171, 172, 173, 174, 175, 176, 245, 246, 248, 249, 299
denitrifying, 27, 37, 168, 170
Denmark, 263
Department of Agriculture, 51, 88, 96, 126
dephosphorylation, 6, 18
deposition, 49, 78, 160, 195
deposits, 9, 53
deprivation, 17, 20
depth, 24, 49, 162, 163, 165, 167, 176, 202, 236, 247, 248, 268, 274, 282, 286
derivatives, 2, 4
desiccation, 51
desorption, 78, 91, 94, 115
destruction, 78, 156, 168, 207
detection, 8, 187, 288
detoxification, 79, 135, 147, 186

developed countries, 278
developing countries, 137, 242, 245
diatoms, 138
diffusion, 3, 18, 24, 40, 47, 48, 50, 103, 206, 235, 239, 240, 241, 258, 285, 312
digestibility, 94
digestion, 61, 112, 164, 281
discharges, 40
diseases, 81, 217, 218, 242, 306
disequilibrium, 30
disinfection, 177, 180, 181, 186, 190
disorder, 191, 250
dispersion, 61, 95, 317
dissociation, 142
distribution, 9, 10, 16, 18, 19, 24, 49, 57, 58, 111, 122, 123, 124, 171, 176, 186, 188, 192, 201, 202, 203, 205, 229, 239, 243, 246, 248, 252, 262, 263, 270, 275, 282, 283, 286, 287, 294, 295
diversity, 51, 56, 57, 58, 59, 61, 63, 64, 65, 66, 67, 68, 69, 70, 71, 73, 74, 76, 82, 112, 113, 129, 148, 150, 168, 171, 172, 199, 201, 203, 205, 206, 207, 209, 210, 218, 226, 232, 236, 270, 271, 272, 273, 274, 316
DNA, 39, 58, 202, 204, 303, 315
DNA polymerase, 303
DOI, 119, 274, 275
dominance, 21, 319
donors, 23, 33, 48
drainage, 96, 242, 245, 250, 286
drought, 103, 123, 253, 261, 285, 300, 303, 305, 308, 313, 314, 316, 318, 319
dry matter, 144, 278
drying, 48, 60, 193, 203, 246, 272, 285
dyes, 79

E

E.coli, 168
earthworms, 57, 61, 63, 64, 66, 69, 70, 85, 98, 111, 112, 113, 124, 126, 132, 211, 212, 213, 216, 217, 218, 226, 228, 230, 247, 250, 254, 259
ecological restoration, 49
ecological roles, 57
ecology, 18, 51, 57, 63, 73, 88, 124, 150, 174, 191, 240, 267, 268, 306, 308
economic downturn, 128
ecosystem, 52, 56, 57, 58, 59, 61, 62, 63, 64, 65, 66, 67, 69, 70, 71, 74, 78, 95, 102, 107, 114, 120, 121, 132, 152, 165, 170, 199, 200, 203, 206, 207, 209, 236, 247, 262, 267, 269, 271, 273, 274, 298, 306, 307, 311, 315, 316, 317, 318
Ecuador, 261
EEA, 292
egg, 218
electrical conductivity, 201, 213
electrodes, 30
electron, 3, 4, 15, 21, 23, 27, 28, 29, 30, 31, 33, 37, 38, 39, 40, 42, 44, 48, 272, 285
electrons, 23, 27, 30, 202
elongation, 8, 9, 11, 15, 47, 49, 50, 53, 81, 214, 227
emission, 23, 33, 38, 53, 160, 238, 252, 265, 293
emitters, 257
encoding, 3, 5, 6, 7, 19
energy, 2, 4, 5, 23, 27, 33, 35, 47, 60, 78, 132, 133, 162, 164, 168, 182, 199, 202, 205, 237, 238, 239, 251, 271, 278
energy consumption, 182
energy input, 162, 164
engineering, 78, 196
enlargement, 137
environment, x, 22, 23, 27, 28, 32, 60, 68, 69, 71, 74, 77, 78, 80, 96, 127, 129, 131, 132, 133, 135, 136, 147, 150, 151, 164, 187, 190, 191, 194, 199, 200, 202, 205, 206, 209, 210, 212, 239, 240, 242, 246, 247, 250, 252, 288, 298, 304
environmental aspects, 312
environmental change, 160
environmental conditions, 18, 47, 53, 169, 199
environmental control, 172
environmental degradation, 85
environmental effects, 36
environmental factors, 102, 140, 156, 159, 167, 247
environmental impact, 22, 23, 130, 200, 209
environmental management, x, 159, 169
environmental protection, 53
environmental quality, 236
environmental stress, 59
environmental threats, 73
enzymatic activity, 101, 112, 172
enzymes, 2, 3, 17, 35, 36, 37, 47, 59, 63, 81, 134, 159, 160, 164, 178, 187, 189, 191, 202, 203, 212, 220, 237, 238, 246, 298, 300, 301, 303, 312, 314
equilibrium, 27, 86, 131, 135, 192, 199, 206, 279
erosion, 201, 202, 242, 243, 252, 261, 263, 267, 269, 281
ethylene, 6, 15, 300
eucalyptus, 214
eukaryotic, 59
Europe, 49
evapotranspiration, 47
evidence, 27, 96, 114, 132, 170, 204, 214, 215, 216, 217, 218, 237, 269, 278
evolution, 51, 171, 173, 202, 290
excretion, 60
explosives, 77
exporter, 279

exposure, 74
extinction, 63, 74
extraction, 51, 60, 142, 175, 208, 280, 281, 282
extracts, 170, 214, 216, 217, 218, 230, 233, 251

F

factories, 287
farm size, 252
farmers, 1, 15, 56, 64, 128, 137, 206, 207, 236, 248, 250, 251, 257, 261, 264, 267, 269, 280
farming techniques, 242
farms, 90, 128, 227, 247, 290
fauna, 56, 57, 58, 59, 60, 61, 62, 63, 65, 66, 67, 68, 70, 95, 218, 252, 254, 281
FDA, 202, 208
feces, 228, 230
feedstock, 213, 214, 215
fermentation, 47, 88, 134, 146, 156, 174
fertility, 22, 36, 55, 57, 61, 62, 63, 64, 67, 69, 73, 74, 82, 83, 85, 86, 90, 94, 102, 104, 117, 119, 120, 121, 125, 130, 137, 147, 159, 160, 168, 170, 173, 200, 205, 207, 211, 219, 227, 231, 235, 236, 242, 245, 246, 247, 250, 251, 252, 253, 254, 255, 257, 258, 259, 260, 261, 263, 264, 265, 268, 274, 277, 280, 281
fertilization, 19, 40, 95, 104, 129, 134, 142, 145, 151, 172, 174, 177, 180, 182, 184, 218, 219, 220, 221, 222, 223, 224, 225, 228, 229, 230, 237, 239, 240, 264, 267, 269, 273, 277, 280, 281, 283, 284, 285, 286, 289, 290, 291, 292, 297, 299, 300, 301, 308, 311, 312, 314, 315, 319
fertilizers, 7, 14, 16, 23, 34, 73, 74, 85, 86, 87, 97, 119, 128, 130, 131, 133, 134, 137, 142, 146, 147, 148, 152, 166, 167, 173, 178, 186, 190, 200, 210, 211, 212, 215, 216, 218, 219, 220, 221, 222, 225, 230, 237, 238, 245, 247, 248, 249, 250, 251, 252, 257, 272, 279, 280, 288, 289, 290, 305
field crops, 173, 231, 277, 285
field trials, 147, 286
financial, 90, 128
financial performance, 90
fire suppression, 95
fires, 95, 114, 272
fitness, 59
fixation, 4, 15, 31, 34, 39, 59, 96, 112, 133, 135, 159, 160, 162, 164, 170, 174, 195, 197, 248, 252, 259, 269, 278, 285, 286, 299, 303
flame, 11
flavonoids, 299
flavour, 248
flooding, 23, 32, 33, 40, 47, 48, 49, 51, 193, 272
floods, 49, 51, 127
flora and fauna, 247
flowers, 80, 87, 214
fluctuations, 22, 45, 73, 189, 191, 252
fluorescence, 187, 192
food, 55, 57, 58, 59, 60, 61, 62, 63, 65, 66, 67, 73, 74, 78, 80, 82, 84, 88, 118, 129, 131, 194, 199, 206, 212, 218, 226, 236, 237, 245, 246, 247, 257, 258, 269, 292, 295, 298
food chain, 62, 78, 80
food habits, 58
food production, viii, 55, 194, 245, 246, 298
food security, 88, 236, 257
food web, 57, 58, 59, 63, 65, 66, 67, 199
forage crops, 7, 237
Ford, 20, 166, 172
forest ecosystem, 75, 113, 164, 263
formation, 2, 6, 18, 32, 33, 36, 37, 38, 42, 47, 54, 55, 56, 57, 64, 66, 79, 85, 88, 120, 130, 132, 135, 136, 142, 152, 178, 182, 184, 186, 189, 190, 191, 197, 200, 203, 204, 232, 238, 240, 245, 246, 254, 285, 287, 288, 299, 317
fragility, 248
fragments, 129, 272
France, 70, 264
free energy, 27
freezing, 203, 272
freshwater, 50, 51, 74
fructose, 11, 12
fruits, 7, 15, 84, 290
fungi, 34, 55, 56, 58, 59, 61, 65, 67, 68, 78, 79, 102, 107, 117, 118, 121, 129, 133, 135, 136, 138, 139, 140, 141, 142, 143, 144, 147, 149, 150, 154, 157, 160, 161, 173, 178, 183, 204, 208, 209, 210, 217, 247, 267, 268, 269, 273, 274, 275, 276, 297, 298, 299, 303, 306, 307, 308, 311, 312, 313, 314, 315, 316, 318
fungus, 59, 69, 103, 111, 115, 125, 140, 142, 143, 144, 145, 147, 148, 153, 178, 182, 185, 189, 194, 316, 317, 318
fusion, 129

G

gas diffusion, 24, 25, 26
gene expression, 6, 315
gene regulation, 278
genes, 3, 5, 6, 7, 15, 17, 18, 53, 217, 299, 315
genetic diversity, 48, 204
genome, 301
genomics, 210
genotype, 191, 206, 214, 220, 225, 226, 277, 279, 308, 314

genus, 59, 94, 97, 142, 143, 161, 171, 181, 183, 270, 286
genus Streptomyces, 183
geography, 132
Germany, 50, 54, 124, 208, 209, 263
germination, 3, 16, 81, 82, 143, 214, 225, 226, 230, 233, 274, 290
Gigaspora margarita, 271, 272
global climate change, 275
global demand, 128
global scale, 22, 87
global warming, 38, 168, 253
Glomus intraradices, 142, 312
glucose, 11, 15, 19, 54, 137, 141, 142
glutamate, 2
glutamic acid, 35
glutamine, 2, 313, 315
glycine, 315
glycolysis, 2
grass, 51, 65, 69, 71, 77, 94, 95, 97, 111, 115, 118, 122, 145, 148, 150, 172, 173, 255, 269
grasses, 77, 94, 95, 97, 102, 115, 123, 170, 176, 242, 317
grasslands, 61, 69, 167, 262, 279
grazers, 102
grazing, 125, 167, 176, 253, 282
Great Britain, 189, 258
Greece, 177, 191, 192, 194, 195
green alga, 39, 45
Green Revolution, 74
greenhouse, 12, 21, 26, 33, 54, 131, 151, 211, 214, 216, 218, 226, 270, 308, 319
greenhouse gases, 26
greening, 252
groundwater, 36, 51, 96, 168, 174, 286
growth factor, 236
growth hormone, 81, 82
growth rate, 59, 81, 181, 187, 220, 227
guanine, 6
guard cell, 3, 16, 17, 20
Guinea, 259

H

habitat, viii, x, 55, 56, 66, 124, 131, 132, 199
habitats, 46, 58, 62, 70, 112, 132, 205
halogen, 286
halogens, 271, 286
hardwood forest, 262
harvesting, 7
hazardous substances, 77, 196
health, 22, 36, 47, 48, 73, 74, 81, 85, 199, 200, 201, 202, 204, 205, 207, 208, 210, 211, 242, 269, 273, 274, 276, 278, 298, 311
health problems, 36
heavy metals, 26, 45, 60, 78, 80, 89, 91, 120, 169, 179, 196, 207, 279, 293
height, 220, 221, 258
heterogeneity, 170, 245, 306
heterosis, 227
heterotrophic microorganisms, 150, 202
higher education, 176
history, 286
homeostasis, 4, 6, 7, 17, 19, 20
hormone, 278
hormones, 53, 133, 298, 299, 300, 301, 303
horticultural crops, 213
host, xiii, 61, 62, 104, 147, 171, 229, 231, 273, 274, 298, 299, 302, 303, 311, 312, 313, 314, 315
hot spots, 75, 76, 118
hotspots, 94, 112, 114, 262
House, 40, 45, 50, 295
human, 39, 58, 64, 74, 87, 89, 95, 129, 132, 207, 212, 236, 237, 254, 267, 269, 272, 282
human health, 207
humidity, 289
humus, 32, 67, 88, 115, 118, 119, 232
hunting, 253
husbandry, 279
hybrid, 220, 222, 308
hydrocarbons, 78
hydrogen, 15, 27, 30, 52, 238, 301, 303
hydrogen gas, 30
hydrogen peroxide, 15, 303
hydrological conditions, 41, 42
hydrolysis, 40, 108, 133, 134, 169, 178, 179, 187, 202, 231, 258
hydrophobicity, 78
hydroponics, 236
hydroxide, 136, 139, 146, 150, 251
hydroxyapatite, 137, 152
hydroxyl groups, 3
hypothesis, 69, 102, 125, 176, 217, 315
hypoxia, 24

I

illumination, 12
imbalances, 240, 268
immobilization, 26, 34, 44, 103, 108, 109, 126, 160, 164, 165, 166, 168, 248, 250, 288
improvements, 216, 220, 225
impulses, 88
in vitro, 143, 147, 154

in vivo, 194
incidence, xii, 156, 208, 217, 218, 227, 228, 277, 279
income, 94
incubation period, 131
indexing, 264
India, 54, 55, 70, 73, 84, 85, 86, 90, 122, 126, 151, 206, 252, 263, 264, 275
indirect effect, x, xi, 46, 57, 60, 71, 159, 160, 211, 217, 218, 313
Indonesia, 131
induction, 15, 217, 218
industrial wastes, 156
industrialization, 22
industries, 78
industry, 33, 128, 251
inefficiency, 280
infection, 121, 123, 125, 217, 299, 308, 317
infrastructure, 280
ingest, 58, 63
ingestion, 112, 117, 119
inhibition, 4, 20, 42, 82, 109, 120, 142, 172, 186, 195
inhibitor, 166, 167, 169, 171, 176, 263
initiation, 30, 81, 204, 217
inoculation, 60, 67, 125, 142, 143, 144, 145, 146, 147, 148, 151, 154, 164, 170, 174, 206, 270, 273, 275, 286, 298, 308, 315, 319
inoculum, 273, 301, 304
inosite, 2
insects, 60, 61, 70, 218
Integrated Nutrient Management (INM), 73, 82
integration, 235
integrity, 76, 199, 201
interaction effect, 314
interaction effects, 314
interchangeability of potassium, 1, 14
interface, 54, 135, 136, 205, 210, 262
interrelations, 56, 63
intervention, 74, 236, 272
intestine, 57, 87
introspection, 88
invertebrates, 55, 56, 57, 58, 61, 63, 64, 67, 121, 152, 175
investment, 280
ion channels, 16, 18, 20
ion transport, 18
ion uptake, 19
ion-exchange, 201
ions, 2, 3, 4, 40, 78, 83, 89, 131, 134, 135, 136, 142, 188, 189, 213, 239, 240, 243, 259, 271, 285, 286
Iran, 297
Ireland, 190, 194
iron, 21, 22, 26, 27, 33, 40, 41, 42, 44, 51, 52, 53, 81, 82, 84, 90, 91, 94, 133, 134, 135, 136, 137, 139, 143, 148, 150, 155, 190, 191, 195, 227, 237, 238, 245, 246, 251, 255, 275, 278, 293, 295, 302, 307, 308
irradiation, 25, 26
irrigation, 33, 74, 169, 173, 186, 239, 257, 263
islands, 74, 274
isotope, 67, 174
issues, 88, 212, 260, 277, 280
Italy, 232, 291
Ivory Coast, 112, 115, 122

J

Japan, 194
Jordan, 121, 152

K

K^+, 5, 6, 8, 11, 14, 15, 16, 17, 18, 19, 20, 130, 143, 144
Kenya, 247, 251, 260, 261
kinetics, 5, 17, 41, 118, 124, 138
Krebs cycle, 2

L

laboratory studies, 115
laboratory tests, 145
lakes, 36, 40
land tenure, 242
landscape, 64, 71, 88, 172, 207, 242, 243, 245, 248, 259, 260, 261, 265
landscapes, 200
lanthanum, 141, 157
Laos, 232
larvae, 218
lattices, 282
leaching, 36, 61, 99, 131, 160, 166, 167, 171, 181, 207, 242, 243, 244, 245, 246, 248, 249, 252, 257, 260, 261, 281, 299, 300, 312, 314
lead, xi, 21, 24, 33, 47, 63, 79, 80, 81, 85, 89, 90, 107, 168, 189, 199, 208, 212, 213, 240, 245, 246, 258, 285
Leahy, 78, 90
leakage, 14
leaks, 245
Lebanon, 294
legume, 77, 85, 90, 196, 251, 252, 269, 270, 275, 299
Lepidoptera, 62
leucine, 202, 208, 220

liberation, 108
life cycle, 31, 236, 278
light, 4, 12, 16, 40, 81, 187, 193, 197, 215, 236, 250, 287, 289, 315
lignin, 63, 69, 164, 165
limestone, 65, 69, 130, 238
lipases, 202
lipid peroxidation, 192
liquid phase, 24, 240
livestock, 87, 124, 229, 246, 253, 255
logging, 206
LSD, 105, 106, 107, 109, 110

M

macrofauna, 55, 56, 58, 61, 62, 63, 66, 67, 111, 112, 119, 121
macromolecules, 3
macronutrients, 18, 31, 99, 194, 223, 229, 235, 237, 250, 268, 278, 289, 300
macropores, 216
magnesium, 10, 11, 16, 81, 82, 143, 192, 237, 239, 242, 243, 251, 255, 263
magnitude, 15, 67, 214, 216, 220, 280
management, 64, 67, 68, 69, 73, 74, 82, 88, 89, 90, 95, 111, 120, 123, 124, 169, 170, 173, 176, 199, 200, 201, 202, 204, 205, 206, 207, 208, 209, 210, 231, 235, 236, 240, 242, 248, 250, 251, 258, 259, 260, 261, 263, 264, 265, 272, 277, 279, 284, 292
manganese, 21, 27, 33, 81, 82, 84, 177, 179, 191, 192, 193, 194, 195, 196, 197, 237, 250, 254, 278, 293, 304, 305, 308
manufacturing, 146
manure, 35, 50, 80, 85, 86, 87, 120, 144, 173, 212, 217, 219, 220, 221, 222, 223, 224, 226, 227, 230, 233, 245, 246, 247, 250, 251, 305
mapping, 90, 257, 260
Maryland, 291
mass, xi, 61, 142, 146, 218, 235, 239, 241, 258, 285, 312, 313
mass loss, 61
materials, 59, 79, 96, 114, 115, 116, 117, 118, 129, 131, 148, 165, 166, 172, 176, 189, 212, 238, 247, 251, 255, 258, 279
matrix, 160, 200
matter, 23, 51, 62, 73, 78, 85, 95, 112, 116, 132, 177, 180, 182, 186, 200, 205, 213, 231, 238, 242, 243, 250, 252, 311, 313
measurement, 25, 30, 47, 50, 208, 260
measurements, 19, 22, 107, 124, 187
media, 9, 82, 102, 137, 143, 156, 206, 211, 214, 215, 216, 218, 225, 227, 228, 229, 233
Mediterranean, 49, 65, 69, 176
Mediterranean climate, 176
medium composition, 9
melon, 8, 156, 171
membrane permeability, 3
membranes, 4, 6, 15
mercury, 80, 261
meristem, 11, 250
mesentery, 87
mesofauna, 55, 58, 60, 61, 63, 67
mesophyll, 4
metabolic disturbances, 1, 82
metabolism, 1, 4, 17, 18, 22, 23, 31, 47, 49, 60, 81, 82, 89, 132, 134, 178, 187, 238, 278, 302, 316, 317
metabolites, 58, 133, 134, 141, 302
metabolized, 80
metal ion, 78, 79, 90, 136
metal ions, 78, 79, 90, 136
metals, 33, 78, 79, 80, 180, 182, 184, 188, 197, 209, 269, 279, 290, 292
meter, 30, 286
Mexico, 112, 262
Mg^{2+}, 130, 143, 192
microbial cells, 25, 116, 168
microbial communities, 60, 165, 170, 215, 226, 229, 272
microbial community, 59, 60, 61, 147, 150, 199, 206, 207, 209, 210, 212, 215, 218, 220, 308, 316
microbial solubilization, x, 127, 129
microbiota, 131, 201, 268
microcosms, 61, 68, 136
microcrystalline, 131
microenvironments, 246
microfauna, 55, 58, 59, 61, 218
microflora, 55, 58, 61, 63, 82, 85, 122, 144, 148, 150, 162, 166, 204, 218, 245, 308
microhabitats, 59, 69
Micronutrient deficiencies, 277, 279
micronutrients, 22, 31, 73, 82, 85, 86, 182, 212, 215, 235, 237, 250, 251, 263, 267, 268, 277, 278, 280, 281, 282, 283, 284, 285, 287, 288, 289, 293, 295, 302, 303, 304, 308
microorganism, 138, 144, 145, 148, 203, 240
microorganisms, 14, 23, 27, 47, 48, 56, 61, 62, 63, 65, 73, 78, 83, 93, 98, 102, 103, 104, 108, 117, 124, 125, 127, 129, 131, 132, 133, 134, 135, 137, 138, 139, 140, 141, 142, 143, 144, 145, 147, 148, 149, 150, 151, 152, 153, 154, 155, 156, 159, 160, 161, 162, 164, 168, 169, 172, 174, 176, 177, 180, 181, 182, 183, 185, 191, 193, 194, 199, 200, 201, 203, 205, 207, 208, 212, 213, 217, 222, 225, 226, 245, 246, 247, 269, 270, 273, 275, 305, 307, 308, 317

migration, 127, 239
mildew, 231
mineralization, 23, 34, 47, 54, 56, 58, 59, 93, 98, 99, 103, 104, 107, 108, 109, 112, 117, 121, 122, 124, 126, 133, 134, 165, 166, 169, 172, 173, 175, 176, 203, 215, 246, 253, 262, 263, 283, 298, 313, 314, 316
Mississippi River, 306
mixing, 111, 117, 132, 202, 216, 220
modelling, 49
models, 149, 220, 269, 305
modifications, 200
moisture, 18, 36, 37, 46, 48, 54, 82, 86, 89, 103, 160, 162, 167, 169, 171, 173, 177, 178, 180, 182, 189, 191, 235, 239, 240, 242, 245, 246, 247, 248, 249, 250, 252, 283, 285
moisture content, 46, 162, 169, 171, 178
molds, 136, 137, 144
mole, 26, 27
molecular mass, 79
molecular oxygen, 24, 30
molecular structure, 212
molecular weight, 23, 134, 145, 153, 181, 189
molecules, 1, 3, 9, 18, 115, 124, 143, 203, 204, 240, 301
mollusks, 56
molybdenum, 82, 238, 246, 254, 278, 286, 291, 292, 294
monosaccharide, 3
Morocco, 142
morphology, 6, 53, 60, 132, 178, 190, 191, 192, 206, 213, 214, 216, 229, 299, 302
mosaic, 197
Moscow, 1, 149
motif, 79
MSW, 261
multicellular organisms, 164
multiple regression, 265
multiple regression analysis, 265
mung bean, 196
municipal solid waste, 228, 233, 251, 255, 260
mycelium, 61, 102, 204, 270
mycorrhiza, 9, 85, 102, 103, 119, 120, 122, 124, 177, 178, 180, 182, 183, 189, 190, 209, 317, 318, 319
mycorrhizas, xii, 59, 209, 210, 267, 268, 269, 270, 276, 306

N

Na^+, 5, 6, 11, 19, 130
NaCl, 11, 270
native species, 62
NATO, 89
natural gas, 33
natural resources, 64, 74, 127
necrosis, 47, 81
negative consequences, 36, 188
negative effects, 40, 47, 67, 188, 286
negative relation, 28
nematode, 68, 102, 218, 226, 232
Netherlands, 43, 49, 52, 53, 90, 171, 228, 231, 259, 260, 265, 291, 305, 307
neutral, 3, 44, 82, 135, 180, 181, 183, 184, 193, 246, 258, 271, 284, 289, 298, 301
New Zealand, 66, 90
N-fixing bacteria, 162, 298, 311, 312, 314
Nicaragua, 260
nickel, 82, 89, 208, 237, 272
Nigeria, 235, 242, 243, 247, 248, 250, 251, 255, 258, 259, 260, 261, 262, 263, 264
Nile, 176
nitrates, 21, 22, 27, 34, 53, 129, 245, 289
nitric oxide, 37
nitrification, 26, 35, 36, 37, 38, 50, 83, 133, 159, 160, 166, 167, 168, 169, 170, 171, 172, 173, 175, 176, 202, 245, 246, 249, 262, 264, 267, 269, 313
nitrifying bacteria, 50, 166, 167, 168
nitrite, 35, 36, 37, 166, 167, 168, 175, 246
nitrogen fixation, 39, 81, 83, 86, 159, 160, 162, 164, 169, 170, 171, 173, 174, 208, 209, 255, 260, 267, 291
nitrogen gas, 38
nitrogenase, 39, 164, 172, 286, 299
nitrogen-fixing bacteria, 39, 59, 142, 145, 148, 162, 169, 171, 174, 176
nitrous oxide, 21, 37, 38, 48, 52, 166, 174, 175
NMR, 121, 176
N-N, 222, 223
Nobel Prize, 74
nodules, 143, 270, 286, 299
North America, 57, 114, 167, 242
Norway, 50, 136, 155
Norway spruce, 50, 136, 155
nucleic acid, 2, 3, 4, 34, 159, 245
nucleotides, 2
nutrient availability, 21, 42, 59, 82, 127, 129, 133, 147, 156, 204, 227, 240, 250
nutrient imbalance, 242
nutrition, 17, 18, 19, 83, 85, 88, 89, 98, 102, 103, 104, 108, 111, 125, 127, 128, 129, 130, 137, 141, 142, 147, 148, 150, 153, 173, 183, 184, 185, 188, 191, 192, 194, 196, 197, 236, 237, 238, 242, 258, 261, 263, 270, 274, 278, 291, 293, 294, 295, 305, 308,317
nutritional imbalance, 279
nutritional status, 83, 94, 109, 308

O

obstacles, 250
oceans, 39, 42
oil, 7, 33, 77, 78, 138, 150, 154, 169, 175, 180, 182, 191, 200, 226, 246, 250, 258, 269, 286, 297, 301
oil production, 302
operations, 250
opportunities, 175
optimization, 78, 141
organ, 3, 19, 180, 189
organelles, 6
organic compounds, 2, 23, 31, 32, 34, 39, 46, 113, 114, 159, 164, 165, 205, 289, 312
organic matter, 22, 23, 27, 33, 44, 47, 49, 55, 56, 57, 58, 59, 61, 62, 63, 65, 69, 74, 82, 83, 85, 94, 95, 96, 97, 101, 103, 112, 115, 116, 121, 123, 130, 133, 135, 159, 160, 166, 169, 174, 175, 177, 180, 182, 186, 187, 189, 192, 200, 202, 204, 208, 209, 212, 213, 215, 219, 225, 230, 237, 239, 240, 242, 246, 247, 250, 252, 253, 255, 260, 268, 271, 280, 281, 283, 284, 287, 288, 298, 305, 313, 314, 315, 317
organic polymers, 165
organic soils, 48, 189
organism, 58, 172, 200
organs, 1, 7, 13, 188, 191, 192
ornamental plants, 80
oscillation, 239
osmotic pressure, 11, 12, 15
overgrazing, 202
overlap, 23, 27, 60, 239
oxalate, 122, 135, 141, 149, 166, 245, 301
oxidation, 2, 27, 28, 30, 34, 35, 36, 38, 42, 45, 48, 51, 53, 82, 83, 132, 133, 135, 148, 167, 169, 173, 178, 181, 183, 187, 189, 202, 207, 212, 288, 293, 304
oxidative stress, 81, 193, 303
oxygen, vii, 5, 21, 22, 23, 24, 25, 26, 27, 28, 33, 39, 40, 47, 48, 49, 50, 79, 133, 137, 162, 174, 238, 240, 246
oyster, 251
ozone, 37, 168

P

Pakistan, 154
parallel, 28
parasite, 218
parasites, 61, 62
parenchymal cell, 14
pasture, 60, 66, 67, 70, 125, 149, 167, 203
pastures, 66, 121, 123, 167, 201, 245, 269, 279, 282, 285
pathogens, 59, 147, 199, 203, 205, 208, 217, 218, 230, 298, 300
pathways, 5, 301
peat, 53, 211, 212, 214, 215, 216, 217, 231, 233
penetrability, 204
PEP, 4
peptides, 2, 79
peri-urban, 253, 255
permeability, 17, 18
peroxidation, 303
peroxide, 15, 20
personal communication, 281, 282, 285, 288, 289
pest populations, 226
pesticide, 88, 228
pests, 82, 87, 147, 199, 217, 231, 242
Petal, 69
petroleum, 77
pH, 5, 14, 22, 25, 28, 29, 30, 36, 37, 38, 40, 42, 44, 45, 46, 48, 51, 52, 73, 79, 82, 83, 96, 99, 104, 112, 120, 130, 131, 133, 134, 135, 137, 143, 145, 146, 151, 167, 168, 177, 178, 180, 181, 182, 183, 186, 188, 191, 192, 193, 201, 203, 205, 208, 216, 235, 242, 243, 246, 247, 254, 255, 258, 262, 264, 270, 271, 284, 285, 286, 288, 290, 300, 301, 307
phenol, 203
phenolic compounds, 302, 303, 306, 307
Philadelphia, 53, 70
Philippines, 88
phloem, 4, 7, 8, 14, 15, 16, 18
phosphate, 4, 5, 25, 26, 32, 40, 42, 44, 50, 51, 52, 53, 83, 85, 86, 94, 119, 121, 122, 124, 125, 127, 128, 129, 133, 134, 135, 136, 137, 138, 139, 140, 141, 142, 143, 144, 145, 146, 147, 148, 149, 150, 151, 152, 154, 155, 156, 157, 167, 176, 182, 184, 205, 207,225, 251, 252, 262, 275, 288, 300
phosphates, 3, 79, 115, 129, 135, 136, 137, 139, 140, 141, 142, 143, 145, 147, 149, 150, 152, 154, 251
phospholipids, 39, 79, 245
phosphorous, 51, 52, 54, 85, 86, 122, 281, 308, 312, 319
phosphorus, 1, 16, 32, 34, 40, 41, 42, 48, 49, 50, 52, 81, 82, 83, 85, 86, 89, 93, 94, 95, 98, 112, 113, 117, 118, 119, 120, 121, 122, 123, 124, 125, 126, 129, 132, 137, 139, 145, 147, 148, 150, 154, 155, 195, 196, 200, 221, 237, 239, 242, 243, 244, 245, 247, 248, 250, 251, 255, 257, 259, 261, 262, 263, 264, 265, 268, 283, 306, 307, 308
phosphorylation, 2, 3, 4, 6, 18
photosynthesis, 4, 7, 13, 17, 31, 32, 39, 42, 45, 47, 50, 51, 81, 82, 187, 194, 195, 197, 237, 238
physical characteristics, 97, 216

physical environment, 250
physical features, 22, 200
physical mechanisms, 215
physical properties, 114, 212, 216, 230, 257, 262, 264
physical structure, 57, 160
physico-chemical parameters, viii, 73
physicochemical properties, 119, 186, 227, 262
physics, 259
physiological, 18, 125, 139, 151, 308
physiology, 1, 2, 16, 52, 91, 132, 192, 193, 209, 210, 229, 278, 306
phytoremediation, 79, 207, 209, 295
pilot study, 264
Pinus halepensis, 154
planets, 87
plant cell turgor pressure, vii, 1
plant diseases, 217, 230
plant establishment, 94
plant growth, 12, 20, 31, 40, 50, 57, 63, 67, 73, 81, 82, 83, 86, 89, 91, 93, 98, 114, 117, 128, 135, 138, 139, 141, 142, 143, 144, 146, 147, 149, 151, 152, 156, 160, 171, 173, 178, 186, 187, 191, 201, 205, 208, 210, 211, 212, 213, 214, 215, 216, 217, 218, 219, 225, 227, 228, 229, 230, 232, 233, 238, 240, 246, 268, 270, 273, 274, 278, 288, 297, 298, 300, 301, 302, 303, 304, 305, 307, 308, 309, 311, 312, 313, 314, 315, 316, 318
plant growth promoting rhizobacteria (PGPR), xii, xiii, 297, 298, 311, 312
plant P-uptake, 93, 98, 117
plant type, 206
plasma membrane, 5, 6, 7, 14, 16, 19, 20, 135, 217
plasticity, 51
platinum, 25, 30, 50
Poland, 21, 49, 50, 53, 179
polarization, 25
policy, 88, 236
policy makers, 88
pollen, 3, 16, 19
pollen tube, 3
pollination, 19
pollutants, 77, 78, 80, 168, 269
pollution, 50, 77, 80, 153, 166, 168, 207, 208, 242, 245
polyamines, 6
polychlorinated biphenyl, 79
polymers, 164
polyphosphates, 147
polysaccharide, 309
pools, 22, 59, 98, 118, 124, 190
population growth, 127
porosity, 24, 46, 144, 212, 216, 240, 242
positive correlation, 42
post-transcriptional regulation, 3
potassium, 1, 2, 3, 4, 5, 6, 7, 8, 9, 10, 11, 12, 13, 14, 15, 16, 17, 18, 19, 20, 82, 86, 89, 115, 130, 131, 133, 134, 136, 139, 143, 144, 145, 146, 147, 148, 149, 151, 153, 154, 155, 156, 157, 237, 238, 239, 245, 247, 248, 251, 255, 257, 261, 280, 301, 305
potassium transmembrane transport, 1
potato, 7, 123, 195, 196
poultry, 85, 86
poverty, 242
precipitation, 40, 44, 45, 81, 96, 97, 98, 162, 173, 288, 302
predation, 62
predators, 58, 62, 63
preservative, 67
prevention, 201, 204, 208
priming, 290
productive capacity, 248
project, 89, 252
prokaryotes, 15, 132
proliferation, 252, 317
proline, 189, 194, 314
protected areas, 95
protection, 77, 82, 151, 178, 189, 191, 201, 205, 278
protein synthesis, 2, 81, 202, 232, 238
proteins, 2, 3, 6, 7, 14, 16, 19, 20, 23, 31, 34, 42, 78, 79, 159, 178, 189, 191, 202, 206, 220, 237, 238, 285, 301, 305, 312
protons, 2, 14, 135, 142
protozoans, 55, 56, 60
pumps, 233
purines, 34
purity, 279
pyrite, 42, 44

Q

quality control, 293
quality improvement, 250
quality production, 19
quantification, 57
quartz, 96, 129, 133, 141, 144, 149, 156
Queensland, 170

R

race, 74, 97
radical formation, 82
rain forest, 125
rainfall, 239, 242, 245, 249, 279, 286
rainforest, 64, 124

rangeland, 95, 175, 263
rape, 140, 155
rare earth elements, 137
reactants, 27
reactions, 2, 3, 4, 15, 26, 27, 28, 32, 35, 39, 47, 82, 83, 131, 132, 133, 135, 137, 147, 178, 187, 195, 203, 287
reactive oxygen, 7, 14, 187, 301
reactivity, 94
reality, 177, 181, 242
recombination, 201
recovery, 82, 145, 173, 199
recycling, 132, 201, 219
redistribution, 6, 13, 17, 62
redox potential, vii, 21, 22, 23, 27, 28, 30, 48, 52, 53, 82, 180, 181, 182, 183, 191, 193, 196, 288
reform, 201
regenerate, 253
regeneration, 151, 235, 242, 247, 250, 251
regrowth, 95
regulations, 18
remediation, 77, 80, 89, 207
reprocessing, 201
reproduction, 278
reproductive organs, 7
RES, 282, 283, 287
reserves, 47
residuals, 32
residues, 34, 55, 56, 63, 70, 77, 141, 159, 165, 168, 170, 173, 200, 251, 255, 263, 264, 315
resilience, 56, 63, 64, 199, 203, 236
resistance, 4, 30, 31, 52, 81, 82, 123, 133, 187, 188, 189, 191, 199, 202, 203, 218, 231, 313, 314
resource allocation, 214
resource availability, 269
resources, 58, 74, 84, 111, 120, 127, 147, 212
respiration, 4, 17, 23, 24, 27, 32, 33, 39, 61, 62, 119, 133, 134, 135, 145, 166, 202, 240
restoration, 48, 50, 73, 88, 207, 208, 210, 255, 258, 267, 268, 271
restoration programs, 271
retardation, 3, 4, 13, 47
reticulum, 7
rhizobium, 269, 273, 298, 299, 311, 312, 314, 315
riboflavin, 304
ribosome, 3
rice field, 33, 265
risk, 31, 33, 51, 212
risks, 290, 295
RNA, 303
root cap, 188
root growth, 14, 25, 124, 187, 189, 196, 202, 206, 214, 217, 231, 238, 241, 242, 264, 283, 300, 309
root hair, 6, 102, 125, 190, 206, 237, 299, 302, 307
root system, x, 17, 61, 119, 168, 177, 183, 185, 187, 188, 189, 190, 191, 205, 206, 255, 283
root zones, 11
roots, 5, 6, 7, 11, 12, 13, 14, 16, 17, 18, 20, 24, 25, 39, 47, 49, 59, 62, 65, 68, 80, 89, 90, 91, 98, 102, 111, 124, 129, 131, 132, 133, 134, 135, 136, 139, 141, 142, 153, 155, 157, 160, 162, 164, 167, 168, 170, 174, 176, 181, 189, 190, 192, 195, 196, 201, 204, 205, 206, 209, 227, 235, 236, 239, 240, 242, 244, 250, 252, 257, 259, 260, 262, 273, 274, 285, 289, 299, 300, 301, 302, 303, 308, 312, 313, 315
roundworms, 60
runoff, 242, 245, 248, 249, 255
Russia, 1, 149

S

safety, 77, 212
salinity, 11, 18, 20, 140, 156, 242, 250, 267, 268, 275, 306, 316
saliva, 117
salts, 179, 237, 238, 281, 289
saturation, 5, 40, 42, 96, 131, 180, 212, 242, 243, 271
savannah, 124
sawdust, 79
scarcity, 16, 248
scavengers, 78
science, 88, 89, 132, 169, 176, 191, 260, 317
Scots pine, 68
seasonal changes, ix, 94, 101, 104, 107, 108, 111, 165
seasonal flu, 107
seasonality, 93, 94, 102, 104, 108, 111, 117, 119, 236
secondary metabolism, 131, 278
secretion, 4, 126, 190
security, 258
sediment, 25, 26, 35, 40, 43, 46, 48, 50, 134
sedimentation, 247
sediments, 31, 39, 41, 43, 49, 50, 52, 54, 170, 279
seed, 20, 81, 86, 90, 143, 214, 237, 238, 284, 285, 290, 292
seeding, 284, 289
seedlings, 9, 11, 81, 90, 141, 143, 157, 194, 195, 196, 197, 231, 232, 246, 252, 259
selenium, 308
senescence, 3, 17
sensing, 19
sensitivity, 6, 77, 185
sensors, 7
services, 64, 121, 129, 132, 152, 273, 274, 317

sewage, 33, 217, 232, 251, 254, 255, 256, 258, 262, 263, 294
shade, 255, 315
shallow lakes, 50
shoot, 15, 47, 50, 53, 84, 124, 178, 184, 185, 191, 196, 206, 214
shoots, 7, 47, 188, 196, 240, 313
shores, 80
shortage, 4, 5, 6, 7, 9, 285
showing, 22, 29, 33, 43, 45, 46, 143, 214, 216, 218, 270
shrubs, 77, 97, 98
signal transduction, 278
signaling pathway, 7, 18, 314
signals, 7
signs, 237, 279, 299, 302
silica, 133, 138
silicon, 134, 136, 140, 143, 193, 194, 196, 237
silver, 30, 155
SiO2, 129
slag, 251
sludge, 51, 78, 217, 232, 251, 254, 255, 258, 261, 262, 263, 293, 294
SO42-, 28, 31, 33, 42, 44
sodium, 4, 5, 6, 11, 20, 51, 237, 251, 255, 308
soil community, 199
soil erosion, 85, 86, 204, 242, 247, 248, 251, 252
soil fertility, 36, 55, 57, 61, 62, 64, 67, 69, 73, 82, 83, 85, 86, 117, 119, 121, 130, 137, 159, 160, 168, 173, 205, 207, 211, 219, 231, 235, 236, 242, 245, 246, 247, 250, 251, 252, 253, 254, 255, 257, 258, 260, 261, 263, 264, 274, 277, 280, 281
Soil microbial activity, vii, 21, 125
soil mining, 235, 242
Soil nutrient depletion, 235
soil particles, 59, 60, 112, 115, 116, 117, 160, 202, 204, 239, 252
soil pollution, 73, 79, 81, 267
soil type, 5, 124, 168, 169, 170, 173, 177, 186, 201, 250, 282, 285
solid waste, 255, 259
solubility, 24, 34, 40, 48, 83, 133, 135, 137, 143, 146, 178, 180, 182, 183, 191, 193, 240, 245, 258, 285, 288, 300, 302, 304
solution, 1, 8, 9, 30, 35, 39, 41, 42, 82, 108, 112, 131, 133, 134, 136, 138, 145, 148, 155, 177, 179, 180, 181, 182, 183, 184, 185, 192, 193, 194, 195, 201, 205, 209, 239, 261, 264, 284, 285, 289, 295
sorption, 38, 52, 94, 99, 104, 112, 115, 116, 118, 123, 212, 265
sorption isotherms, 99
sorption process, 115
South Africa, 65
South America, 114, 115, 117, 123, 273
sowing, 16, 88, 220, 290
soybeans, 279, 289, 319
Spain, 67, 69, 70, 124, 209, 211, 261, 293, 307, 318
speciation, 131, 293
species richness, 56, 62
spectroscopy, 121
spiders, 62, 71
spirituality, 88
spore, 109, 110, 111, 143, 219, 271, 272, 274, 299
Spring, 165, 167
stability, 30, 45, 46, 52, 56, 59, 63, 64, 68, 89, 124, 181, 199, 201, 203, 204, 209, 216, 242, 252
stabilization, 28, 64, 80, 204, 230, 253
standard error, 221, 222, 223, 224
starch, 3, 7, 17
starvation, 4, 6, 14
state, 1, 2, 3, 21, 22, 23, 24, 25, 26, 27, 28, 29, 30, 31, 35, 36, 40, 41, 46, 47, 48, 95, 131, 132, 207, 219
states, 34, 42, 45
stele, 10
sterile, 136, 203
stigma, 255
stoichiometry, 306
stomata, 3, 6, 300
storage, 7, 13, 15, 18, 19, 49, 56, 62, 112, 119, 201, 202, 229, 253, 272
stratification, 68
stress, 6, 20, 47, 48, 52, 58, 63, 75, 82, 103, 135, 192, 194, 195, 199, 202, 209, 238, 268, 269, 273, 275, 285, 298, 300, 302, 303, 305, 306, 307, 308, 313, 314, 316
stress factors, 47
structure, 2, 4, 5, 46, 50, 55, 56, 59, 61, 63, 64, 66, 68, 69, 79, 86, 94, 112, 128, 143, 147, 150, 170, 181, 190, 199, 201, 202, 204, 205, 206, 207, 208, 209, 210, 212, 216, 218, 220, 226, 229, 232, 238, 248, 252, 270, 272, 273, 276, 298, 303, 306, 308, 312, 319
sub-Saharan Africa, 235, 236, 245, 247, 248, 253, 257
substrate, 4, 27, 28, 46, 47, 59, 60, 68, 111, 122, 162, 164, 171, 204, 209, 213, 214, 215, 218, 220, 226, 227, 229, 240, 247
succession, 68, 69, 120, 319
sucrose, 3, 5, 7, 8, 9, 10, 11, 12, 14, 15, 16, 17, 19, 20
sugar beet, 1, 7, 11, 12, 13, 14, 19, 20, 90, 172
sugarcane, 1, 4, 17, 79, 164, 170
sulfate, 21, 22, 27, 32, 33, 40, 42, 43, 44, 50, 90
sulfur, 41, 42, 45, 82, 280, 281, 290, 301, 308
sulphur, 50, 82, 85, 135, 237, 238, 239, 251, 255

suppression, 131, 199, 217, 218, 233
surface area, 129, 213
surface layer, 26
surface tension, 289
surfactant, 88, 91
surfactants, 78
survival, 47, 50, 121, 236
susceptibility, 108
suspensions, 45, 54, 148
sustainability, viii, 55, 69, 128, 147, 205, 227, 232, 264, 265
Switzerland, 16, 295
symbiosis, 39, 94, 104, 111, 117, 120, 123, 125, 178, 191, 229, 269, 273, 298, 299, 305, 307, 308, 311, 312, 314, 315, 316, 317, 318
symptoms, 81, 184, 186, 187, 189, 281, 283, 285
synergistic effect, 86, 184
synthesis, 1, 2, 3, 4, 5, 7, 15, 36, 48, 50, 82, 95, 102, 108, 141, 194, 202, 237, 239, 258, 286, 290
Syria, 263

T

target, 21, 189
taxa, 55, 59, 61, 62, 64
taxa of invertebrates, viii, 55
techniques, 112, 141, 147, 200, 204, 208, 235, 246, 247, 251, 258, 277, 281, 290
technologies, 78, 147, 258
technology, 16, 77, 199, 207, 209, 211, 251, 252, 272, 277, 284, 289, 290, 317
temperature, 5, 24, 27, 28, 30, 36, 37, 40, 46, 48, 49, 96, 98, 129, 131, 133, 137, 160, 162, 167, 168, 169, 173, 182, 183, 187, 189, 195, 236, 240, 241, 246, 247, 249, 250, 252, 259, 261, 264, 279, 283, 285, 289, 307
terrestrial ecosystems, 60, 63, 69, 132
testing, 248, 281
texture, x, 25, 99, 101, 104, 123, 177, 179, 180, 181, 182, 183, 201, 202, 204, 207, 248, 250, 271
Thailand, 260, 262, 265, 293
thermal treatment, 207
thorium, 79
tissue, 11, 12, 18, 20, 25, 47, 164, 197, 227, 278
titanium, 271
tobacco, 80, 89, 194, 195, 197, 218, 232, 306
topology, 74
toxic metals, 209
toxic substances, 23, 205
toxicity, 44, 50, 51, 52, 81, 88, 152, 167, 175, 177, 178, 180, 184, 185, 186, 187, 188, 189, 191, 192, 193, 194, 195, 196, 197, 242, 247, 268, 270, 292
trace elements, 82, 83, 197, 237, 278, 279, 288, 292
traits, 204, 270, 317
transcription, 15, 217
transformation, 22, 32, 35, 66, 77, 90, 94, 103, 104, 108, 109, 113, 129, 133, 151, 181, 196, 238, 246, 291
transformation processes, 113
transformations, 32, 45, 83, 94, 104, 119, 123, 125, 133, 169, 178, 240, 245, 246, 247, 252, 267, 269, 281, 317
translocation, 3, 4, 7, 13, 16, 17, 80, 89, 97, 192, 194, 269, 304
transpiration, 189
transport, 1, 2, 3, 4, 5, 6, 7, 13, 14, 15, 16, 18, 19, 20, 24, 25, 47, 49, 60, 178, 184, 187, 188, 190, 191, 238, 239, 240, 244, 259, 294, 305, 306, 316, 317, 319
transport processes, 2
transportation, 212, 244
treatment, 11, 15, 33, 86, 87, 144, 145, 230, 241, 254, 290
tricarboxylic acid, 2, 134
tricarboxylic acid cycle, 2
Trinidad, 70
tropical dry forest, 119
tropical forests, 122
tropical savannas, 94, 114, 120, 123
tundra, 61, 260
turgor, 1, 2, 3, 4, 9, 11, 14, 15, 19, 20
Turkey, 89
turnover, 17, 18, 55, 56, 59, 113, 120, 121, 159, 160, 201

U

UK, 65, 66, 67, 70, 122, 194, 210
ultrastructure, 196
UNESCO, 57, 122
United States (USA), 16, 17, 18, 20, 49, 51, 53, 67, 68, 89, 96, 119, 126, 137, 171, 175, 176, 195, 259, 260, 276, 318
universal gas constant, 27
uranium, 79
urban, 40, 74, 127, 253, 255, 259, 292
urea, 34, 82, 173, 251
urine, 86, 167, 253
USDA, 96, 99, 126
Uzbekistan, 159, 176

V

vacuole, 2, 6, 8, 9, 10, 14, 239
valence, 39

Valencia, 95, 96, 97, 103, 121, 122
vanadium, 237
variables, 74, 201, 243, 269, 291
variations, 102, 200, 236, 246, 250, 270, 277, 279, 282
varieties, 137, 172, 184, 214, 225, 233, 247, 248, 303
vascular bundle, 313
vegetables, 7, 257, 290
vegetal tissue, 188
vegetation, 21, 35, 40, 44, 47, 49, 50, 73, 95, 107, 111, 172, 174, 252, 272
vegetative cover, 202
vegetative stages, 289, 290
Venezuela, 93, 94, 97, 98, 119, 122, 123, 124, 125
Venus, 17
vinasse, 140, 154
vitamin C, 7, 231
vitamins, 238
volatilization, 95, 166, 240, 245, 248, 249, 255

W

Washington, 51, 88, 260, 292
waste, 40, 78, 129, 138, 145, 146, 148, 154, 211, 212, 213, 218, 219, 226, 229, 231, 232, 254, 255, 261, 263, 294
waste management, 129
waste water, 40, 255
water quality, 49, 51, 52, 129
watershed, 98
West Africa, 69, 124, 258, 259
Western Australia, 131, 132, 149, 151, 156
wetlands, 33, 40, 48, 50, 51, 80
wetting, 26, 48, 119, 202, 203, 246, 272, 289
wheat germ, 292
Wisconsin, 195
World Health Organization (WHO), 278, 294, 295
worms, 118, 132

X

xylem, 4

Y

yeast, 78, 137, 138, 140, 141, 143, 146, 154, 155
yield, 1, 4, 7, 19, 81, 84, 86, 90, 125, 138, 140, 145, 148, 150, 154, 155, 173, 175, 206, 214, 216, 219, 220, 221, 222, 226, 227, 228, 229, 230, 231, 238, 241, 245, 247, 248, 250, 256, 259, 261, 262, 263, 268, 277, 278, 279, 282, 284, 285, 286, 289, 292, 297, 298, 301, 302, 303, 304, 305, 308, 311, 312, 316, 319

Z

zinc, 81, 82, 84, 89, 136, 137, 149, 195, 196, 227, 237, 238, 246, 250, 278, 280, 284, 292, 293, 294, 295, 308
zinc oxide, 238
zinc sulfate, 308